DARK MATTER IN THE UNIVERSE

INTERNATIONAL ASTRONOMICAL UNION
UNION ASTRONOMIQUE INTERNATIONALE

DARK MATTER IN THE UNIVERSE

PROCEEDINGS OF THE 117TH SYMPOSIUM OF THE
INTERNATIONAL ASTRONOMICAL UNION
HELD IN PRINCETON, NEW JERSEY, U.S.A.,
JUNE 24-28, 1985

EDITED BY

J. KORMENDY

*Dominion Astrophysical Observatory, Herzberg Institute of Astrophysics,
Victoria, B.C., Canada*

and

G. R. KNAPP

*Department of Astrophysical Sciences, Princeton University,
Princeton, New Jersey, U.S.A.*

SPRINGER-SCIENCE+BUSINESS MEDIA, B.V.

Library of Congress Cataloging in Publication Data

International Astronomical Union. Symposium
    (117th: 1985: Princeton, N.J.)
Dark matter in the universe.

    At head of title: International Astronomical Union = Union astronomique
internationale.
    Sponsored by IAU Commission 28 (Galaxies) et al.
    Includes index.
    1. Interstellar matter–Congresses. 2. Galaxies–Congresses. 3. Stars–Cong-
resses. I. Kormendy, J. (John) II. Knapp, G. R. III. International
Astronomical Union. IV. International Astronomical Union. Commission 28
(Galaxies) V. Title.
QB790.I573 1985   523.1'12   86–21905

ISBN 978-90-277-2357-4   ISBN 978-94-009-4772-6 (eBook)
DOI 10.1007/978-94-009-4772-6

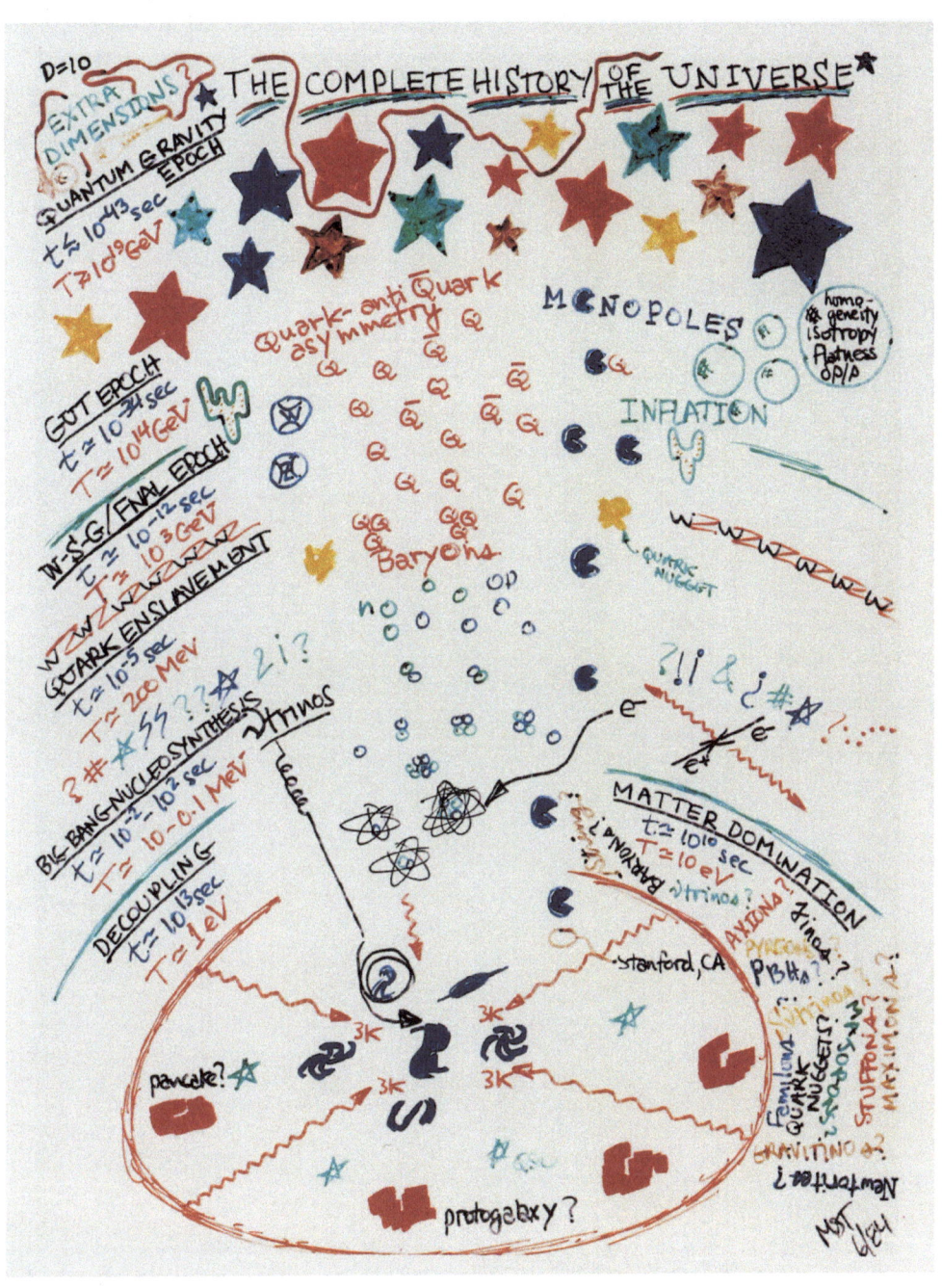

THE COMPLETE HISTORY OF THE UNIVERSE

(See pp. 449 - 450)

TABLE OF CONTENTS

(* indicates review paper)

POSTER PAPERS NOT SUBMITTED FOR THE PROCEEDINGS

HOT AND COLD DARK MATTER   S. Achilli, F. Occhionero and
    S. Scaramolla

A STATISTICAL APPROACH TO LARGE-SCALE STRUCTURE
    J. M. Bardeen, J. R. Bond, N. Kaiser and A. Szalay

DWARF GALAXIES AND THE MASSIVE HALO PROBLEM
    A. Bosma, J. Kormendy and J. Souviron

GALAXIES WITH LOW SURFACE BRIGHTNESS DISKS
    A. Bosma, J. M. van der Hulst and E. Athanassoula

FORMATION OF GALAXIES AND DARK MATTER HALOS
    R. Carlberg, G. Lake and C. A. Norman

AN X-RAY AND OPTICAL STUDY OF THE CLUSTER OF GALAXIES A754
    D. Fabricant, T. Beers, M. Geller and M. Kurtz

STRUCTURE OF THE GALACTIC HALO   F. D. A. Hartwick

A SINGLE-MASS MODEL FOR ω CENTAURI   B. Jarvis

AN EXPERIMENT TO DETECT GALACTIC AXIONS
    A. C. Mellissinos, J. Rogers, W. Wuensch, H. Halama,
    A. Prodell, P. Thompson and W. B. Fowler

SUPERCLUSTERS AND PANCAKES   M. Mijic

RADIAL VELOCITIES OF CARBON STARS IN THREE DWARF SPHEROIDAL GALAXIES
    P. Seitzer and J. Frogel

KINEMATICS AND DYNAMICS OF THE SYSTEM OF BLUE HORIZONTAL BRANCH
    FIELD STARS IN THE OUTER GALACTIC HALO   J. Sommer-Larsen

ISOTHERMAL AXION PERTURBATIONS   M. S. Turner and D. Seckel

PREFACE

This is the first time that the International Astronomical Union has held a symposium on objects of totally unknown nature. In fact, M. Rees has pointed out that the mass of the individual particles that make up the dark matter is unknown to > 70 orders of magnitude. Since dark matter appears to make up ~ 90 % of the mass of the Universe, it presents us with one of the most fundamental problems in astrophysics. IAU Symposium 117 on *Dark Matter in the Universe* was held on June 24 - 28, 1985. Our hosts were Princeton University and the Institute for Advanced Study, which together form one of the most active centers of work on the dark matter problem. There were ~ 190 participants from 16 countries. These proceedings include the 31 review and invited papers, 72 of the 85 poster papers, and the two general discussions.

The idea that the Universe might contain much more mass than we see in gas, stars and their remnants has been with us for over 50 years. In 1933, F. Zwicky pointed out that the Coma Cluster could be in equilibrium at the large observed velocity dispersion only if a great deal of unseen matter were present. However, in the absence of other evidence, the idea of "dark matter" was not widely pursued. Then in the mid-1970s it became clear that rotation curves of galaxies do not have the expected Keplerian declines at large radii; they stay flat as far out as we can observe them. The evidence became compelling when rotation curves were measured to very large radii. Other techniques of mass measurement were developed, based, e. g., on the confinement of X-ray halos; these also implied large amounts of DM. At about the same time J. Ostriker and P. J. E. Peebles pointed out that DM halos in galaxies could explain the observed stability of disks against the formation of bars. Also, large amounts of DM could close the Universe, an appealing possibility that was beginning to look unlikely as the inventory of visible matter was improved. The simultaneous appearance of a variety of observational and theoretical arguments resulted in a general acceptance of the existence and importance of DM. Questions about the amount of DM remained, and some degree of healthy skepticism about the existence of DM persists today. But a scientific revolution had begun.

Until recently we knew little more than that DM appears to exist; we had little systematic information about its properties. Only in the past several years have we progressed to the point that we can measure DM density distributions. For example, with accurate rotation curves extending over large ranges in radius, we can try to decompose the effects of visible and dark matter to measure DM density profiles. Already some regularities in DM behavior have

turned up. We need to look for more. For example, we need to look
for correlations of structure parameters similar to the Faber-Jackson
relation for visible matter. Then we can look at the astrophysics of
DM in more rigorous detail. In addition, there is growing evidence
that some DM is not baryonic. If this consists of elementary
particles created in the early Universe, then the DM problem is
intimately related with current work in fundamental physics. In
particular, Grand Unified Theories suggest the existence of various
particles that would necessarily contribute importantly to the
mass density of the Universe if they exist and have mass. Then DM
observations and particle theories provide interesting constraints
on each other. By the mid-1980s progress had become very rapid. The
time had come to hold an IAU Symposium. The Scientific Organizing
Committee (SOC) wanted to provide an opportunity for people in the
very diverse fields now involved in DM work to get together. We did
not expect to solve the problem of the composition of DM. However,
we hoped that a successful meeting would stimulate further work and
suggest the directions that would be most productive.

In planning the meeting, the SOC tried to emphasize a number of
key issues; these are introduced in papers by S. Faber and V. Rubin.
To allow the reviewers to be as thorough as possible, it was decided
that all contributed papers would be posters. Similarly, the
reviewers were given as many pages in the proceedings as possible to
allow them to write papers that would be useful references. Posters
are included as one-page abstracts. Long discussion periods were
scheduled to allow participants to explore the issues; the session
chairmen were M. Schmidt, M. Roberts, E. Salpeter, P. Shapiro,
P. Schechter, A. Dekel, C. Norman, D. Lynden-Bell, G. Steigman,
J. Peebles and S. Tremaine. The discussions are included in full,
but were edited to reduce their length. We have tried to preserve
the content and flavor of what was said; of course any errors that
were introduced are our responsibility.

IAU Symposium 117 was sponsored by IAU Commission 28 (Galaxies)
and co-sponsored by Commissions 33 (The Galaxy), 47 (Cosmology)
and 48 (High Energy Astrophysics). We are grateful to the IAU for
its support. Financial support was also provided by the Department
of Astrophysical Sciences at Princeton University, the Dominion
Astrophysical Observatory, the Institute for Advanced Study and
De Luxe Travel of Princeton.

A large number of people contributed to the success of the
meeting. We thank the members of the SOC for their help with the
planning. In addition, P. Schechter, A. Toomre and S. Tremaine
gave very helpful advice at the time the program was defined.
Much of the planning took place at a workshop on Dynamics Within
Galaxies held at the Weizmann Institute of Science in 1984. JK
thanks M. Milgrom and the Institute staff for their hospitality. In
Princeton, the symposium was run by a very capable Local Organizing
Committee; we thank them very much for their hard work. V. Nixon
and W. Tankins provided invaluable assistance at registration.
We are grateful to the students who handled the microphones and
discussion sheets: S. Brown, W. Ewell, N. Katz, H. M. Lee, T. Quinn,

S. Ratcliff, M. Richmond, M. Rupen, B. Ryden, T. Statler, M. Vietri
and M. L. West. We also thank M. Beahn, A. Frost, and S. Marguarat
of the Princeton University Audiovisual Dept., who showed the slides
and taped the discussions. Part of the editing of the discussions
was carried out during a visit to the Institute for Advanced Study;
JK is grateful to J. Bahcall and the IAS staff for their hospitality.
At Princeton University, Wrae Tankins typed several papers and the
early drafts of the discussions. It is a pleasure to thank her for
her fast, careful and always cheerful work. First drafts of the
Name and Object Indices were typed at the DAO by Y. Hurwitz. It
is a special pleasure to thank Mary Ker Kormendy for her dedicated
editorial help; without this, it would have taken much longer to
complete the manuscript. Finally, we thank the many contributors
to the meeting, who presented papers, who contributed to lively
discussions, and who puzzled, speculated and agonized over the nature
of dark matter in the Universe.

J. Kormendy

G. R. Knapp

SCIENTIFIC ORGANIZING COMMITTEE

J. Audouze, France

J. N. Bahcall, USA

B. F. Burke, USA

J. Einasto, USSR

K. C. Freeman, Australia

J. Kormendy, Canada (Chairman)

D. Lynden-Bell, England

J. P. Ostriker, USA

V. C. Rubin, USA

R. Sancisi, The Netherlands

M. Schmidt, USA

LOCAL ORGANIZING COMMITTEE

J. N. Bahcall

E. Blakey

D. B. Hortenbach

G. R. Knapp

E. L. Turner (Chairman)

J. B. Turner

M. Wisnovsky

LIST OF PARTICIPANTS

M. AARONSON
Steward Observatory
University of Arizona
Tucson, AZ 85721
USA

L. AGUILAR
Department of Astronomy
University of California
Berkeley, CA 94720
USA

T. S. VAN ALBADA
Kapteyn Astronomical Institute
Postbus 800
9700 AV Groningen
The Netherlands

C. ALCOCK
6-207 MIT
Cambridge, MA 02139
USA

E. ATHANASSOULA
Observatoire de Marseille
2, Place Le Verrier
13248 Marseille Cédex 04
France

J. AUDOUZE
Institute d'Astrophysique
98 bis, Boulevard Arago
F-75014 Paris
France

R. BACON
Observatoire de Genève
Ch. des Maillettes 51
CH-1290 Sauverny
Switzerland

J. BAHCALL
School of Natural Sciences
Institute for Advanced Study
Princeton, NJ 08540
USA

N. BAHCALL
Space Telescope Science Institute
Homewood Campus
Baltimore, MD 21218
USA

J. BARDEEN
Physics Department FM-15
University of Washington
Seattle, WA 98195
USA

J. BARNES
School of Natural Sciences,
Institute for Advanced Study
Princeton, NJ 08540
USA

J. BEKENSTEIN
Physics Department
Ben Gurion University
Beersheva 84105
Israel

J. BERGERON
Institute d'Astrophysique
98 bis, Boulevard Arago
75014 Paris
France

S. VAN DEN BERGH
Dominion Astrophysical Observatory
5071 W. Saanich Road
Victoria, BC V8X 4M6
Canada

E. BERTSCHINGER
Astronomy Department
University of California
Berkeley, CA 94720
USA

J. J. BINNEY
Department of Theoretical Physics
1 Keble Road
Oxford OX1 3NP
United Kingdom

G. BLUMENTHAL
Lick Observatory
University of California
Santa Cruz, CA 95064
USA

P. C. BOESHAAR
Department of Physics
Rider College
P.O. Box 6400
Lawrenceville, NJ 08648
USA

A. BOSMA
Observatoire de Marseille
2, Place Le Verrier
13248 Marseille Cédex 4
France

F. BOUCHET
IAP et École Polytechnique
98 bis, Boulevard Arago
75014 Paris
France

B. F. BURKE
26-335, Physics
MIT
Cambridge, MA 02139
USA

D. BURSTEIN
Department of Physics
Arizona State University
Tempe, AZ 85281
USA

R. H. BUSS, Jr.
Campus Box 391 APAS
University of Colorado
Boulder, CO 80309
USA

C. CANIZARES
37-501 Physics
MIT
Cambridge, MA 02139
USA

C. CARIGNAN
Kapteyn Laboratorium
Postbus 800
9700 AV Groningen
The Netherlands

R. CARLBERG
York University
4700 Keele Street
Downsview
Ontario M3J 1P3
Canada

B. C. CARNEY
Department of Physics and Astronomy
Phillips Hall 039A
University of North Carolina
Chapel Hill, NC 27514
USA

B. CARR
School of Mathematical Sciences
Queen Mary College
Mile End Road
London E1 4NS
United Kingdom

S. CASERTANO
School of Natural Sciences
Institute for Advanced Study
Princeton, NJ 08540
USA

A. CAULET
Astronomy and Astrophysics
University of Chicago
5640 S. Ellis Ave
Chicago, Il 60637
USA

G. CHINCARINI
Department of Physics and Astronomy
University of Oklahoma
Norman, OK 73019
USA

M. CRÉZÉ
Observatoire de Besançon
41 Ave. de l'Observatoire
25000 Besançon
France

R. A. DALY
Department of Astronomy
Boston University
725 Commonwealth Avenue
Boston, MA 02215
USA

B. DATTA
Indian Institute of Astrophysics
Sarjapur Road
Bangalore 560 034
India

A. DAVIDSEN
Department of Physics
Johns Hopkins University
Baltimore, MD 21218
USA

R. L. DAVIES
N.O.A.O.
P.O. Box 26732
Tucson, AZ 85726
USA

M. DAVIS
Department of Astronomy
University of California
Berkeley, CA 94720
USA

A. DEKEL
Department of Physics
The Weizmann Institute of Science
Rehovot 76100
Israel

A. DRESSLER
Mt. Wilson and las Campanas
   Observatories
813 Santa Barbara St.
Pasadena, CA 91101
USA

J. EINASTO
Tartu Astrophysical Observatory
202444 Toravere
Estonia
USSR

R. D. EKERS
NRAO, P.O. Box O
Socorro, NM 87801
USA

S. M. FABER
Lick Observatory
University of California
Santa Cruz, CA 95064
USA

A. C. FABIAN
Institute of Astronomy
Madingley Road
Cambridge CB3 OHA
United Kingdom

D. FABRICANT
Center for Astrophysics
60 Garden Street
Cambridge, MA 02138
USA

S. M. FALL
Space Telescope Science Institute
Homewood Campus
Baltimore, MD 21218
USA

J. E. FELTEN
Laboratory for Extraterrestrial
   Physics
Code 697, Goddard Space Flight
   Center
Greenbelt, MD 20771
USA

K. FREESE
P-236
Center for Astrophysics
60 Garden Street
Cambridge, MA 02138
USA

K. C. FREEMAN
Mt. Stromlo Observatory
Private Bag
Woden P.O., ACT
2606 Australia

C. S. FRENK
Astronomy Centre
University of Sussex
Brighton BN1 9QH
United Kingdom

D. GERBAL
D.A.F., Observatoire de Meudon
92195 Meudon Cédex
France

O. E. GERHARD
Max Planck Institut f. Astrophysik
Karl Schwarzschild Str. 1
D-8046 Garching bei München
Federal Republic of Germany

G. GILMORE
Institute of Astronomy
Madingley Road
Cambridge CB3 0HA
United Kingdom

J. H. VAN GORKOM
N.R.A.O., P.O. Box 0
Socorro, NM 87801
USA

J. R. GOTT
Department of Astrophysical
   Sciences
Princeton University
Princeton, NJ 08544
USA

R. F. GREEN
N.O.A.O.
P.O. Box 26732
Tucson, AZ 85726
USA

E. J. GROTH
Physics Department
Princeton University
Princeton, NJ 08544
USA

E. GUENDELMAN
Center for Theoretical Physics
MIT
Cambridge, MA 02139
USA

J. E. GUNN
Department of Astrophysical
   Sciences
Princeton University
Princeton, NJ 08544
USA

A. GUTH
Center for Theoretical Physics
MIT
Cambridge, MA 02139
USA

F. D. A. HARTWICK
University of Victoria
Department of Physics
P.O. Box 1700
Victoria, BC V8W 2Y2
Canada

J. HEISLER
6-206 Physics
MIT
Cambridge, MA 02139
USA

L. HERNQUIST
Department of Astronomy
University of California
Berkeley, CA 94720
USA

J. HEWITT
26-349 Physics
MIT
Cambridge, MA 02139
USA

L. HOFFMAN
Department of Physics
Lafayette College
Easton, PA 18042
USA

Y. HOFFMAN
Department of Physics
University of Pennsylvania
Philadelphia, PA 19104
USA

J. P. HUCHRA
Center for Astrophysics
60 Garden Street
Cambridge, MA 02138
USA

P. HUT
School of Natural Sciences
Institute for Advanced Study
Princeton, NJ 08540
USA

S. IKEUCHI
Tokyo Astronomical Observatory
Mitaka, Tokyo 181
Japan

G. ILLINGWORTH
Space Telescope Science Institute
Homewood Campus
Baltimore, MD 21218
USA

T. ISHIZAWA
Department of Astronomy
University of Kyoto
Kyoto 606
Japan

B. J. JARVIS
Cerro Tololo Interamerican
  Observatory
Casilla 603
La Serena
Chile

C. JOG
Department of Astronomy
P.O. Box 3818
University of Virginia
Charlottesville, VA 22903
USA

J. JONES
NORDITA
Blegdamsvej 17
DK-2100 København Ø
Denmark

B. JONES
NORDITA
Blegdamsvej 17
DK-2100 København Ø
Denmark

N. KAISER
Institute of Astronomy
Madingley Road
Cambridge CB3 OHA
United Kingdom

A. J. KALNAJS
Mt. Stromlo Observatory
Private Bag
Woden P.O., A.C.T.
2606 Australia

S. KENT
Center for Astrophysics
60 Garden Street
Cambridge, MA 02138
USA

G. R. KNAPP
Department of Astrophysical
  Sciences
Princeton University
Princeton, NJ 08544
USA

H. KODAMA
Department of Physics
University of Tokyo
Bunkyo-Ku
Tokyo 113
Japan

J. KORMENDY
Dominion Astrophysical Observatory
5071 W. Saanich Road
Victoria, BC V8X 4M6
Canada

L. KRAUSS
Lyman Laboratory of Physics
Harvard University
Cambridge, MA 02138
USA

P. C. VAN DER KRUIT
Kapteyn Laboratorium
Postbus 800
9700 AV Groningen
The Netherlands

N. KRUMM
Physics Department
University of Cincinnati
Cincinnati, OH 45221-0011
USA

C. LACEY
Department of Astrophysical
   Sciences
Princeton University
Princeton, NJ 08544
USA

M. LACHIÈZE-REY
Service d'Astrophysique
CEN-Saclay DPhG-SAp
91191 Gif-sur-Yvette Cédex
France

G. B. LAKE
School of Natural Sciences
Institute for Advanced Study
Princeton, NJ 08540
USA

R. B. LARSON
Astronomy Department
Yale University
New Haven, CT 06511
USA

D. LATHAM
Center for Astrophysics
60 Garden Street,
Cambridge, MA 02138
USA

B. LIEBERMAN
Center for Theoretical Physics
MIT
Cambridge, MA 02139
USA

F. J. LOW
Steward Observatory
University of Arizona
Tucson, AZ 85721
USA

V. N. LUKASH
Space Research Institute
USSR Academy of Sciences
Profsojusnaya 84-32
117810 Moscow
USSR

D. LYNDEN-BELL
Institute of Astronomy
Madingley Road
Cambridge CB3 OHA
United Kingdom

J. MADSEN
Institute of Astronomy
University of Aarhus
DK-8000 Aarhus C
Denmark

E. MALUMUTH
Department of Physics and Astronomy
Rutgers University
Piscataway, NJ 08854
USA

G. MAMON
Physics Department
New York University
4 Washington Place
New York, NY 10003
USA

L. MARTINET
Geneva Observatory
CH 1290 Sauverny
Switzerland

R. D. MATHIEU
Center for Astrophysics
60 Garden St.
Cambridge, MA 02138
USA

J. MCCLINTOCK
37-521
Center for Space Research
MIT
Cambridge, MA 02139
USA

J. MCDOWALL
Institute of Astronomy
Madingley Road
Cambridge CB3 0HA
United Kingdom

A. MELOTT
Astronomy and Astrophysics
University of Chicago
5640 South Ellis Ave
Chicago, IL 60637
USA

D. MERRITT
Department of Astronomy
University of California
Berkeley, CA 94720
USA

M. MIJIC
452-48, Caltech
Pasadena, CA 91125
USA

M. MILGROM
Department of Nuclear Physics
The Weizmann Institute of Science
Rehovot 76100
Israel

J. MOODY
Institute for Theoretical Physics
University of California
Santa Barbara, CA 93106
USA

R. A. NOLTHENIUS
Steward Observatory
University of Arizona
Tucson, AZ 85721
USA

C. A. NORMAN
Space Telescope Science Institute
Homewood Campus
Baltimore, MD 21218
USA

F. OCCHIONERO
Intituto Astronomico
University of Rome
Via G.M. Lancisi 29
00161 Rome
Italy

J. P. OSTRIKER
Department of Astrophysical
   Sciences
Princeton University
Princeton, NJ 08544
USA

B. E. PACZYŃSKI
Department of Astrophysical
   Sciences
Princeton University
Princeton, NJ 08544
USA

P. J. E. PEEBLES
Department of Physics
Princeton University
Princeton, NJ 08544
USA

R. C. PETERSON
Apt. D22
3636 North Campbell
Tucson, AZ 85719
USA

J. PRIMACK
Physics Department
University of California
Santa Cruz
CA 95064
USA

P. J. QUINN
130-33 Caltech
Pasadena, CA 91125
USA

M. J. REES
Institute of Astronomy
Madingley Road
Cambridge CB3 OHA
United Kingdom

D. O. RICHSTONE
Department of Astronomy
University of Michigan
Ann Arbor, MI 48109
USA

M. S. ROBERTS
NRAO
Edgemont Road
Charlottesville
VA 22901
USA

H. J. ROOD
School of Natural Sciences
Institute for Advanced Study
Princeton, NJ 08540
USA

L. RUDNICK
School of Physics and Astronomy
University of Minnesota
116 Church St., SE
Minneapolis, MN 55455
USA

V. C. RUBIN
Department of Terrestrial Magnetism
Carnegie Inst. of Washington
5241 Broad Branch Road, NW
Washington, DC 20015
USA

E. E. SALPETER
Newman Laboratory of Natural
   Science
Cornell University
Ithaca, NY 14853
USA

R. SANCISI
Kapteyn Laboratorium
Postbus 800
9700 AV Groningen
The Netherlands

R. L. SANDERS
Kapteyn Laboratorium
Postbus 800
9700 AV Groningen
The Netherlands

C. L. SARAZIN
Department of Astronomy
P.O. Box 3818
University of Virginia
Charlottesville, VA 22903
USA

P. SCHECHTER
Mt. Wilson and Las Campanas
   Observatories
813 Santa Barbara St.
Pasadena, CA 91101
USA

M. SCHMIDT
105-24 Astronomy
Caltech
Pasadena, CA 91125
USA

M. SCHWARZSCHILD
Department of Astrophysical
   Sciences
Princeton University
Princeton, NJ 08544
USA

D. SECKEL
Fermilab, MS 209
P.O. Box 500
Batavia, IL 60510
USA

P. SEITZER
N.O.A.O., P.O. Box 26732
Tucson, AZ 85726
USA

J. SELLWOOD
Kapteyn Laboratorium
Postbus 800
9700 AV Groningen
The Netherlands

P. SHAPIRO
Department of Astronomy
University of Texas
Austin, TX 78712
USA

N. A. SHARP
N.O.A.O., P.O. Box 26732
Tucson, AZ 85726
USA

W. L. SHUTER
Department of Physics
University of British Columbia
Vancouver, BC V6T 1W5
Canada

J. SILK
Department of Astronomy
University of California
Berkeley, CA 94720
USA

M. SKRUTSKIE
Space Sciences Building
Cornell University
Ithaca, NY 14853
USA

J. SOMMER-LARSEN
Niels Bohr Institute
Blegdamsvej 17
DK-2100 København Ø
Denmark

L. S. SPARKE
Institute of Astronomy
Madingley Road
Cambridge CB3 OHA
United Kingdom

D. N. SPERGEL
Center for Astrophysics
60 Garden Street
Cambridge, MA 02138
USA

L. STAVELEY-SMITH
Nuffield Radio Astronomy
   Laboratories
Jodrell Bank, Macclesfield
Cheshire SK11 9DL
United Kingdom

A. STEBBINS
Department of Astronomy
University of California
Berkeley, CA 94720
USA

G. STEIGMAN
Bartol Research Foundation
University of Delaware
Newark, DE 19716
USA

M. STRUBLE
Department of Astronomy and
   Astrophysics, E1
University of Pennsylvania
Philadelphia, PA 19104
USA

L. L. STRYKER
Department of Terrestrial Magnetism
Carnegie Inst. of Washington
5241 Broad Branch Road, NW
Washington, DC 20015
USA

Y. SUTO
Department of Physics
University of Tokyo
Bunkyo-Ku
Tokyo 113
Japan

A. SZALAY
Fermilab, MS 209
P.O. Box 500
Batavia, IL 60510
USA

P.A. THOMAS
Institute of Astronomy
Madingley Road
Cambridge CB3 OHA
United Kingdom

S. D. TREMAINE
C.I.T.A.
McLennan Physical Lab.
University of Toronto
60 St. George Street
Toronto, Ontario M5S 1A1
Canada

G. TRINCHIERI
Center for Astrophysics
60 Garden Street
Cambridge, MA 02138
USA

W. TUCKER
Center for Astrophysics
60 Garden Street
Cambridge, MA 02138
USA

R. B. TULLY
Institute for Astronomy
University of Hawaii
2680 Woodlawn Drive
Honolulu, HI 96822
USA

E. L. TURNER
Department of Astrophysical
   Sciences
Princeton University
Princeton, NJ 08544
USA

M. S. TURNER
Fermilab, MS 209
P.O. Box 500
Batavia, IL 60510
USA

J. A. TYSON
AT&T Bell Laboratories
ID-316, Murray Hill
NJ 07974
USA

J. P. VADER
Astronomy Department
Yale University
New Haven, CT 06511
USA

S. VEERARAGHAVAN
Department of Astronomy
University of California
Berkeley, CA 94720
USA

E. T. VISHNIAC
Astronomy Department
University of Texas
Austin, TX 78712
USA

W. D. WATSON
Loomis laboratory of Physics
University of Illinois
1110 W. Green Street
Urbana, IL 61801
USA

S. D. M. WHITE
Steward Observatory
University of Arizona
Tucson, AZ 85721
USA

B. WHITMORE
Space Telescope Science Institute
Homewood Campus
Baltimore, MD 21218
USA

D. WILKINSON
Department of Physics
Princeton University
Princeton, NJ 08544
USA

T. B. WILLIAMS
Department of Physics and Astronomy
Rutgers University
Piscataway, NJ 08854
USA

H. VAN WOERDEN
Kapteyn Laboratorium
Postbus 800
9700 AV Groningen
The Netherlands

J. P. WRIGHT
National Science Foundation
Extragalactic Astronomy
Washington, DC 20550
USA

W. WUENSCH
Department of Physics and Astronomy
University of Rochester
Rochester, NY 14627
USA

R. WYSE
Department of Astronomy
University of California
Berkeley, CA 94720
USA

A. YAHIL
Astronomy Program
State University of New York
Stony Brook, NY 11794
USA

W. H. ZUREK
Theoretical Astrophysics
T-6, MS B288
LANL, Los Alamos
NM 87545
USA

LIST OF PARTICIPANTS

H. Jan WARNER
Kapteyn Laboratorium
Postbus 800
9700 AV Groningen
The Netherlands

J. P. WRIGHT
National Science Foundation
Extragalactic Astronomy
Washington, DC 20550
USA

R. RUFFINI
Department of Physics and Astronomy
University of Rochester
Rochester, NY 14627
USA

C. WILL
Department of Physics
University of California
Berkeley, CA 94720
USA

A. WITT
Astronomy Program

B. SPEED
Theoretical Astrophysics
LANL, Los Alamos
NM 87545
USA

DARK MATTER: KEY ISSUES

Sandra M. Faber
Lick Observatory
University of California
Santa Cruz, CA 95064
USA

ABSTRACT. Key outstanding questions regarding dark matter are formulated, as a backdrop to the upcoming discussions at this meeting. A major issue involves how many species of dark matter there are, and whether both baryons and non-baryons are implicated in cosmological dark matter. How is dark matter distributed relative to baryons on all scales? Are voids really empty? And finally, is there high-amplitude structure in the matter distribution of the universe on scales ~100 Mpc, and, if so, how can it be accounted for in terms of known, plausible physical processes?

INTRODUCTION. Two years ago marked the golden anniversary of Fritz Zwicky's landmark study of the Coma cluster (Zwicky 1933). The results provided dramatic evidence for dark matter that is still among the best we have. The way to final acceptance of dark matter was not smooth, however; the intervening fifty years were marked by incessant discussion and controversy that called into question every facet of the problem - from data, to theory, to the basic laws of physics. In astronomy, only the quasar redshift debate approached the "missing mass" controversy in sheer intensity, but QSOs were resolved to most astronomers' satisfaction in a much shorter period of time. And arguments over missing mass are not over yet.

Looking back, it is interesting to reflect on how utterly paralyzing dark matter was in those days for any rational attempt to model the dynamics of the universe. The turnaround since then has been remarkable. Now regarded as at least a decent working hypothesis, dark matter has turned from nemesis to powerful ally: initially invoked merely to bind galaxies and clusters of galaxies, it is now called upon to form them and, by some, even to close the universe itself. These achievements, though substantial, ought not to impress us unduly. If one is allowed to play freely with nine-tenths of the mass of the universe, miracles are not too much to expect!

After fifty years, it is fitting that we take stock to review how far we have come with dark matter (hereafter, often DM) and what we know and don't know about it. To that end, I have organized a few key ques-

1

J. Kormendy and G. R. Knapp (eds.), Dark Matter in the Universe, 1–16.
© *1987 by the IAU.*

tions to introduce the following speakers. Let us start with what is still the most important and most basic question:

1. ARE WE SURE YET THAT DARK MATTER EXISTS?

Conservative sceptics can still attack inadequacies in the observational data and analyses. More radical sceptics can attack the law of gravity itself, as Dr. Milgrom will explain to us later. Actually, the basic astronomical data have not increased or altered much in recent years. Perhaps I am too pessimistic, but it is my feeling that, short of an actual detection in the laboratory, most of the direct evidence for DM is already in. What we will see in the future, I predict, is a slow testing over the years, as people work patiently to fit DM into the wider context of physics, astronomy, and cosmology. This indirect process has already started, with encouraging early results, and will doubtless continue for a long while to come.

Let us therefore pass by this basic issue and rather regard dark matter as a "decent working hypothesis," as suggested above. In this spirit, the next most vital question is:

2. HOW MANY SPECIES OF DARK MATTER ARE THERE?

2.1 Do we need baryonic dark matter?

In all, DM has been claimed to exist in five types of structures that differ greatly in size: the solar neighborhood, dwarf galaxies, large galaxies, groups and clusters, and superclusters. By well known arguments (White and Rees 1978), the last four seem to require that DM consist of material that entered into a dissipationless state before galaxies collapsed. The existence of DM in the solar neighborhood, by contrast, is usually assumed to require dissipation during or after the formation of the Milky Way, although this has not yet been rigorously shown.

The existence of DM in the solar neighborhood from the Oort-Bahcall analysis (Bahcall 1984) thus points strongly to a dissipative and hence probably baryonic component in dark matter. Fortunately, within the context of solar-neighborhood astrophysics, viable baryonic candidates are not hard to find. As Larson points out (this conference), plausible changes in the local IMF and history of star formation could increase the white-dwarf component in Bahcall's models considerably. Recall also that the local density, ρ_{tot}, derived in the models is directly proportional to the square of the assumed scale height of the stellar tracers, and thus to their assumed absolute magnitude: $\rho_{tot} \propto L$ (tracer). The assumed luminosities of the F dwarfs and K giants used as tracers must surely introduce uncertainties of at least a few tens of percent. Bahcall has also pointed out other difficulties, including how the dark component is assumed to be distributed with height. It is important to try to analyze critically the uncertainties in the Oort-Bahcall estimate to see if, after all, there is any troubling discrepancy with a plausible baryonic origin for the local DM.

Even if problems do remain, many people including myself would argue that, because dark matter in the solar neighborhood is probably dissipational, it can have little or no connection with cosmological dark matter, which is pretty clearly dissipationless. Others (e.g., Schramm and Freese 1985) are not convinced, arguing that, if we can establish the existence of dark baryons in the solar neighborhood, by continuity arguments this is powerful evidence for additional baryons in galaxy halos. Schramm and Freese point out that $\Omega_b \, h_{50}^2 \lesssim 0.01$ based on baryons that are detected in galaxies, whereas Big Bang nucleosynthesis requires $\Omega_b \, h_{50}^2 \gtrsim 0.03$. They thus argue for extra dark baryons outside galaxies and have shown that, within the errors, <u>all</u> DM in galaxy halos could be baryonic without violating nucleosynthesis constraints, provided $\Omega_{DM}^{TOT} \lesssim 0.2$ in total.

An alternative, equally interesting interpretation is that there are indeed extra baryons in the universe, but they are not DM around galaxies -- they are unseen baryons in voids associated with galaxies that failed to form, i.e., biased galaxy formation. I confess outright that I am a strong believer in biased galaxy formation and regard it as a natural consequence in most hierarchical clustering scenarios. If biasing operates on large enough scales, it could obviously profoundly alter present notions about the large-scale distribution of matter in the universe.

Either way, "missing baryons" in the universe are important. It is therefore vital to review carefully the above limits on $\Omega_b \, h^2$ to assess how soft they are. The estimate of baryons detected in galaxies is now especially out of date. It is based simply on the mean luminosity density of Zwicky galaxies without regard to changing baryonic M/L with Hubble type or luminosity. The baryonic component of individual galaxies has also likely been overestimated, owing to inclusion of DM within the optical radii. Extra baryons in hot gas in X-ray clusters have not been allowed for either. The whole calculation is clearly ripe for more careful reconsideration.

2.2 Do we need non-baryonic dark matter?

There are two prime reasons for wanting non-baryonic DM: belief in inflation, which strongly implies $\Omega \approx 1$ (if $\Lambda = 0$), and the need to make galaxies by the present epoch without violating $\delta T/T$ in the microwave background (Bond and Efstathiou 1984, Vittorio and Silk 1984). Failing any known mechanism to generate galaxy-sized isothermal perturbations in a purely baryonic universe, the $\delta T/T$ argument seems fairly firm. There is admittedly an alternative picture of galaxy formation in which pregalactic perturbations are generated from hydrodynamic processes from winds and supernova explosions in an early generation of stars (Ostriker and Cowie 1981). These first stars must simply be posited, however, and the model also cannot explain structure on very large scales, the energy requirements being too great. The more usual unified approach, which attempts to explain all structures from galaxies to superclusters as arising from the same underlying DM fluctuation spectrum, is simpler and more elegant. I am therefore inclined to believe strongly in non-baryonic DM for this reason.

With the door now open, the next critical question is whether there
is so much DM that $\Omega = 1$. The main impetus for this comes from inflation -- a largely esthetic argument still not considered compelling by
most astronomers. Until recently, I was myself deterred from $\Omega = 1$ by
the great ages measured for globular clusters: several workers (see
review by Sandage and Tammann 1983) had converged with great unanimity
on the value 17 ± 2 b.y., whereas, even with H_0 as low as 50 km s^{-1} Mpc^{-1},
the age of the universe with $\Omega = 1$ must be under 14 b.y. An important
change in this picture has now occurred, as Vandenberg (1985) has re-
considered cluster ages taking into account the fact that oxygen, the
most abundant metal, is up by as much as $+1.0$ dex relative to Fe in
metal-poor stars. Vandenberg's new ages are close to 14 b.y., in good
agreement with $\Omega = 1$. This, coupled with some possible extra, uncounted
mass in voids due to biased galaxy formation, makes $\Omega = 1$ an attractive
astronomical option for the first time.

3.   WHAT IS ρ_{DM}/ρ_{BARY} ON GALACTIC AND SUBGALACTIC SCALES?

Let us turn now to the distribution of dark matter relative to baryons
in the universe. A precise understanding of this distribution is surely
one of our most powerful clues to the identity of dark matter. It is
clear that dark matter and baryons are radially separated on scales
smaller than galaxies, where dissipation operates. The simplest picture
is that all galaxies as a whole had initially the same ratio of baryons
to DM, perhaps altered later by processes like tidal stripping of dark
halos, ram-pressure stripping of baryons, ejection of baryons by supernovae, etc. It has been usual to assume a   constant initial ratio
(White and Rees 1978, Faber 1982a, Gunn 1982, Blumenthal et al. 1984,
Dekel and Silk 1985), but, as this assumption is clearly fundamental,
solid evidence is badly needed to confirm it.
     It being virtually impossible to obtain an accurate measure of the
total DM associated with any one galaxy, it is probably realistic to
hope at this stage only to obtain indisputable evidence that every galaxy contains at least some dark matter. This has already been established for spirals over a wide range in luminosity. For ellipticals, the
detection of dark matter is more tentative. We have a very recent report
by Jean Brodie and John Huchra (Brodie and Huchra 1985) of a high halo
velocity dispersion in the globular clusters around M87. More ellipticals might be studied in this way, but there will always be some ambigu-
ity caused by the unknown anisotropy, β, of the globular cluster velocity
ellipsoid (Binney 1982). This is why X-ray studies of ellipticals are
so important, because β must be identically zero for gas in pressure
equilibrium. Hot gas in the galaxy potential well thus gives an unambiguous mass distribution and also the true shape of the potential well,
which can be compared to the shape of the visible isophotes. All this
is possible provided both the gas density and temperature profiles are
accurately known and the X-ray maps have sufficient angular resolution.
The first images from Einstein (Forman, Jones, and Tucker 1985) hint
strongly at dark matter around E's, but really adequate data will not be
available until AXAF. This will be one of the most important applications of this satellite.

Until then, the only other evidence we have for dark matter around E's is indirect: flat rotation curves in disks associated with large spheroids (e.g., the Sombrero [Bajaja et al. 1984], the polar-ring S0 A0136-080 [Whitmore et al. 1982], and the HI-rich S0 NGC 4203 [Burstein and Krumm 1981]), plus the simple fact that groups and clusters dominated by E's and S0's always show the strongest dynamical evidence for dark matter. Analysis of the total baryon content of the latter (Blumenthal et al. 1984) in fact suggests a ratio of dark matter to baryons that is similar to spiral-dominated groups, within the quite considerable errors of measurement.

In addition to E's vs. spirals, there is also the major question of DM in little galaxies vs. big ones. This is a powerful test of elementary-particle models of dark matter that in principle could rule out neutrinos (Tremaine and Gunn 1979, Aaronson 1983, Lin and Faber 1983). Very small dwarf spheroidals are particularly attractive objects in which to search for dark matter, as their exceedingly low surface brightness (Bingelli et al. 1984) may reflect an abnormally low baryon content, perhaps due to some form of ram-pressure stripping (Lin and Faber 1983) or supernovae-driven gas loss (Dekel and Silk 1985). If the baryon deficit is large enough, one may even find a marked and unambiguous DM excess within the optical boundaries of the galaxy, a degree of excess that seems never to occur in larger spirals. According to this argument, dwarf spheroidals with exceptionally low surface brightness like Ursa Minor and Draco could show higher M_{DM}/M_{BARY} and M/L than brighter systems such as Fornax, and that is generally what the data are showing (Aaronson, this conference). However, doubts about stellar radial velocities due to stellar pulsation and binary motion are not yet fully resolved and will not be until a few more years of monitoring are available.

In gas-rich dwarf irregulars, comparable baryon loss has evidently not occurred, and M_{DM}/M_{BARY} in the inner regions might be closer to the one-to-one ratio typical of spirals. An average over the inner parts may therefore not show any clear DM excess, even though there may be much dark matter in the galaxy. In these dwarfs, as in ordinary spirals, one may be forced to infer dark matter from the shape of the rotation curve in the outermost regions. In many of these objects, the gas extends far beyond the optical boundaries, but rotation is often weak, random motions are significant, and the dynamics are difficult to analyze. Nevertheless, better 21 cm images will clearly be a very important tool in studying dark matter in dwarf galaxies in the near future (Kormendy, this conference).

Although the mass distributions of large spirals are better understood than E's or dwarf galaxies, major questions about dark matter still remain. The main goal at present is to study the total mass distribution with radius and decompose it into baryonic and DM components. This process can be criticized on several levels. On the lowest, there is concern that, in the bulge-dominated regions of early-type galaxies, the observed rotation curves may not represent true circular velocities. The prototype example is again the Sombrero, in which the rotation velocity of only 100 km s^{-1} at 1 Kpc indicates a local M/L_B of only 0.5

(Steiman-Cameron 1984), which is quite a lot lower than expected for an old bulge stellar population (Faber and Gallagher 1979). M/L_B also rises dramatically outward, finally reaching 5.0 at 10 Kpc, a reasonable value for bulge-type stars. Perhaps the ionized gas used to trace the inner rotation curve exhibits large random motions in addition to rotation, or perhaps the rotational motion is somehow impeded and slowed by a reservoir of slowly-rotating, million-degree plasma in the bulge, like that seen in E's (Bajaja et al. 1984). The Sombrero is admittedly an extreme example because its bulge is so large. However, if the effect exists generally in the bulge-dominated regions of spirals it could systematically distort attempts to decompose the DM and baryonic components versus radius.

A further, higher objection is that decomposition is never possible without at least one additional assumption. Up to now, this has usually been the "maximum disk" assumption applied to the baryons or an isothermal sphere assumption for the dark matter. Neither of these is well justified. The ratio M_{BARY}/L for disks is not known well enough to justify the maximum disk, whereas recent theoretical results (Blumenthal et al. 1985, Barnes 1985, Gunn and Ryden 1985) indicate that the DM density may be far from isothermal owing to baryonic compression during infall.

If these problems can be solved, for example, by following a new approach suggested by Athanassoula and Bosma based on Toomre's q-index (Freeman, this conference), the overall goal is clear and important: to determine ρ_{BARY} (r) and ρ_{DM} (r) versus Hubble type and other parameters. There is suggestive evidence (Tinsley 1981) that dissipative baryonic infall may be systematically larger in early Hubble types compared to late types. This might be discernible from accurate mass measurements and could play an important role in theories for the origin of Hubble types (Faber 1982a,b).

4.   WHAT IS THE LARGE-SCALE STRUCTURE OF THE UNIVERSE, AND CAN DARK
     MATTER ALONE ACCOUNT FOR IT?

We now come to what I believe is currently the most critical problem in cosmology: the nature and origin of perturbations in the universe on large scales. The issues here are equally observational and theoretical. Observationally, there is still a severe lack of reliable data, although preliminary evidence for 100 Mpc-scale structure has come from galaxy redshift surveys (e.g., the Boötes void [Kirshner et al. 1984] and the Perseus-Pices supercluster [Giovanelli, Haynes, and Chincarini 1983]) and from significant large-scale amplitude in the cluster-cluster correlation function (Bahcall and Soneira 1983). Surveys to study large-scale structure are afflicted currently by two difficult problems: nearby 100-Mpc structures cover large angles on the sky, and it is hard to maintain strict uniformity of data in the face of variable Galactic extinction, seasonal variations in observing conditions, and different equipment in the northern and southern hemispheres. A more fundamental difficulty is that, with current sensitivity, we see out far enough to sample only a few 100-Mpc-sized volumes. We therefore do not know yet whether a largé void such as Boötes is a rare event. To remedy this will unfortunately require extensive redshift surveys at faint levels.

We also need to question what the distribution of L_* galaxies, the subject of virtually all surveys so far, really tells us about the underlying total matter distribution. As noted above, biased galaxy formation may help to form large-scale voids and clumps and is a natural accompaniment to several scenarios, but it also introduces a major new degree of freedom in interpreting the data. It could imply that galaxies do exist in voids but that they are systematically smaller and/or of lower surface brightness than the familiar ones that populate the nearby Local Supercluster. Magnitude- or diameter-limited catalogs would systematically undersample such objects. Deeper, more careful searches using a variety of techniques are needed to answer the question: does a void really mean no baryons, no dark matter, both, or neither?

A related issue is how much weight should be placed on the standard galaxy-galaxy correlation function, ξ_{gal} (Peebles 1980), which has come to be accepted as a key test of all clustering models. As a number-density-weighted index, ξ_{gal} badly underestimates the correlation contribution on short scales from high-density cluster cores, where the stellar (M/L) is low and where additional baryons are present in the form of hot gas. On large scales, the missing contribution by baryons in voids due to biasing could likewise be important. In view of these uncertainties, it is perhaps unwise to view mismatches between models and observations too critically, as is sometimes done.

With regard to theory, there are also problems of practice and principle. There are two main methods so far for estimating the amplitude of ξ_{gal} derived from any of the common fluctuation spectra: N-body simulations and simple linear evolution of the initial density fluctuation spectrum. Both indicate strongly that ξ_{gal} derived from any of the common density fluctuation spectra plus random phases should go negative beyond about 30 Mpc (Dekel 1985). It has been proposed that the observed high amplitude of the cluster-cluster function, ξ_{clus}, on large scales is simply a "super-correlation" effect due to looking at 2-σ or 3-σ peaks in the Abell clusters (Kaiser 1984, Politzer and Wise 1984). This picture also predicts, however, that $\xi_{clus} \sim n^2 \xi_{gal}$, where $n\sigma$ is the average overdensity of Abell clusters. Thus ξ_{clus} should also go negative at 30 Mpc, in contrast to the observations, which show it to be positive out to 100 Mpc (Bahcall and Soneira 1984).

A major question is whether it is possible to cure this problem without abandoning random phases simply by modifying the fluctuation spectrum slightly on scales near 100 Mpc. Recent work (Dekel 1984; Barnes, Dekel, Efstathiou, and Frenk 1985) has obtained good agreement with observation by adding a bump, or discontinuity, to the spectrum at this location, such as might result from a hybrid scenario with two types of DM particles. The critical question, not yet fully explored, is whether such a modification is compatible with the microwave background $\delta T/T$ limits. It may be that the real solution requires abandoning random phases, for example, by invoking large-scale strings (Vilenkin 1981). Several groups are now looking into this prospect.

We also need to question whether ξ_{gal} is really the optimum function for characterizing large-scale structure. It is not clear yet exactly how it is that ξ_{gal} tends to small values beyond 10 Mpc while the void-cluster pattern appears to have much higher amplitude on longer length scales. This suggests that ξ_{gal} is not especially well tuned to the particular structure of voids in our universe, as indeed mathematically it need not be. Perhaps it might be better to develop an alternative statistic keyed to voids, with particular regard to their sizes and shapes.

To summarize, it is not clear at this time what the true amplitude is of matter fluctuations in the universe on 100 Mpc scales and, if large, how these fluctuations may be generated from known, plausible physical processes. With strong observational constraints placed by the microwave background, galaxy counts, and radial velocity surveys, plus strong interest generated by the obvious connections to the early universe, large-scale structure is likely to remain one of the most lively and productive areas of observational cosmology in the near future.

5. HOW HAS DARK MATTER SHAPED THE STRUCTURE OF GALAXIES?

The influence of DM on the visible parts of galaxies began early and lasted through what may be idealized as three phases: initial formation; a period of isolated, self-contained evolution; and any later inter-actions. During the first phase, the central question is whether DM gravity controlled the gravitational collapse of galaxies, at least initially. If DM is the reason why galaxies formed early without viola-ting the microwave background, the answer to this question must be "yes." If so, there are two further questions: what was the resultant angular momentum spectrum of galaxies, and what did a forming protogalaxy look like -- was it a centrally concentrated, rather symmetrical blob, or was it a collection of smaller lumps, each one collapsing simultaneously on its own and developing its own substructure? If the latter, we might have to add dynamical friction to the list of dissipational processes that shaped the structure of visible galaxy cores.

To understand the phase of isolated evolution, we need to know what is special about the dynamics of two-component galaxies in which at least one component, the dark halo, may be triaxial. A number of inter-esting phenomena have been suggested, including angular momentum exchange between the baryons and halo, stabilizing or destabilizing of disk orbits at certain radii, and disk warps induced by triaxial halos or stabilized by spherical ones. The timescale for halo-baryon interactions is es-pecially important. Binney (this conference) speculates that transfer of angular momentum between disk and halo is efficient and that mis-alignment between the two can be only short-lived. He even suggests that gradual accumulation of a disk by infall can, by an adiabatic-invariant process, actually rotate the angular momentum vector of the original bulge to cause alignment between the two, as is observed. Conversely, instability timescales in the outer parts may be quite long, with the result that polar orbits in a flattened or slightly triaxial potential may be stable or at least quite long lived (Steiman-Cameron and Durisen 1982). The nature of stable orbits in tumbling potentials is also rather different from that in static potentials (Steiman-Cameron and Durison 1984; David, Steiman-Cameron, and Durison 1984). The net result is that it is proving quite a bit more difficult to infer the shape of the DM potential from the orientation and morphology of gas and dust lanes than was once hoped.

During the third, or interactive, phase, DM influences the frequency and nature of galaxy-galaxy interactions. Large DM halos considerably increase the cross-section for galaxy-galaxy collisions (White and Sharp 1977), and merger and coalescence are quite efficient whenever two halos strongly interpenetrate, provided the mutual orbital energy is low or negative (White 1978). Recent N-body simulations by Frenk et al. (this conference) suggest that, with a cold DM spectrum, merging of galaxy-sized halos continues to the present. This raises the question of whether galaxies have on average always continued to grow in size, even until now. On the other hand, perhaps the halos do continue to merge, but the luminous cores stay separated for long times. If binary galaxies and small groups have long lifetimes, this must indeed be what happens. It would be important to reassure ourselves that this is physically possible, perhaps using realistic two-component N-body simulations of binaries and small groups. On the other hand, if the luminous portions of galaxies continue to grow in size, too, this is an effect that could be searched for using lookback observations with Space Telescope and other instruments.

6. WHAT NEW AVENUES LOOK PROMISING FOR THE NEAR FUTURE?

In conclusion, we should mention a few new approaches to dark matter that have not been much discussed up to now but which may yield important insights in the near future. Most of these are well represented in the agenda of this conference. To me, the most interesting is gravitational lenses. As Ed Turner emphasizes (this conference), it is striking that an obvious lens candidate is apparent in only one of the known lenses. The situation is highly reminiscent of the original "missing mass" problem, in which a gravitational field was clearly present, but the matter causing it was invisible. Thus, lenses could be telling us that there are major mass concentrations in the universe that are not centered on visible galaxies. This would certainly cause a profound shift in our current picture of dark matter.

A second promising area is lookback studies to cosmological redshifts. The possibility of an increase in mean galaxy luminosity as a function of time owing to mergers was mentioned above. A related phenomenon is an evolution in the amplitude and slope of the galaxy-galaxy correlation function. A pioneering step in lookback studies of ξ_{gal} has been taken in a recent paper by Koo and Szalay (1984) based on galaxy counts, but the definitive treatment will have to await deep redshifts from the biggest telescopes.

Finally, there is the great arena of particle physics, to which astronomers look for theoretical inspiration and perhaps even experimental confirmation of dark matter. There are several possibilities for direct detection, including the measurement of a non-zero neutrino mass, watching axions interact with a magnetic field, or detecting phonons as DM particles collide with a crystal lattice at the weak interaction rate. These and other interesting possibilities will doubtless come to light in the next days of discussion.

I would like to thank my colleagues at Santa Cruz, George Blumenthal, Joel Primack, and Visiting Professor Avishai Dekel, for many helpful discussions.

This work was partially supported by NSF grant AST 82-11551.

REFERENCES

Aaronson, M. 1983, Ap.J. Lett., 266, L11.

Bahcall, J. 1984, Ap.J., 287, 926.

Bahcall, N.A., and Soneira, R. 1983, Ap.J., 270, 20.

Bajaja, E., van der Burg, G., Faber, S.M., Gallagher, J.S., Knapp, G.R., and Shane, W.W. 1984, Astron. Ap., 141, 309.

Barnes, J. 1985, private communication.

Barnes, J., Dekel, A., Efstathiou, G., and Frenk, C. 1985, Ap.J., 295, 368.

Binggeli, B., Sandage, A.R., and Tarenghi, M. 1984, A.J., 89, 64.

Binney, J.R., 1982, Ann. Rev. Astron. Astrophys., 20, 399.

Blumenthal, G.R., Faber, S.M., Flores, R., and Primack, J.R. 1986, Ap.J. in press.

Blumenthal, G.R., Faber, S.M., Primack, J., and Rees, M. 1984, Nature, 311, 517.

Bond, J.R., and Efstathiou, G. 1984, Ap.J. Lett., 285, L45.

Brodie, J., and Huchra, J. 1985, private communication.

Burstein, D., and Krumm, N. 1981, Ap.J., 250, 517.

David, L.P., Steiman-Cameron, T.Y., and Durisen, R.H. 1984, Ap.J., 286, 53.

Dekel, A. 1984, Ap.J., 284, 445.

Dekel, A. 1985, private communication.

Dekel, A., and Silk, J. 1985, preprint.

Faber, S.M. 1982a, Astrophysical Cosmology: Proc. Study Week on
    Cosmology and Fundamental Physics, eds. Brück, H.A., Coyne, G.V.,
    and Longair, M.S. (Vatican: Pontifical Scientific Academy),
    pp. 191-218.

Faber, S.M. 1982b, ibid., pp. 219-232.

Faber, S.M., and Gallagher, J.S. 1979, Ann. Rev. Astron. Astrophys., 17,
    135.

Forman, W., Jones, C., and Tucker, W. 1985, Ap.J., 293, 102.

Giovanelli, R., Haynes, M., and Chincarini, G. 1983, as quoted by
    Oort, J.H., Astron. Astrophys. 21, 373, 1983.

Gunn, J.E. 1982, ibid., pp. 233-260.

Gunn, J.E., and Ryden, B. 1985, poster paper this conference and
    private communication.

Kaiser, N. 1984, Ap.J. Lett., 284, L9.

Kirshner, R.F., Oemler, A., Schechter, P.L., and Shectman, S.A. 1984,
    Ap.J.,

Koo, D.C., and Szalay, A.S. 1984, Ap.J., 282, 390.

Lin, D.N.C., and Faber, S.M. 1983, Ap.J. Lett., 266, L17.

Ostriker, J.P., and Cowie, L.L. 1981, Ap.J. Lett., 243, L127.

Peebles, P.J.E. 1980, The Large-Scale Structure of the Universe,
    (Princeton: Princeton University Press), Chap. 3.

Politzer, H.D., and Wise, M.B. 1984, Ap.J. Lett., 285, L1.

Sandage, A.R., and Tammann, G. 1983, in Large-Scale Structure of the
    Universe, Cosmology, and Fundamental Physics, eds. G. Setti, and
    L. Van Hove (Geneva: ESO-CERN), p. 127.

Schramm, D., and Freese, K. 1985, preprint.

Steiman-Cameron, T. 1984, Master's thesis, Indiana University.

Steiman-Cameron, T., and Durisen, R.H. 1982, Ap.J., 257, 94.

Steiman-Cameron, T., and Durisen, R.H. 1984, Ap.J., 276, 101.

Tinsley, B.M. 1981, Mon. Not. Roy. Astron. Soc., 194, 63.

Tremaine, S.D., and Gunn, J.E. 1979, Phys. Rev. Lett., 42, 407.

Vandenberg, D. 1985, colloquium delivered at U.C. Santa Cruz.

Vilenkin, A. 1981, Phys. Rev., D24, 2028.

Vittorio, N., and Silk, J. 1984, Ap.J. Lett., 285, L39.

White, S.D.M. 1978, Mon. Not. Roy. Astron. Soc., 184, 195.

White, S.D.M., and Rees, M. 1978, Mon. Not. Roy. Astron. Soc., 183, 341.

White, S.D.M., and Sharp, N.A. 1977, Nature, 269, 395.

Whitmore, B.C., Schweizer, F., and Rubin, V.C. 1982, Bull. A.A.S., 14, 643.

Zwicky, F. 1933, Helv. Phys. Acta., 6, 110.

DISCUSSION

STEIGMAN: I'd like to make some comments relating to the issue of dark
baryons. I think it is probably unfair to argue that there is a
discrepancy between $\Omega = 0.01$ in visible matter and what is predicted by
Big-Bang nucleosynthesis. I think the prediction goes down to about
0.01 and up to about 0.15, so there's a big range mostly connected with
the uncertainty in the Hubble parameter. I'm also very pleased that
you're emphasizing the importance of the hot gas. It's a point that
Schramm and I have made over the years, that probably most of the
baryons in the Universe are dark by the conventional astronomer's
definition. They just happen to be shining in x-rays. And for that
reason I would urge everyone here to distinguish between dark baryons
and dark matter. When you ask, for example, for the ratio of baryons
to dark matter, remember that some of the dark matter is clearly in the
form of baryons.

RUBIN: Sandy, let me answer your question about the Sombrero galaxy,
although the question is a general one. The evidence that we don't
observe infall in the gas is that along the minor axis you see no
velocity other than the systemic velocity. This seems to be true
generally. In fact, in one specific case where we do see a normal
rotation curve and rather peculiar things happening in the inner
regions of the galaxy, things that I thought might be evidence for
infall, the most plausible geometry says that the gas is moving the
other way.

FABER: What do you think, then, about the resulting M/L ratios for the
stars? Do you agree with the basic premise that in the Sombrero, at
least, you would find a rather small value of M/L?

RUBIN: There are certainly things going on in the inner regions of
galaxies that we don't understand. But in the work that we have done
we are led to the conclusion that dark matter is important on very
small scales, and that the dynamics in the inner parts of galaxies are
not being controlled by the luminous stars.

FABER: I think it's fair to say that in no galaxy does one see the
rapid rise to very large velocities and the decline farther out that
one might have expected from the mass distribution derived from the
light using conventional M/L ratios.

RUBIN: I agree.

DRESSLER: The surprisingly low rotation speed of gas in the central
disk of the Sombrero galaxy is also found by Fillmore, Boroson and
Dressler in several other early-type spirals. The gas rotation curves
consistently fall below the velocity predicted from the stellar
kinematics. We suggest that the gas is on non-circular orbits in these
bulge-dominated regions, as would occur, for example, if it were being
shed by stars in the spheroid.

FABER: You make the very interesting point that we have an independent estimate of the mass present from the <u>stellar</u> kinematics; this is in rough agreement with the usual estimates for an old stellar population but in strong disagreement with masses derived from the emission-line gas. That is exactly the point I wanted to make, but I didn't realize that there existed good data on galaxies other than the Sombrero.

FREEMAN: To qualify what Dressler just said: There are other ways of probing the potential in the inner parts of systems like the Sombrero - one can use the kinematics of the stars themselves. Kormendy and Illingworth have measured the rotation velocity and stellar dispersion in the inner parts of the Sombrero, and Jarvis has made simple dynamical models with circular rotation. These reproduce the stellar kinematics beautifully, both the rotation and the dispersion, and not just with radius but also up off the plane. Jarvis finds an M/L ratio of about 7.

P. QUINN: A comment about dark matter in ellipticals. Spirals have the advantage that we can make excellent assumptions about the orbits in their outer parts and so can say how much mass is out there. For ellipticals, anisotropy makes it not so simple, as you pointed out. However, the shells seen around some ellipticals can be used as test particles. Hernquist and I have just made an extensive study of the kinematics of shells. Shells are a complicated phenomenon, but we are confident that they are telling us a lot about the distribution of dark matter in at least some galaxies.

FABER: Yes, you're right.

E. TURNER: When you conclude from the microwave background limits that dark matter must play a dominant role in galaxy formation, do you discount the hydrodynamic, explosive galaxy formation theories or do these provide an escape from that conclusion?

FABER: Yes, in a sense I am discounting the hydrodynamic approach, since the origin of the explosive seeds has always seemed to me <u>ad hoc</u>. However, there is also a problem in matching the microwave background limits on cluster-sized masses and above. On these scales it is not clear that hydrodynamic effects are strong enough to create structure.

SCHECHTER: Your other firm conclusion was that there is clear evidence for non-baryonic dark matter. Scanning the meeting schedule, I am discouraged by the lack of any discussion of the constraints imposed on baryonic dark matter by the microwave background measurements. I hope the experts in this area will tell us more about it. Because I, for one, am loath to admit that the dark matter must be non-baryonic.

USON: The theoretical implications of the microwave background observations that Dave Wilkinson and I have made rest on a very strong assumption, which is that the initial perturbations had a Gaussian distribution. If this assumption is relaxed, the limits could go up by as much as a factor of 2.5 (for rather pathological initial conditions). Even then, isothermal fluctuations would still be allowed for $0.1 < \Omega < 0.3$. (Although these fluctuations are currently not popular, they could become fashionable again.) We have increased the size of our sample but the results are not yet available.

DEKEL: When considering constraints from fluctuations in the microwave background, we should consider the possibility of smearing by reionization. The energy may come from Population III stars or from galactic explosions, and can smear out fluctuations with scales smaller than 7°. Thus, this can not be used as a strong argument for non-baryonic dark matter.

FABER: I'm not familiar with the details, but I thought that the 7° limit requires high ionization until rather recent epochs, and consequently very high energy demands that are hard to satisfy.

LAKE: Just following up on Schechter's comment. There's a gap at the low end between the amount of material visible in the stars (including the stuff confined to the disk which is measured by the Oort limit) and the baryonic limit. However, there's an overlap between the value of Ω derived from large-scale surveys and the largest amount you measure dynamically. So the only real gap is the gap between stars and baryons, not between stars and dark matter.

OSTRIKER: The question has been raised in your talk and in prior questions as to whether diffuse (gaseous) baryons could contribute significantly to the mass density. The situation has not changed in any favorable way over the last decade. Cold gas is severely restricted by the Gunn-Peterson test. Very hot gas is limited by the x-ray background and by the Zel'dovich-Sunyaev effect on the microwave background. If the gas were smoothly distributed and kept at 0.1 to 0.3 keV a value of Ω_b of 0.1 would be possible. There are physical problems in keeping gas at this temperature and conflicts with pressure in the Lα clouds. Such gas will, in any case, soon be observable using the Helium Gunn-Peterson test.

STEIGMAN: In recent work, Henriksen, Mushotzky and Cowie studied the Coma and Perseus clusters out to about 3 Mpc radius; they argue that something like 30 to 40% of the binding mass on that scale is in hot gas. For some other Abell clusters, they quote even larger fractions. If that evidence were to stand up - and I don't know it well enough to judge it - then are we all in trouble trying to hide nine times as much non-baryonic mass as baryonic mass.

FABIAN: Field and I have just written a paper on the x-ray background. We get Ω_b = 0.2 or 0.3 and in fact can fit the x-ray background better than a black body spectrum fits the microwave background.

BURKE: For what range of Hubble constants?

FABIAN: The scaling is $H^3\Omega^2$=constant. We're using $H=50$ km s^{-1} Mpc^{-1}.

OSTRIKER: The energy requirements of such a picture make explosive galaxy formation seem pitifully unimaginative (laughter).

FABIAN: The gas has to be in energy equipartition with the microwave background, like the gas in our Galaxy.

FELTEN: I would like to ask Andy Fabian to clarify what the x-ray background tells us about Ω_b. My understanding is that the intergalactic medium could emit the x-ray background if it were very hot ($T > 10^8$ K) and if its density were $\Omega_b \approx 0.2 - 0.3$. On the other hand, if we attribute the x-ray background to something else, then we could set T much lower, but still high enough to avoid the Gunn-Peterson constraint (say $T \sim 10^5$-10^6 K). We could then allow Ω_b to be as large as unity without violating any observation. Jerry Ostriker suggested that this isn't possible, but the arguments against it may not be conclusive.

FABIAN: You are quite right as regards the x-ray background. (Guilbert and I did allow about 15% of the background to originate in a power-law spectrum from point sources.)

FABER: The consensus seems to be that we can have at least $\Omega_b = 0.1$ and possibly as much as $\Omega_b = 0.2$ or 0.3 in diffuse hot gas. That is easily enough to be cosmologically significant, for if we apply the customary ratio of dark matter to baryons of ~ 10, we could easily have enough total matter to close the universe.

Dark Matter in the Galactic Disk

John N. Bahcall
Institute for Advanced Study
Princeton, New Jersey 08540

ABSTRACT. The Poisson and Vlasov equations are solved self-consistently for realistic Galaxy models which include multiple disk components, a Population II spheroid, and an unseen massive halo. The total amount of matter in the vicinity of the Sun is determined by comparing the observed distributions of tracer stars, samples of F dwarfs and of K giants, with the predictions of the Galaxy models. Results are obtained for a number of different assumed distributions of the unseen disk mass. The major uncertainties, observational and theoretical, are estimated. *For all the observed samples, typical models imply that about half of the mass in the solar vicinity must be in the form of unobserved matter.* The volume density of *unobserved* material near the Sun is about $0.1 M_\odot pc^{-3}$; the corresponding column density is about $30 M_\odot pc^{-2}$. This so far unseen material must be in a disk with an exponential scale height of less than 0.7 kpc. If the unseen material is in the form of stars with masses less than $0.1 M_\odot$, then the nearest such object is about 1 pc away and has a proper motion of more than 1 arcsecond per year.

1. Introduction

The main results that I wish to convince you of are (Bahcall 1984a,b):

$$0.5 \leq \frac{\rho_{unobserved}(0)}{\rho_{observed}(0)} \leq 1.5 \qquad (1)$$

and

$$z_{scaleheight} \leq 0.7 kpc \qquad (2)$$

The first equation says that the amount of unobserved material in the vicinity of the Sun is between 0.5 and 1.5 times the already observed material. The second equation says that the exponential scale height of the unobserved material, if it is a single population, must not exceed 0.7 kpc. Thus about half of the matter in the vicinity of the Sun is in the form of unseen disk material which has a scale height of less than 0.7 kpc. The unseen material that is inferred from galaxy rotation curves at large galactocentric distances and from applying the virial theorem to groups and clusters of galaxies may not be the same as the unobserved disk matter.

17

J. Kormendy and G. R. Knapp (eds.), Dark Matter in the Universe, 17–31.
© *1987 by the IAU.*

As we have just heard from Sandy Faber and as Mike Turner will tell us in more detail, the unobserved material at large galactic radii and in clusters of galaxies is often discussed by particle physicists in terms of dissipationless particles (various 'inos'), while the unseen disk material is presumably dissipational. It is possible that the unseen material in the disk consists of stars or planets that are not massive enough to burn hydrogen and hence are of too low a luminosity to have been detected by searches carried out so far.

If you are willing to take the results in equations (1) and (2) on faith, you can doze through the rest of the talk without missing very much.

Before we get down to the justification of the main results, I want to remind you of a bit of the history of this subject because this perspective may have a special significance for the participants in this symposium. Oort's (1932,1960) early studies of the total amount of matter in the solar vicinity led to what may have been the first astronomical suggestion of a large "missing mass." Nevertheless, most of the explanations for missing matter that will be discussed at this conference do not account for the unseen matter near the Sun.

2. The Method

The method of weighing the matter in the local neighborhood that I have used, and which Oort pioneered, can be summarized as follows. A detailed model of the observed matter (in stars, gas, and clouds) is constructed from all the available observations. In addition, the density distribution and velocity dispersion of a set of tracer stars perpendicular to the galactic plane is taken from published measurements. Theoretical models are then computed for the expected distribution of tracer stars in different gravitational potentials (mass distributions). The amount of matter that is actually present in the Galaxy is determined by comparing the observed and computed distributions.

The problem is similar to computing the distribution of an isothermal atmosphere (since for the tracer stars of interest the velocity dispersion changes much more slowly with height above the plane than does the density). Clearly, the more matter there is close to the plane. the more quickly will the density fall off with height above the plane.

The availability of modern computers has made possible important improvements in the theoretical analysis of this problem at the same time that better observational samples of tracer stars have been obtained. I have taken advantage of these developments to sharpen the determinations of the total amount of matter in the solar vicinity, using more realistic Galaxy models and more accurate theoretical solutions. I have solved numerically the combined Poisson and Vlasov equations for the gravitational potential of Galaxy models consisting of realistically large numbers of individual isothermal disk components in the presence of a massive unseen halo. Most previous calculations were carried out without requiring self-consistency between the Poisson and Vlasov equations. For example, in Oort's work the equations were solved separately. In the solutions that I will discuss, the distribution functions that solve Vlasov's equation for the observed matter and the tracer stars also depend on the potential that appears in Poisson's equation and generate, through their associated densities, the mass densities in Poisson's equation. I have carried out the calculations with different assumptions about the unseen matter and have compared the results with the observed number densities of F dwarfs and K giants versus height above the plane, assuming that the F dwarfs and K giants are reasonably faithful tracers of the total gravitational potential. Because the solutions are obtained with the aid of a computer, I can make more quantitative estimates of the errors by varying all of the parameters and by trying many different models.

Incidentally, the work of Oort and other previous investigators referred only to the equivalent of equation (1) above. The derivation of equation (2) requires combining the studies of the motion perpendicular to the plane with knowledge of the Galaxy rotation curve.

3. The Input Data

Table 1 summarizes the relative amounts of the observed mass components, and their velocity dispersions (i. e., temperatures) that were derived - using data from many sources - by Bahcall and Soneira (1980) and by Hill, Hilditch, and Barnes (1979), often referred to as the B&S and the HHB Galaxy models. The models contain many observed disk components (typically 14) whose characteristics are determined by local measurements, a Population II spheroid inferred from faint star counts, different models for the unobserved disk components, and an unseen massive halo whose normalization is fixed by the solar rotation velocity. The mass fractions are defined in terms of the total *observed* mass density (in stars, gas, and dust), i. e.,

$$A_i = \frac{\rho_i(0)}{\rho_{obs}(0)}. \tag{3}$$

I use the difference between the results obtained with the B&S and the HHB Galaxy models as one measure of the uncertainty. The two models are similar, since the luminosity function of the disk stars is reasonably well determined (see Wielen 1974) over much of its range. The B&S and HHD models mainly differ in the mass density assigned to white dwarfs and to the interstellar matter. In both cases, I have made use of more recent determinations. For example, fewer white dwarfs are observed at faint absolute magnitudes than had been expected on the basis of earlier theoretical estimates. I have used in the B&S model the observed number density | Green (1980), Liebert, Dahn, Gresham, and Strittmatter (1979) | down to $M_V = 17.2$ and a white dwarf mass of 0.6 $M_\odot$. I have also adopted the value for interstellar matter density that has been estimated by Spitzer (1978), which is consistent with the recent value inferred by Sanders, Solomon, and Scoville (1984). This value is rather larger than the interstellar matter density that was used by HHB.

Previous theoretical studies of the total amount of matter in the vicinity of the Sun have been limited mainly to simplified Galaxy models with one or, at most, a few disk components and no spherical component. The previous solutions were also limited either by what was tractable analytically or by assuming a numerical form for the total matter density that was independent of the potential. Since I have access to a VAX computer, I have calculated numerical models with many different sets of input data and a several assumptions about how the unseen material is distributed. I estimate the major uncertainties in the determination of the distribution of unseen matter by comparing an extensive collection of theoretical models with the available data.

4. The Simplest Model for the Unseen Material

Since we haven't yet observed the unseen material, we don't know how it is distributed. Therefore we have to try different models for the unseen material to see how the results depend upon our assumptions.

Table 1

The Galaxy Model for Observed Components[a]

Component (1)	B & S Mass Fraction (A_i) $(M_\odot \ pc^{-3})$ (2)	$< v_z^2 >^{1/2}$ $(km \ s^{-1})$ (3)	HHB Mass Fraction (A_i) $(M_\odot \ pc^{-3})$ (4)
Main Sequence Stars:			
$M_V <$ 2.5 mag	0.021	4	0.038
2.5 mag $\leq M_V \leq$ 3.2 mag	0.015	8	0.019
3.2 mag $\leq M_V \leq$ 4.2 mag	0.031	11	0.033
4.2 mag $\leq M_V \leq$ 5.1 mag	0.035	21	0.034
5.1 mag $\leq M_V \leq$ 5.7 mag	0.025	20	0.023
5.7 mag $\leq M_V \leq$ 6.8 mag	0.037	17	0.036
	0.0358	8	
	0.0626	13	
$M_V \leq$ 6.8 mag	0.0536	15	0.0262
	0.0626	20	
	0.0834	24	
Subgiants and Giants	0.016	20	
White dwarfs	0.052	21	0.185
Atomic H and He			0.287
	0.469	4	
Molecular H and dust			0.083
Spheroid	0.001	100	......
Total	0.0958	...	0.108

[a]Disk luminosity functions and velocity dispersions from Wielen (1974).

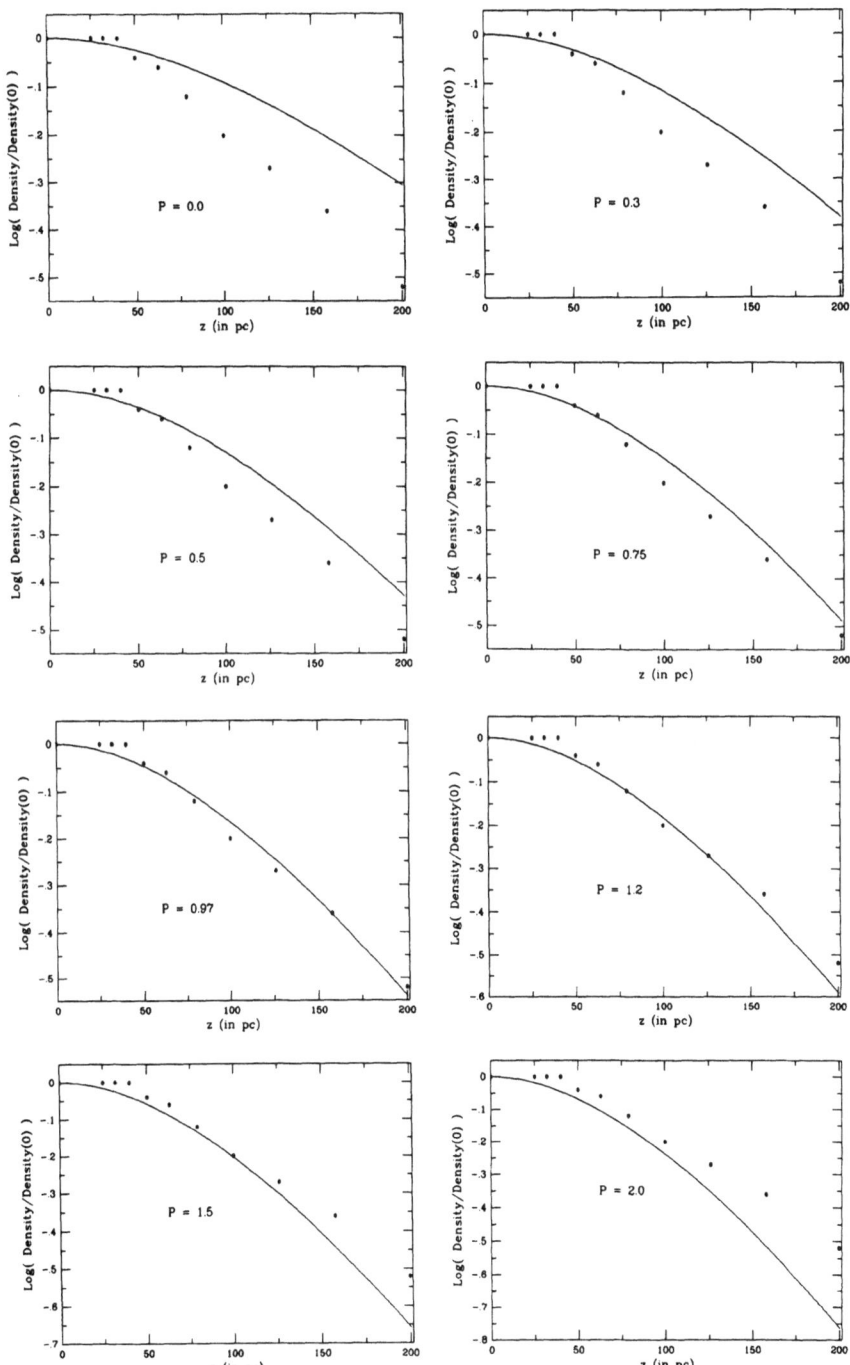

Figure 1. Comparison of measured versus computed number densities of F stars. The measured densities are taken from the work of Hill *et al.* (1979). The mass in unobserved material is assumed to be proportional to the mass in observed material, stellar and interstellar, with proportionality constant P.

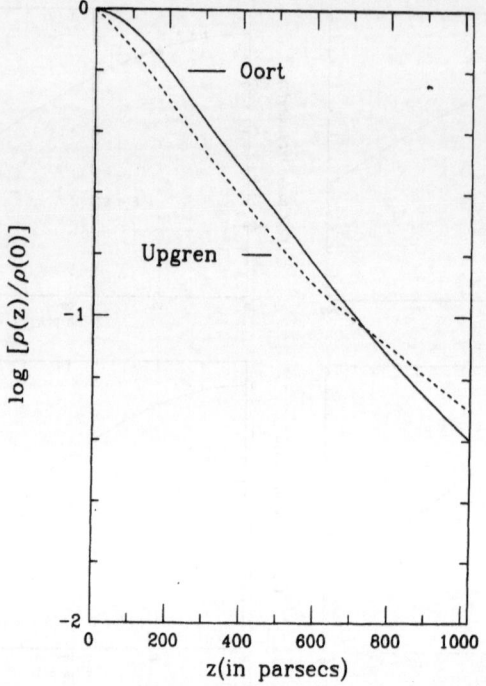

Figure 2. The comparison of the Oort (1960) and Upgren (1962) density distribution using the average visual magnitude and absorption adopted in Bahcall (1984b).

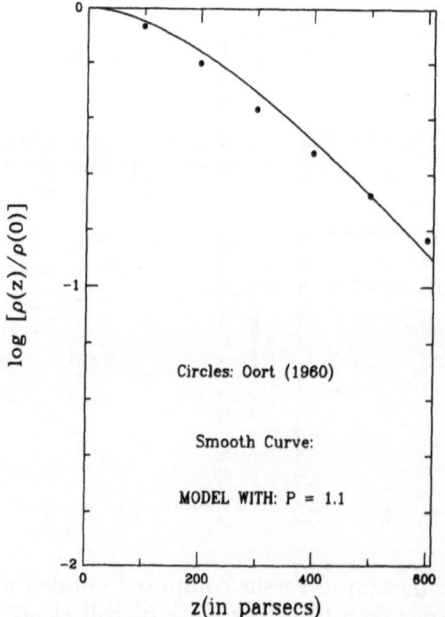

Figure 3. The best-fit model for the Oort K-giant data using the simple scale model defined in section III of this paper.

There is one model which is uniquely simple and is characterized by only one parameter, the overall scale factor, P, between observed and unobserved material. In this illustrative model, the unobserved mass density in every component, i , is proportional to the observed mass density in the same component,

$$(\text{Observed})_i \equiv P \times A_i, \qquad\qquad (4)$$

and the unobserved and observed velocity dispersions for the ith component are equal. Of course, this is only one of the many different models that have been explored.

Figure 1 is a chi-by-eye illustration of why one needs missing matter in the disk. I compare in this figure the measured star densities of Hill, Hilditch, and Barnes (1979) with a sequence of models computed assuming that the scale factor P = 0.0, 0.3, 0.5, 0.75, 0.97, 1.2, 1.5, and 2.0. You can judge for yourself the improvement in the agreement between model and observation as the amount of material is increased from no unobserved material (P = 0.0), through the best fit (P = 0.97), to a worsening of the fit at large ratios of unobserved to observed matter (up to an unacceptable P = 2.0). For small values of P, the observed distribution of F stars falls off more rapidly than does the calculated distribution. Therefore, we have to add additional unseen matter to pull down the calculated curve. A formal statistical treatment (Bahcall 1984a) of the fit gives for this case $P = 0.97 \pm 0.23$. Does that agree with your chi-by-eye assessment of the uncertainty? Incidentally, the flatness of the observed distribution within the first 40 pc (the first three data points in Figure 1) is an artifact of the way that Hill *et al.* reduced their data and does not reflect any real observational constraint on the shape of the distribution at small heights above the plane.

Figure 2 shows the agreement between the number densities for the Oort(1960) and the Upgren(1962) samples of K giants. Figure 3 shows the best fit for the Oort data, again with the simple proportional model.

5. Other Models and Equation (1)

I have explored many possible models for the distribution of unobserved material. I have calculated , e. g., models in which the unobserved material has a small velocity dispersion (like the interstellar material), has a distribution like the older stars (e. g., like the white dwarfs or K giants), is distributed like all the observed stars (ignoring the interstellar material), or has the maximum scale height consistent with the Galaxy rotation curve.

Table 2 gives the ratio of unobserved to observed mass density for twenty-eight detailed models (see Bahcall 1984b for a description of these models) that fit the observed distribution of K giants. The models represent numerical solutions of the combined Poisson-Vlasov equation for different input parameters, as well as for several assumptions about the distribution of the unobserved disk material. There are separate columns referring to the observed K-giant samples of Oort (1960) and to the Upgren (1962) K giant density distributions. For both the volume and the column density, the typical best-fit model has, for the Oort densities, about equal amounts of unobserved and observed material. For the Upgren densities, the typical best-fit model has about 40% more unobserved than observed matter. *These averages are only illustrative since at most one of the models considered for the distribution of unseen matter can be correct.* Similar results are obtained by comparing theoretical models to the observed sample of F dwarfs (Bahcall 1984a).

TABLE 2

RATIO OF UNOBSERVED TO OBSERVED DISK MATERIAL

Row[a] (1)	Oort Densities		Upgren Densities	
	$\dfrac{\rho_{unobs}(0)}{\rho_{obs}(0)}$ (2)	$\dfrac{\sigma_{unobs}}{\sigma_{obs}}$ (3)	$\dfrac{\rho_{unobs}(0)}{\rho_{obs}(0)}$ (4)	$\dfrac{\sigma_{unobs}}{\sigma_{obs}}$ (5)
1..............	1.1	1.1	1.5	1.6
2..............	1.6	1.6	2.1	2.1
3..............	0.6	0.6	1.0	1.0
4..............	0.9	0.9	1.3	1.4
5..............	1.3	1.3	1.8	1.8
6..............	1.3	1.1	1.8	1.6
7..............	0.6	1.2	0.8	1.6
8..............	0.7	0.7	1.0	1.0
9..............	0.4	0.7	0.5	1.0
10.............	2.4	0.5	2.6	0.5
11.............	1.5	0.3	2.2	0.5
12.............	0.6	2.5	0.7	3.2
13.............	1.1	1.1	1.5	1.6
14.............	1.5.	1.5	2.0	2.0
Average....	1.1	1.1	1.5	1.5

[a]Disk luminosity functions and velocity dispersions from Wielen (1974).

I conclude that *a typical best-fit model implies that about half of the disk material at the solar position has not yet been observed.* This conclusion, which is sunmmarized in equation (1), is in qualitative agreement with the previous major studies (see e. g., Oort 1932, 1960, Hill 1960, Woolley and Stewart 1967, Lacarrieu 1971, and Hill, Hilditch, and Barnes 1979), although I find a larger ratio of unobserved to observed matter than in some of the earlier analyses. The present investigation establishes more firmly and specifically the existence of unobserved disk material. The added confidence in the results arises because: 1) more realistic Galaxy models are used; 2) the Poisson and Vlasov equations are solved self-consistently; 3) improved (and more homogeneous) observational data are utilized; and 4) many theoretical models are compared with the observations in order to estimate the uncertainties.

6. But...

I do not want to sound too satisfied, however. There is no modern data sample of K giants; the samples that I have been forced to use are a quarter of a century old! The stars are very bright (apparent magnitudes less than 10) so that it would be very easy to get a much improved sample with modern techniques, using spectroscopic observations to assure that the population was homogeneous with height above the plane. The velocity dispersions of both the K giants and the F dwarfs could be improved with modern radial velocity techniques. The absolute magnitude of the tracer stars should be redetermined using Hipparcos as well as the soon-to-be-published Yale parallax catalogue.

The largest identifiable source of uncertainty in the Oort limit is the unknown form of the distribution of unseen matter (see the last row of Table 9 of Bahcall 1984b). In the future, it should be possible to constrain sharply the distribution of unseen matter by requiring consistency with observations of several carefully selected samples of tracer stars with different scale heights.

7. The Rotation Curve and Equation (2)

The unseen material must be mostly in a disk form, i.e., be dissipational. If all of the material were in a relatively round halo, then the rotation velocity at the solar position would have to be as large as 500 $km \ s^{-1}$. For a given local volume density of unseen mass, the total amount of mass required in a round halo is larger than the amount of mass needed in a disk by about the ratio of the galactocentric distance of the Sun to the disk scale height, i.e., by more than an order of magnitude. The largest scale height of the unseen disk material that is consistent with the solar rotation velocity is 0.7 kpc (see row 12 of Tables 5 and 6 of Bahcall 1984b). I determined this value by making a succession of models in which the unseen material had a progressively larger vertical velocity dispersion. For each model I required that the predicted distributions of tracer stars fit the observations of F dwarfs and K giants and also be consistent with the observed (220 km/s) rotation velocity at the solar galactocentric position. The maximum allowed vertical velocity dispersion is 40 km/s.

8. What is it?

If the missing material is in the form of stars that are not massive enough to burn hydrogen ($M < 0.1 M_\odot$), then the nearest such brown dwarf is probably less than a parsec away and has a proper motion of more than an arcsec per year. Brown dwarfs of the required number density might be detected in future dedicated large area surveys for very red, high proper motion objects. If the unseen material has a typical mass like that of Jupiter, the nearest such object would be about 0.2 pc from the Sun, moving with a proper motion of order 5 arcseconds per year. Such remarkable objects might be discoverable with IRAS.

Moti Milgrom will describe at this conference the work he has done in constructing a theory of modified dynamics to account for missing matter (see, for example, Milgrom 1983 and references in the talk on this subject in the present volume). The modification that Milgrom has proposed implies that there should be missing matter in the z-direction, as well as in the plane of the Galaxy, and that this motion should be described by the simple scale model discussed in section IV of the present talk. In fact, the modified dynamics model predicts that $P \sim 1$, in agreement with the results summarized in equation (1) and in Table 2.

It may be that this agreement is just a coincidence. On the other hand, the value of P could easily have been 0.1 or 10.0. I think we should be alert to the possible significance of the fact that the order of magnitude is the same for the "missing mass" inferred locally from the z-motion and globally from arguments about galactic halos. In most conventional models there must be two different explanations for the "missing mass". Perhaps there is a deep connection between the disk and the halo missing mass and they are both manifestations of the same phenomenon (which of course may have nothing to do with modified dynamics).

9. The Harmonic Approximation

Professor Einasto will report at this conference on the important investigations of Professor G. G. Kuzmin on the problem of determining the total amount of matter at the solar vicinity. The references to this work are mainly in the Russian literature and are contained in Professor Einasto's report. The Soviet work is not as well known in the West as it should be and therefore I will comment on the approximations that are involved.

The basic assumption that is made by Professor Kuzmin and his collaborators is that the gravitational potential is quadratic in height above the galactic plane, i. e.,

$$\phi(z) \approx 2\pi G \rho_{total}(0) z^2 \qquad (5)$$

and therefore that the motion of the tracer stars is harmonic. The fractional error in the potential that is caused by this approximation is

$$\frac{\Delta \phi}{\phi} \approx -\pi G \rho_{total}(0) z^2 / 3\sigma_{total}^2 , \qquad (6)$$

where σ_{total} is a characteristic velocity dispersion for all of the matter (observed and unobserved). Suppose that we want to measure the total matter density using a set of tracer stars with a known velocity dispersion, σ_{tracer}. We need to have an accurate solution of Poisson's equation for at least two exponential scale heights of the tracer stars, i. e., we must have a solution that is valid for $z \leq z_{tracer}$, where:

$$z_{tracer}^2 = \sigma_{tracer}^2 / \pi G \rho_{total}(0) . \qquad (7)$$

Let ϵ be the maximum allowed fractional error in the potential and insert equation (7) into equation (6). Then the velocity dispersion of the tracer stars must satisfy

$$\sigma_{tracer} \leq (\epsilon/0.1)(\frac{\sigma_{total}}{10 \; km/s})5 \; km/s. \qquad (8)$$

In addition, the sample of tracer stars must be relaxed and homegeneous. I do not know of any sample of observed stars that satisfies simultaneously equation (8) and the other requirements.

This work was supported in part by the National Science Foundation grants PHY-8217352 and by NAS8-32902.

References

Bahcall, J. N. 1984a, *Ap. J.* **276**, 169.
Bahcall, J. N. 1984b, *Ap. J.* , **287** , 926.
Bahcall, J. N. and Soneira, R. M. 1980. *Ap. J. Suppl.* **44**, 73.
Green, R. F. 1980, *Ap. J.* **238**, 685.

Hill, E. R. 1960, *Bull. Astr. Inst. Netherlands* **15,** .
Hill, G., Hilditch, R.W., and Barnes, J.V. 1979, *M.N.R.A.S.* **186,** 813.
Lacarrieu, C.T. 1971, *Astron. and Astrophys.* **14,** 95.
Liebert, J., Dahn, C. C., Gresham, M., and Strittmatter, P. A. 1979,
      *Ap. J., * **233,** 226.
Milgrom, M. A. 1983, *Ap. J.* **270,** 371.
Oort, J. H. 1932, *Bull. Astr. Inst. Netherlands* **6,** 249.
Oort, J. H. 1960, *Bull. Astr. Inst. Netherlands* **15,** 45.
Sanders, D. B., Solomon, P. M. and Scoville, N. Z. 1984, *Ap. J.* **276,** 182.
Spitzer, L. 1978, *Physical Processes in the Interstellar Medium*
      (New York: Wiley).
Upgren, A. R. 1962, *A. J.* **67,** 37.
Wielen, R. 1974, *Highlights of Astronomy,* **Vol. 3,** 395, ed. Contopoulos, G.,
      (Dordrecht: D. Reidel).
Woolley, R. and Stewart, J. M. 1967, *M.N.R.A.S.* **136,** 329.

DISCUSSION

LAKE: If you make the mass of the dark objects large enough, is there
any chance that bound triples can masquerade as binaries? Have you
actually done the dynamics of these things, and do you find reasonable
masses for the stars?

J. BAHCALL: In many of the cases the stars in binaries with 0.1 pc
separations have identical radial velocities that are stable to better
than 0.5 km s^{-1}. These measurements are by D. Latham.

LAKE: But do you know enough about them to calculate a consistent mass
function? Is there a chance that they could be bound triples with
something of say 10 M_Θ?

J. BAHCALL: All that is required for my argument is that they exist,
because they are disrupted so easily. There is no evidence for binary
motion. They are so fragile that you can't put anything else in them.

OSTRIKER: You've convinced us that there is difficult-to-observe
matter locally and that the amount is not negligible. But before we
can speculate on what it might be, it would be useful for you to remind
us what you've included in the <u>observed</u> matter, so that we don't try to
add that in. For example, how many faint companions and how many white
dwarfs of what luminosity are already in your observed sample?

J. BAHCALL: I integrated the Green-Liebert luminosity function for
white dwarfs as far as it goes and used a mass of 0.6 M_Θ for a typical
white dwarf. For the interstellar medium, I include HI, HII, He, He$^+$
and H$_2$ (found from CO measurements using Solomon's conversion factor -
Spitzer has a very similar estimate based on the reddening of nearby
stars). I think that here I've erred by putting in too much - there's
nothing like that amount within 100 or 200 pc of the Sun; we are in a
hole. With respect to main-sequence dwarf stars, I integrated the
Wielen luminosity function broken down into 11 components and used the
individual velocity dispersion for each component. I have separated the
subgiants and giants from the main sequence stars. The spheroid
contributes 0.001 M_Θ pc^{-3}.

OSTRIKER: And what fraction of the mass is assumed to be in normally
unobserved companions?

J. BAHCALL: That's 25% of the 0.044 M_Θ pc^{-3}. Different prescriptions
change that number by ± 10%. However, it makes a difference of only
~ 0.002 M_Θ pc^{-3} in the final answer.

SANDERS: You said that you put in the observed white dwarfs, but you
mentioned that you did not include the number of white dwarfs that you
expect from stellar evolution. What would that number be?

J. BAHCALL: Wiedemann's estimate is two or three times bigger than the number I quoted. This is more than has been observed, but it's not a terribly big contribution.

SCHWARZSCHILD: You mentioned that Larson has a possible picture for increasing the number of white dwarfs. Could Larson be called on?

LARSON: If you stick to conventional models of galactic evolution, the latest one by Beatrice Tinsley predicts an amount of mass in white dwarfs which is something like a third of the mass in other ordinary forms (lower main sequence stars, interstellar gas and so on). So in answer to Martin's question: If you are willing to increase this number by revising the model of galactic evolution, you only need a factor of 3 change to get the amount of mass in remnants to be equal to the amount of ordinary matter in other forms. I have a poster paper which suggests a way of doing this. It uses a model in which the initial mass function is not a Salpeter power law, but is bimodal. In that kind of model you can quite easily get enough mass in remnants of 2, 3 or 4 $M_\odot$ stars to give you the factor of 3 increase.

I have another comment, as to whether the dark matter could be very low mass stars. I don't think there's any suggestion of this in the available data. It has been clear for a long time that there are few stars known that have $M < 0.1$ $M_\odot$. There may be very few out there. The latest discussions of the luminosity function of faint stars are two extensive preprints, one by Scalo and one by Poveda. They agree that there's a peak at around 0.2 or 0.5 $M_\odot$, followed by a fairly steep decline toward lower masses. This luminosity function implies a negligible mass in stars less massive than ~ 0.1 $M_\odot$. To have a lot of stars with masses less than 0.1 $M_\odot$, the luminosity function would have to be quite remarkable, with another peak at very faint luminosities.

J. BAHCALL: I think you've correctly summarized the general opinion. I am less confident of these conclusions than are some of my colleagues. Isn't a peak between 0.1 and 0.01 $M_\odot$ just as likely as one between 0.5 and 2 $M_\odot$? I guess the general answer to Martin's question is that Larson's paper makes it theoretically acceptable to consider bimodal distributions. For me, that was the main point of his paper, which I liked very much.

SCHMIDT: Your statement was that if the unseen mass is included in the Luyten luminosity function, the slope has at least to be between 0.01 and 0.05 or so?

J. BAHCALL: Yes. I know that this conflicts with your result and with a number of others. I'm not absolutely convinced, however.

FABER: A question for Larson. When you were describing conventional models and the amount of mass in remnants, was the star formation rate uniform or was it slightly declining as a function of time?

LARSON: Uniform.

FABER: And is that still your feeling now?

LARSON: Such evidence as there is suggests that it has been uniform, but that evidence only applies to a restricted range of masses (~ 1 $M_\odot$). Maybe the stars that formed at earlier times were more massive, a suggestion first developed seriously by Maarten Schmidt.

PACZYNSKI: I understand that all of your models assume for simplicity that the distribution of the ISM in the galactic plane is smooth. Yet there's this hole near the Sun, ~ 100 pc across. Obviously, this doesn't affect the behavior of stars at z distances of 600 pc; these stars average over a large area of the disk. However, when you are closer to the disk, the existence of the hole might be felt. How does it affect the slope of the decrease in stellar density with z?

J. BAHCALL: Well, as an example, I made an extreme model containing no ISM, and it was within the range of acceptable models. One can also ask questions like: How massive would a molecular cloud at the edge of the hole have to be to have an effect? The answer is a very high mass, $\sim 10^8$ $M_\odot$.

WHITE: All of your models assume a Gaussian velocity distribution for each component and a velocity dispersion which is independent of height above the galactic plane. How well do the observational data support this independence for the F stars and the K giants on which you rely most heavily? Specifically, are there sufficient stars nearby and at z greater than one scale height to be sure that the dispersions of the two samples do not differ by more than 30%?

J. BAHCALL: The K giant velocities that I have studied are consistent with a Gaussian distribution and a constant velocity dispersion to the accuracy of the measurements. This work is shown in Table 1 and Figure 1 of Ap. J., 287, 926 (1984). An essential and new element in this analysis is to use only normal-metallicity disk objects. The fractional change in the velocity dispersion was < 10% over the entire distance surveyed. The F-star data are less extensive but are also consistent with a constant velocity dispersion within the heights I have studied. For both samples, I have calculated the maximum effect allowed due to departures from isothermality, and found a change of only a few percent of the local matter density.

P. QUINN: Can you constrain the scale length of the dark matter in the disk? If not, could the scale length be very large and affect the shape of the rotation curve?

J. BAHCALL: I can't constrain the disk scale length from my arguments. When we found an upper limit of 0.65 kpc for the scale height of the dark material, Soneira and I used a disk scale length of 3.5 kpc. If we use van der Kruit's larger value of 5.5 kpc, that limit rises to 0.7 kpc.

TREMAINE: I guess I'm still a little confused about how much you have
to stretch the conventional picture to get rid of the missing mass in
the disk. I can easily imagine that you have an extra 0.01 or 0.02 $M_\odot$
pc^{-3} in molecular gas and about the same amount in unobserved white
dwarfs. That would bring the observed density of 0.11 $M_\odot$ pc^{-3} up to
0.14 $M_\odot$ pc^{-3}. This seems to be getting within a standard deviation or
two of the Oort limit.

J. BAHCALL: You have to get the observed mass up to 0.2 $M_\odot$ pc^{-3}.

TREMAINE: OK. Let me phrase it another way. How embarassed would you
be if all this went away?

J. BAHCALL: It won't go away. I'm confident of that.

MATHIEU: A comment on open star clusters. To date the dynamics of
three clusters varying in age from a few times 10^7 to a few times 10^9
years have been studied. Two of these, M11 and M67, are discussed in a
poster paper. In all three cases, the agreement of the observed
velocity dispersion with that predicted from the stellar spatial
distribution is good. There is no evidence for dark matter in these
systems. However, due to mass segregation, the observed velocity
dispersions are not sensitive to objects with masses less than 1 $M_\odot$.
So if we consider the possibility that the disk dark matter consists of
objects formed in typical star-forming regions and that the open
clusters are a microcosm of the disk population, then these results
suggest that the dark matter is in objects with masses less than $\sim$ 1
$M_\odot$.

J. BAHCALL: That is an important comment. It is consistent with the
argument based on the existence of wide binaries.

SHUTER: I have the impression, which could be confirmed by an analysis
of the Bell Labs ^{13}CO Galactic Plane survey, that there is a missing
mass problem of comparable magnitude in Giant Molecular Clouds.
The mass estimated from their dynamics is a factor of 2-3 greater than
that estimated from the CO column density. If this is the case, the
unobserved material must have relatively small random velocity to be
captured in molecular clouds. It must therefore have a small scale
height, perhaps comparable to that of CO clouds in the solar
neighborhood, which is $\sim$ 85 pc.

J. BAHCALL: Very interesting.

EXTREME M DWARFS AND BROWN DWARFS IN A DEEP CCD SURVEY

P. Chikotas Boeshaar[1], J. A. Tyson[2], and P. Seitzer[3]

[1]Rider College
[2]AT&T Bell Laboratories
[3]National Optical Astronomy Observatories

Two years ago, Seitzer and Tyson began a program of 4-meter prime focus CCD observations at CTIO, with the aim to develop techniques for imaging and photometry to the theoretical limit of that telescope and overall 52% efficiency: 26.6 J mag, 26 R mag, 25 I mag. Color-magnitude plots show that the limiting magnitudes for detection are 28 J mag, 27 R mag, and 25.3 I mag. Each of our CCD fields covers about 12 sq. arcmin., with total exposure time about 7000 sec. in each of the three bands (see Boeshaar and Tyson 1985). Processed images are put on a VAX 11/780 and the FOCAS v.3.2 automated detector and classifier is run. In order to determine the probability of detection and photometric errors as a function of magnitude, FOCAS is rerun many times on artificial images made by adding in real star and galaxy images, dimmed many magnitudes, at random locations on the CCD sky frame. Over 2000 objects are detected in every high galactic latitude CCD field, less than 50 of which are classified as stars. Most of the faint galaxies are very blue (due to evolution), making it possible to search for infrared excess stars in the presence of so many galaxies.

An analysis of six CCD fields indicates no evidence for a dynamically significant disk or halo population of extreme M dwarfs. We surveyed 128 pc^3 in the R and I bands for stars of 0.1 $M_\odot$ (M_v=16). If we assume 0.5 to 1.5 times the locally observed mass density (0.1 $M_\odot$ per pc^3) is in the form of these stars (Bahcall 1985), we should have seen 20-60 such red dwarfs. One candidate was found.

We do not wish to address at present the question of evolution effects on the detectability of objects below the end of the hydrogen burning main sequence greater than 2-3 billion years old. There exist some variation in the theoretical estimates of the evolution of "brown dwarfs"; furthermore, great uncertainty exists regarding the observationally determined absolute magnitude vs. mass relation for stars at the very bottom of the main sequence. For stars of 0.08 $M_\odot$ (M_v=18-19 mag), we surveyed 87 pc^3 in the R and I bands, and would have expected to find 20-50 of the intrinsically faintest M dwarfs for a dynamically significant disk population. Possibly one was found.

The I-band stellar number counts reveal 20 times fewer counts than would be expected for a barely dynamically significant halo composed of a distribution of masses (0.15-0.09 $M_\odot$) rather than stars of a single mass and luminosity. Other halo models would predict even more M dwarfs.

Bahcall, J. N., 1985, Bull. Am. Astron. Soc. 17, 581.
Boeshaar, P. C., and Tyson, J. A., 1985, Astron. J. 90, 817.

J. Kormendy and G. R. Knapp (eds.), Dark Matter in the Universe, 32.

Kinematics of the Galactic Inner Spheroid

G. Gilmore*, R. Wyse**
*Institute of Astronomy, Cambridge
**University of California, Berkeley

Analysis of the detailed photometric, kinematic and chemical properties of stellar populations constrains the formation history of the Galaxy. We have completed a photometric survey and initiated a spectroscopic survey, obtaining radial velocities and abundances for volume complete samples of spheroid dwarfs in situ, to distances of a few kpc. Three fields under study are those for which Chiu (Ap.J.Suppl. 1980) obtained proper motions - SA 57 (NGP), SA51 (anticenter field) and SA68. Two of these fields are on the sun - Galactic center meridional plane (SA57 and SA51) so that (U,V) and (V,W) components of space motion respectively may be derived on the basis of the proper motions alone, once distances have been obtained. Our initial distance estimates are from Chiu's photometry and population classes, which are based on the position of the star on the reduced proper motion diagram.

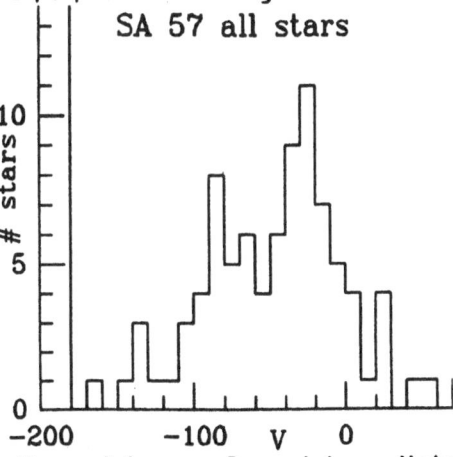

As may be seen from the figure, there is a peak at V-velocities $\sim$ 80 - 100 kms^{-1} with respect to the sun. The stars in this peak also have a U and W velocity dispersion intermediate between the thin disk and extreme spheroid. The other features of the UVW distributions are compatible with a centrifugally supported thin disk, after taking account of asymmetric drift and differential rotation, plus a non-rotating extreme spheroid, seen in the last velocity bin in the figure.

Chiu however did not allow for the existence of an intermediate population when classifying the stars. It is more self consistent to derive distances iteratively by adopting an initial disk main sequence classification for all stars, and modelling population changes by a suitable abundance gradient. Our results shown in the Figure are robust, since there is no qualitative change in the UVW distributions using the latter approach. The importance of Chiu's sample is that though a proper motion sample it is magnitude limited and hence free of kinematic bias. The results found here are in good agreement with those found in Eggen's kinmatically selected proper motion sample, and in spectroscopically selected samples (Gilmore & Wyse, A.J. 1985). More detailed information about the kinematics and metallicity of the inner spheroid, and about the Galactic potential, will be available once we analyse the extensive radial velocity dataset we have obtained for our photometrically defined samples.

J. Kormendy and G. R. Knapp (eds.), Dark Matter in the Universe, 33.

RELATIVELY DARK MATTER: THE LOCAL MASS DENSITY OF STELLAR REMNANTS

Richard Green
Kitt Peak National Observatory

The Palomar-Green Survey produced statistically complete samples of hot white dwarfs and subdwarfs, from which the local mass density of these objects can be determined. The luminosity function of hot hydrogen-atmosphere white dwarfs (DA's) was recently re-determined by Fleming, Liebert, and Green (1985, Ap. J., submitted). The contribution of each of the 353 objects to the local space density was calculated from the ensemble of limiting magnitudes and a spectroscopically or photometrically derived absolute magnitude. The local surface density of white dwarfs with $M_V < 12.75$ (log $L/L_\odot > -3.1$; $T_E > 9000$ K) is 0.32 ± 0.03 pc^{-2}. Assuming an exponential disk with scale height 250 pc, we derive the differential luminosity function in the figure; the total volume density is 0.65 ± 0.06 per 1000 pc^3. He atmosphere degenerates add another 20%.

The cooling time to this limiting luminosity is modeled to be $\sim 1.2 \times 10^9$ yrs. In $\sim 1 \times 10^{10}$ yrs, degenerates can cool to log $L/L_\odot < -4.5$, log $T_E < 4000$ K. At this limit, there is an apparent deficit of cool degenerates detected. Assuming a constant local birthrate over this entire time interval leads to a total density for all degenerates of 6.5 per 1000 pc^3 or 3.2 pc^{-2}. With 0.6 $M_\odot$ per degenerate core, the mass density is 3.9 $M_\odot$ per 1000 pc^3 or 1.9 $M_\odot$ pc^{-2}. This value is 2.2% of the Oort limit of 0.18 $M_\odot$ pc^{-3}, an insignificant contributor. Halo degenerates contribute $\sim 1\%$ of the local density.

Do hot subdwarfs add significantly to the mass density? It can be seen from the figure that hydrogen atmosphere subdwarfs (sdB's) exceed the white dwarfs in cumulative counts by almost a factor of two. Their volume density is estimated by assuming that the coolest ones make the largest density contribution, and adopting for them $M_V = +5.2$. Using Downes' galactic plane survey objects and the PG galactic pole objects, we find $Z_0 = 300$ pc to match $\rho \sim 2 \times 10^{-6}$ pc^{-3}. Although an order of magnitude higher than previous estimates, this value represents only 0.3% of the white dwarf mass density.

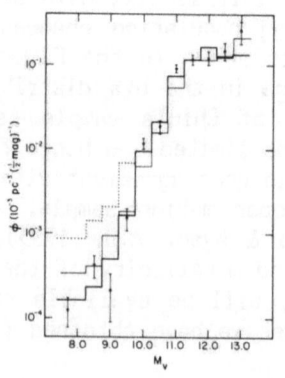
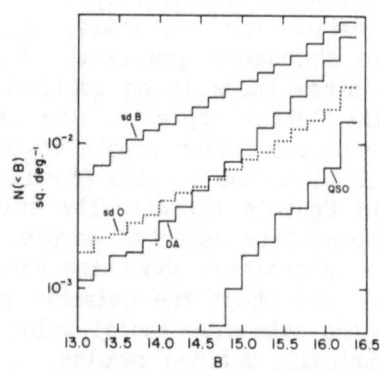

J. Kormendy and G. R. Knapp (eds.), Dark Matter in the Universe, 34.

THE DYNAMICS OF OPEN CLUSTERS

Robert D. Mathieu
Harvard-Smithsonian Center for Astrophysics

Many of the "observations" of dark matter involve the application of the virial theorem or more sophisticated dynamical models to bound gravitational systems. Such studies often adopt a certain mass function and assume that the system is in a state near equipartition. In many cases though the data are not sufficient to determine the mass function or justify the equipartition assumption. Open clusters provide excellent case studies of systems which are relaxed and which have a large observable stellar-mass range, approaching an order of magnitude in the nearest young clusters. As a result open clusters have recently been the subject of extensive dynamical study.

A comprehensive review has recently been given by Mathieu (1985). The essential results are two-fold. First, open clusters do show substantial mass segregation. Furthermore, the stellar surface-density profiles (as a function of stellar mass) are well-fit by multi-mass equipartition King models. The mass segregation is thus consistent with these clusters being in equipartition. Secondly, the observed velocity dispersions agree well with the predictions of the dynamical models; there is no need to invoke dark matter in these clusters to explain large internal motions. However, dark matter in the form of 1 $M_\odot$ objects or less would not have been detected if the total mass in such dark matter were less than the observed cluster mass.

A new result not shown in Mathieu (1985) is given in Fig. 1, where we show the radial distribution of the 1.2 $M_\odot$ single stars in M67 and the spectroscopic binaries with 1.2 $M_\odot$ primaries. Notice the marked central concentration of the binaries and the excellent agreement with the theoretical density profile for a 2 $M_\odot$ component in equipartition. Presumably any dark matter in the form of similarly massive objects would be distributed in a similar fashion.

Mathieu, R.D. 1985, Dynamics of Star Clusters, (eds. J. Goodman and P. Hut), p. 427, (Dordrecht: Reidel).

Fig. 1

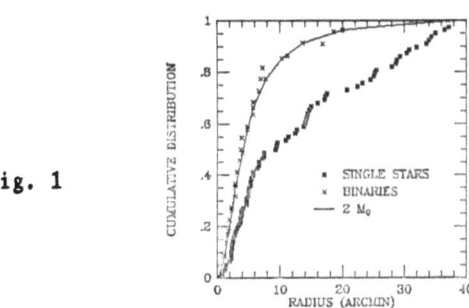

J. Kormendy and G. R. Knapp (eds.), Dark Matter in the Universe, 35.

SEEN AND UNSEEN MATTER IN THE GALACTIC DISC

Annie Robin, Olivier Bienaymé, Michel Crézé
Observatoire de Besançon
41 bis avenue de l'observatoire
F-25044 Besançon Cedex
France

ABSTRACT. We use a Galaxy model including three components for the 'seen' stellar matter (Robin, Crézé, 1985). A scenario of star formation and evolution in the disc constrains the age distribution within each spectral type and luminosity range. The model fits available star counts. We add the 'seen' interstellar matter with local density 0.04 $M_\odot$ pc^{-3} and scale height 140 pc. Thus the total observed local density is 0.085 $M_\odot$ pc^{-3}.

We add ingredients (corona + bulge) to fit a typical flat rotation curve (Caldwell, Ostriker, 1981). We get a minimum variance with a corona characterized by central density 0.0794 $M_\odot pc^{-3}$ and core radius 3.147 kpc, and a bulge mass of 0.129 10^{11} $M_\odot$.

We compute a series of potentials through numerical integration of Poisson equation based on above density laws plus arbitrary Unseen Mass Discs (UMD) with density in the range 0 to 0.25 locally and scale heights from 250 to 3000pc.

We derive model density distribution through the Boltzman equation for a sum of isothermal components. Our Galaxy model provides directly the age distribution for any spectral type selection and then associates a velocity dispersion with each age. We compare computed and observed $\rho(z)/\rho(0)$ for F5-F8 stars (Hill, Hilditch and Barnes, 1979) and for K giants (Oort's and Upgren's densities rescaled by Bahcall (1984)). The age-velocity distribution mixture in spectral type limited samples directly follows from our galaxy model. So we can fit observed density data up to distances as large as 1kpc from the plane just playing with the UMD.

CONCLUSION. An Unseen Mass Disc cannot be avoided. No acceptable fit can be obtain with UMD local density smaller than 0.15 $M_\odot$ pc^{-3}. The scale height should be larger than 1000pc. Smaller scale heights can be accepted with larger local densities.

REFERENCES.
Bahcall, J.N., 1984, Astrophys. J. **287**, 926.
Caldwell, J.A.R., Ostriker, J.P., 1981, Astrophys. J. **251**, 61.
Hill, G., Hilditch, R.W., and Barnes, J.V., 1979, M.N.R.A.S. **186**, 813.
Robin, A., Crézé, M., 1985, Astron. Astrophys. in press.

36

J. Kormendy and G. R. Knapp (eds.), Dark Matter in the Universe, 36.
© *1987 by the IAU.*

KINEMATICS OF NEARBY FV STARS

W.L.H. Shuter and W.N. Stocker
University of British Columbia

The line-of-sight velocity field has been determined for all 80 FV stars within 15 degrees of the North Galactic Pole (NGP) with measured and constant velocities, and measured distances, from the catalogs of Hill et al. (1976, 1982). These data are a subset of the data used in a determination of K_z by Bahcall (1984). Procedures were as described by Goulet and Shuter (1984). The velocity field is depicted in Figure 1, and the magnitude of the velocity residuals is plotted in Figure 2.

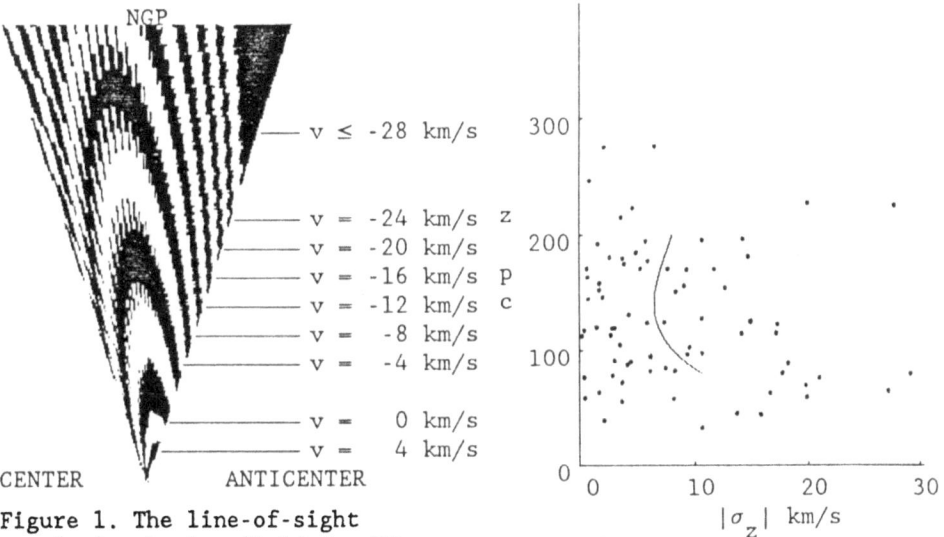

NGP

——— v ≤ -28 km/s   300

——— v = -24 km/s   z
——— v = -20 km/s   200
——— v = -16 km/s   p
——— v = -12 km/s   c
——— v = -8 km/s
——— v = -4 km/s   100

——— v =   0 km/s
——— v =   4 km/s

CENTER   ANTICENTER   0
                                  0   10   20   30
                                       $|\sigma_z|$ km/s

Figure 1. The line-of-sight vertical velocity field for FV stars in the center-anticenter direction referred to its own LSR. The Sun is at the bottom of the cone which has a height of 400 pc.

Figure 2. A plot of the magnitude, σ_z, of the velocity residuals versus height, z, above the Sun. The solid line represents a best fit cubic polynomial.

If the FV stars were in statistical equilibrium all velocities plotted in Figure 1 should be zero. Instead, there is a marked kinematic compression. If the velocity residuals plotted in Figure 2 were isothermal, there should be no variation of σ_z with z. No clear trend exists.

REFERENCES

Bahcall, J.N.: 1984, Astrophys. J., 287, p. 926.
Goulet, T., and Shuter, W.L.H.: 1984, in "Local Interstellar Medium", eds. Kondo, Y. Bruhweiler, F.C. and Savage, B.D. (NASA), p. 319.
Hill, G. et al.: 1976, Mem. R. Astron. Soc., 82, p. 69.
Hill, G., Barnes, J.V., and Hilditch, R.W.: 1982, Publ. D.A.O., XVI, p. 111.

J. Kormendy and G. R. Knapp (eds.), Dark Matter in the Universe, 37.

THE LOCAL GALACTIC ESCAPE VELOCITY[1]

Bruce W. Carney
Department of Physics and Astronomy
University of North Carolina
Chapel Hill, North Carolina 27514
USA

David W. Latham
Harvard-Smithsonian Center for Astrophysics
60 Garden Street
Cambridge, Massachusetts 02138
USA

ABSTRACT. From a new photometric and spectroscopic survey of high proper motion stars, combined with previously published work, we find that the local value of the escape velocity from the Galaxy exceeds 500 km s^{-1}. This gives direct dynamical evidence that the total Galactic mass exceeds the mass inside the solar orbit by a factor of at least five.

1. INTRODUCTION

Stars in the solar neighborhood moving on circular orbits about the Galactic center have Galactic rest-frame velocities, V_{RF}, of about 220 km s^{-1} (Gunn, Knapp, and Tremaine 1979, hereafter GKT). If the total mass of the Galaxy were contained within the solar Galactocentric distance, R_O, the Galactic mass would be 1.0×10^{11} $M_\odot$, assuming R_O = 8.5 kpc (GKT), and the local value of the escape velocity would be 311 km s^{-1}. The total mass of the Galaxy is, however, much larger, as has been shown by three different methods.

First, the Galaxy's observed rotation curve is flat or rising out to distances approaching $2R_O$ (Blitz, Fich, and Stark 1982), so that its dynamical mass is at least twice that inside R_O. The Galaxy thus resembles other disk galaxies with flat or rising rotation curves, as shown by optical data (Rubin et al. 1978, 1982; Burstein et al. 1982; Burstein and Rubin 1985) and suggested by radio data (Bosma 1981a,b).

---

[1] Some of the observations reported here were obtained with the Multiple Mirror Telescope, a joint facility of the Smithsonian Institution and the University of Arizona.

J. Kormendy and G. R. Knapp (eds.), Dark Matter in the Universe, 39–50.

Second, radial velocities of distant globular clusters and of the Galaxy's retinue of dwarf spheroidal galaxies suggest a total mass approaching 10^{12} $M_\odot$ (Hartwick and Sargent 1978; Lynden-Bell et al. 1983; Peterson 1985). Here, unfortunately, the sample is small, and a relatively large uncertainty results from the unknown orbital eccentricities of the systems. The poor accuracy of the radial velocities has been a factor in the past, but is no longer the dominant source of uncertainty. In a similar context, Hawkins (1983,1984) has deduced a Galactic mass exceeding 10^{12} $M_\odot$, based on the (as yet) uncon- firmed distance and velocity of his RR Lyrae candidate star R15.

Finally, the space velocities of nearby stars may be used to determine lower limits to the local escape velocity. For example, the wide co-moving pair HD 134439/40 has been known for over half a century to be such an extreme velocity system. The recent trigonometric para- llax (Russell 1977), together with its very large proper motion and high radial velocity yield $V_{RF} > 400$ km s^{-1}. More recently, Sandage's (1969) very high radial velocity for the high latitude dwarf G64-12 yield $V_{RF} > 400$ km s^{-1}.

In this paper we use a sample of nearby stars with high space velocities to investigate the local velocity of escape from the Galaxy. Although this approach is vulnerable to errors in the distance deter- minations, the use of stars allows us to sample a larger number of high-velocity objects than the second method. In the sections below we discuss the data sources that have yielded extreme-velocity stars, the determinations of V_{RF}, and the limits that we can set on the local escape velocity and the total mass of the Galaxy.

2. DATA SOURCES

2.1 A New Survey of High-Velocity Stars

We have recently completed the first phases of a photometric and spect- roscopic study of over 900 F, G, and K stars selected from the Lowell Proper Motion Catalog (Giclas et al. 1971, 1978). In brief, the stars all have proper motions exceeding 0.2 arc s yr^{-1}, and in most cases have two independent measures, the second being that of Luyten (1979 a,b; 1980 a,b). The V magnitudes range from 7.0 to 16.3 mag (the Lowell Catalog limit). UBV photometry exists for the entire sample, including our own 1225 measures for 867 stars. Digital spectrograms have been obtained for all the stars as well, using photon-counting Reticons to record 45 A of a single echelle order centered near 5200 A at high dispersion (2.2 A mm^{-1}) and high resolution (10 km s^{-1}) but at low to moderate signal-to-noise. Radial velocities with a typical accuracy of ± 0.7 km s^{-1} per observation were derived using digital cross-corre- lation techniques. As of July 1985 we had measured approximately 3400 velocities of the stars in our proper-motion sample.

2.2 Previous Surveys

Although our new survey is large and unbiased in metallicity, we chose to enlarge the sample significantly by adding stars from several other studies of high-velocity stars. We have therefore included in our analyses all high-velocity F, G, and K dwarfs of which we are aware whose proper motions exceed 0.2 arc s yr^{-1}. The major sources are the studies of Sandage (1964, 1969, 1981), the compilation of Eggen (1964), Saio and Yoshii (1979), and a private file maintained by one of us (BWC). These additional samples contain about 300 stars, all with published UBV photometry and radial velocities.

Eggen (1976) obtained UBV photometry for stars near the South Galactic Pole, and Carney and Peterson (1985a) studied 27 of the fainter metal-poor stars. JHK photometry was obtained to confirm the photometric parallaxes, and low-resolution spectrograms were taken to eliminate white dwarfs and measure radial velocities. Eggen (1969 his Table 2) also published a list of possible extreme velocity stars, with V_{RF} ranging from 400 to 2000 km s^{-1}. Included in his list are 23 stars blue enough (B-V < 1.0 mag) for reliable photometric parallaxes (see section 3.1 below). Carney and Peterson (1985b) have studied 18 of these stars utilizing Vby and JHK photometry, and low-resolution spectroscopy. One star has V_{RF} > 400 km s^{-1}.

3. KINEMATICS

To determine space velocities, it is necessary to know the distance for each star in order to convert proper motions into tangential velocities. To correct each space velocity to the galactic rest frame we must remove the small contribution due to the solar peculiar motion and the large contribution due to θ_O, the circular velocity of the Local Standard of Rest (LSR):

$$(V_{RF})^2 = U^2 + (V + \theta_O)^2 + W^2. \tag{1}$$

θ_O plays a second role in our work, because the ratio of the total Galactic mass, M_{TOT}, to the mass within the LSR orbit, M_{LSR}, will depend on V_{ESC}/θ_O, where V_{ESC} is the local escape velocity measured in the Galactic rest frame. In the simple case of a flat rotation curve extending from R_O to some abrupt cut-off at R_{LIM}, we have

$$(V_{ESC}/\theta_O)^2 = 2 \ln (M_{TOT}/M_{LSR}) + 2$$

$$= 2 \ln (R_{LIM}/R_O) + 2 \tag{2}$$

3.1 Photometric Parallaxes

We have used UBV colors to derive photometric parallaxes for the dwarf stars in our survey. In the absence of interstellar reddening, the B-V color of a dwarf is determined primarily by its temperature and composition. In Figure 1a we show absolute magnitude, M_V, versus color, B-V, for members of the Hyades cluster, after removal of binaries and binary candidates (Carney 1982). The halo dwarfs with accurate trigonometric parallaxes are also shown (Carney 1979b). The large separation

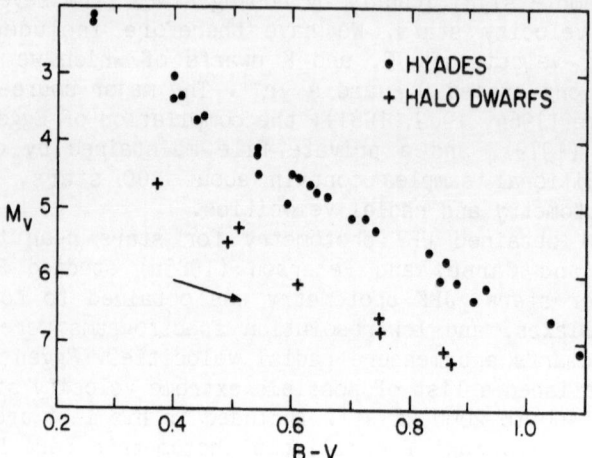

Fig. 1a. The main sequences for the Hyades and for the metal-poor halo dwarfs in the B-V versus M_V color-magnitude diagram.

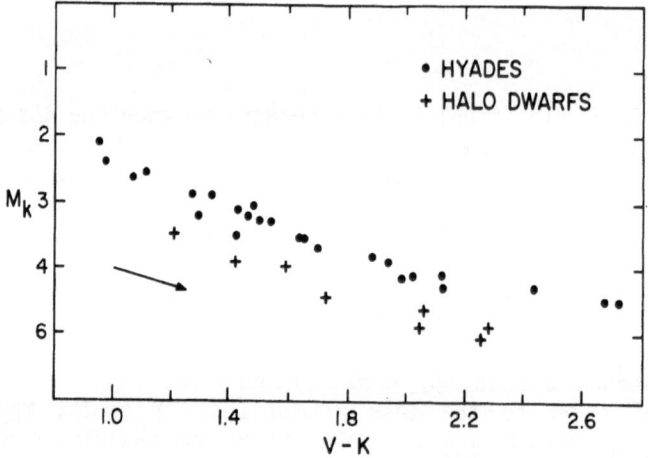

Fig. 1b. The main sequences for the Hyades and for the metal-poor halo dwarfs in the V-K versus M_K color-magnitude diagram.

between the Hyads and the halo dwarfs illustrates one of the main difficulties in determining M_V for stars with various compositions. If we assume that the disk and halo populations have identical helium abundances, or that helium correlates (or anticorrelates) with metallicity in a well-behaved way, we can interpolate between the two sequences as a function of metallicity. With UBV photometry, metallicity affects both M_V (through opacity) and B-V (through blanketing). This last point is illustrated by Figure 1b, where we use metallicity-insensitive V-K (Carney 1983) instead of B-V. The gap between the Hyads and the halo dwarfs has been more than halved because the redder pass-bands are much less affected by blanketing. However, since we rely primarily on UBV data, we must establish a reliable metallicity-based interpolation scheme that does not require infrared K measurements. We have chosen to work with the normalized ultraviolet excess, $S(U-B)_{0.6}$ (hereafter S) as defined by Sandage (1969). Its relationship to [Fe/H] has been calibrated by Carney (1979a). Carney and Latham (1985a) discuss this procedure in some detail, and find that by interpolating B-V using S^2, a typical accuracy of $\pm$ 0.2 mag in M_V (corresponding to an accuracy of $\pm$ 10 percent in the distance) can be achieved.

3.2 Reddening

The first-order effects of reddening are shown for E(B-V) = 0.1 mag by the arrows in Figures 1a and 1b. On the U-B versus B-V color-color diagram the reddening vector is steeply inclined to the stellar locus; reddening increases a star's ultraviolet excess and thus lowers its estimated metallicity and luminosity. Thus if the effects of reddening are ignored, the values derived for the star's space velocity will be too low. We can ignore reddening and still set lower limits to V_{ESC}. However, since we are also interested in the upper limits, we have estimated E(B-V) for each star.

For stars lying above or below the Galactic plane by more than 10 degrees, we used the reddening maps of Burstein and Heiles (1982) and assumed the dust is distributed smoothly with a scale height of 150 pc. For lower-latitude nearby stars, we used the map of Paresce (1984); for those lower-latitude stars more distant than 150 pc, we determined E(B-V) versus distance using published spectral types and UBV photometry for early-type stars within 1 or 2 degrees of each program star, then the revised δ and M_V values were iterated a few times. We discuss our results with and without these reddening corrections in sections 4.1 and 4.2 below.

3.3 Galactic Rotation

GKT determined Θ_0 = 220 km s^{-1} using 21 cm data, and Arp (1985) utilized Local Group radial velocities to estimate Θ_0 = 250 km s^{-1}. If we can select a halo population that has no net rotation in the galactic rest frame, then we can determine Θ_0 by measuring the mean motion of the sample with respect to the LSR. Several such studies have already been done. Woolley and Savage (1971) found a value of Θ_0 = 224 $\pm$ 25 km s^{-1} using 79 field RR Lyraes, while Pier (1984) obtained 272 $\pm$ 41 km s^{-1} using 150 field halo AB stars. With a much smaller sample (21) of blue horizontal branch stars, Sommer-Larsen and Christensen (1985) found Θ_0 = 172 $\pm$ 40 km s^{-1}. For a subsample of 46 metal-poor globular clusters, Frenk and White (1980) found a value of 190 $\pm$ 35 km s^{-1}, while Zinn's (1985) sample of 81 such clusters yielded Θ_0 = 170 $\pm$ 23 km s^{-1}, compared to 225 $\pm$ 52 km s^{-1} for the sample of 33 clusters more than 7 kpc from the Galactic center. Using a sample of 107 southern metal-poor ([Fe/H] < -1.4) dwarfs and giants, Norris et al. (1985) found Θ_0 = 182 $\pm$ 20 km s^{-1}. To these studies, we now add two more.

First, we have obtained over 300 radial velocities for 82 of the red giants with [Fe/H] < -1.5 in the studies of Norris et al. (1985) and Bond (1980). This all-sky kinematically unbiased (all stars were originally identified in objective prism surveys) sample of 175 metal-poor stars yields θ_O = 205 ± 23 km s^{-1}.

Second, we have used our own, kinematically biased, database. If we use only stars with absolute U or W velocities greater than θ_O, we will have a kinematically-selected spheroidal population. The 205 such stars in our survey yield θ_O = 231 ± 8 km s^{-1}.

The mean of the various determinations of θ_O for stellar halo population is θ_O = 227 ± 7 km s^{-1} (with our results superceding those of Norris et al. 1985). It appears that some of the globular clusters may belong to a slightly different population, that the local metal-poor halo population has no obvious net rotation, and that θ_O lies between 220 and 230 km s^{-1}.

4. RESULTS

4.1 A Lower Limit to V_{ESC}

In Figure 2a we show a histogram of rest-frame velocities that exceed the Keplerian escape velocity, assuming no reddening (the shaded parts represent stars not contained in our new survey). From this plot we conclude that V_{ESC} must be at least 400 km s^{-1}, and may be as high as 550 km s^{-1}. When reddening corrections are included, the space velocities of several stars increase substantially, as can be seen by comparing Figures 2a and 2b. Unfortunately the stars with the largest space motions in Figure 2b are also the stars for which the reddening corrections are the largest and the most uncertain. If we restrict ourselves to stars with modest reddening, we get the sample plotted in Figure 3. For each star a line connects the velocities calculated with

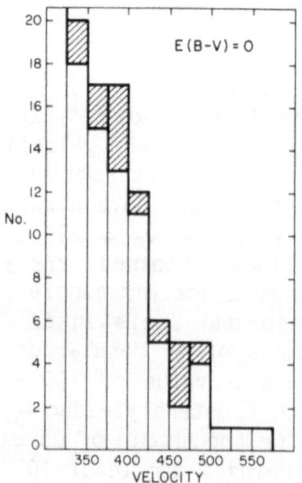 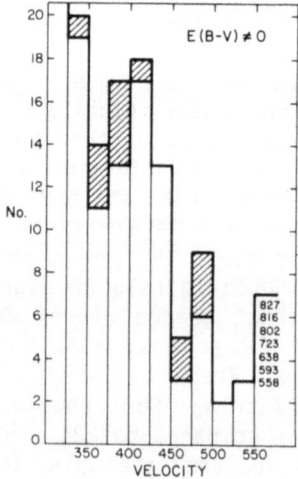

Figs. 2a and 2b. Histograms of the numbers of stars versus Galactic rest-frame velocity. The left plot assumes no reddening, while the right plot includes reddening corrections. The stars in the shaded areas are from previous studies.

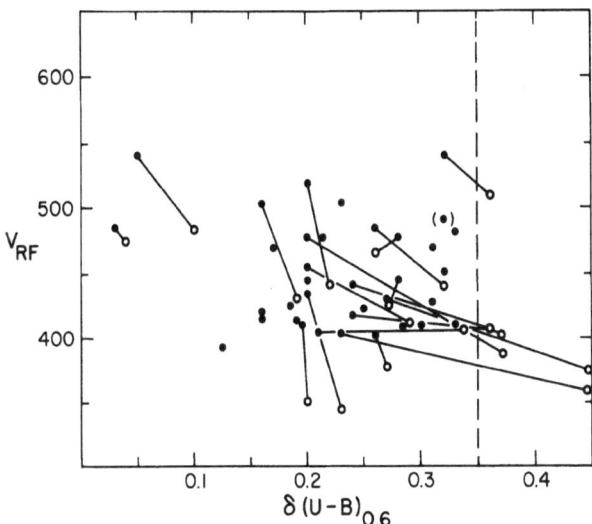

Fig. 3. V_{RF} versus metallicity for stars with modest reddening. For each star a line connects the velocities calculated with and without reddening corrections.

---

and without reddening. The dashed line at $\delta = 0.35$ mag represents the limit expected for extremely metal-deficient stars. Stars to the right of this limit may have a blue companion or may not have been sufficiently dereddened. Note that none of these stars have space velocities greater than 550 km s^{-1}, but a few exceed 500 km s^{-1}. We now consider whether any of these stars have had their velocities inflated by uncertainties in their photometric parallaxes and proper motions.

For the uncertainty in the photometric parallax we adopt $\pm$ 0.3 mag, and for the proper motion errors we use the difference between the Lowell and Luyten values. If we convolve the uncertainties due to reddening, photometric parallax, and proper motion, and eliminate stars with velocity uncertainties larger than $\pm$ 70 km s^{-1}, we are left with the 15 stars plotted in Figure 4. We conclude that the lower limit to V_{RF} is about 525 km s^{-1}.

4.2 An Upper Limit to V_{ESC}?

Two questions arise when we try to estimate the upper limit to V_{ESC}. First, have we determined V_{RF} reliably for all our extreme-velocity stars? Regrettably, the answer is yes for only 15 of the 55 stars likely to have $V_{RF} > 400$ km s^{-1}. More work must be done on the remaining 40 stars to see if stars with V_{RF} much larger than 500 km s^{-1} can be confirmed.

Second, how complete is our sample? Have we worked to faint enough limits to reach the highest-velocity stars? When we began the observational work, we chose a V magnitude limit of approximately 14.0 mag. At this limit we expected a significant fraction of the faintest

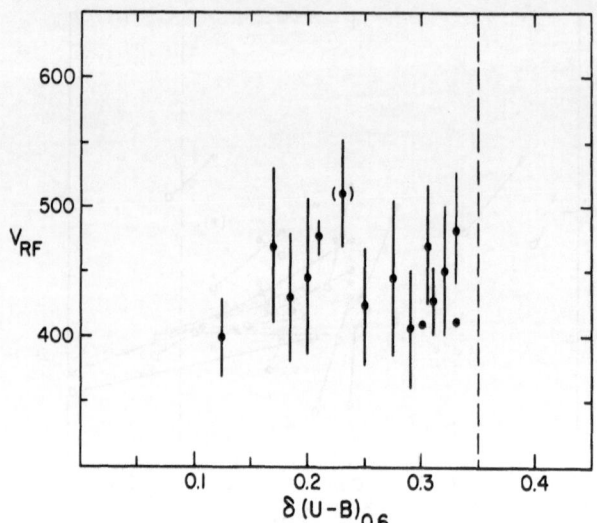

Fig. 4. V_{RF} versus metallicity for stars with modest reddening. The error bars include uncertainties in reddening, photometric parallax, and proper motion. Only the 15 stars with velocities uncertain by less than ± 70 km s^{-1} are shown.

stars would prove to be cool white dwarfs. When this expectation was not borne out, we extended the survey to the limit of the Lowell Catalog. In the deeper sample the fraction of white dwarfs does rise dramatically as the faint limit is approached; all five stars with V > 15.5 are white dwarfs. Thus we doubt that extending the survey to a fainter limit using the NLTT Catalog is likely to uncover very many extreme velocity stars.

Our survey does not extend much south of declination -10 deg, except for the South Galactic Pole work and results from previous studies mentioned above in section 2.2. Eggen's list of candidates for extreme-velocity stars consists mostly of southern stars, but only one has proven to actually have an extreme velocity. We conclude that our future work should focus on improving the velocity determinations for our 40 stars with uncertain but potentially extreme velocities.

4.3 Implications for the Total Mass of the Galaxy

If we adopt V_{ESC} = 525 km s^{-1}, equation (2) suggests that M_{TOTAL}/M_{LSR} = 5.0 for θ = 230 km s^{-1} (or 6.4 for 220 km s^{-1}). The total Galactic mass is thus likely to be in excess of 5×10^{11} M$_{\odot}$. This mass is similar to the result derived from the radial velocities of distant Galactic satellite systems (Peterson 1985). Thus these results give direct dynamical evidence for a large total mass of the Galaxy.

5. FUTURE WORK

We plan to refine the space velocity determinations for our highest-velocity stars in four ways. First, we will obtain more radial velocities to eliminate the possible error introduced by a large-amplitude binary and to indicate which stars are spectroscopic binaries and therefore may have their photometry contaminated by a companion. Second, we will obtain JHK (and possibly uvby) photometric data to confirm the metallicities (via Δm_1^* - Carney 1983) and especially the photometric parallaxes. Third, we hope to determine metallicities by comparing our observed spectra with theoretical spectra recently calculated covering the wavelength range 5150 - 5250 A for a grid of model atmospheres with T_{eff} = 4750, 5000 ..., 6500 K, [M/H] = 0.0, -0.5, ..., -3.0, and gravities appropriate for old main sequence stars. This effort should provide us with better metallicities, especially for the extremely metal-deficient stars, where the photometric metallicities have so little leverage. Finally, we will obtain moderate-resolution spectrograms at high signal-to-noise to measure the strengths of the narrower diffuse interstellar features at 5780, 5797, and 6195 A. Since these features correlate well with E(B-V) (Herbig 1975), we should be able to reduce significantly the uncertainties due to reddening.

We have identified many stars whose origins lie in the outermost Galactic halo, yet are momentarily near enough and bright enough for detailed study. Just as comets are used to study the primordial record of the outermost parts of the Solar System, so we plan to use our new high-velocity stars to probe the chemical history of the Galaxy's outer halo.

We thank Bob Stefanik, Bob Davis, Ed Horine, Jim Peters, Skip Schwartz, Jon Morse, and Dick McCrosky for help with the radial velocity observations. This work has been supported by NSF grant AST-8312842 to the University of North Carolina.

REFERENCES

Arp, H. C. 1985 Astron. Astrophys., in press.
Blitz, L., Fich, M., and Stark, A. A. 1982 in IAU Symp. No. 87, Interstellar Molecules, B. H. Andrew, ed., Reidel, Dordrecht p. 213.
Bond, H. E. 1980 Astrophys. J. Suppl. 44, 517.
Bosma, A. 1981a Astron. J. 86, 1791.
Bosma, A. 1981b Astron. J. 86, 1825.
Burstein, D. and Heiles, C. 1982 Astron. J. 87, 1165.
Burstein, D. and Rubin, V. C. 1985, preprint.
Burstein, D., Rubin, V. C., Thonnard, N., and Ford, W. K., Jr. 1982 Astrophys. J. Lett. 225, L107.
Carney, B. W. 1979a Astrophys. J. 233, 211.
Carney, B. W. 1979b Astrophys. J. 233, 877.
Carney, B. W. 1982 Astron. J. 87, 1527.
Carney, B. W. 1983 Astron. J. 88, 623.

Carney, B. W. and Latham, D. W. 1985a, in IAU Coll. No. 88, Stellar
    Radial Velocities, A. G. D. Philip and D. W. Latham, eds., L.
    Davis Press, Schenectady, p. 139.
Carney, B. W. and Latham, D. W. 1985b, in preparation.
Carney, B. W. and Peterson, R. C. 1985a, in preparation.
Carney, B. W. and Peterson, R. C. 1985b, in preparation.
Eggen, O. J. 1964 Royal Obs. Bull. No. 84.
Eggen, O. J. 1969 Astrophys. J. Suppl. 19, 31.
Eggen, O. J. 1976 Astrophys. J. Suppl. 30, 351.
Frenk, C. S. and White, S. D. M. 1980 Monthly Notices Roy. Astron.
    Soc. 193, 295.
Giclas, H. L., Burnham, R., Jr., and Thomas, N. G. 1971 Lowell Proper
    Motion Survey, Northern Hemisphere, Lowell Obs., Flagstaff.
Giclas, H. L., Burnham, R., Jr., and Thomas, N. G. 1978 Lowell Obs.
    Bull. No. 164.
Gunn, J. E., Knapp, G. R., and Tremaine, S. D. 1979 Astron. J. 84,
    1181.
Hartwick, F. D. A. and Sargent, W. L. W. 1978 Astrophs J, 220, 453.
Hawkins, M. R. S. 1983 Nature 303, 406.
Hawkins, M. R. S. 1984 Monthly Notices Roy. Astron. Soc. 206, 433.
Herbig, G. H. 1975 Astrophys. J. 196, 129.
Latham, D.W. 1985 in IAU Coll. No. 88, Stellar Radial Velocities,
    A. G. D. Philip and D. W. Latham, eds., L. Davis Press, Schenec-
    tady p. 21.
Luyten, W. J. 1979a The NLTT Catalogue Vol. I, Univ. Minn., Minnea-
    polis.
Luyten, W. J. 1979b The NLTT Catalogue Vol. II, Univ. Minn., Minnea-
    polis.
Luyten, W. J. 1980a The NLTT Catalogue Vol. III, Univ. Minn., Minnea-
    polis.
Luyten, W. J. 1980b The NLTT Catalogue Vol. IV, Univ. Minn., Minnea-
    polis.
Lynden-Bell, D., Cannon, R. D., and Godwin, P. J. 1983 Monthly Notices
    Roy. Astron. Soc. 204, 87P.
Norris, J., Bessell, M. S., and Pickles, A. J. 1985 Astron. J. 58, 463.
Paresce, F. 1984 Astron. J. 89, 1022.
Peterson, R. C. 1985 Astrophys. J., in press.
Pier, J. R. 1984 Astrophys. J. 281, 260.
Rubin, V. C., Ford, W. K., Jr., and Thonnard, N. 1978 Astrophys. J.
    Lett. 225, L107.
Rubin, V. C., Ford, W. K., Jr., Thonnard, N., and Burstein, D. 1982
    Astrophys. J. 261, 439.
Russell, J. 1977 Astron. J. 82, 293.
Saio, H. and Yoshii, Y. 1979 Publ. Astron. Soc. Pacific. 91, 553.
Sandage, A. 1964 Astrophys. J. 139, 442.
Sandage, A. 1969 Astrophys. J. 158, 1115.
Sandage, A. 1981 Astron. J. 86, 1643.
Sommer-Larsen, J. and Christensen, P. R. 1985 Monthly Notices Roy.
    Astron. Soc. 193, 295
Woolley, R. and Savage, A. 1971 Royal Obs. Bull. No. 170.
Zinn, R. 1985, preprint.

DISCUSSION

MATHIEU: (1) What galactic mass model did you use for your total mass determination? How does the galactic mass implied by your high-velocity stars compare to that determined from the Blitz rotation curve? (2) To what extent can your results be affected by undetected photometric binaries?

CARNEY: (1) Our model is very simple. We assume a flat rotation curve beyond the solar circle to some limiting radius, R_{lim}, in which case M_{tot}/M_{LSR} is equal to R_{lim}/R_{LSR}. Thus if $V_0 = 220$ km s^{-1} and $V_{esc} = 500$ km s^{-1}, $R_{lim} \sim 5$ R_{LSR} or ~ 40 kpc. This extends further than Blitz's rotation curve, which is also essentially flat, and so implies even more mass. (2) Regarding binaries, our spectra rule out single- and double-lined binaries with periods shorter than a few years. Also, ubvy and JHK photometry can rule out pairs whose components differ by one magnitude or more in M_V. We cannot yet rule out equal components with moderately wide separations, but with Hal McAlister and others we've begun speckle interferometry observations of the brighter stars. In any event, should one star later prove to be two, the correction would increase the distance, hence the tangential velocity, hence the inferred escape velocity.

OSTRIKER: While your lower limits have got to be right, because you see the stars, I wonder about the upper limits. First, a statistical point. For example, suppose you wanted to determine the tallest possible person by measuring the height of everyone in this room. You'd probably not get it right. But you would get a better estimate by measuring everyone in Princeton. And your estimate would keep going up as the sample size increases. In your present sample you may just run out of stars. There is also a physical point. Imagine that the dark halo extends half way to M31, i.e., has a radius of 250 kpc, but suppose that the Galaxy has a hard edge with no stars beyond 40 kpc. Then all of the stars you see will have fallen at most from 40 kpc to us. You would only be measuring that much difference in the potential, which you would think gives the escape velocity. So there is always going to be a question about the distribution of stars you can sample compared with the distribution of matter.

CARNEY: I agree, although some stars may come from further out than 40 kpc. Even if the Galaxy has a hard edge, it has companions which have been stripped, and some of their stars have fallen from distances greater than 40 kpc.

OSTRIKER: Statistically you would expect very few of them to be picked up in your survey.

CARNEY: Very few, at least, of the stars coming from the systems we now see.

TREMAINE: I'd like to emphasize the point you're making. It seems to me that since you've selected the stars so heavily, it's possible that they include a population of stars wandering through the Local Group that could even have been stripped from M31. When I do the estimates very roughly, I find that there's a reasonable chance that a sample of your size might contain a few such stars. So, if I might make the opposite argument to Jerry, it's conceivable that these stars are not telling us much about the escape speed from the Galaxy but are telling us something about the kinematics of the Local Group.

CARNEY: That's perhaps why some of the metal-rich stars with high velocities are interesting.

LYNDEN-BELL: If Scott is correct, there should be a velocity bias, or anisotropy, of ~100 km s^{-1} toward M31.

AARONSON: What percentage of your distant halo giants are turning out to be binaries?

CARNEY: Of the halo dwarfs and giants in our programs with three or more radial velocities obtained over one- to three-year baselines, 15% are velocity variables. The true halo binary fraction is, of course, somewhat higher.

FABER: Can you redo the analysis using only the radial velocities?

CARNEY: Yes, and some of the stars, those with small error bars, are dominated by the radial velocity signature. For the star with V = -585 km s^{-1}, in particular, you can use the radial velocity alone to get a rest-frame velocity of ~410 km s^{-1}. There are also some stars which are leading the LSR - pulling away from it at substantial velocities. The halo red giants themselves don't unfortunately give particularly interesting numbers, but there are a few stars which suggest values of ~400 km s^{-1}.

WHITMORE: You commented that the Mg lines fell outside your wavelength coverage for the high velocity stars. At what velocity does this occur? Could it cause a selection bias, since it may be easier to measure the velocity when the Mg line is present than when it is outside your wavelength range?

CARNEY: The cutoff is at about 500 km s^{-1}. The correlation technique used to measure the velocities can still measure the velocity fairly well even when the Mg lines are not present.

RICHSTONE: In principle, your sample also gives you information on the velocity ellipsoid for the stars in the halo - for example, you know that for stars within 30 kpc, it's not very tangential.

CARNEY: We have chosen not to estimate the velocity ellipsoid for the proper-motion-selected sample just because the kinematic bias is so extreme. We have computed it for the halo red giants and find the normal proportions of 2:1:1 between the radial, tangential and axial velocity dispersions.

CONSTRAINTS ON THE DARK MATTER FROM OPTICAL ROTATION CURVES

Vera C. Rubin
Department of Terrestrial Magnetism
Carnegie Institution of Washington
5241 Broad Branch Road, N. W.
Washington D.C. 20015

ABSTRACT. From the observed rotation curves of Sa, Sb, and Sc spiral galaxies, it is possible to deduce a dozen constraints on the nonluminous matter in spirals. Within the optical image, the dark matter is less concentrated than the luminous, and contributes about 1/2 of the mass, for spirals of all Hubble types and luminosities.

1. INTRODUCTION

The determination of rotation curves for spiral galaxies has been a fruitful industry for the past decade. High dispersion spectrographs on large telescopes, coupled with electronic enhancement of incoming photons, has made it possible to derive accurate emission line rotation velocities for the optical disks of Sc's, Sb's and even for some Sa's. From a systematic study of the dynamical properties of about 60 relatively nearby Sa, Sb, and Sc field spirals, Rubin and colleagues (1985 and references therein) have discussed dynamical properties as a function of various galaxy parameters. Virtually all of this work has been done photographically.

   Within the past few years, a dramatic change in observing techniques has been introduced. Replacing the photographic plate with a CCD detector permits accurate digital subtraction of weak nightsky emission features and hence velocity measurements of weaker galaxy emission; it also presents simultaneously an accurate record of the variation of emission line strengths across the galaxy disk. Rotation curves can be obtained for galaxies with redshifts as great as z=0.05 and probably even 0.1. (I hope that some day I will be amused at the conservative nature of this prediction). Dynamical properties of entire new classes of galaxies can be studied.

   I show in Figure 1 an image of a portion of the compact group Hickson 88, and a spectum of Hi88a. Note the entanglement of the [NII] 6583A line with the nightsky line. Such confusion is easily eliminated, now that I measure CCD spectra at a computer terminal with a new facility developed by Kent Ford, based in part on software supplied by Schechter, Tonry, and Boroson. Velocities measured from the galaxy plus

J. Kormendy and G. R. Knapp (eds.), Dark Matter in the Universe, 51–65.

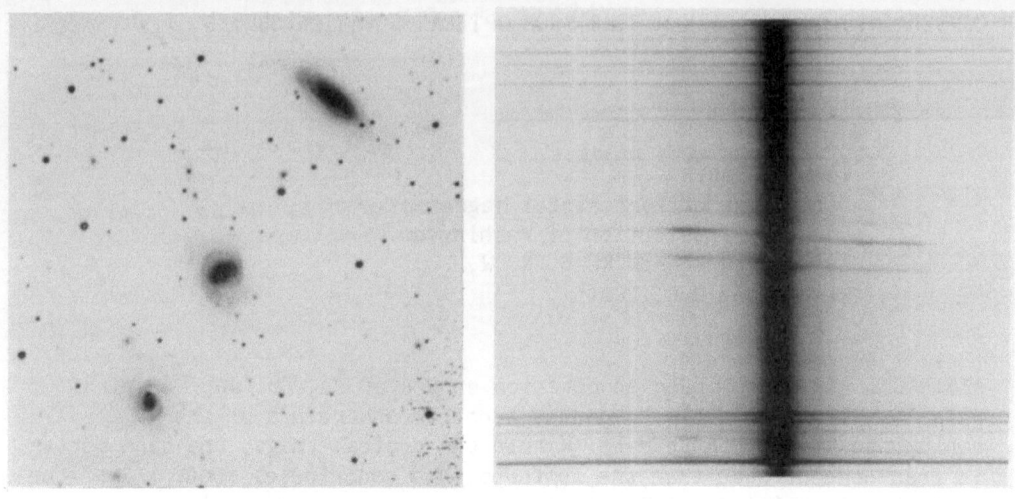

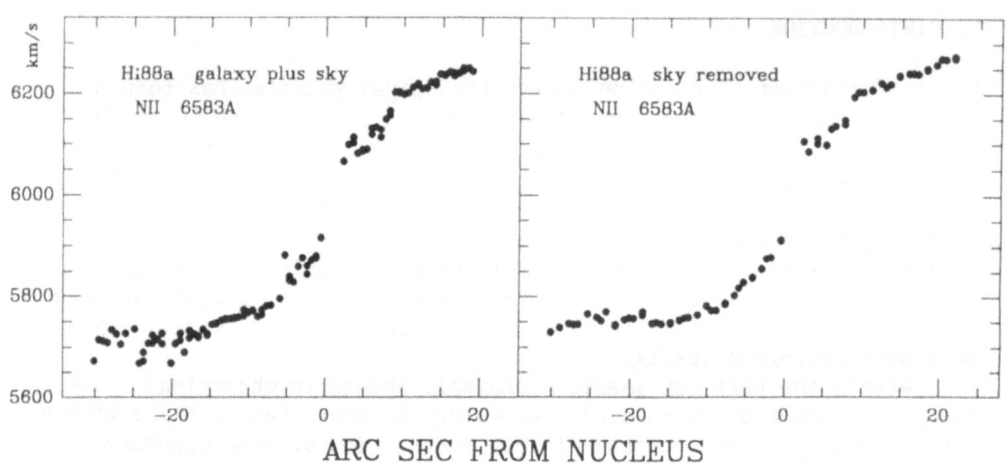

ARC SEC FROM NUCLEUS

Figure 1. Upper Left: The inner region of the compact group Hickson 88, from a KPNO 36-inch CCD + R (Mould filter) frame. Hi88a is the NE spiral. Upper Right: Hα and [NII] lines in the spectrum of Hi88a, from a Palomar 200-inch spectrum recorded on a CCD; wavelength increases down. Original dispersion and scale are 0.8A/pix and 0.59"/pix; exposure 6000 sec. The nightsky line between Hα and [NII] (right) merges with the low velocity [NII] (left). Lower: Measures of [NII] velocities, with and without sky subtraction, found by computer fits to the continuum levels and emission line shapes. Generally, 2 or 3 columns are summed in making the fits. Note the reduced scatter and the increased extent after sky subtraction. Additional accuracy, especially in the central region, can be gained by removing the galaxy continuum.

sky frame and for the galaxy minus sky (Figure 1) attest to the decrease in scatter when the sky spectrum is removed. Increased velocity accuracy offers us the capability of asking and answering questions, for example, about subtle alterations in dynamics arising from environmental causes.

While these techniques offer great promise for the immediate future, in this paper I want to probe what we have learned in the past. Rather than discuss rotation curves as a tool for studying galaxy dynamics, I want to describe what rotation curves tell us about the nonluminous matter in spirals. Specifically, my aim is to compile a list of the properties of the dark matter, as they are inferred from the observations, generally optical, of the dynamics of galaxies, generally spirals.

It seems to me remarkable that rotation curves for galaxies of a variety of Hubble types and a wide range of luminosities exhibit such a simplicity of form. This in turn implies a simplicity of mass distribution. It seems equally remarkable that a comparison of the rotation curves, or mass forms, with optical properties makes it possible to place a dozen or so contraints on the dark matter, something we have so far detected only by its gravitational effect. The observations from which these properties are inferred are already known to many of you, but here the emphasis will be slightly different. Due to space limitations, the less controversial points will be mentioned only briefly; literature references are incomplete, but can be traced from papers cited.

2. WHERE IS THE DARK MATTER?

2.1. It is clumped around spiral galaxies

Since the early work of Zwicky (1933) and Smith (1936), evidence has existed for nonluminous matter in clusters of galaxies. Only in the past decade has the study of field spirals demonstrated that nonluminous matter is a property of isolated spirals as well (Bosma 1978). For a galaxy with a flat rotation curve, the mass interior to R grows linearly with R, $M(R) = kV^2R$, and the density ρ falls as R^{-2}; for many spirals where the outer rotation curve is slowly rising rather than flat, the density falls more slowly, $\rho \sim R^{-1.7}$. At the isophotal radii of massive spirals, the density has fallen only to about 10^{-25} or 10^{-26} gm/cc which is 4 orders of magnitude higher than the closure density of the universe. Thus field galaxies with their attendant halos represent density peaks in the distribution of matter in the universe.

If density continues to fall as R^{-2}, it requires halos of a few Mpc extent, which is about one-half the average distance between galaxies, to have sufficient mass to close the universe. There is presently no evidence that space is filled with these large halos.

2.2. It is less concentrated than the luminous matter

In a disk galaxy, the surface brightness of luminous mass falls exponentially, while mass density falls more slowly, nominally as R^{-2}. Thus the ratio of M(total)/M(lum) increases significantly with

increasing radial distance across a galaxy disk. Resolution of spirals
into component parts by Carignan and Freeman (1985; see other papers
this volume, especially Bosma, Athanassoula, and van der Hulst) shows
that M/L(local) increases by a factor of about 100 across the 10 kpc
disks of late-type spirals.

2.3. It is more extended than the luminous matter

It is a rare circumstance that permits the measurement of optical
velocities beyond the optical disk; SO galaxies with polar rings offer
such an opportunity. For a few of these galaxies (Schweizer et al. 1983;
Whitmore et al. this volume) velocities have been measured over a ring
whose spacial extent is several times that of the disk it girdles. In
all cases, the constant ring velocity matches that in the disk. Hence
mass continues to rise linearly with radius to a distance several times
that of the disk isophotal radius. For polar ring galaxies, the dark
matter extends beyond the galaxy disk.

For some spirals, HI gas is significantly more extended than optical
luminosity. Almost without exception, measured HI velocities show flat
rotation curves to the Holmberg radius (Bosma 1981, Sancisi 1983), beyond
which warps become prominent and rotation velocities uncertain. For NGC
3198 (van Albada et al. 1985), accurate 21 cm velocities exist to radii
almost twice the Holmberg radius. Cumulative M/L values increase from
unity at R(25) to 2 at R(Holmberg) to 4 at the limits of HI observations.
In all of these special cases with extensive measurements, the dark halo
extends well beyond the optical galaxy.

3. WHAT IS THE FORM OF THE DARK MATTER DISTRIBUTION?

3.1. Gravitational potential is more nearly spherical than flat

SO galaxies with polar rings offer the very unique possiblity to sample
the 3-dimensional form of the gravitational potential of a disk galaxy.
The velocity of particles in the ring at polar distance R can be
compared with the velocities of particles in the disk at radial distance
R. For four polar ring galaxies (Schweizer et al. 1983; Whitmore et al.
this volume) the ratio V(ring)/V(disk) is unity to within 10%, implying
a nearly spherical form for the potential. Moreover, the constant
velocities in the ring which match and extend those in the disk offer
evidence that the density distribution of the dark matter is not
discontinuous at the limits of the optical disk, at least for these
objects.

3.2. Forms of rotation curves, and hence forms of mass distributions,
are unrelated to galaxy morphology

I show in Figure 2 two sets of rotation curves, plotted in log V, log R,
coordinates to emphasize their forms. The lower three curves come from
an Sa, Sb, and Sc galaxy whose bulge-to-disk ratios range over a factor
of 40. NGC 2639, an almost diskless Sa, and NGC 801, an almost bulgeless

Figure 2. Lower: Rotation curves of similar form for an Sa, Sb, and Sc galaxy; bulge-to-disk ratios differ by a factor of forty among these three. Upper: Rotation curves of similar form for 3 other galaxies, whose form differs from that below. The form of the rotation curve is not closely correlated with the Hubble type or luminosity.

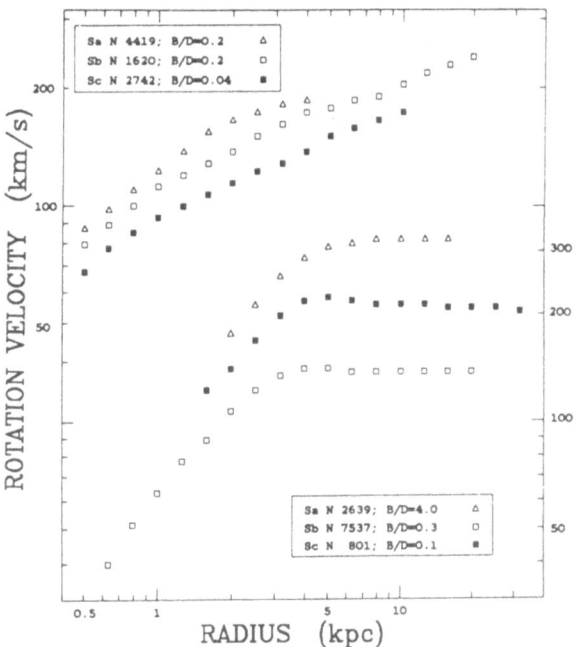

Sc, both exhibit a similar form for their rotation curve. The three upper rotation curves exhibit a similar form among themselves, but a very different one from the lower set. Thus, galaxies of similar morphology can have very different rotation curves, while galaxies of the same morphology can have very different rotation curves. Galaxy morphology is not a useful predictor of the distribution of mass within a galaxy. This is a surprising result, but a fairly well-founded one.

3.3. Rotation curves, and hence forms of mass distribution, come only in a limited variety

It is a remarkable fact that the forms of the rotation curves of galaxies, in distinction to their amplitudes, show a notable sameness, when scaled to appropriate units in radius and velocity. Rotation curves cover only a small fraction of the V,R plane; velocities at moderate and large R range only from slightly rising to slightly falling. Burstein and Rubin (1985) show that this simplicity transforms to an equivalent simplicity in integral mass distributions; these are useful in discussing overall galaxy properties. When we assume that the relation between the rotation curve and the projected mass distribution is the same for all galaxies, then $M(R) = kV^2R$ is the mass interior to R. We can classify the forms of mass variations by examining a plot of M(R) vs R.

In Figure 3 we superpose plots of the mass interior to R vs R for 20 galaxies (11 Sc, 5 Sb, 4 Sa), each identified by a letter. Radii and masses are scaled to give minimum scatter. All of these galaxies exhibit a similar form of rotation curve and hence mass distribution, even though they differ markedly in morphology and in luminosity. From the restricted

Figure 3. Upper: Mass
interior to R plotted vs R
for 20 galaxies; each
coordinate is scaled to
produce minimum scatter. The
resulting curve we call mass
type I. Lower: Mean lines for
each of the three integral
mass distributions, plotted
together to emphasize the
difference in curvature
between the mass forms, and
the continuity of forms among
the types.

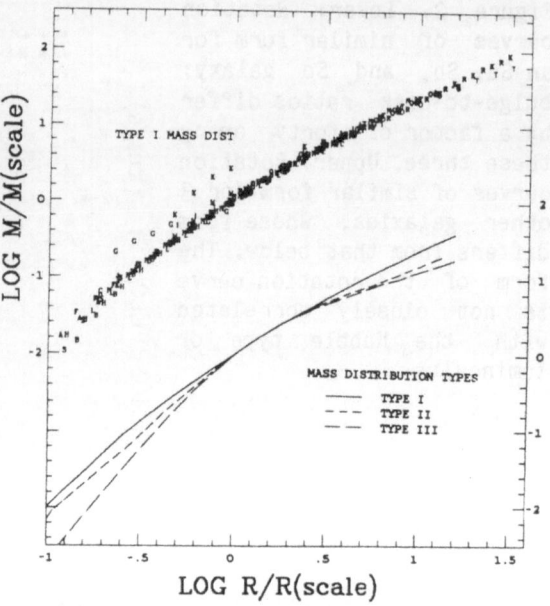

LOG R/R(scale)

continuum of mass forms, we have identified 3 principal ones, shown
schematically in the lower part of the Figure. Mass type II contains 2
Sc, 4 Sb, and 5 Sa spirals; type III contains 1 Sc, 5 Sb,and 4 Sa's.
Twelve spirals have intermediate forms, and 7 have complex (uncertain)
forms. Although a larger number of Sc galaxies are of mass type I, there
is a selection of all Hubble types among all mass types.

　　Mass forms are differentiated by the amount of curvature on a [log
M(R), log R] plot; they illustrate the progression in the form of the
rise of the rotation curve relative to the radius at which the rotation
curve becomes nearly flat. For most spirals, maximum curvature in the
rotation curve occurs within one or two radial scale lengths, generally
of order 10" or 20" for the galaxies we observe. In the limit that the
rise is unresolved, the mass form is a straight line. Thus it is
significant that many virtually bulgeless Sc galaxies have a faster rise
(i.e., type I) than the large-bulged galaxies of mass type III.

3.4. Dark matter contributes to the galaxy potential on all scales

There is little controversy that dark matter dominates the galaxy
potential on large scales. Here we question the smallest scales on which
the dark matter is significant. The already-classic procedure for
discussing the distribution of mass in spiral galaxies (Kaljnas 1983,
van Albada et al. 1985, Bahcall and Casertano 1985, Carignan and Freeman
1985) is to decompose the galaxy into a nucleus, bulge, disk, and halo
and deduce the relative mass for each, subject to the condition that
each component separately satisfy the observed luminosity properties,
and that all components sum to predict a gravitational potential which
produces the observed rotation curve. In practice, the rise of the
rotation curve is attributed solely to the luminous disk mass, producing

what van Albada et al. have called the "maximum disk." For example, in NGC 3198, the halo mass contributes negligible mass interior to one scalelength (2.7 kpc = 60"). I would now like to argue, on admittedly tenuous grounds, that although the contribution of the dark matter is not dominant at small radii, it is present here too.

If gas in a spiral disk has noncircular motions anywhere, these are likely to exist near the nucleus. However, optical observations give only minimal evidence for such. In many galaxy spectra, emission lines can be traced smoothly as close as one or two arc sec to the nucleus, typically 100 or 200 pc along the major axis (and of order 500 pc along the minor axis) for our sample. The width and character of the lines do not change as they approach the nucleus; minor axis measures give little evidence for radial motions. (In practice, a slight slope along the adopted minor axis is taken as evidence of a small error in the adopted line of nodes, which can have an uncertainty of several degrees due to complex galaxy morphology).

If we accept the conclusion that the disk gas is in circular orbit within a few hundred pc of the nucleus, then how are we to interpret the similar observed mass form at small radii (Figures 2 and 3) for galaxies of very different nuclear morphology, or the steep velocity rise for Sc galaxies of minimal bulge? One simple explanation is that the contribution of the dark matter is already signficant by 0.5 or one kpc. More detailed optical observations might help settle this question; spectra of nuclear regions of the nearest galaxies from the space telescope will be crucial. To date, most radio observations are unable to resolve velocities at small R. Thus while the evidence is not overwhelming, I think we may learn that dark matter is a nontrivial component of nuclear regions of spiral galaxies.

4. HOW MUCH DARK MATTER MAKES A SPIRAL GALAXY?

4.1. The amounts of dark and luminous matter are related

Within the sphere defined by the radius of its optical disk, a spiral galaxy has equal parts luminous matter and dark matter. This proportion holds for galaxies whose optical luminosities differ by as much as a factor of 100. These conclusions come both from statistical studies, and from mass decomposition of a few individual galaxies. This incredible concordance of luminous and nonluminous matter must offer one of the primary clues as to the nature of the dark mass.

I show in Table I the dynamical mass-to-luminosity ratios which are derived from the rotation curves of the field spirals (Rubin et al. 1985). Within each Hubble type, there is a good correlation of dynamical mass with optical luminosity over the entire observed luminosity range. A comparison of the derived M/L values with M/L values predicted (Larson and Tinsley 1978) from luminous mass in stars plus gas shows that for each spiral type, the dynamical (i.e., luminous plus dark) mass exceeds the luminous mass by a factor of about two. Hence the mass in dark matter equals the mass in luminous matter interior to R(25).

For those galaxies whose mass distributions have been deconvolved

Table 1. Ratio of Dark-to-Luminous Matter in Spiral Galaxies

Hubble Type	M(R25)/L Dynamical	n	(M/L)stars Tinsley Larson	(M/L)stars plus gas	(M/L)dynamical / (M/L)stars,gas
Sa	6.1+0.7	11	3.1	3.1	2.0
Sb	4.5+0.4	22	2.0	2.1	2.1
Sc	2.6+0.2	20	1.0	1.2	2.2

into component parts (NGC 247, 300, and 3109, Carignan and Freeman 1985; NGC 891, Bahcall 1983; NGC 3198, van Albada et al. 1985; NGC 4564, Casertano 1983a; NGC 5907, Casertano 1983b; see other contributions this volume), the same 1:1 ratio of dark-to-luminous matter is derived. This conclusion is discussed more fully by Bahcall and Casertano (1985). Finally, Bahcall (1984; see also Oort 1960) has shown that the mass required in the galactic disk in the solar neighborhood to account for the observed z-motions of stars exceeds by a factor of two that mass which is observed. However, there is as yet no evidence that this dark disk material is related to the more extended dark halo matter.

4.2. M(dark+lum) has an upper limit of order $10^{12} M_\odot$

It is exceptional for a spiral mass interior to R(25) to exceed $10^{12} M_\odot$. I have obtained rotation curves for two spirals for which Giovanelli et al. (1982; 1985) had discovered exceedingly high rotational velocities,

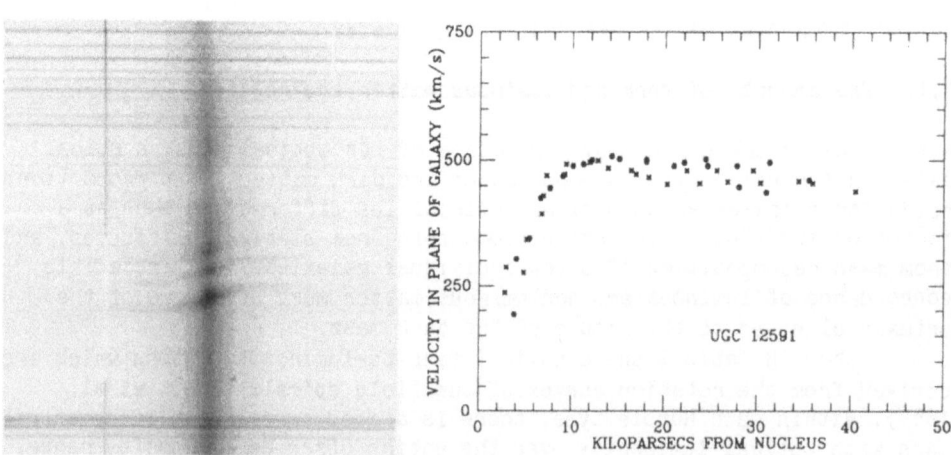

Figure 4. Left: Spectrum of UGC 12591 from a Palomar 200-inch CCD frame. Dispersion and scale 0.8A/pix and 0.59"/pix; integration 5400 sec. Hα emission is here bracketed by [NII]. Right: Rotation velocities for UGC 12591, the most rapidly rotating spiral disk known, from Hα and [NII]; different symbols refer to opposite sides of the galaxy.

Figure 5. Distribution of mass in spiral galaxies for which we have determined rotation curves. The normal spiral galaxies of highest mass are UGC 12591 (SO/a, Vmax = 500 km/s), NGC 669 (Sab, Vmax = 363 km/s), and UGC 2885 (Sc, Vmax = 304 km/s, but R(25) is over 100 kpc).

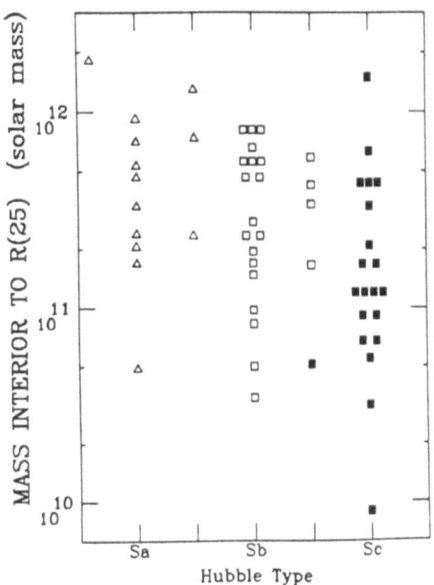

NGC 669 [Sab, V(max)=363 km/s] and UGC 12591, [SO/a, V(max)=500 km/s]. The rotation curve for UGC 12591 is shown in Figure 4. Even though its rotation velocity exceeds by 25% that of the next highest spiral, [IC 724, V(max)=374 km/s], the mass of UGC 12591 is not significantly in excess of 10^{12} M☉ (Figure 5). A lower limit for spiral masses is probably not well established from our sample.

5. WHAT ROLE DOES THE ENVIRONMENT PLAY IN DETERMINING SPIRAL DYNAMICS?

5.1. Statistics of forms of rotation curves, and hence mass forms, differ for field and cluster spirals

It seems well established that field galaxies are embedded in massive, extended halos composed of nonluminous matter. What happens to halos of galaxies in clusters? Are they modified, or indeed, never given the opportunity to form, in the denser cluster environment? In an attempt to see if the dynamics of spirals in clusters differs from those of field spirals, Burstein, Whitmore, and Rubin are analyzing rotation curves of about 20 spirals in clusters: Peg I, Cancer, Hercules, and about 10 from the southern Dressler (1980) cluster DC1842-63 (Figure 6).

Individually, the rotation curves for cluster galaxies show the characteristic turnover and nominally flat outer portions that are observed for field spirals. Statistically, however, there is a difference in the distribution of galaxies among the three mass types, which we believe is environmentally produced. Of the 18 cluster galaxies with well defined mass types, 7 are Sc, 4 Sb, 3 Sa or Sab, 2 S..., and 2 distorted; a distribution not notably different from the field spirals. But whereas 50% of the field spirals show mass type I, the type with the

Figure 6. The central region of the southern Dressler cluster DC1842-63, from a Las Campanas 100-inch plate. Baked 103aO+GG385 filter, exposure 2 hours.

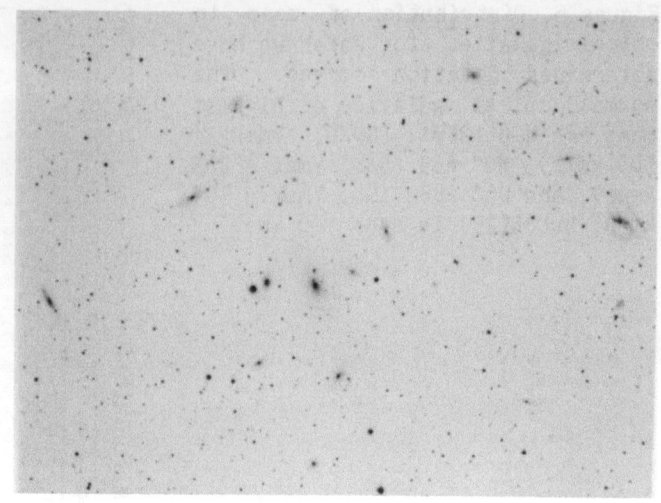

most rapid mass increase in the outer parts, none of the cluster spirals do. Conversely, while only 20% of field spirals are mass type III, the type with slowest mass increase, 50% of the cluster spirals are type III. Thus, cluster galaxies do not exhibit the slowly rising outer portions which are so characteristic of field spirals, but have flat or slowly decreasing velocities, evidence of a more rapid decrease in mass density with radius. A likely explanation is that the denser cluster environment has modified the distribution of dark matter surrounding the spirals which still exist there. In an effort to examine halo properties of spirals in a yet denser environment, we are presently obtaining rotation curves for galaxies in Hickson (1982) compact groups, where the galaxy surface density is as high as it is in the centers of the large clusters. Ultimately, mass forms may be an important clue to the evolutionary and environmental history of spiral galaxies.

5.2. What role does "dim matter" play in spiral galaxies?

In concluding this discussion of the properties of matter that we cannot see, I would like to call attention to the matter that we can almost not see. I refer to the low surface brightness "dim matter" which is not visible on the Palomar Sky Survey prints, the images from which many of us formed our ideas of what a galaxy is. On the Sky Survey prints, UGC 10205 appears as a bulge crossed by an indistinct absorption lane, sufficiently isolated and sufficiently similar to the Sombrero galaxy that we observed it spectroscopically in our sample of normal Sa galaxies. New CCD images (Figure 7) illustrate that there is more to UGC 10205 than appears on the Survey prints. At moderate light levels, UGC 10205 resembles a classic Sa galaxy seen edge-on; at lower light levels it reveals faint features like the shells and ripples which Malin and Carter (1980) and Schweizer and Ford (1984) have pointed out in Ellipticals and SO's. And at the faintest levels, barely 3% above sky, UGC 10205 shows streamers and features characteristic of Arp galaxies.

Until we understand the role of this dim matter, which we can see, in the formation and evolution of spirals, the properties of the dark matter will be difficult to illuminate.

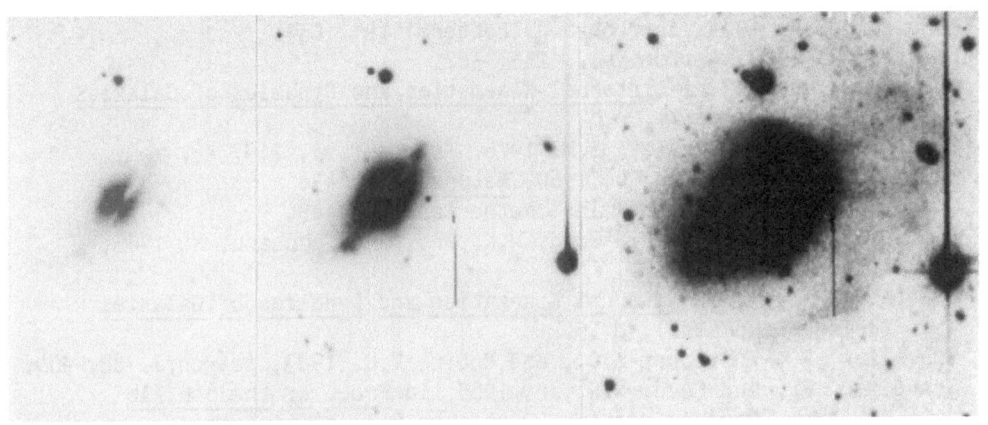

Figure 7. UGC 10205 from a KPNO 36-inch CCD + R (Mould filter) frame, showing successively lower surface brightness features. While at the highest light levels the galaxy resembles a fairly normal Sa, at low light levels it appears embedded in an extensive envelope, from which streamers and other faint features emerge.

ACKNOWLEDGEMENTS AND OTHER

I acknowledge with thanks the valued support of my observing colleagues, Kent Ford, Deidre Hunter, Paul Schechter, and Brad Whitmore and thank them especially for putting up with me on cloudy nights. Much of the work discussed here was done in collaboration with them and with Dave Burstein and Norbert Thonnard. Thanks also to the Directors of Cerro Tololo, Kitt Peak, Las Campanas and Palomar Observatories for observing time. Finally I wish to announce a Vmax Prize consisting of a $100 bill to the first astronomer who observes a spiral galaxy with a rotation velocity of 600 km/s (after correction for 1+z) or greater.

REFERENCES

Bahcall, J.N. 1984, Astrophys.J. 287, 926.
Bahcall, J.N. 1983, Astrophys.J. 267, 52.
Bahcall, J.N., and Casertano, S. 1985, Astrophys.J. 293, L7.
Bosma, A. 1978, Thesis, Rijksuniversiteit te Groningen.
Bosma, A. 1981, Astron.J. 86, 1825.
Burstein, D., and Rubin, V.C. 1985, Astrophys.J. 297, October 15.
Carignan, C., and Freeman, K.C. 1985, Astrophys.J. 294, 494.
Casertano, S. 1983a, Thesis, Pisa.
Casertano, S. 1983b, Mon.Not.Roy.Ast.Soc. 203, 735.

Dressler, A. 1980, Astrophys.J.Suppl. 42, 565.
Giovanelli, R., Haynes, M.P., and Chincarini, G.L. 1982, Astrophys.J. 262, 442.
Giovanelli, R., Haynes, M.P., Rubin, V.C., and Ford, W.K.,Jr. 1985, Astrophys.J., submitted.
Harrison, E.R. 1974, Astrophys.J.(Letters) 191, L51.
Hickson, P. 1982, Astrophys.J. 255, 382.
Kaljnas, A.J. 1983, in Internal Kinematics and Dynamics of Galaxies (Reidel:Dordrecht), p.87.
Larson, R.B., and Tinsley, B.M. 1978, Astrophys.J. 219, 46.
Malin, D.F., and Carter, D. 1980, Nature 285, 643.
Oort, J.H. 1960, Bull.Astr.Inst.Netherlands 15, 45.
Rubin, V.C., Burstein, D., Ford, W.K.,Jr., and Thonnard, N. 1985, Astrophys.J. 289, 81.
Sancisi, R. 1983, in Internal Kinematics and Dynamics of Galaxies (Reidel:Dordrecht), p.55.
Schweizer, F., Whitmore, B.C., and Rubin, V.C. 1983, Astron.J. 88, 909.
Schweizer, F., and Ford, W.K.,Jr. 1985, in Proc. of the 8th IAU Regional European Meeting, (Springer:Berlin), in press.
Smith, S. 1936, Astrophys.J. 83, 23.
van Albada, T.S., Bahcall, J.N., Begeman, K., and Sancisi, R. 1985, Astrophys.J. 295, 305.
Zwicky, F. 1933, Helv.Phys.Acta. 6, 110.

DISCUSSION

DRESSLER: Your point 3.2 that mass distributions are not correlated
with galaxy morphology seems overstated. A plurality of your Type I
mass distributions are Sc galaxies, while Sa's dominate the Type III's.
Although it is true that all morphologies are present for each type,
there is apparently a strong trend. To me this suggests that the
morphological type criteria are not quite correct, that there might be
a "Platonic" morphological type that would correlate much better with
mass distribution. I am further concerned that the correlation of mass
distribution with environment is really a manifestation of this
correlation with morphological type.

RUBIN: There are 4 or 5 Sb's and 4 or 5 Sa's in each of mass types I,
II and III. You are correct that the majority of Sc galaxies have mass
type I, but I am still impressed with the fact that each mass type
contains Sa, Sb and Sc galaxies. As far as the cluster galaxies go,
they are not principally early types. Most of the cluster galaxies we
observed are Sb's or Sc's.

BURSTEIN: If one accepts our result on environmental differences in
mass types, one has the following unclosed circle: a) morphology of
spirals is correlated with environment; b) mass type is correlated with
environment; but c) morphology is uncorrelated with mass type! How
does the environment distinguish between luminous mass and total mass
in the formation process of a spiral galaxy?

MILGROM: I wonder if it wouldn't be more convenient to use the density
distribution. After all, the mass distribution is an integral of the
density distribution, and so has little freedom to vary. It must start
out at zero and go linear more or less asymptotically. So it is no
surprise that all of the mass types look similar. If I understood
correctly, the distinction between the types comes down to whether the
rotation curve is flat or slightly rising.

RUBIN: There is also a difference in the inner parts. A rotation
curve which is slightly rising in the outer parts is always rapidly
rising in the inner parts. A rotation curve which has a shallow rise
in the inner parts has a slight fall in the outer parts. You don't see
any rotation curves which are crossovers between these two cases. So
some mass types which are possible are not observed.

KORMENDY: For me, the most uncomfortable thing you said was your point
3.4 that dark matter is important at all radii. I think that this
corollary would follow: The dark matter could probably not even be
close to isothermal. If it were, and it had a density high enough to
be important near the center, then (given typical bulge and disk scale
lengths) the mass of the disk would be negligible. This would be
uncomfortable, because then you couldn't make density waves, which we
know are there. So I'd like to ask the people who make models: Is it
possible, if you start with an isothermal halo and then embed within it

a much smaller, high-density concentration, to pull the dark matter enough so that it becomes non-isothermal but important at all radii?

FABER: I don't think we fully know the answer. The best information we have comes from the adiabatic assumption, because that allows us to look at many different baryonic mass distributions. The adiabatic calculations suggest that at just about the degree of infall required to give a typical specific angular momentum, the final dark matter density equals the baryonic density within the optical radius. The amount of dark matter present before the infall was much smaller. That is, because of the squeezing effects, you have four or five times as much dark matter within the optical radius after infall as you did before. Further down toward the core, the dark matter density is smaller than the baryonic density. By the time you get to the core of an elliptical galaxy, the dark matter is really not important.

LYNDEN-BELL: I do not agree with Kormendy's point, for the following reason. It seems to me that what matters is the core radius. Just saying that something is isothermal doesn't tell you its core radius. Galaxies have roughly V = constant a long way in, and in principle you can make isothermal models with density $\propto r^{-2}$ all the way in to the middle. Then the dark matter would be important even at the center. So I think your statement is based on the idea that dark matter must be isothermal but with a large core radius. You didn't say that.

KORMENDY: That's true. I'm not used to thinking about singular isothermals because elliptical galaxies looked at carefully and dark matter distributions measured by decomposing rotation curves have all had cores. Even if dark matter can exist as singular isothermals, you can't afford to make the dark matter completely dominant at any radius where there is a disk. Otherwise, if the disk is massless, you can't make density waves.

PEEBLES: One could have the dominant mass in a spheroid and still have instabilities in the disk.

KORMENDY: Not if the disk is totally non-self-gravitating.

LYNDEN-BELL: Even if the disk is totally non-self-gravitating.

PEEBLES: Then some mechanism other than gravity would have to be making spiral arms.

LYNDEN-BELL: That's right (laughter).

FELTEN: Some of your observational points might be used to argue for or against alternative Milgrom-type theories of dynamics. For example, your result that mass types show strong environmental effects in clusters might be surprising in a Milgrom theory, in which there is no dark matter and all the matter is concentrated and tightly bound. Perhaps Milgrom will comment on this, now or later. Have you given any thought to whether some of these points can be used for or against unconventional dynamics?

RUBIN: No.

YAHIL: I want to go back to a question which Sandy posed this morning. What happens at small radii to the spheroidal component? If you take a reasonable mass-to-light ratio, e.g., $M/L \geqslant 1$, then you predict a higher velocity than is observed. Now, your calculations of the integral mass are made by assuming that $M \propto V^2 R$. So it may be possible to resolve Sandy's paradox if you took the flatness of the disks into account. Also, what if there is a hole in the center of the disk?

RUBIN: It is hard to believe that disks have holes, because you observe emission lines right across their centers with no discontinuity. Observations of the Na D lines also coincide exactly with observations of the emission lines. I, too, have worried about rotation curves in the inner parts of galaxies – how accurately we know them, whether they reflect circular motions, etc. This last spring, Gallagher and Hunter took a large-scale, high-dispersion echelle plate of the inner regions of NGC 3198. Within the errors (~ 10 km s^{-1}) this very detailed rotation curve falls exactly in the HI rotation curve. This doesn't tell us that the motions are circular. But this very accurate rotation curve fits on the family curve. So if there are non-circular motions, they are fooling us, because they are similar from galaxy to galaxy and because the minor axis spectra reveal no peculiar motions.

STEIGMAN: Comparing the luminous mass to the total mass is comparing something derived from observations to something derived from theory. Am I supposed to believe the factor of two difference, or am I supposed to say, "Great agreement!"?

RUBIN: You're supposed to believe the factor of two difference. That is, if you build a galaxy out of stars, you get $M/L = 3$. If you measure its dynamics, you get $M/L = 6$.

DIFFERENCES IN MASS DISTRIBUTION FOR FIELD AND CLUSTER SPIRAL GALAXIES

David Burstein   and   Vera C. Rubin
Ariz. State Univ.   DTM - CIW

Our group has now obtained rotation curves for 80 spiral galaxies, Hubble types Sa through Sd. As described in Rubin et al. (Ap. J. 289, 81; 1985), the forms of these rotation curves are similar for all Hubble types. Given this observational fact, we have chosen to analyze the mass distributions for these galaxies under the assumption that the mass distributions for all spirals can be described by the same three-dimensional form, here taken to be spherical for simplicity. The mass distribution forms for 71 of these galaxies can be placed into a simple classification scheme based on the curvature of mass distribution form in a log(radius) - log (integral mass) diagram. The three most common mass forms among this continuum are termed Types I, II and III, the forms of which are displayed below (see also the discussion by Rubin elsewhere in this Symposium).

The forms of mass distribution show no correlation with galaxy size, luminosity or mass density, and little correlation with Hubble type (see Burstein and Rubin, Ap. J., Oct. 1985 for details). However, we do find an environmental dependence of Mass Type: Of the 54 field galaxies with good Mass Types, 27 are of Type I (50%) and 11 are of Type III (20%). Of the 18 analagous galaxies in clusters (ranging in Hubble type from Sb to Sd/Irr), none are of Type I (0%) and 9 are of Type III (50%).

The details of the correlation of Mass Type with environment will be discussed in a future paper.

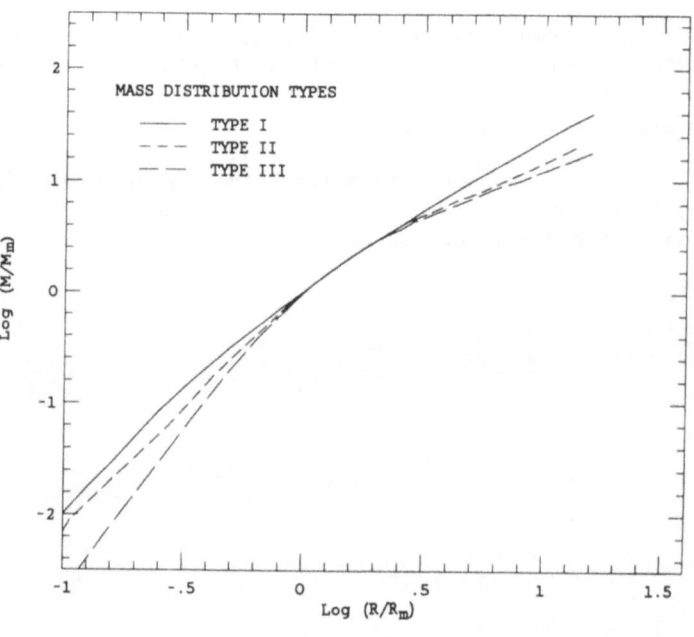

MASS DISTRIBUTION TYPES

——— TYPE I
- - - TYPE II
— — TYPE III

Log (M/M$_m$)

Log (R/R$_m$)

J. Kormendy and G. R. Knapp (eds.), Dark Matter in the Universe, 66.

HI ROTATION CURVES OF GALAXIES

R. Sancisi and T. S. van Albada
Kapteyn Astronomical Institute
Groningen University
The Netherlands

ABSTRACT. The observational evidence on the discrepancy between the mass distribution in galaxies derived from HI rotation curves and that derived from the distribution of light is reviewed. In the outer parts the discrepancy is such that in some galaxies there is at least three times as much dark matter as luminous matter. This is a direct consequence of the nearly constant circular velocity far beyond the edge of the visible part of the galaxy, as derived from the motion of HI. The discrepancy is clearly present already near the edge of the visible disk (R_{25}). In the inner regions, i.e. inside approximately 2.5 disk scale-lengths, no dark matter is required, but its presence can not be ruled out. There is no evidence for a dependence on galaxy luminosity or morphological type. These results suggest a strong coupling between luminous matter and dark matter within individual galaxies, and among galaxies as well. Finally attention is drawn to the large-scale asymmetries in the outer parts of galaxies and to possible implications for the vertical distribution of dark matter.

1. INTRODUCTION AND OBSERVATIONAL REQUIREMENTS

The main issue we want to address in this paper is a quantitative determination of the discrepancy between the distribution of luminous matter in galaxies and the distribution of matter derived from the kinematics of the gaseous component. For this purpose neutral hydrogen observations are crucial as they make it possible to trace the kinematical properties of galaxies far beyond the optical disk. In the last decade HI rotation curves have been derived for a large number of systems of various morphological types and luminosities (see e.g. Bosma 1981; Carignan and Freeman 1985). They have been used to calculate total masses and mass distributions, and have revealed significant dis-crepancies between the luminous and the dynamical mass (see e.g. Faber and Gallagher 1979). The majority of these studies, however, were limited by low angular resolution in the inner parts, poor signal/noise ratio in the outer parts, and occasionally large-scale deviations from axial symmetry.

J. Kormendy and G. R. Knapp (eds.), Dark Matter in the Universe, 67–81.

At present a second generation 21-cm line study for a number of carefully selected objects is in progress. The increased sensitivity of these new observations makes it possible to reach column densities of about 1×10^{19} cm^{-2} for resolutions of ~1 arcmin. The selection criteria are straightforward: the galaxies must have extended, unperturbed HI disks, and be sufficiently inclined to the line of sight - somewhere between 50 and 80 degrees - so that possible effects of warping in the outer parts are small and can be unambiguously corrected for. A wide range of morphological types and luminosities should be covered.

The method generally followed to derive rotation curves from the observed velocity fields is to represent the hydrogen disk by a number of circular rings, each ring being characterized by an inclination angle, a position angle in the plane of the galaxy, and a circular velocity. The orientation parameters, and their variation with radius, are determined from the velocity field. For good cases such as NGC 3198 and 2403 (see below) this can be done with high precision: $\sigma_i \simeq 1^0$ and $\sigma_{pa} \simeq 1^0$. These kinematical inclination and position angles generally agree well with those derived from optical images. The quality of the fit with the derived orientation parameters and rotation curve is assessed by inspecting the residual (model - observed) velocity field. Values of the 'apparent' circular velocity accurate to about 2 or 3 km/s are generally obtained from such fitting procedures. But irregularities and asymmetries in the velocity field often produce larger fluctuations, of order 5 to 10 km/s. A requirement that may be used to obtain an estimate of the true uncertainty in the circular velocity is that rotation curves derived separately for the two sides of a galaxy are symmetrical. For a useful rotation curve the two sides should agree to within 5 - 10 km/s. With the high sensitivity of present HI observations this condition, and not S/N, is the real limitation; it restricts the choice of objects to study. The angular and velocity resolutions which can be obtained with the VLA and Westerbork synthesis radiotelescopes are normally adequate. The rotation curves derived from optical observations (see e.g. Rubin et al 1985 and references therein), which reach out to at most R_{25}, are an essential complement in the central regions where beam smearing effects or HI deficiences can be a severe limitation to the radio observations.

For galaxies viewed edge-on a more sophisticated analysis, using models of the position-velocity maps, is necessary for a precise determination of circular velocities. But, for such large inclinations the possibility of deviations from circular motion and/or HI deficiencies at the line of nodes can never be completely ruled out, and the velocity determination is bound to remain somewhat uncertain. Nevertheless, edge-on galaxies offer a clear advantage for studying also the z-distribution of the gas and of the total mass (e.g. van der Kruit 1981).

2. RESULTS

A number of these newly determined HI rotation curves are shown in Figure 1. They have been reduced to the same linear scale. Morphological types of the galaxies shown vary from early Sb to Sdm; the luminosities

cover a range of 260 : 1. Two curves (NGC 2403 and 3198) extend to more
than twice the optical size and to many (9 and 11) disk scale lengths. In
comparison, UGC 2885, which is well known for its huge dimensions and
flat rotation curve (Rubin et al. 1980), does not extend in HI much
beyond R_{25} (Roelfsema and Allen 1985), i.e. about 4 disk scalelengths.

All curves remain approximately flat out to the last measured
point. They are not, however, entirely featureless. The curve of NGC
4565 shows a slow but steady decline over the outer half of the system.
Inside 10 kpc the curve is not well determined; it may be significantly
affected by deficiency of gas or by non-circular motions. A drop-off by
about 20 km/s is also observed in NGC 5907 beyond the optical edge at
about 18 kpc. On one side the gas extends farther out to about 45 kpc,
i.e. beyond the symmetrical part (R < 32 kpc) shown in Figure 1. In this
tail the line-of-sight velocities appear to drop off towards the
systemic velocity in a way similar to that found in the southern tail of
NGC 891 (Sancisi and Allen 1979; see also Fig. 10). The asymmetries in
these cases are suggestive of non-circular motions. A hint of a decline
of V_{cir} beyond R_{25} and a flattening at large radii is also noticeable in
NGC 3198. NGC 2403 and UGC 2259 show slightly rising rotation curves.

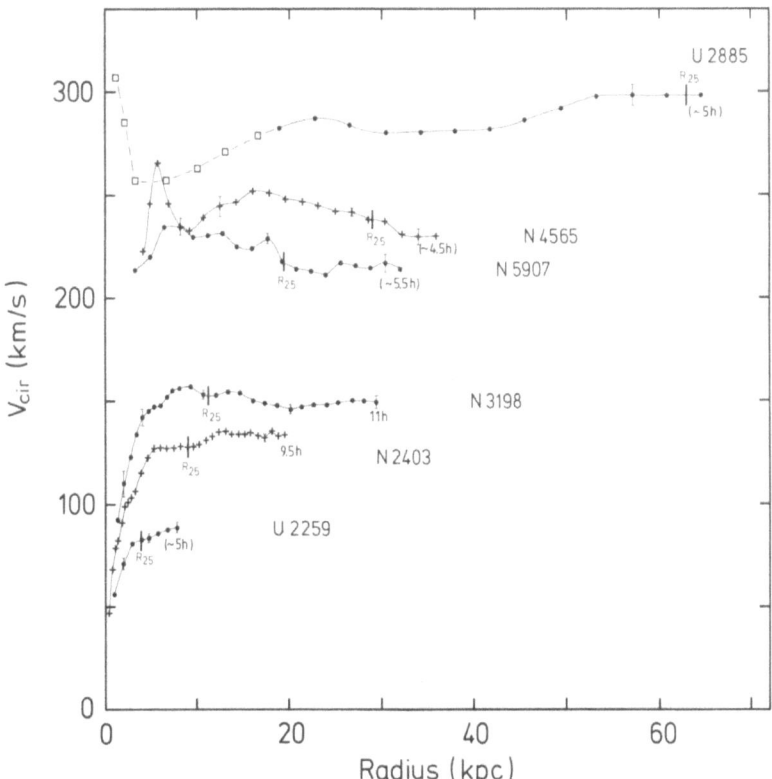

Figure 1.   HI rotation curves for a number of spiral galaxies. (Linear
scale based on H = 75 km/s/Mpc.) The optical radius, R_{25}, and the number of disk scalelengths, h, at the last measured point are indicated.

Possibly these features are pointing at some significant property of the mass distribution, and perhaps are related to the overall distribution of luminous matter as they occur over the region of the disk or close to its edge. At present we will, however, investigate only the first order flat behaviour of rotation curves, with particular attention to the cases of NGC 3198 and 2403 where the curves remain flat far beyond the optical image. The results for these two galaxies are summarized in Table 1.

Table 1. Parameters for NGC 2403 and 3198.

NGC	2403	3198	
Distance	3.25	9.2	(Mpc)
Disk scalelength, h	2.1	2.7	(kpc)
R_{25}	8.5	11.2	(kpc)
R_{max} (HI)	20	30	(kpc)
R_{max}/h	9.5	11	
V_{max}	135	157	(km/s)
M_{HI}	3.2	4.8	($10^9 M_\odot$)
M_{total}	7.9	15.4	($10^{10} M_\odot$)
L_B	0.79	0.86	($10^{10} L_{B\odot}$)
M_{total}/L_B	10	18	($M_\odot/L_{B\odot}$)
M_{disk} (max)	1.9	4.1	($10^{10} M_\odot$)
M_{halo} ($R<R_{25}$)	1.3	1.9	($10^{10} M_\odot$)
M_{disk}/L_B	≤ 2.4	≤ 4.7	($M_\odot/L_{B\odot}$)
M_{dark}/M_{lum} ($R<R_{25}$)	≥ 0.8	≥ 0.5	
M_{dark}/M_{lum} ($R<R_{max}$)	≥ 3.2	≥ 2.7	

NGC 3198. This is clearly the prototype of a spiral satisfying the criteria mentioned in section 1. The HI (Begeman 1985) has been traced out to about 2.7 R_{25}, corresponding to ~ 11 disk scalelengths (Figure 2). The velocity field shows large scale regularity and symmetry. The inclination angle is sufficiently large for a non-ambiguous and precise determination of the circular velocity. The largest correction for changes in inclination in the outer parts is less than 5 km/s. The estimated uncertainties in the rotational velocities in the outer parts are also of the order of 5 km/s and come entirely from the small asymmetry between the two sides of the galaxy. HI circular velocities in the inner region are fully confirmed by recent optical observations (V.C. Rubin 1985, private communication). In Figure 2 (bottom) the observed rotation curve is compared with the model rotation curve calculated from the photometric profile (top) of Wevers (1984), with the assumption of a constant M/L ratio. The value of the latter has been chosen such as to maximize the disk mass while matching the HI rotation curve. It is clear that the observed curve can be accounted for by the visible disk inside about 6 kpc (~ 2.2 scalelengths), but beyond the

peak of the model curve the discrepancy between the observed and the
expected curves becomes increasingly larger.

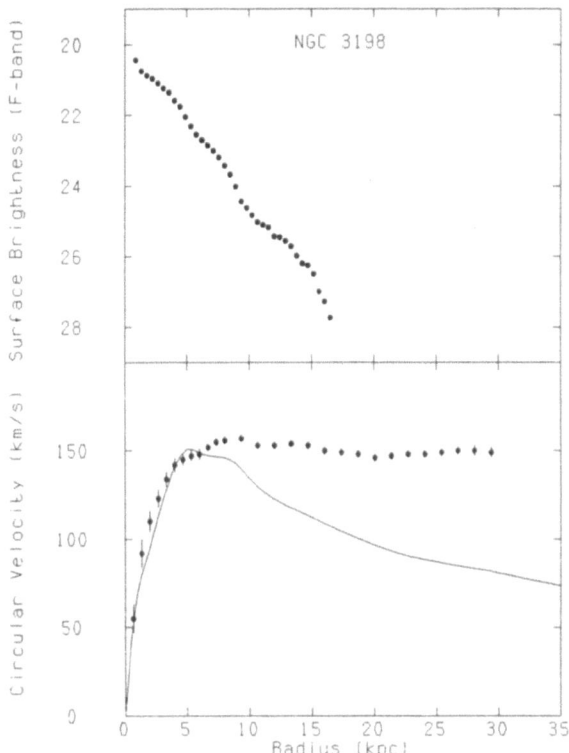

Figure 2.   Top: Luminosity profile of NGC 3198 from the photographic
surface photometry of Wevers (1984). The scalelength of the disk is 2.7
kpc. Bottom: HI rotation curve (dots with error bars, Begeman 1985), and
curve representing the circular velocity of light and gas (solid line).
The light contribution has been computed from the luminosity profile by
assuming that M/L is constant with radius; maximization of the disk mass
while matching the observed rotation curve gives $M/L_B = 4.0$ (stars
only). The gas contribution, which includes a correction for helium, is
negligible inside 10 kpc.

The conventional explanation for such a discrepancy is the presence
of a dark, more or less spherical halo. The amount of dark matter inside
the last point of the rotation curve, at 30 kpc, is at least a factor
three larger than the amount of luminous matter. The maximum M/L_B ratio
for the disk is 4.7 (gas + stars, Table 1), but any lower value, in
combination with an appropriate halo, would also be consistent with the
HI curve. Such a disk-halo model for NGC 3198 has been discussed in
detail by van Albada et al. (1985); see Figs. 3 and 4. The core radius
of the distribution of dark matter lies between 1.7 and 12.5 kpc.

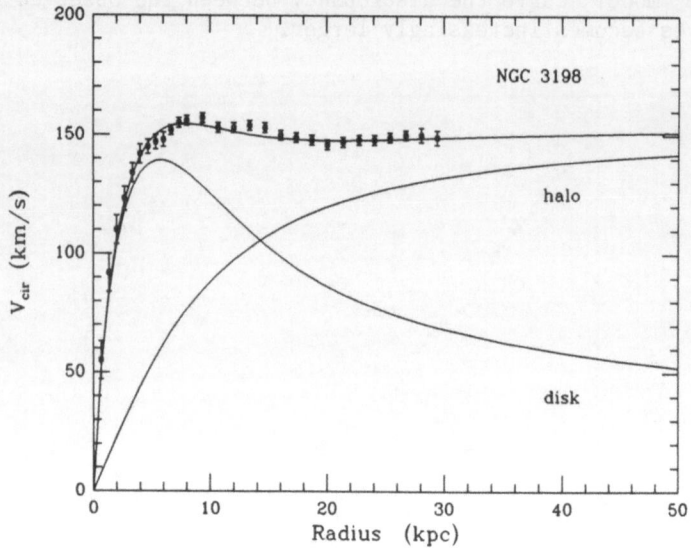

Figure 3.   Fit of exponential disk with maximum mass and halo to the observed rotation curve for NGC 3198 (dots with error bars) from van Albada et al. (1985). The scalelength of the disk has been taken equal to that of the light distribution (2.7 kpc). The maximum circular velocity of the disk has been somewhat reduced with respect to that in Fig. 2 to allow a halo with a non hollow core.

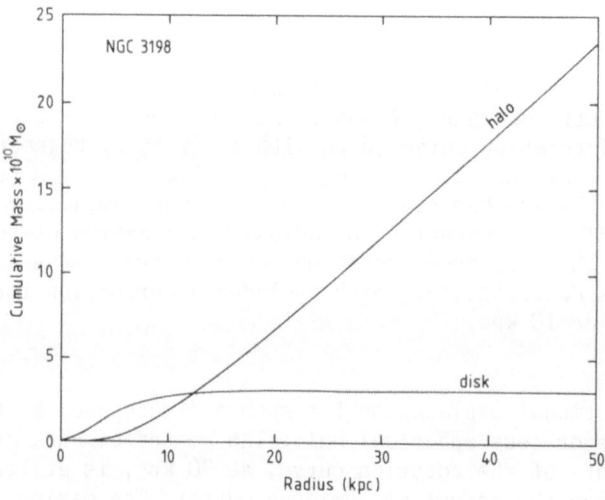

Figure 4.   Cumulative distribution of mass with radius for the exponential disk and halo with maximum disk mass shown in Figure 3. (van Albada et al. 1985).

NGC 2403. The rotation curve of this galaxy, showing the symmetry
between the two sides, is given in Fig. 5. An analysis of the mass
distribution similar to that for NGC 3198 has been made. The luminosity
profile (Wevers 1984) has a clear exponential shape (Figure 6 top). A
maximum disk fit to the HI rotation curve is shown in Figure 6 (bottom).
A large discrepancy between dynamical mass and mass expected from the
luminosity profile with − maximum − constant M/L_B shows up at radii
larger than ~ 4.5 kpc (≃ 2.2 h). The M/L_B for the disk is very low,
less than 2.4. If the gas contribution is also taken into account the
M/L_B for the stellar component becomes less than 1.9, suggesting that
the true disk mass can not be much smaller than the maximum disk mass
(see Larson and Tinsley 1978). The dark mass inside the last measured
point at ~ 2.3 R_{25} (= 9.5 h) would then be about 3 times as large as
the luminous mass (stars + gas).

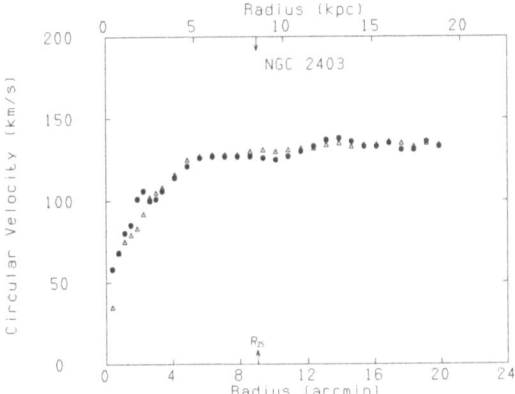

Figure 5.
HI rotation curve
of NGC 2403 (Begeman
1985). The receding
and approaching sides
(triangles and dots)
are given separately
to show the symmetry
of the galaxy.

NGC 4565 and 5907. The rotation curves expected from the photometric
profiles (van der Kruit and Searle 1981, Jensen and Thuan 1982) for
these nearly edge-on systems (with the assumption of a disk with
constant and maximum M/L) show large deviations from the HI rotation
curves in Fig. 1. The model for NGC 5907 is shown in Fig. 7; for NGC
4565 see article by Casertano et al. (this volume). Contrary to what has
been found for NGC 3198 and 2403 a deviation is present also in the
inner parts. For NGC 4565 this may be explained at least partly by the
presence of a bulge component. In NGC 5907 the large M/L value of 9
derived for the maximum disk case, which is consistent with the large
M/L value of van der Kruit and Searle, may point at the presence of a
stellar component not represented in the photometric profile (i.e.
hidden by the dust) or at some selective absorption affecting the
profile inside 15 kpc. At any rate, the question of the luminosity
distribution in nearly edge-on systems (i ≳ 85°) with a dust layer,
should be further investigated before concluding on a real mass
discrepancy. In the outer parts the discrepancy between dynamical and
luminous mass seems indisputable and similar to that found in the less
inclined galaxies discussed above.

The observed 20 km/s drop-off in the rotation curve of NGC 5907 at 18 kpc (≈ 3 h) does not correspond to the drop-off of the curve predicted from the observed light profile and may therefore not be related to the truncation effect studied by Casertano (1983), who based his conclusion on an approximate model for the light distribution with a sharp truncation.

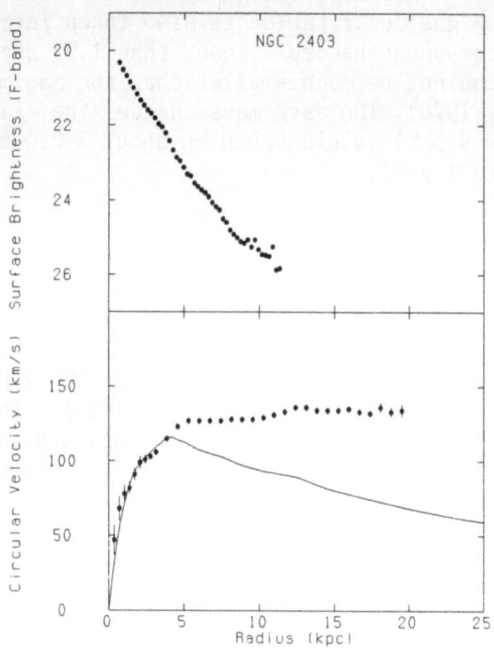

Figure 6.   Top: Luminosity profile of NGC 2403 from the photographic surface photometry of Wevers (1984). The scale length of the disk is 2.1 kpc. Bottom: HI rotation curve (dots with error bars, Begeman 1985) and curve representing the circular velocity of light and gas (solid line). The light contribution has been computed from the luminosity profile by assuming that M/L is constant with radius; a value of 1.9 has been used (cf. Figure 2). The contribution of gas to the circular velocity inside 5 kpc is negligible.

Low luminosity galaxies. For dwarf irregular galaxies the observational situation is much less satisfactory than for the higher luminosity systems discussed above. Although rotation always appears to be the dominant form of motion in these systems, the HI distribution and kinematics show large-scale irregularities and asymmetries. The optical picture itself is too irregular to be of use for the definition of the geometrical parameters (inclination angle, line of nodes and center of mass). To our knowledge, it has not been possible to draw any firm conclusion on mass discrepancies and on the amount of dark matter in dwarf irregulars from existing HI observations.

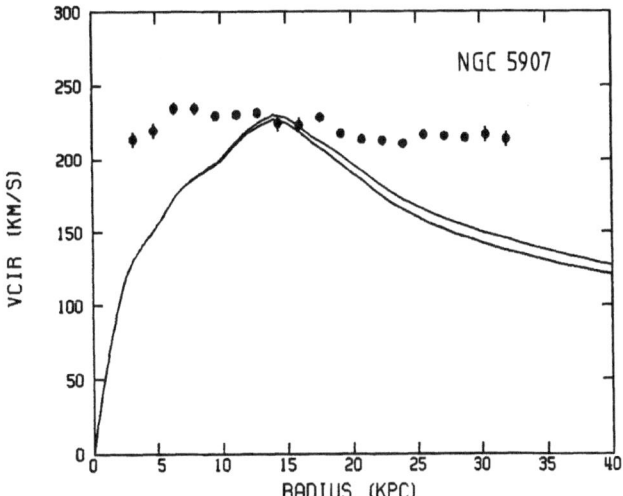

Figure 7.   HI rotation curve of NGC 5907 (dots with error bars) and
rotation curves (solid lines) representing the distribution of light
(lower) and light + gas (upper). The curve for the light has been
computed from the light profiles of van der Kruit and Searle (1981) by
assuming that M/L is constant and equal to 9 (maximum disk case). Note
discrepancy in the inner region between observed and calculated rotation
curves (see text).

      An interesting and more promising category of objects are galaxies
of low luminosity and more regular morphology (type Sd–Sm), and circular
velocity between 50 and 100 km/s. The galaxies studied by Carignan and
Freeman (1985) fall in this class. Another example, shown in Figure 1,
is UGC 2259, which was selected for a 21 cm study by Carignan, Sancisi
and van Albada (1985, see also paper in this volume). The HI in this
galaxy extends out to about 2.3 R_{25}, has overall regular shape, and its
velocity field shows large-scale axial symmetry. The rotation curve
rises slowly out to the last measured point. No surface photometry is
available yet to allow a detailed analysis such as made for NGC 3198 and
2403. Estimates based on total luminosity and size (R_{25}) indicate the
same kind and amount of discrepancy in the outer parts as found in other
galaxies. This is consistent with the results obtained by Carignan and
Freeman (1985) in three late-type spirals from less detailed HI
observations but with complete photometric information.

3. DISCUSSION

a) Conclusions from HI observations.

      The main question we have attempted to address is: how large is the
mass discrepancy in disk galaxies, and how does it depend on location
within a galaxy, on luminosity, and morphological type? According to the
conventional hypothesis such mass discrepancy is taken as a measure of

the amount of dark matter in the system. From the observations discussed above the following picture emerges (cf. Table 1, see also discussion by Bahcall and Casertano 1985).

1) There is a large discrepancy between dynamical and luminous mass. It begins well inside the optical disk, certainly around 2.5 disk scale-lenghts and possibly even closer in, and increases with radius.

2) For the actual ratio of dark-to-luminous matter (M_{dark}/M_{lum}) only certain limits can be set. It could be as low as zero inside 2.2 h, but it is at least as large as 0.5 inside R_{25}. This is based on the maximum disk fit to the rotation curves of NGC 2403 and 3198. The maximum M/L_B values for the disks are between 2 and 5 (somewhat lower if the gas contribution is taken into account). These values seem plausible when compared to the value for the solar neighborhood (Bahcall 1984) and compared to those predicted by stellar evolution models (Larson and Tinsley 1978), and could be close to the true ones.

3) Outside R_{25} non-luminous matter clearly becomes dominant. The ratio M_{dark}/M_{lum} inside the last measured point, out to 20 to 30 kpc (~ 10 disk scalelengths) can be as large as 3.

4) The ratio M_{dark}/M_{lum} does not seem to vary much with total luminosity. This is indicated by a comparison of low luminosity galaxies (Carignan and Freeman 1985; Carignan, Sancisi and van Albada 1985) with the higher luminosity objects discussed above; see also Bahcall and Casertano (1985). Over a range of one hundred in luminosity the variation may not be more than a factor 2 or 3, with no clear systematic trend.

b) The disk-halo dilemma.

The basic assumption made in the above analysis is that of constant M/L for the disk. The 'halo' is the additional component necessary to explain the rotation curve: the halo mass depends on the assumed value for the disk mass. Examination of two extreme possibilitites - a dominating disk and an insignificant disk - in the two best studied objects NGC 3198 and 2403 (Table 1) may elucidate the question and consequences of the disk-halo ratio.

(i) <u>Dominating disk</u> inside R_{25} (maximum disk, Fig. 8). M_{dark}/M_{lum} would be somewhat less than 1 inside R_{25} with 2 < $(M/L)_{disk}$ < 5. This would provide a natural explanation for the Tully-Fisher relation, since the maximum observed circular velocity is uniquely related to the amount of luminous matter. However, the <u>conspiracy</u> between disk and halo, required to produce a flat rotation curve, would remain unresolved.

(ii) <u>Insignificant disk</u>, with maximum circular velocity say 0.5 times the maximum observed circular velocity (Fig. 9). In this case M_{dark}/M_{lum} ≃ 5 inside R_{25}, and $(M/L)_{disk}$ about 1. There is now no problem of conspiracy of disk and halo as the halo accounts for the flat part of the rotation curve. But a fixed ratio M_{dark}/M_{lum} inside R_{25} is required to explain the Tully-Fisher relation. The large contribution of the presumably 'hot' halo to the total mass inside R_{25} would inhibit the formation of two-armed spiral structure, in contrast to the observed situation for NGC 3198 and 2403. (From the dynamical point of view this is a strong argument against the insignificant disk case.)

There may be an intermediate case between that of the insignificant disk – leading to the rather low values of $(M/L)_{disk}$ and to the other difficulties – and that of a dominating disk – with the puzzle of the disk-halo conspiracy – that minimizes these problems. To some extent however the difficulties will remain. The <u>alternative</u> solution would be that of a disk with M/L increasing with radius and a functional form of $M/L(r/h)$ similar for all galaxies. The z-thickness of such a disk may also increase with radius.

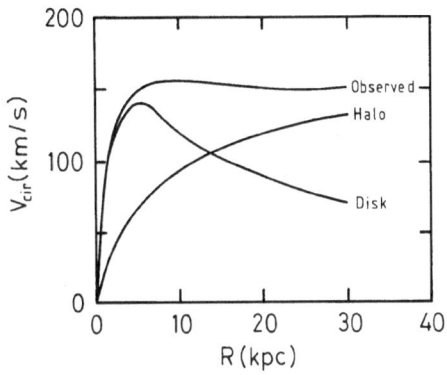

Figure 8.
Diagram showing the decomposition of an observed rotation curve into the separate contributions by disk and halo for the maximum disk case.

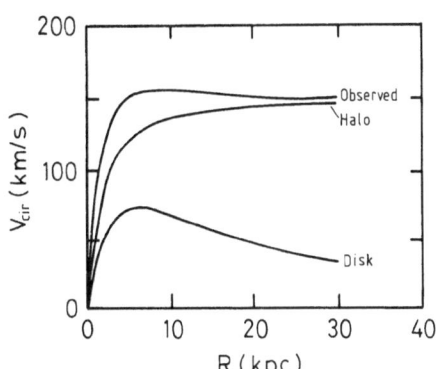

Figure 9.
Same as Figure 8 for an insignificant disk.
$(M_{disk} = 0.25 \times M_{disk\ max})$

c) The outer asymmetries.

Above we have emphasized the requirement of axial symmetry for galaxies to derive their rotation curves, and argued that large-scale deviations from symmetry are the real observational limitation. Yet, galaxies do have asymmetries in their HI density distributions and kinematics (cf. e.g. Baldwin, Lynden-Bell and Sancisi 1980). Although the assumption of circular motion may no longer be valid in such cases, the velocity structure may still be used to obtain information on the mass distribution.

Here we present an example of such asymmetries occurring in the outer parts of two edge-on galaxies (NGC 891 and 5907; cf. Fig. 10), which illustrates the limitations in deriving rotation curves at large

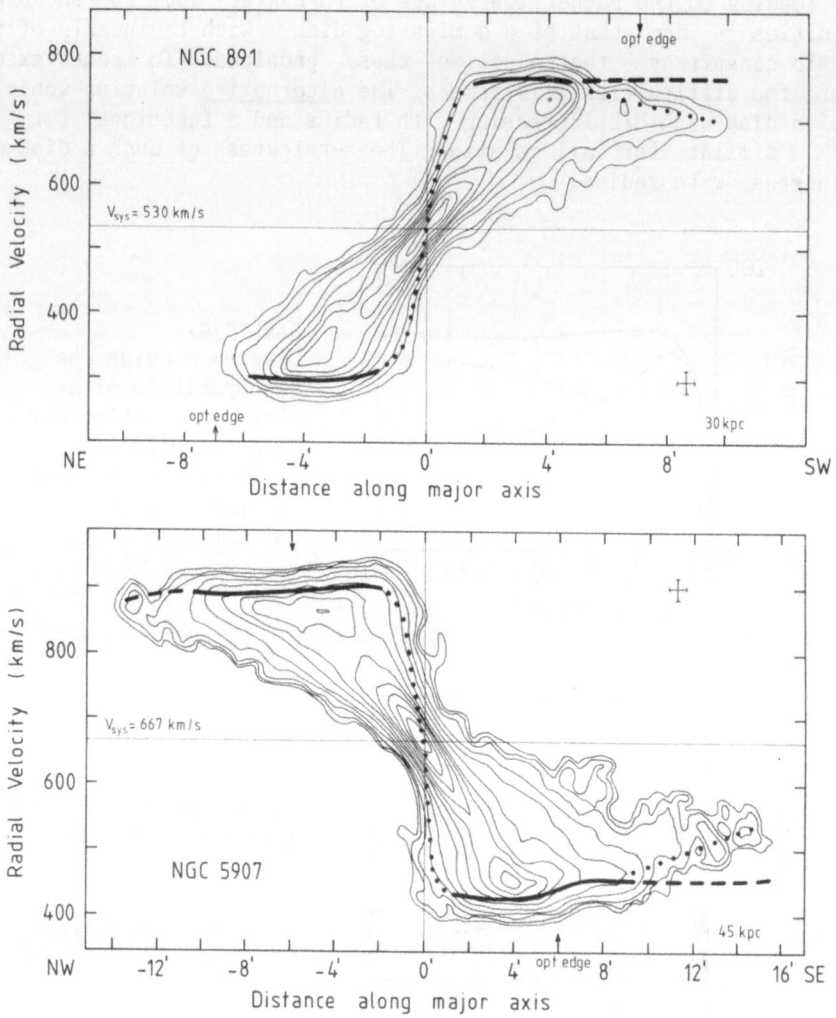

Figure 10.   Maps of the edge-on galaxies NGC 891 and 5907 showing the distribution of HI brightness temperature along the major axis, after integration in the perpendicular direction, as a function of heliocentric radial velocity. The contours for NGC 891 are: 0.5,1.6,2.6,5.2 and so on with equal steps of 2.6, for NGC 5907: 0.1,0.3,0.6,1.5,3.0, 6.0, etc. with steps of 3.0 ; units are 10^2 K x arcsec. Crosses indicate angular and velocity resolutions (FWHP). The rotation curves for the symmetrical parts are shown as thick solid lines; in the inner regions (dots) they are rather uncertain. In the tails extending far outside the optical image on the right side of the figure, where the HI radial velocities re-approach the systemic velocities, both the flat extrapolation of the rotation curve (dashed) and the close to Keplerian drop-off (dotted) are shown.

radii and at the same time may provide a clue to the true z-distribution of dark matter. The observations show a picture of an inner system, which is ordered and symmetrical and for which a rotation curve can be derived, changing to an outer disordered and more irregular distribution. In particular: (i) the HI shows a tail on one side, (ii) the velocities in the tail drop back to the systemic velocity and (iii) the HI in the tail remains close (although warped in one of the two objects) to the plane of the disk out to 2 to 3 optical radii. The velocity drop-off, which is close to Keplerian, does not necessarily imply a finite mass distribution. It may also be reconciled with a flat rotation curve in combination with an HI tail displaced from the line of nodes. Or, perhaps, the galaxy is lopsided and the HI moves on eccentric orbits.

   Clearly, no firm conclusion can be reached on the mass distribution and on the presence of dark matter in the outermost parts of these systems. But, assuming that dark matter is present, the asymmetrical HI distribution and the location of the tail close to the plane of the disk, may hint at a flattened distribution of dark matter in the outer parts. This is based on the following considerations. The gas in the tail is rotating differentially with relatively short orbital periods of the order of 1×10^9 yrs. If it were part of a primordial, asymmetrical configuration, the survival of the asymmetry would be difficult to explain (cf. Baldwin et al. 1980). If, on the other hand, the gas has been accreted recently and yet has been able to find the plane of the disk at large radii (2 to 3 times optical), this may imply the presence far out of a flat dark system, more or less coplanar with the inner disk.

   We thank K. Begeman for the use of his rotation curve data prior to publication and for his assistance with the preparation of figures. We are grateful for the hospitality at The Institute for Advanced Study, where part of this review was prepared, and where we benefited from discussions with S. Casertano, N. Kylafis, and M. Schwarzschild.

REFERENCES

Bahcall, J.N. 1984, Astrophys. J. 287, 926.
Bahcall, J.N., and Casertano, C. 1985, Astrophys. J. 293, L7.
Baldwin, J.E., Lynden-Bell,D., and Sancisi, R. 1980, Mon. Not. R. Astr. Soc. 193, 313.
Begeman, K. 1985, in preparation.
Bosma, A. 1981, Astron. J. 86, 1791.
Carignan, C., and Freeman, K.C. 1985, Astrophys. J. 294, 494.
Carignan, C., Sancisi, R., and van Albada, T.S. 1985, in preparation.
Casertano, S. 1983, Mon. Not. R. Astr. Soc. 203, 735.
Faber, S.M., and Gallagher, J.S. 1979, Ann. Rev. Astron. Astrophys. 17, 135.
Jensen,E.B., and Thuan, T.X. 1982, Astrophys. J. Suppl. 50, 421.
Larson, R.B., and Tinsley, B.M. 1978, Astrophys. J. 219, 46.
Roelfsema, P.R., and Allen, R.J. 1985, Astron. Astrophys. 146, 213.

Rubin, V.C., Ford, W.K.Jr., and Thonnard, N. 1980, Astrophys. J. 238,
       471.
Rubin, V.C., Burstein, D., Ford, W.K.Jr., and Thonnard, N. 1985,
       Astrophys. J. 289, 81.
Sancisi, R., and Allen, R.J. 1979, Astron. Astrophys. 74, 73.
van Albada, T.S., Bahcall, J.N., Begeman, K., and Sancisi, R. 1985,
       Astrophys. J. 295, 305.
van der Kruit, P.C., and Searle, L. 1981, Astron. Astrophys. 95, 105.
van der Kruit, P.C. 1981, Astron. Astrophys. 99, 298.
Wevers, B.M.H.R. 1984, Ph.D. Thesis, Groningen University.

DISCUSSION

OSTRIKER: When you look at M/L ratios integrated out to an optical
radius, you get numbers that are moderate, like M/L ~ 10. But since
the light is all in the inner parts, what happens if you work out the
local M/L in the outer regions? I.e., if you estimate the added mass
and the added light at around the radius of the HI disk, what sort of
numbers do you get? Is M/L as big as 1000?

SANCISI: By R_{25} the integrated amounts of light and dark matter are
comparable. If you go to greater radii, the local values of M/L become
very large, certainly larger than 100.

CARIGNAN: In the case of NGC 3109 (Carignan and Freeman 1985) the
local M/L varies from 2 at the center to 10^3 at the last measured
velocity point.

PARTRIDGE: Vera Rubin mentioned a couple of cases where you have
background quasars shining through the disks of spiral galaxies. Can
you get information from these cases about the rotation curve at very
large distances, or do asymmetries kill you?

SANCISI: Well, we very much rely on having fully sampled 3-dimensional
data to identify large-scale deviations from circular motion, such as
asymmetries, warps, etc. Now if you have one point far out, even if it
is on the major axis, you don't know what sort of inclination
correction to use - the gas could be orbiting in a different plane in
the inner and outer parts. So although such an observation provides
useful information, such as how far the disk extends, it provides only
a very uncertain measure of the rotation velocity.

FABER: You might know more about asymmetries if you had observed more
face-on objects. So what are the statistics of the phenomenon? How
many galaxies have been observed at very low surface brightnesses?

SANCISI: I would say that the majority of galaxies have lopsided
forms. But a real statistical study remains to be done. Warps and
asymmetrics occur in the outer parts of galaxies, and only by
understanding the physics and state of motion of the gas there can you
use it to measure rotation curves. This may in the end prove to be
impossible.
     I also worry that gas in edge-on galaxies may not lie on the line
of nodes. So even if it is on circular orbits, it may not give the
rotation velocity. In the well-known case of the galaxy NGC 1961, for
example, we know that the gas is not along the major axis.

KINEMATICS AND WARP IN NGC 4565

S. Casertano [1]. R. Sancisi [2] and T. S. van Albada [2]
[1] Institute for Advanced Study, Princeton, NJ 08540 - USA
[2] Kapteyn Laboratorium. Postbus 800, 9700 AV Groningen - The Netherlands

We present preliminary results of the analysis of WSRT observations of the edge-on spiral NGC 4565 (Sancisi 1985). The neutral hydrogen layer of this galaxy extends to $1.3 R_{25}$, permitting to trace the rotation curve outside the optical edge of the system, and is warped in the outer parts.

The rotation curve is obtained by constructing model velocity profiles at different positions along the major axis, and matching them to the observed profiles. The uncertainties in the derived velocities are of about $5\,km/s$. The relevant features of the rotation curve are a steep inner maximum and a steady slow decline of the rotation velocity in the outer parts.

In the inner parts, the rotation velocity quickly rises to reach a maximum value of about $260\,km/s$ at about $5\,kpc$ from the center. After the inner maximum, the velocity shows a sharp, faster-than-keplerian drop of about $30\,km/s$. If real, this feature would imply a nearly homogeneous, highly flattened (axis ratio < 0.5) nuclear mass component, comprising at least 20% of the disk mass within one disk scale length. These characteristics do not agree with the properties of the observed bulge. Alternatively, the inner maximum may be only apparent. due to either non-circular motions or irregularities in the HI distribution.

At larger radii the rotation velocity exhibits a broad maximum ($v = 250\,km/s$ at $17\,kpc$). past which it slowly declines to about $225\,km/s$. The amount of the velocity drop implies that a good fraction of the total mass within the last measured point is probably in a non-luminous form.

The disk of NGC 4565 shows a warp both in the HI and in the optical image (van der Kruit and Searle 1981). At first sight, the warps in the two components do not coincide, in that the plane of the gas appears to become tilted *before* the stars do. However, a detailed model of the warp. made possible by the high resolution and quality of the maps, shows that the difference can be entirely due to projection effects. The model successfully reproduces the observed height above the galactic plane on both sides of the galaxy with a unique tilted-ring model. The tilt is rather abrupt, with the inclination increasing from 0 to 10 degrees in less than one disk scalelength, and then probably leveling off.

A more detailed description of our analysis is in preparation.

REFERENCES

Sancisi 1985, in preparation.
van der Kruit, P. C. and Searle, L. 1981, *Astr. Ap.*, **95**, 105.

J. Kormendy and G. R. Knapp (eds.), Dark Matter in the Universe, 82.

MASS/LUMINOSITY RATIOS FROM ROTATION OF HI IN SO GALAXIES

Wim van Driel and Hugo van Woerden
Kapteyn Astronomical Institute, Groningen, The Netherlands

At Westerbork, we have mapped the distribution and motions of HI in about twenty gas-rich SO and SO/a galaxies, with resolutions of order 30 arcsec. Gas-rich SO's often have most of their HI in an outer ring, with diameters between 0.9 and 2.5 times the optical (D_0). Assuming circular shape and motions for this ring, radius R, and a spherical distribution of matter, we derive the inclination and rotation speed of the ring, the amount of mass $M_T(R)$ interior to it and the corresponding ratio M_T/L_B^0 to blue luminosity.

Gas-rich SO/a's often have (partly) filled HI disks, and SBO/a's tend to have broad outer rings or sets of arms around big central holes

Figure 1 shows 4 rotation curves so obtained. The SO NGC 4203 has an inner and outer HI ring (Van Woerden et al. 1983), which together give a flat rotation curve. The disk and the broad rings in NGC 3900 (SArO$^+$) 3941 (SBsO/a) and 5101 (RSBO/a), respectively, give flat, rising and falling rotation curves, though the latter is uncertain because of its low inclination. Figure 2 shows rotation speeds V_{rot} at outermost (radius R_{max}) in 13 SO and SO/a galaxies. There is only a marginal trend for V_{rot} to increase with R_{max}, due to 2 dwarfs and one supergiant galaxy. The trend (and the fit) is, however, consistent with the relation between diameter A(0) and corrected profile width $\Delta V(0)$ found by Shostak (1978).

A plot of M_T/L_B^0 vs. linear HI extent R_{max} shows no significant trend. Figure 3 does suggest a trend for M_T/L_B^0 to increase with the ratio R_{max}/R_{opt}, where $R_{opt} = \frac{1}{2}D_0$. At any rate, our data suggest a wide range of mass/luminosity ratios in SO galaxies.

The Westerbork Radio Observatory is operated by the Netherlands Foundation for Radio Astronomy, with financial support from ZWO. WvD acknowledges support by ASTRON, and HvW a NATO Research Grant (RG 098-82).

References
Shostak, G.S. 1978, Astron.Astrophys. 68, 321
Van Woerden, H., van Driel, W. and Schwarz, U.J. 1983, IAU Symp.100, p.99

J. Kormendy and G. R. Knapp (eds.), Dark Matter in the Universe, 83–84.
© *1987 by the IAU.*

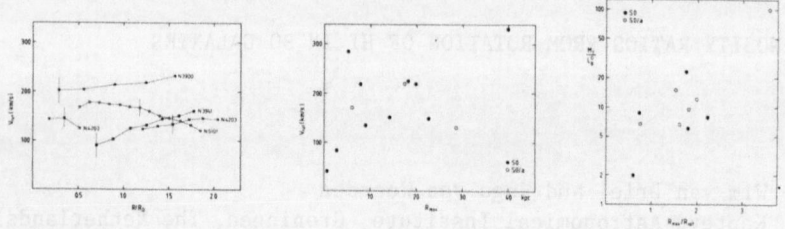

Figure 1 (left): Rotation curves for 4 S0 and S0/a galaxies.
Figure 2 (middle): Rotation speed vs. HI extent for 13 S0's and S0/a's.
Figure 3 (right): Mass/luminosity ratio vs. HI extent/optical size.

MASS DETERMINATIONS AND DARK MATTER AT INTERMEDIATE SCALES

J. P. Ostriker
Princeton University Observatory
Princeton, NJ 08544

1.   HISTORICAL BACKGROUND

The issue of "dark matter" in astronomy is extremely confusing.
Difficulties exist on two levels.   First there are the, in principle,
straightforward scientific questions of measurement.   A certain region
of space is studied, and by some technique, the mass within it is
determined.   Separately the energy output in some wavelength band from
the region is measured and then, with due allowance for distance
uncertainties, a "mass-to-light" ratio is determined.   These
measurements are difficult, with the results affected both by small
number statistical uncertainties (as when using globular clusters to
determine the mass of the galactic halo), measurement errors (as with
binary galaxies), and systematic questions of interpretation (as with
X-ray emitting gas around galaxies).   Ultimately, with patience and
skill these problems have been reduced and, as we shall see in
subsequent sections of this report, there exists moderate agreement
among observers concerning the large mass ($\sim 10^{12}$ $M_\odot$) and high
mass-to-light ratio ($M/L_B > 100$ $M_\odot/L_\odot$) for material integrated over
distances in the range (30 kpc $< r <$ 300 kpc) from the centers of giant
galaxies.

But there is another level of confusion which is purely semantic,
is less defensible, and where little improvement is apparent.   This
occurs when observed values of M or M/L are translated into statements
about "missing matter" or "dark matter".   In fact most of the detected
optical light is from giant stars contributing very little to the mass
of stellar systems.   And most of the "observed stellar mass" reported
is from low mass 0.1 $< M/M_\odot <$ 0.6 normal stars which contribute almost
nothing to the observed flux and whose presence is simply assumed on
the basis of a presumed analog to the solar neighborhood.   This mass is
estimated from the observed light times an assumed value of (M/L).
Thus, the "observed mass" is really the implicit product of an observed
light and an assumption.   Then, from the dynamically determined mass,
one subtracts off this inferred (or assumed) stellar mass, calling the
residual material "dark matter".   Given this procedure, identical

J. Kormendy and G. R. Knapp (eds.), Dark Matter in the Universe, 85–93.

observations of M and L can give rise to wildly and meaninglessly
different estimates for the amount of "dark matter" in the system.
While it is sometimes useful to have such estimates, they can only be
interpreted if presented with a detailed description of the assumptions
under which they were derived.

The use of the term "missing mass" is even more peculiar since
even the sign of this quantity is often unknown. It is used both to
describe matter known to be present from dynamical studies (i.e., "dark
matter") but not seen or inferred from the visual light, and also used
for the opposite case of mass not observed dynamically but whose
presence is expected on the basis of some other argument such as a
preference for a certain form of inflationary cosmology. Thus one does
not know if the "missing matter" in a given case is a positive or
negative entry on the dynamical ledger. Fortunately, the phrase is
being abandoned with increasing use instead of the slightly preferable
term "dark matter".

Historically, the intermediate mass range has been very important
in clarifying our understanding of the dark matter problem. As early
as 1937 Zwicky pointed out that the mass-to-light ratio for clusters of
galaxies was much larger (× 100) than estimates for the solar
neighborhood with only a relatively small part of the difference
attributable to difference in types of stars found in the two
environments. The amount of local dark matter required to balance the
books in the solar neighborhood using the Oort analysis was much
smaller; it seemed either to be an entirely different phenomenon or
merely a reasonable accounting error given the inaccuracy of the
measurements on which it was based. On the other hand, Zwicky's
anomalous result was so odd as to indicate the necessity of a
"cosmological" explanation, or to deny the validity of the virial
method when applied to clusters. It required the relatively recent
measurements at intermediate scales, detailed in the next section, to
show that a mass-to-light ratio gradually increasing with radius is a
common characteristic of stellar systems. It can be understood (though
not uniquely so) by an ever greater admixture of dark matter with
increasing radius as we proceed from the Oort to the Zwicky scale.

But before reviewing the modern results, it is worth pointing out
that several careful early studies had indicated an increasing (M/L) at
intermediate scales. In our own galaxy estimates of the mass based on
the rotation curve were for, well known reasons, limited to radii less
than the sun's galactocentric orbit. Most other galaxies were simply
too far away for spectroscopic observations of extended parts. But the
Andromeda Nebula had been studied to fairly large galactocentric radius
by Babcock in thesis work (1939), Wyse and Mayall (1942) and in an
extremely prescient paper by Schwarzschild (1954) who noted that the
rotation curve of that galaxy was apparently flat, implying a
mass-to-light ratio increasing rapidly with radius. He also observed
that there was a trend with size from globular clusters having
mass-to-light ratios of order unity through spiral and elliptical

galaxies to the giant clusters of galaxies like Coma which Zwicky (1937) had discovered had mass-to-light ratios in the range $10^{2.5}$-10^3 $M_\odot/L_\odot$.

The early work certainly indicated that the mass-to-light ratio of large systems was much greater than that in the solar neighborhood or in globular clusters but a good deal of skepticism remained. The observations were admittedly fragmentary. Could they be in error? Or, if correct, could the interpretation, based normally on the virial theorem, be seriously fallacious? In the two decades after Schwarzschild's paper, evidence accumulated in favor of "dark matter". It was summarized in a paper by Ostriker, Peebles and Yahil (1974), which contained no new work, but rather sought to display the substantial body of data indicating that conventional estimates of the mass and radius of galaxies might be severe underestimates.

Those authors reviewed the evidence primarily from local group spirals for mass at intermediate distance scales. Tidal limits of local dwarf spheroidals, which give $(M(r)/r^3)$ at perigalacticon (Hodge 1966), indicated a surprisingly large interior mass but the result depended on the statistics of small numbers, the poorly known masses of the test objects and orbital undertainties. The radio rotation curve of M31 (Roberts and Rots 1973) confirmed the earlier optical evidence for flatness but did not extend very far. Binary galaxies studied by several investigators gave conflicting results. The strongest individual piece of evidence reviewed had been brought forward by Kahn and Woltjer (1959) who noted the fact that M31 was approaching our Galaxy. This innocent observation, pertaining to two objects which are presumably unaffected by forces other than gravity and initially taking part in the expanding Hubble flow, indicated an attracting mass in the system far in excess of the assumed stellar mass. The important result has sometimes been called the "timing argument", since it depends on achieving the velocity reversal in a given time period. With then current observational numbers, assigning the mass of unknown origin to the two galaxies in rough proportion to their luminosities, a total mass of 5×10^{12} $M_\odot$ was obtained or about 2×10^{12} $M_\odot$ for our galaxy out to a distance of 300 kpc. In a recent re-examination of the problem, prompted by Sandage's (1986) analysis of local group velocities, I found that the timing argument is quite sensitive to assumptions about the distribution of the attracting mass. For example, if it is all placed at the center of mass of the system, the total mass is reduced by a factor of nine, to values even less than obtained from the individual rotation curves.

Einasto and co-workers (1974) at the same time obtained similar results on the basis of similarly fragmentary evidence. One could summarize the situation at that time, a decade ago by saying that all the evidence pointed to assigning a mass per giant spiral galaxy of order 10^{12} $M_\odot$ within $10^{5.5}$ pc of the galactic center, but that none of the evidence was very good.

2.   RESULTS OF THE DECADE 1975-1985

2.1. Binary Galaxies and Satellites of Galaxies

There were two new and important surveys in this period by Turner (1976) and Peterson (1979) using, respectively, optical and radio data samples large enough, $\sim 10^{1.5}$ -10^2 galaxies, to reduce some of the statistical uncertainties. Despite careful efforts, misidentifications (due to projection effects creating "optical binaries"), isolation from clusters, velocity errors and a host of other difficulties make the analysis of this data prone to serious uncertainty. The best analysis of this data to date with the most current observations and analytical techniques by White et al. (1983) determined for this data base the result

$$M(r) = 1.3 \times 10^{12} \ (r/100 \ \text{kpc})$$

independent of H_0 with an uncertainty of approximately 30% -50% . It is interesting to compare this result with the original findings of Turner and Ostriker (1977) and Peterson (1979), the former obtained 2.2 $\times 10^{12}$ at 270 kpc separation and the latter quotes 1.0×10^{12} $M_\odot$ per galaxy at a separation of 130 kpc. Thus, surprisingly all results are in quite good agreement. Phrased in terms of mass-to-light ratio, the binary results give

$$(M/L_B) \approx (70 \pm 20)h^1 \times (M/L_B)_\odot$$

within a radius of about 100 h^{-1} kpc. This is smaller than the (M/L) ratios for clusters, but significantly larger than the (M/L) ratios obtained from rotation curves of similar galaxies. Davis and Peebles (1983) taking an alternative statistical approach using the two-and three-particle distribution functions, which does not depend on isolation, determined a typical mass for an L_* spiral galaxy of 2 $\times$ 10^{12} $M_\odot$ in conformity with the binary and local timing results.

My re-analysis of the observations, summarized above, does not differ significantly from the earlier review by Faber and Gallagher (1979).

2.2   Satellites of Galaxies

Hartwick and Sargent (1978) analyzed the orbits of a group of distant globular clusters to infer the mass of the Galaxy interior to the group. Since only radial velocities are available (and tidal limits only eliminate the possibility of an extremely radial distribution of orbits), the assumed degree of orbital eccentricity affects their results as does statistical noise from the small sample of 11 objects. They found $M(r) = (8 \pm 2) \times 10^{11}$ $M_\odot$ within r = 50 kpc for an isotropic orbital distribution.

Recently Peterson (1985) obtained improved spectroscopic velocity measurements for these clusters with several additional distant objects measured.   Using the   analytical methods of Lynden-Bell et al. (1983) she obtains   a mass of $(5.1 \pm 3.1) \times 10^{11}$ $M_\odot$ for isotropic velocities within a galactocentric distance of about 80 kpc.   Since, the satellites   are   very   diffuse   and   could   not   survive   a   close galactocentric passage, an alternate solution with   high eccentricity orbits excluded was made; this gave a mass four times larger.   In order to improve further these measurements, larger samples will   be required such as distant R-R Lyrae stars.   But either proper motion data, or an extremely detailed density distribution for the test particles will be required before the uncertainties due to unknown orbital eccentricity can be lessened.

2.3 X-Ray Halos about Massive Galaxies

Luminous elliptical cluster galaxies emit thermal X-ray bremsstrahlung at high enough rates to allow determination of both temperature and density radial profiles.   The best studied case is M87 where Fabricant et al. (1980) compute a mass, based on hydrostatic equilibrium of 2 × 10^{13} $M_\odot$ within 230 kpc of the galaxy center.   Binney and Cowie (1981) found only 5 × 10^{11} $M_\odot$ within r = 100 kpc for that galaxy, but the inconsistency is apparent not real, since these authors also compute a total mass of 2 × 10^{13} $M_\odot$ within r = 230 kpc but attribute most of it to the cluster.   The mass-to-light ratio of the material between 100 and 200 kpc in any case exceeds   100 $(M/L)_\odot$ whether it is regarded as galaxy or cluster material.   It will be interesting to see if, when equivalently high quality data is available for "field" galaxies, the results are similar.   Preliminary results by Forman, Jones and Tucker (1986) and analysis by Muzhotsky (1985) for more isolated ellipticals indicate total masses of order 5 × 10^{12} $M_\odot$ within 100 kpc of the center of these early type systems, with (M/L) ratios far in excess of 10^2 in the region 30-100 kpc containing most of the mass.

2.4 Gravitational Lensing

There are several ways that beams of light to background objects passing through galactic halos can be used to probe for the existence of dark matter.   If there are point-like masses in the beam, there can be a significant amplification of the brightness of some background stellar objects--an effect studied by Canizares (1982) and Vietri and Ostriker (1983) for quasars.   Alternatively, extended background objects like galaxies will appear slightly crescent shaped, a phenomenon noted by Russell (1937) and investigated recently by Tyson et al. (1984).   Suffice to say that these methods, while potentially powerful and completely independent of those using mass points as the test particles, give, at present, conflicting and highly insecure results.

2.5 Theory of Merging Systems

Similarly, extended massive halos will promote rapid merging of stellar systems, especially in groups with low velocity dispersion (relative to the internal velocity dispersion within galaxies). Simulations by Barnes (1985) and Mamon (1985) indicate higher rates of merging for galaxies with substantial halos than are permitted by the observation that thin spiral discs imply merging has been at most a few percent effect for most galaxies. Once again the evidence is too fragmentary for firm conclusions to be drawn.

3. SUMMARY AND CAVEATS

The bulk of the evidence seems to indicate that within a volume between spheres of radio $10^{4 \cdot 5}$ pc and 10^5 pc surrounding a normal giant spiral of luminosity $L_* = 1.5 \times 10^{10} L_\odot$ there is typically found $2 \times 10^{12} M_\odot$ with a mass-to-light ratio exceeding $10^{2 \cdot 5} (M/L_B)_\odot$. For ellipticals of the same luminosity the mass found is typically twice as much.

However Yahil (1977) found that, although a statistically secure (M/L) ratio might be definable, there was no correlation detectable between M and L, even when allowance was made for the variation due to galactic types. Peterson (1979) puzzlingly found no correlation between (M/L) and galaxy pair separation and White et al. (1983) found no correlation between Δv and Δr or L in the best analyzed binary data.

Thus, although the presence of substantial amounts of dark matter in the outskirts of galaxies seems to be established, it is not at all clear, at this time, how well bound or even how well correlated the "light" and "dark" components are. It seems attractive to this observer to consider galaxy formation and the development of halos as two relatively separate phenomena with the halos accumulating around galaxies at late times. Then environmental influences will produce the apparent irregularities which we see.

REFERENCES

Babcock, H. W. 1939, Lick Obs. Bull., 19, 41.
Barnes, J. 1985, M.N.R.A.S., 215, 517.
Binney, J. and Cowie, L. L. 1981, Ap. J., 247, 464.
Canizares, C. 1982. Ap. J., 263, 508.
Einasto, J., Kaasik, A. and Saar, E. 1974, Preprint #1 of Tartu
    Observatory.
Faber, S. and Gallagher, J. 1979, Ann. Rev. Astron. & Astroph., 17,
    135.
Fabricant,, D., Lecar, M. and Gorenstein, P. 1980, Ap. J., 241, 552.
Forman, W., Jones, C. and Tucker, 1986, Ap. J. (in press).
Hartwick, F. D. A. and Sargent, W. L. W. 1978, Ap. J., 221, 512.
Hodge, P. W. 1966, Ap. J., 144, 869.
Huchra, J., Lathan, D. and Davis, M. 1983, M.N.R.A.S., 203, 701.

Kahn, F. D. and Woltjer, L. 1959, Ap. J., 130, 105.

Lynden-Bell, D., Cannon, R. D. and Godwin, P. J. 1983, M.N.R.A.S., 204, 87P.

Mamon, G. 1985, Ph.D. Thesis, Princeton University.

Muzhotsky, R. 1985, private communication.

Ostriker, J. P., Peebles, P. J. E., and Yahil, A. 1974, Ap. J. (Lett.), 193, L1.

Davis, M. and Peebles, P. J. E. 1983, Ap. J., 267, 465.

Peterson, S. D. 1979, Ap. J., 232, 20.

Peterson, R. 1985, Ap. J., 297, 309.

Roberts, M. S. and Rots, A. H. 1973, Astron. & Astroph., 26, 483.

Russell, H. N. 1937, Scientific American, 84, 76.

Sandage, A. 1986 (preprint).

Schwarzschild, M. 1954, A. J., 59, 273.

Turner, E. L. 1976, Ap. J., 208, 304.

Turner, E. L. and Ostriker, J. P. 1977, Ap. J., 217, 24.

Tyson, T., Valdes, F., Jarvis, J. F. and Mills, A. P. 1984, Ap. J., (Lett.), 281, L59.

Vietri, M. and Ostriker, J. P. 1983, Ap. J., 267, 488.

Wyse, A. B. and Mayall, N. J. 1942, Ap. J., 95, 24.

Yahil, A. 1977, Ap. J., 217, 27.

Zwicky, F. 1937, Ap. J., 86, 217.

DISCUSSION

PEEBLES: Jerry, you don't want too much irregularity in the process of accumulating halos or you won't get the remarkable uniformity of rotation-curve shapes that Vera Rubin showed us. It would be difficult to get this if you added the components stochastically.

OSTRIKER: You're right.

YAHIL: The second problem with making the halo after the disk is that you need it to stabilize the disk.

OSTRIKER: I wouldn't necessarily make the halo after the disk, because both components accrete fairly late. The models by Gunn and others in which the disk is produced late are interesting because there do not seem to be many stars in the solar neighborhood which are much older than the Sun. This suggests that these ideas may be right.

LAKE: What is disk formation late with respect to?

OSTRIKER: The formation of the spheroid. I have in mind a more-or-less straightforward one-parameter sequence of spheroids or ellipticals which are made by the Divine Hand at a redshift of 10-20. Then, later on, depending on the environment and other circumstances, the spheroids accrete gas, sometimes, and halos.

PACZYNSKI: I am confused about this 10^{12} $M_\odot$ business. In all of the diagrams for pairs of galaxies which you showed us there was a linear relationship between the mass within R and R; i.e., $M(R) \propto R$, with rather little scatter. The diagrams suggested that the circular velocity is roughly the same for all these galaxies. Yet you know that for spirals V_c varies from 70 to 500 km s^{-1}. How do you reconcile these two statements?

OSTRIKER: My own guess - and we can't tell from the observations as they presently exist - is that at large distances from the Sombrero you would find V_c decreasing, while at large distances from a dwarf elliptical you'd find V_c increasing.

PEEBLES: Didn't you show averages over the range of V_c from 70 to 500 km s^{-1}, rather than individual cases?

OSTRIKER: Yes, I did show averages. But I'm also suggesting that there is less variation in V_c at large radii than at small radii.

SANDERS: Earlier you mentioned that only satellites and binaries would be seriously affected by dynamical friction. But now you're suggesting that the two members of a binary pair might be swimming in a common halo. Is it an embarrassment that we see binaries at all?

OSTRIKER: Ed Turner and I looked at that in a paper about five years ago. We calculated the merging rate and found that there's no problem.

TURNER: It's just a case of steady-state flow.

OSTRIKER: Let me stress that point, because the same thing arises with the growth of cD galaxies. You have to worry about the continuity equation. Given the shape of the two-particle correlation function, you keep on forming new binaries and so can withstand some merging without changing the observed distributions.

SANDERS: And you preserve the two-point correlation function?

OSTRIKER: Yes. But that's not the constraint. The constraint is that you don't mess up the disk too much.

GUNN: Since there is a very nice correlation between mass and luminosity or rotation velocity and luminosity on the scales that the 21-cm and optical rotation curves sample, would you comment on the continued believability of the statement that on your scales, which are really not all that much larger, L and M seem to be uncorrelated? Isn't this really a problem?

OSTRIKER: It doesn't strike me as a problem because of the timescales. I can easily believe that during the formation process, all of the inner parts were magically formed with a constant ratio of dark matter to baryons. But when other material accreted later, there were additional effects, like competition from other galaxies. So I don't see why L and M have to be correlated.

GUNN: But don't you have to do something drastic to the rotation curves between ~20 kpc, where they are well observed, and 50 kpc, where they go to hell?

OSTRIKER: No. Because the amount of dark matter you need inside the visible galaxy is less than or comparable to the baryonic mass, whereas the amount you need at larger radii can be ten times that amount. So the amount in the center doesn't have much influence on the amount outside.

THE MASS OF THE BINARY GALAXIES NGC 4038/39 (THE "ANTENNAE")

J.M. Mahoney, B.F. Burke
Massachusetts Institute of Technology
and J.M. van der Hulst
Westerbork Radio Observatory

The binary galaxies NGC 4038/39 have extended filamentary arms
generated by tidal interactions (Toomre and Toomre, Ap. J. 178; 623,
(1972)(TT)). The velocity field was determined by HI observations taken
with the VLA (a facility operated by the NRAO under contract with the
NSF), and the combined velocity and morphological information was used
to constrain the allowed orbital parameters, halo characteristics, and
dynamical friction. TT-type calculations were carried out with central
masses and rings of test particles, and the calculated results compared
with the data. Using disk orientations derived from optical data (Rubin
et al., (1970) Ap. J. 160 81), and solving for the six remaining orbital
parameters, central potential softening constant (representing the halo),
and frictional relaxation time, a good fit between the model and the
radio data was found. The best model is shown in Figure 1, and is
superimposed on an HI column density map in Figure 2. The orbit is
well-determined, and must be nearly parabolic; the pair are interacting
for the first time, and if the galaxies have extensive massive halos
much larger than their discs, then their tidal arms would be shorter
and stubbier than observed. More limited halos are allowed; each galaxy
could have up to 80% of its total mass in a halo, but the halos cannot
be much larger than the discs. A halo several times larger than the
disc, with 10 to 20 times the disc mass, is not permitted by the data.

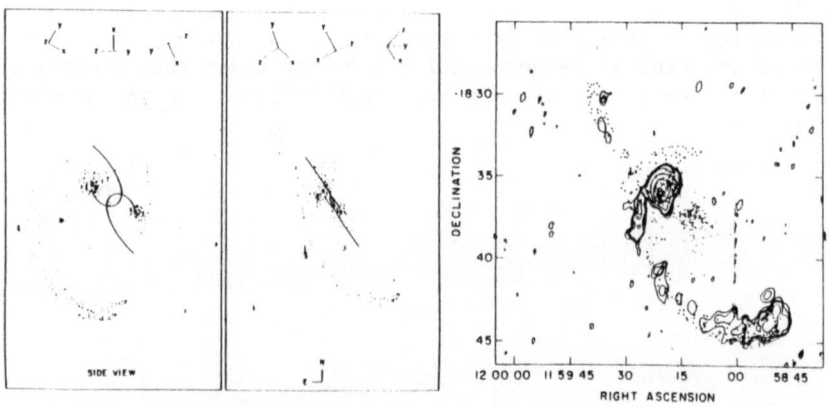

Figure 1: Best-fit model cal-   Figure 2: Hydrogen density con-
culation for NGC 4038/39   tonrs compared to model

J. Kormendy and G. R. Knapp (eds.), Dark Matter in the Universe, 94.

RADIAL VELOCITIES OF REMOTE GLOBULAR CLUSTERS

Ruth C. Peterson
Fred Lawrence Whipple Observatory, Smithsonian Institution

ABSTRACT. Radial velocities good to 25 km/s have been measured for the remote globular clusters Eridanus, NGC 2419, Palomar 3, Palomar 4, Palomar 5, Palomar 14, and Palomar 15. Spectra with a resolution of 3.0 A were recorded for 4 to 10 stars at a time in each cluster, using an aperture plate with the KPNO 4m telescope, R-C spectrograph, Simmons camera, and baked IIIaJ plates. Radial velocities for each star were determined from the position of the night-sky emission feature near 4360A with respect to several strong atomic features in the vicinity.

For two remote systems, Eridanus and Palomar 14, the radial velocity transformed to the galactic rest frame is large enough to suggest a total mass for the Galaxy of 10^{12} $M_\odot$, and a comparable mass is indicated from the average rest-frame velocities of the remote systems. A full presentation of the measurements and results will appear in the October 1, 1985, issue of the Astrophysical Journal.

J. Kormendy and G. R. Knapp (eds.), Dark Matter in the Universe, 95.

THE TROUBLE WITH BINARY GALAXIES

N.A.Sharp
Kitt Peak National Observatory

The statistical analysis of magnitude and center-of-mass velocity data for binary galaxies will **never** provide an adequate estimate of the mass within the orbit, and will therefore never furnish useful information about the mass-to-light ratios of galaxies. There are three main reasons. Firstly, it is very difficult to find a sample which is both objectively selected and dynamically isolated, and in those cases with obvious tidal distortions it is not clear which two points of the confused systems can be taken to be the centers of mass. Secondly, there are several different ways to produce a mass estimator from the available data, and it is not clear which estimator corresponds most closely to what is normally thought of as the gravitational mass of the system. Finally, those mass estimates which include it as a variable depend quite strongly on the orbital eccentricity, and any analysis method is necessarily least sensitive to just this parameter. Therefore, variation even within a single estimator is both considerable and only weakly constrained.

In addition, there are subsidiary observational problems. The comparison between data from different observers demonstrates that optical redshifts are not known accurately enough to permit a believable analysis, even were such a thing possible. (A secondary result is that the claimed 72 km s^{-1} periodicity, although present in the data with high statistical significance, cannot be a property of the physical pairs themselves, and must be some kind of artefact. Regrettably, this approach provides no clues as to just what could be causing such an unusual error.) Although velocities derived from radio observations are more accurate, samples of radio-observed binaries have other serious problems. Pairs with angular separations less than the beam size must be separated in velocity by more than their mean velocity width to be unambiguously resolved. The other unconfused pairs are those separated by more than the beam size. However, since surveys are almost always flux limited, this angular restriction introduces a strong correlation between the linear separation and the absolute luminosity. These two extra selection effects are difficult to incorporate into any method of analysis.

Quoted values of masses and mass-to-light ratios for binary galaxy samples are not reliable estimates of the real physical properties of the sample. However, they **are** acceptable as ways of intercomparing different studies, as long as the same estimator is used in each case. It is very important to be aware of the large differences caused simply by using different definitions. If, at some future time, some members of any sample can be studied in a way which gives unambiguous physical masses, then, and only then, can the statistical results be correctly normalized. Such physically meaningful results can be derived by the careful dynamical simulation of pairs for which both kinematical and morphological data are available. This is a much better way to use binary galaxies to study mass distributions, and more observational effort should be expended in this direction.

J. Kormendy and G. R. Knapp (eds.), Dark Matter in the Universe, 96.
© *1987 by the IAU.*

EVIDENCE FOR DARK MATTER IN GALACTIC SYSTEMS

Marc Davis
Depts. of Physics and Astronomy
University of California
Berkeley, California 94720 U.S.A.

ABSTRACT

The evidence for dark matter in binaries and groups of galaxies is very strong, and is seen in all recent observational studies. Measurements of mass in galactic systems is possible on scales ranging from 50 kpc using virial analysis of binary galaxies to 15 Mpc using Virgocentric infall analysis. The Ω estimates derived from these studies are generally consistent with $\Omega < 0.2$, with a fairly weak trend toward larger Ω estimates on larger scales. However, measurements of the galaxy distribution in the IRAS catalog yields a dipole anisotropy consistent in direction with the microwave dipole anisotropy, suggesting that the local galaxy distribution is responsible for the microwave velocity. This will eventually provide the most reliable estimate of Ω, and is likely to result in a value somewhat larger than previous estimates on smaller scales. Study of the velocity field around large clusters in cosmological n–body experiments provides a useful guide for understanding the limitations of the spherically symmetric models of Virgocentric infall. We point out a number of biases that could affect the existing Virgocentric flow studies.

1. INTRODUCTION

Discussions of dark matter in groups and clusters of galaxies is an old and venerable subject that dates back at least to Zwicky's (1933) comments on the missing mass problem in the Coma cluster. One way to discuss the dark matter problem is in terms of M/L ratios. The observed B_o luminosity density of galaxies is known within approximately 20% uncertainty to be $L_B \approx 1.1 \times 10^8 \, h^2 \, L_\odot/\mathrm{Mpc}^3$ ($H_0 = 100 \, h \, \mathrm{km \, s^{-1} \, Mpc^{-1}}$) (Davis and Huchra, 1982). The cosmological density parameter Ω can then be expressed as

$$M/L_B \approx 2300 h\Omega \, M_\odot/L_\odot$$

Since M/L ratios of stellar populations range from 1-10, "dark matter" must be supplied to explain any measurements that imply $\Omega > 0.005$.

J. Kormendy and G. R. Knapp (eds.), Dark Matter in the Universe, 97–110.

In this paper I shall briefly review evidence for dark matter in groups of galaxies and in Virgocentric flow studies. I shall focus on results postdating the detailed review of Faber and Gallagher (1979). Recent binary galaxy results were presented in the oral version of this presentation but have been transferred to J. Ostriker's report (this volume). In section 2 I shall discuss the measurement of relative peculiar velocity of pairs of galaxies and the continuity of its behavior on scales from binary orbits to clusters of galaxies. The status of the virial analysis of groups of galaxies has not greatly changed in the last few years, and was reviewed last year by Geller (1984); I shall limit the discussion of section 3 to one new study by Nolthenius and White. Much recent activity has focused on the measurement of the anisotropy of the Hubble flow around the Virgo supercluster, which gives a mass estimate on a scale of 15 Mpc and so is sensitive to matter that perhaps is not clustered on smaller scales. The subject has been reviewed by Davis and Peebles (1983a) and new results are reviewed in section 4. Section 5 discusses a new sample of galaxies; the IRAS galaxy list, which can be used in an improved version of large scale mass determination. Finally, section 6 describes recent n-body studies designed to test the reliability of Virgocentric mass determinations.

2. THE RELATIVE VELOCITY OF PAIRS OF GALAXIES

The median velocity difference of well isolated late-type binary galaxies is about 100 km s^{-1} (White *et al.*, 1982; Schweizer, 1985). Binary galaxies that are less isolated have an RMS velocity difference of 200 km s^{-1}. These "binary" galaxies are presumably influenced by the gravitational field of additional neighbors, and the distribution function of their velocity differences is in fact indistinguishable from that of all close galaxy pairs $(r_p < 100 \ h^{-1}$ kpc) in the CfA Survey (Davis and Peebles, 1983b).

The measured rms peculiar velocity of all pairs is a smooth function of the projected separation from the scale of 20 h^{-1} kpc to 5 h^{-1} Mpc (Davis and Peebles, 1983a; Bean *et. al.*, 1983); over the measured range the RMS velocity difference of all pairs is 200-350 km s^{-1}. On larger scales it is very difficult to separate the effects of Hubble expansion from peculiar velocity and the dispersions become model dependent.

The smooth behavior of the relative velocities must be a significant clue to the nature of the dark matter on these scales, although we do not yet understand its implications. If galaxies trace the mass and if the two point correlation function $\xi(r) \propto r^{-\gamma}$ then relative velocities should scale as $\langle v_{21}^2(r) \rangle \propto \Omega r^{2-\gamma}$ (Peebles, 1980). Since γ is observed to be 1.8 we expect a slowly rising behavior for $v_{21}^2(r)$, as observed in the largest sample studied to date. With the proportionality constants in place, the derived density estimate is (Davis and Peebles, 1983b) $\Omega \approx 0.2e^{\pm 0.4}$.

However, this cosmic virial theorem result actually depends on the ratio of a nearly divergent moment of the three point correlation function ς and on the assumption that the three point function can be accurately written as a product function of the two point correlations, $\varsigma_{123} = Q(\xi_1\xi_2 + \xi_2\xi_3 + \xi_1\xi_3)$, which is consistent with observations for $Q \approx 0.8$. Furthermore one wonders how well galaxies can be expected to trace the mass, particularly on small scales. The smoothness of $\langle v_{21}^2(r_p)\rangle$ and the derived Ω estimate suggests dark halos around galaxies extend a considerable distance, of order 300 h^{-1} kpc, or that most of the measured mass is associated with numerous lower luminosity objects clustered with the larger, brighter galaxies which dominate the observed samples.

3. GROUPS OF GALAXIES

A complementary type of analysis to the above statistical approach is virial analysis of individual groups of galaxies. At least two separate group analyses have been performed on the CfA catalog. Press and Davis (1982) selected groups on the basis of crossing time and reported a linear trend of M/L on group size, in the range 50 kpc $<$ hr$_G$ $<$ 2 Mpc. Unfortunately the larger size groups were frequently contaminated by bogus outliers and the strong trend must be discounted for hr$_G$ $>$ 300 kpc. This contamination did not occur in the fake catalogs drawn from n–body models which were used to calibrate the method.

Huchra and Geller have selected groups on the basis of overdensity, using a neighbor sphere that varies with distance to account for the selection effect. They find an enormous scatter in M/L estimates (a factor of 10^4!) and argue the data is too noisy to discern any trends of M/L versus anything. This work has been recently reviewed by Geller (1984). More recently Nolthenius and White (1986) have extended this analysis and have made a detailed comparison to n–body models of cold dark matter dominated universes (Davis et al., 1985). They compared groups drawn from the CfA catalog to groups taken from model universes with Ω = 0.2 and Ω = 0.3 in which galaxies trace the mass, and to a model with Ω = 1.0 in which galaxies are a biased mass tracer. In the unbiased models, one expects no trend of M/L with scale size and none is observed. The biased models are expected to show a trend of M/L with size, but it is very weak and is completely consistent with the CfA data. Nolthenius and White argue that the $\Omega = 1$ biased model gives the overall best fit to the data; if one insists that galaxies must trace the mass, then Ω = .1 or .2 is indicated, in complete agreement with the statistical approach.

4. VIRGOCENTRIC FLOW

The Virgo supercluster presents a unique opportunity to measure mass on a really extended scale, our distance to the center of Virgo, r $\approx$ 15 Mpc. These studies combine some measure of density fluctuations, such as the mean overdensity $\bar{\delta}$ within our radius to Virgo, with a peculiar velocity derived from the anisotropy of the microwave background radiation or from anisotropy of the local Hubble flow. The mean galaxy overdensity has been measured to be $\bar{\delta} \simeq 2.0 \pm$

0.2 (Davis and Huchra, 1982). This value supercedes previous higher estimates of $\bar{\delta}$ which were based on smaller, shallower redshift catalogs.

One can derive an estimate of Ω with the additional assumption of spherical symmetry for Virgo, which in linear theory gives the mean infall velocity V_v averaged on a sphere at radius r from the cluster center as

$$V_v/Hr = \frac{1}{3}\bar{\delta}\Omega^{0.6}. \tag{1}$$

Alternatively one can relax the assumption of spherical symmetry and use the linear perturbation theory result (Peebles, 1980).

$$\vec{v}_p = \frac{2}{3}\left(\frac{\vec{g}}{H\Omega}\Omega^{0.6}\right), \tag{2}$$

where $\vec{g}$ is the net peculiar gravitational acceleration measured by an observer with velocity $\vec{v}_p$. Unfortunately the real Virgo cluster is both non–linear and non–spherical, and there may be systematic errors that bias Ω estimates from these analyses, a topic to be discussed below in section 6.

The observation of primary relevance for Virgo flow analysis is the velocity inferred from the microwave dipole anisotropy, V_μ. In the local group center of mass the measured velocity is $V_\mu = 600\pm50$ km s^{-1}, directed toward $\ell = 270°$, b $= 30°$, some 45° from the direction of the Virgo cluster (Fixsen et al., 1983; Lubin et al., 1983). The fundamental question here is to understand what fraction of the microwave anisotropy is induced by the local supercluster, and what is responsible for the balance.

A number of recent studies strongly suggest that the entire microwave anisotropy is likely to be induced by matter distributed within 50 h^{-1} Mpc of us, a region largely dominated by the local supercluster. In particular, Aaronson et al. (1985) measure a dipole Hubble flow anisotropy of 800$\pm$200 km s^{-1} in a set of 10 clusters arranged around the sky, and at distances of 40–100 h^{-1} Mpc. The direction and amplitude is consistent with the microwave results and implies that we are moving relative to the clusters which themselves appear to be at rest in the comoving frame. This result confirms previous, less secure Hubble anisotropy studies of Hart and Davies (1982) and de Vaucouleurs and Peters (1984). It furthermore is consistent with theoretical prejudices that the primordial power spectrum of density perturbations is the Zel'dovich constant curvature spectrum in which case the peculiar velocity field will be dominated by perturbations of wavelength short compared to 100 h^{-1} Mpc (Peebles, 1980; Davis, 1985).

Another set of observations is limited to determination of the component of our infall velocity directed toward Virgo. Tonry and Davis (1981) argued that the Faber–Jackson luminosity–velocity dispersion correlation for E and S0 galaxies distributed within cz < 8000 km s^{-1} leads to an infall estimate consistent with this component of microwave anisotropy (420 km s^{-1}). Dressler (1984) also used the Faber–Jackson relation to measure the relative distance modulus of elliptical galaxies in Coma and Virgo and took special precaution to maintain apertures of constant projected metric size for the velocity dispersion measurements. His derived infall velocity was 230 ± 80 km s^{-1}. The source of this discrepancy with the Tonry–Davis result is presently unresolved.

The best local studies of Virgo flow use distances derived independently of redshift to attempt to fit the observed Hubble anisotropy to a non-linear spherical flow model centered on the Virgo cluster (e.g. Aaronson et al., 1982). These studies are usually confined to galaxies with cz < 3000 km s^{-1}, and typically yield infall estimates of 200–300 km s^{-1}, substantially less than the Virgo component of microwave anisotropy.

If Virgo itself induces this relatively small infall velocity then the entire Virgo supercluster must be moving at some 400 km s^{-1}, roughly in the direction of the Hydra–Centaurus cluster $(\ell \sim 285°$, b $\sim 25°)$. Hydra–Centaurus has a redshift cz ≈ 3000 km s^{-1}, so by equation (1) with fixed Ω the mean overdensity toward Hydra–Centaurus should be $\bar{\delta}_{HC} > 1$, compared to the $\bar{\delta}_{Virgo} \approx 2$, in order to induce this velocity. If this were the case, we should expect to readily observe a hemispheric anisotropy of the galaxy counts extending beyond 15th magnitude toward this direction that should rival the anisotropy of the Shapley Ames catalog induced by the Virgo supercluster. No such anisotropy exists.

It is important to keep in mind that the Virgo flow test is not a measure of the mass of the central region of the Virgo or Hydra-Centaurus clusters, but of all the matter overdensity within a radius of our distance to the cluster center. It is unrealistic to imagine that Hydra–Centaurus, clusters virtually unrecognizable in full–sky maps of the galaxy distribution, would have more influence on our peculiar velocity than the net effect of the Virgo supercluster. Although most studies point toward a low infall velocity toward Virgo, and a corresponding low estimate of density Ω (< 0.25), these conclusions are unsettling because they do not readily explain the bulk of the microwave anisotropy, which would appear to have been generated by matter within 50 h^{-1} Mpc of us, yet neither by Virgo nor by Hydra–Centaurus, the only known superclusters within this distance and in this general direction.

5. THE IRAS GALAXY CATALOG

The above confusion is likely to be resolved soon by a dramatic result that has emerged from the IRAS point source catalog. In this catalog it is a straightforward procedure to separate stars from galaxies on the basis of IR color. Meiksin and Davis (1985) (see also Yahil, this conference) have generated a galaxy catalog from the IRAS database, which is ideal for measurement of the anisotropy of the local galaxy distribution. Given a catalog of galaxies complete over the entire sky, the net measured dipole of the surface density distribution, weighting galaxies either by number or by flux, is proportional to the net peculiar gravitational acceleration acting on us. In linear perturbation theory, one would expect this dipole direction to agree with the direction of the microwave dipole anisotropy. The proportionality constant between the surface density dipole of IRAS sources and the gravitational acceleration $\vec{g}$ can be readily inferred once redshifts are available for the sample. Of course the method assumes these objects trace the underlying mass distribution on these large scales.

The IRAS galaxy catalog is uniquely suited for this type of whole sky analysis. The catalog generated by Meiksin and Davis contains 6730 sources, selected to have flux $f_{60\mu} > .6$ Jy and to have $f_{60\mu}/f_{12\mu} > 3$. We estimate the catalog is contaminated by less than 10 stars, although some galactic cirrus may be masquerading as external galaxies. These contaminating sources will be readily weeded out in followup optical studies. The resulting catalog is displayed in figure 1. We have eliminated all sources with $|b| < 10°$, and have excised additional sections of the sky around prominent HII regions, which are associated with excessive quantities of cirrus. The empty slices through the middle of each hemisphere are the two small regions not surveyed by IRAS. Our catalog has been generated from 9.55 steradians in a uniform fashion, which would be an impossible task for an optically selected catalog because of galactic extinction.

The most conspicuous clusters apparent in figure 1 are the Virgo and Perseus superclusters. None of the clusters are especially prominent because of the dilution in projection by background galaxies. We estimate this catalog samples to a depth of order 50–100 h^{-1} Mpc, well beyond the local supercluster. The IRAS galaxies are preferentially late type spirals undergoing active star formation, and we believe these should be a fair, nearly random sample of all late type spirals. Certainly the IRAS galaxies will underrepresent regions dominated by E and S0 galaxies in the centers of clusters, but most galaxies, and we suspect most mass, are not associated with the high density regions underrepresented by spirals. In any event, if galaxies are a biased mass tracer, it is likely that spirals more closely trace the mass distribution than do the early type galaxies. The large scatter in infrared to optical flux or in infrared flux to mass can dilute, but will not erase the anisotropy. For the purposes of the anisotropy study, we are indifferent to all details of the IR properties of the selected galaxies. We are interested only that the flux limit and other sampling considerations be uniform across a large region of the sky, and that the IR properties of bright galaxies are a random and fair sample of all galaxies of their morphological type.

Given the above prelude, we measured the dipole anisotropy of the sample as a whole, and after subdividing it into 4 independent subsamples based on quartiles of 60μ flux. Details are provided by Meiksin and Davis (1985). The overall dipole anisotropy is only 4.1% of the mean surface density, but the top quartile subsample, comprised of the brightest sources which should be statistically the closest to us, has an anisotropy of 7%. The anisotropy directions of the 4 independent samples are shown in figure 1, along with the microwave anisotropy direction and the direction of anisotropy seen in the CfA analysis (Davis and Huchra, 1982) which was generated from an optically selected catalog with $|b| \geq 40°$. The center of the Virgo cluster is also indicated. The major result to note is that the IRAS anisotropies for the top 3 quartiles all point very close to the microwave anisotropy, $\theta \sim 22\text{--}29°$, and about 35 degrees from Virgo. The faintest quartile deviates in direction, but this subsample is the most affected by galactic cirrus and other problems. Our anisotropy results agree very well with those of Yahil *et al.* (1985), whose analysis of the IRAS data is quite independent and different from ours.

Examination of figure 1 shows no prominent density peak in the direction of the anisotropy. The dipole is truly the vector sum of all the clusters and voids seen in the catalog, and the low density regions in figure 1 play as significant a role in repelling us as the clusters do in attracting us. The robustness of the anisotropy implies it is dominated by the local galaxy distribution well sampled in all flux quartiles. The fact that the IRAS anisotropy is relatively close in direction $(\theta \sim 30°)$ to the CfA anisotropy, which necessarily presumed no anisotropy was generated from matter in the region $|b| < 40$, suggests that the same clustering dominates both samples. Complete redshift information is available for the CfA analysis, and as shown by Davis and Huchra (1982), the peculiar gravity $\vec{g}$ of that sample is completely dominated by the Virgo supercluster. Quantitatively the gravity anomaly measured in the CfA catalog should generate a peculiar velocity $|v_p| \simeq 670\Omega^{.6}$ km s^{-1}. The additional anisotropy detected by the IRAS sample will probably raise the expected peculiar velocity to $\approx 900\ \Omega^{.6}$ km s^{-1}. This expected velocity should be compared directly with the microwave velocity and in turn implies a high value of density, $\Omega \approx 0.5$. When redshifts are available for a subset of the IRAS galaxies (e.g. the brightest quartile) this density estimate will be better quantified. Since it is the largest scale mass estimate conceivable and is the least model dependent it should become the most reliable measure of the true cosmological density.

6. CLUSTER FLOW IN N–BODY SIMULATIONS

There remains an explanation for the serious discrepancy of the low infall velocity measured in the local Hubble flow studies with the larger peculiar velocity indicated by the microwave and IRAS anisotropies. As indicated above, I seriously doubt if the center of the Virgo cluster is moving anywhere at a speed approaching 400 km s^{-1}. A possible explanation for the discrepancy could be some inadequacy of the spherical models used to measure the local Hubble anisotropy.

To study the accuracy of the spherical models, Villumsen and Davis (1985) have examined cluster formation in cosmological n–body simulations. Since the clusters are sliced out of large models, their exterior boundary conditions are much more realistic than the usual assumption of spherical symmetry. The initial conditions of the models studied by Villumsen and Davis were either power law models, $\delta_K^2 \propto k^n$, n $= -1, -2$, or of the cold dark matter type (Blumenthal and

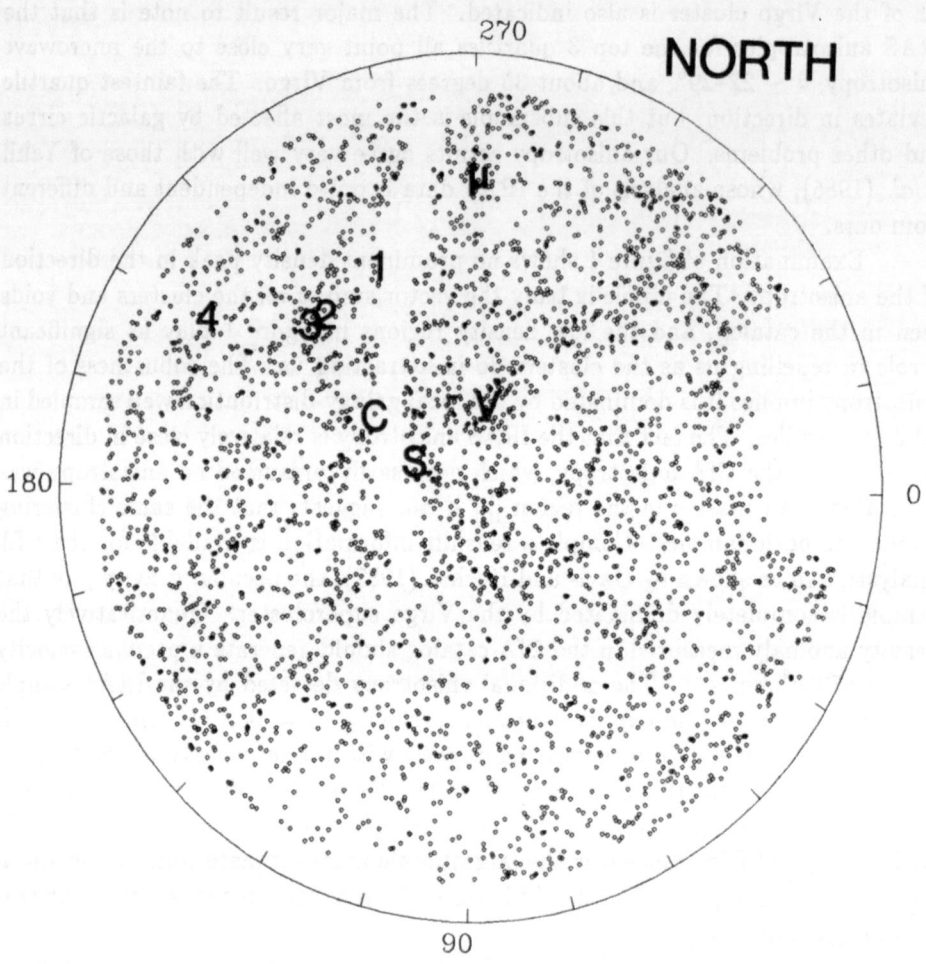

Figure 1a: Equal area projection of the north galactic hemisphere of the IRAS Galaxy catalog of Meiksin and Davis (1985). 3476 galaxies are plotted. The microwave anisotropy direction is indicated by the μ, the Virgo cluster center by V, the Shapley-Ames luminosity anisotropy by S, the CfA dipole anisotropy direction by C, and the dipole directions of the four independent flux quartiles of the IRAS sample by 1–4.

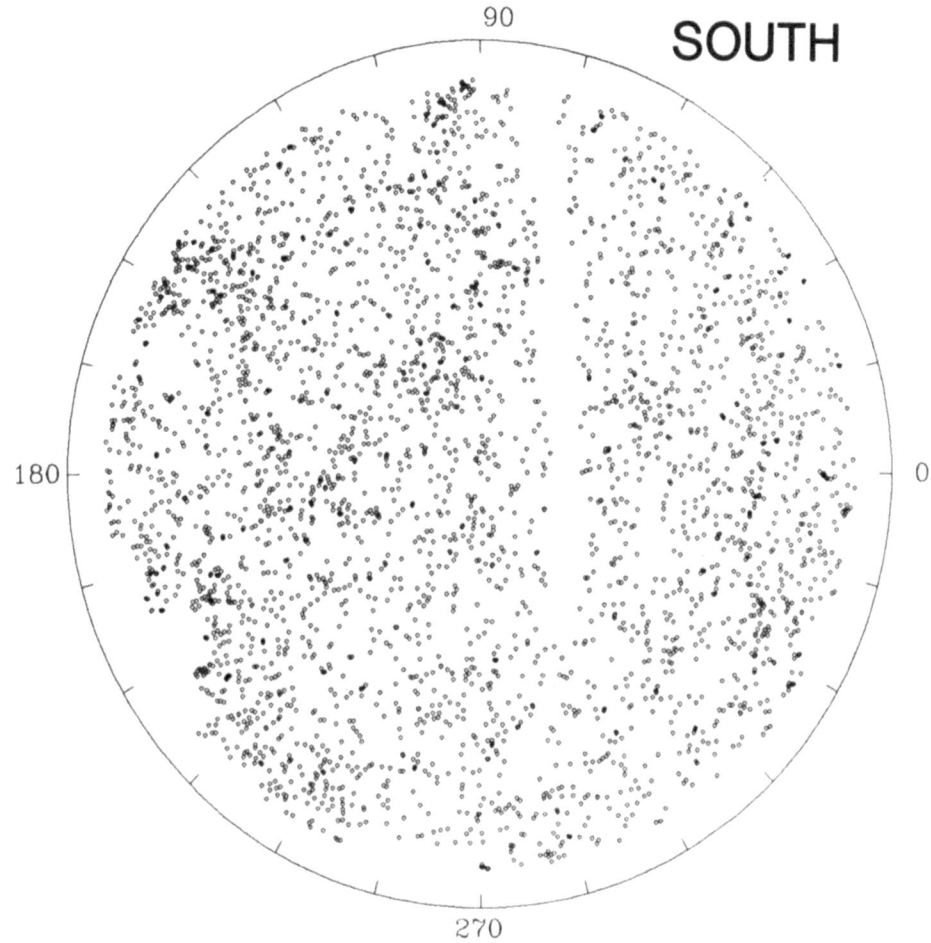

Figure 1b: The same as figure 1a, for the south galactic hemisphere. 3254 galaxies are plotted. Perseus is centered at $l = 150°$, $b = -13°$.

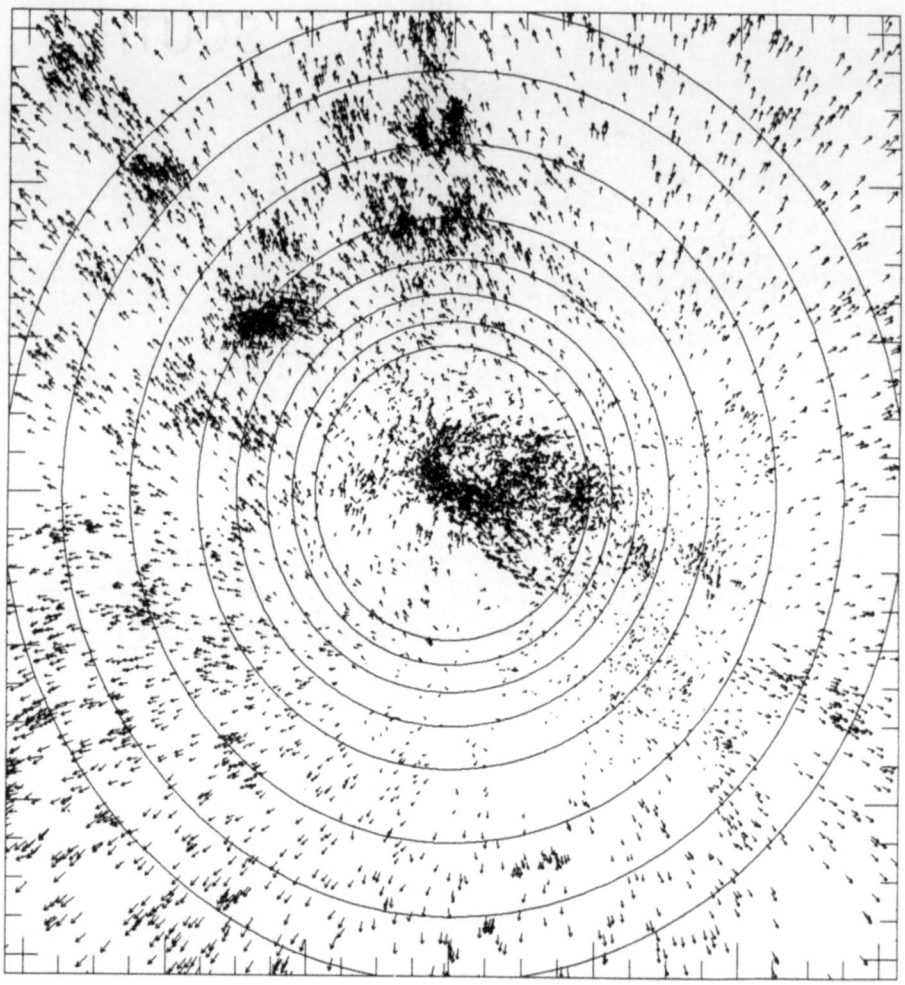

Figure 2: The flowfield around a cluster in a cosmological n–body simulation with cold dark matter, random phase, initial conditions. The arrows denote proper velocity. The circles are draw at $\bar{\delta} = 2^{\frac{n-1}{2}}$, n = 1,...,8. Both small scale and large scale deviations from a spherically symmetric radial flowfield are apparent. The degree of subclustering and asphericity is typical of these models.

Primack, 1983). All models were for $\Omega = 1$ and assumed random phase initial conditions, although some had enhanced power in the fundamental wave. The qualitative nature of the departures from spherical symmetry are independent of initial conditions in the models, although the exterior tidal fields are larger for the models with more power on large scales, as expected.

Clusters were selected from the models if their surface of $\bar{\delta} \simeq 2$ occurred at a distance from cluster center $r \sim 2\text{-}3\ r_o$, where r_o is the correlation length $\xi(r_o) = 1$. This approximates the situation for Virgo, presuming galaxies trace mass. We also attempted to select clusters reasonably isolated from their neighbors. An example of a flowfield around a cluster is shown in figure 2. There are a number of features in common among all these clusters which I mention here. Details are provided in Villumsen and Davis (1985).

1. The clusters are never spherical or isolated; they are usually triaxial. They are always subclustered and perturbed by adjacent clusters.

2. The velocity field, when averaged over spherical surfaces, fits the non–linear flow model quite well for $\bar{\delta} < 3$ but the mean infall is systematically low for increasing $\bar{\delta}$ and is 25% below the spherical model prediction at the $\bar{\delta} = 6$ surface.

3. Substantial (50%) quadrupole and higher order distortions to the velocity field are frequently present in the clusters which can seriously affect infall and mass estimates based on spherical models. The asphericity of the flow field can bias the estimation of mass by a factor of 2-3.

4. These distortions in the velocity field correlate very poorly with the inertia tensor of the interior mass distribution, indicating they are largely influenced by the exterior mass distribution.

5. In spite of the serious local deviations from the spherical models, even on the $\bar{\delta} < 3$ surfaces, the local peculiar velocity field is extremely well aligned with the local force field. On the $\bar{\delta} = 2$ surface, the mean cosine of the angle between the local force and velocity is 0.9, an average deviation of 25°. The constant of proportionality is close to that expected by equation (2). This result applies only for those points not imbedded in dense subclumps.

Several important lessons for application to the Virgocentric flow problem can be drawn from this study:

1. The aspherical flow fields observed in the models can seriously bias the measured mean infall even on the $\bar{\delta} = 2$ surface unless data is drawn from the full 4π steradians of the sphere surrounding Virgo. In practice this is rather difficult to arrange. Not only are the galaxies preferentially distributed on the supercluster plane, but the galaxies near us on the front side of the cluster have more weight than those on the backside of the cluster. Whether the bias increases or decreases the mean infall velocity depends on the exterior mass distribution; perhaps it can be calculated from the IRAS catalog.

2. What is actually measured in the Hubble anisotropy studies is the gradient of the peculiar flow field; solid body rotation would of course be undetectable. A good deal of weight to the measured infall comes from galaxies within our Virgocentric radius (but outside the triple value zone). However the n–body models show that the mean infall drops below the spherical prediction for higher δ, diminishing the spatial gradients in the peculiar velocity field. This is a clear bias that will result in an underestimate of the true infall velocity. It could be dealt with by deleting all galaxies located within the $\bar{\delta} = 3$ surface, but this will likely result in larger statistical errors for the infall.

3. Fortunately the deviations from spherical symmetry seem to obey linear perturbation theory (equation (2)), which provides the best possible measure of Ω. The 25° deviation of the local "gravity" direction seen by IRAS from the microwave anisotropy direction is fully consistent with the behavior of the n–body models.

7. SUMMARY

Observations of binary galaxies suggest they have an extended mass distribution but are consistent with $\Omega = 0.05$ on scales of ~ 50 h^{-1} kpc. Groups and clusters of galaxies are consistent with $0.1 < \Omega < 0.2$ on scales of less than a few Mpc, whether analysed by the cosmic virial theorem or by use of the virial theorem on individual groups. This is consistent with calibrations against fake catalogs drawn from n–body models. However the data is also consistent with $\Omega = 1$ if galaxies are a biased mass tracer.

The general consensus seems to be that the infall velocity to Virgo is small, again consistent with $\Omega < 0.2$. However this result cannot readily explain the bulk of the microwave dipole anisotropy and it is very unlikely that Hydra–Centaurus is the accelerator. It is more probable in my opinion that the bulk of the microwave anisotropy has been induced by material loosely connected to the local supercluster ($cz < 3000$ km s^{-1}). The comparison to n–body results suggest that the spherical infall models used to measure mass in the Virgo cluster are not terribly accurate and may be biased. However, the models do indicate that an excellent cosmological test is possible by comparing local forces to local velocities. The close agreement in direction of the IRAS and microwave dipole anisotropies is very encouraging and suggests that an accurate Ω estimate at large scales is close at hand. The resulting estimate is likely to be larger than measured on smaller scales.

In brief, estimates of Ω are probably not consistent on different scales, and there is no existing data that contradicts the notion that galaxies are a biased mass tracer and that we live in an $\Omega = 1$ universe.

REFERENCES

Aaronson, M., Huchra, J., Mould, J. *et al.* 1985, *Ap.J.*, submitted.

Bean,A.J., Efstathiou,G., Ellis,R.S., Peterson,B.A., and Shanks,T.,1983, *M.N.R.A.S.*,,2

Blumenthal, G. P. and Primack, J. P. 1983, in *Fourth Workshop on Grand Unification*, ed. H. A. Weldon, P. Langacker, and P. J. Steinhardt, (Boston:Birkhausen), p. 256.

Davis, M. 1985, in *Proceeding of the Inner Space-Outer Space Workshop*, Fermilab, 1984, (ed. by M. Turner and R. Kolb, Univ. of Chicago Press).

Davis, M. and Huchra, J. 1982, *Ap.J.* **254**, 425.

Davis, M. and Peebles, P. J. E. 1983a, *Ann. Rev. of Astron. & Astrophys.* **21**, 109.

Davis, M. and Peebles, P. J. E. 1983b, *Ap.J.* **267**, 465.

Davis,M., Frenk,C.S., Efstathiou, G., and White,S.D.M.,1985,*Ap.J.*, **292**,371.

de Vaucouleurs, G. and Peters, W. L. 1984, *Ap.J.* **287**, 1.

Dressler, A. 1984, *Ap.J.* **281**, 512.

Faber, S. M. and Gallagher, J. S. 1979, *Ann. Rev. Astron. & Astrophys.* **17**, 135.

Fixsen, D. J., Cheng, E. S. and Wilkinson, D. T. 1983, *Phys. Rev. Lett.* **50**, 620.

Geller, M. J., in *Clusters and Groups of Galaxies*, IAU Trieste meeting.

Hart, L. and Davies, R. D. 1982, *Nature* **297**, 191.

Lubin, P. M., Epstein, G. L. and Smoot, G. F. 1983, *Phys. Rev. Lett.* **50**, 616.

Meiksin, A. and Davis, M. 1985, *A.J.*, submitted.

Nolthenius, R. and White, S. 1986, (abstract in this volume).

Peebles, P. J. E. 1980, *The Large Scale Structure of the Universe*, Princeton Press.

Press, W. H. and Davis, M. 1982, *Ap.J.* **259**, 449.

Schweizer, L. 1985, Ph.D. thesis, UC Berkeley.

Tonry, J. and Davis, M. 1981, *Ap.J.* **246**, 680.

Villumsen, J. and Davis, M. 1985, in preparation.

White, S., Huchra, J., Latham, D. and Davis, M. 1983, *M.N.R.A.S.* **203**, 701.

Yahil, A., Walker, D. and Rowan-Robinson, M. 1985, preprint.

Zwicky, F. 1933, *Helv. Phys. Acta* **6**, 110.

DISCUSSION

TULLY: I would like to make two points. First, with respect to the
Hydra-Centaurus structure: I have a poster paper which shows the
distribution of large-scale structure around us. On that diagram is
something called Virgo-Hydra-Centaurus. This is a unit of five Abell-
class clusters (since it is in the southern hemisphere, only one of
them actually appears in Abell's catalogue). One of these clusters,
the Centaurus Cluster, is almost certainly about twice the size of
Virgo. So I don't at all discount that it has the overdensity you were
talking about. It is a major feature which extends about 50 - 60 Mpc,
although unfortunately it is badly cut up by the galactic plane.
    Second, with regard to your statement about transverse peculiar
velocities: That is something I definitely do see. I describe the
Local Supercluster as an environment made of clouds of galaxies, and I
have gotten distances, independent of redshift, for many of these
clouds. The picture is quite complex, with significant transverse,
non-Virgocentric velocities.

OSTRIKER: In the latter part of your talk, when you were discussing
recent results, you convinced us that galaxies don't fairly represent
the mass distribution. But in your earlier discussion, you described
the virial method for getting the mass distribution, which seemed to
indicate that galaxies are good tracers of the mass.

DAVIS: Well, the first part of the discussion was based on an argument
which depends not on the two-point correlation function but on a very
difficult integral of the three-point correlation over the two point
correlation which may not have the simple scaling that I put down. In
fact, we've never been able to get that behavior in the numerical
simulations, try as we might.

GALAXY LUMINOSITY FUNCTIONS, M/L RATIOS, AND CLOSURE OF THE UNIVERSE

James E. Felten
NASA Goddard Space Flight Center and University of Maryland

Data on the luminosity function (LF) of galaxies are reviewed and compared on a common magnitude system and with common assumptions. The result of Kirshner et al. (Astron. J. 88, 1285, 1983) is chosen as a best guess. Departures from this "standard LF" for specific galaxy types and environments (clusters, groups, field) are discussed briefly. If the Galactic absorption is $A_B = 0.2$ csc$|b|$ and the solar absolute magnitude is $M_{B\odot} = 5.48$, this LF leads to a mean luminosity density $\mathcal{L} = 2.4 \times 10^8 h$ $L_\odot$ Mpc^{-3} on the B_T system, or about $1.4 \times 10^{-2} h$ "Galaxies" Mpc^{-3}. The mean M/L ratio needed to give critical cosmological density ($\Omega_0 = 1$) is then 920h in solar units on the B_T^0 (face-on) system. [The latter number can be larger by a factor ~ 2 if a different LF is used, and larger still on different systems and with different assumptions (see Table).] The "weighed" M/L on this system is $\sim (250 \pm 50)h$ in clusters, but it is smaller in binaries and small groups. Estimates of the weighed (clumped) Ω_0 vary, but it is definitely < 1. Comparison with constraints imposed by inflation and nucleosynthesis suggests that we distinguish at least two "dark-matter" problems: (1) What is the weighed mass, and how is it distributed? It contributes an $\Omega_0 \sim 0.1-0.5$ and could be all baryonic if at the lower end of this range. (2) Is there additional matter, more smoothly distributed and probably nonbaryonic, which brings Ω_0 up to unity? For details see Comments on Astrophys., in press.

COMPARISON OF M/L RATIOS ON VARIOUS SYSTEMS
AND WITH VARIOUS ASSUMPTIONS

Magnitude system	If $A_B = 0$		"Face-on"	If $A_B = 0.2$ csc$|b|$		"Face-on"
	$B(0)$	B_T	B_T^0	$B(0)$	B_T	B_T^0
"Weighed" M/L (Great clusters)	$(480\pm100)h$	$(380\pm80)h$	$(300\pm60)h$	$(400\pm80)h$	$(310\pm60)h$	$(250\pm50)h$
Critical M/L to give $\Omega_0 = 1$, according to LF of:						
Kirshner et al.	1960h	1540h	1220h	1470h	1160h	920h
Davis & Huchra	2530h	1990h	1580h	1900h	1500h	1190h
Ellis	4430h	3480h	2770h	3330h	2620h	2080h

J. Kormendy and G. R. Knapp (eds.), Dark Matter in the Universe, 111.

ABOUT THE COMA CLUSTER (Progress Report)

D. Gerbal, Observatoire de Meudon
G. Mathez, Observatoire de Meudon
A. Mazure, Observatoire de Toulouse
E. Salvadore-Solé, Universidad de Barcelona

The study of the dynamics of the Coma Cluster is of interest for several reasons. First, there exists a great deal of observational information about the cluster, including data on morphology, magnitude, color and redshift for the galaxies, and reasonably detailed x-ray data for the hot gas. Second, the present dynamical state of the cluster is reasonably well-defined. In addition, the segregation of the more luminous ($\equiv$ massive) galaxies towards the cluster center shows that two-body relaxation effects are well-advanced (Capelato et al. 1980). The profile of velocity dispersion with radius shows that in the outer parts of the cluster the galaxy velocities are non-isothermal (des Forêts et al. 1984). There is, however, evidence of continuing dynamical evolution. The velocity field of the galaxies at large distances from the center of the cluster suggests continuing infall (Capelato et al. 1982), and two sub-condensations are located in the inner regions (Mazure and Proust 1986). A new dynamical analysis for the cluster is being carried out in two stages. First, a relaxed model with a wide mass spectrum (c.f. Inagaki 1980) is fitted to the data. The contribution of the intergalactic gas is taken into account. With H_O = 75 km/sec/Mpc, the total mass within a 3° radius of the center is $\sim 1.5 \times 10^{15}$ M_Θ, of which $\sim$ 30% is in the intergalactic medium, and $M/L \sim 75$ M_Θ/L_Θ. The ratio of specific energies of the galaxies and the gas is $\sim$ 1.1, i.e., there is no scale-height problem (these results are described more fully by Gerbal et al. 1986). A second "model independent" analysis using the profiles of the galactic density and velocity dispersion gives the radial dependence of the galactic mass, the gas mass and also gives the total mass, which is found to be $\sim$ 1.1 $\times 10^{15}$ M_Θ within 3° (Gerbal et al. 1984).

References

Capelato, H., Gerbal, D., Mathez, G., Mazure, A., Salvadore-Solé, E. and Sol, H. 1980, Ap.J. 241, 521.
Capelato, H., Gerbal, D., Mathez, G., Mazure, A. and Salvadore-Solé, E. 1982, Ap.J. 252, 433.
des Forêts, G., Dominguez-Tenreiro, R., Gerbal, D., Mathez, G., Mazure, A. and Salvadore-Solé, E. 1984, Ap.J. 280, 15.
Gerbal, D., Mathez, G., Mazure, A. and Morin, J.L. 1984, in "Clusters and Groups of Galaxies" ed. F. Mardorissian, G. Giurcin and M. Mezzetti, Astrophy. Space Sciences Lib. 111, 147 (D. Reidel Co.)
Gerbal, D., Mathez, G., Mazure, A. and Salvador-Solé, E. 1986, Astron. Astrophys. (in press).
Inagaki, S. 1980, P. Ast. Soc.Japan. 32, 213.
Mazure, A. and Proust, D. 1986, preprint.

112

J. Kormendy and G. R. Knapp (eds.), Dark Matter in the Universe, 112.

ESTIMATING THE MASSES OF GALAXY GROUPS: ALTERNATIVES TO THE VIRIAL
THEOREM

J. Heisler and S. Tremaine
Massachusetts Institute of Technology, Cambridge, MA

J. Bahcall
Institute for Advanced Study, Princeton, NJ

Group masses are almost always computed using the virial theorem.
Bahcall and Tremaine (1981) studied the reliability of mass estimates
based on the virial theorem for the case of test particles orbiting
a central mass. They found that virial theorem estimates were both
inefficient and biased. We have extended the work of Bahcall and
Tremaine by exploring alternative mass estimators to the virial theorem
for self-gravitating systems.

We present three alternatives to the virial theorem for estimating the
masses of groups of galaxies. The projected mass estimator uses the
mean value of $V_i^2 |\vec{R}_i|$, where V_i and $|\vec{R}_i|$ are the radial velocity and
projected separation of galaxy i from the group center. The other two
methods rely on the average and median of $(V_i - V_j)^2 |\vec{R}_i - \vec{R}_j|$ over all
pairs of galaxies in the group. These three estimators and the virial
theorem estimator are tested using a series of N-body simulations and
Monte Carlo realizations of Michie models. No one mass estimator per-
forms significantly better than the others. From the numerical simulations we find that for all four estimators 75% of the mass estimates
lie within $10^{0.25}$ of the correct value for groups with 5 members (N=5)
and within $10^{0.15}$ for N=10. The values obtained from the different
mass estimators are all strongly correlated.

We use the estimators to calculate the masses and mass-to-light ratios
of nearby groups catalogued by Huchra and Geller (1982). Our work
confirms the work of Huchra and Geller implying that there is a large
amount of dark mass in groups on scales of ~700 kpc. The median M/L
is about $10^{2.8}$ M_Θ/L_Θ for groups with three members and for H_0=100 km/s/
Mpc. Most of the width in the M and M/L distributions in this catalog
arises from statistical uncertainties. The width does not necessarily
reflect a real distribution of the mass-to-light ratio or contamination
by unrelated galaxies ("interlopers") which appear projected onto the
group. The measured M/L ratios can only be improved by measuring more
radial velocities. The dominant source of uncertainty in determining
M/L arises from the possibility that the galaxy distribution does not
trace the mass distribution. However, the corrections for this effect
are strongly model-dependent, and we have chosen not to include them
in this paper.

REFERENCES
Bahcall, J., and Tremaine, S. 1981. Astrophys. J. 244, 805.
Huchra, J.P., and Geller, M.J. 1982. Astrophys. J. 257, 423.

J. Kormendy and G. R. Knapp (eds.), Dark Matter in the Universe, 113.

BIASES IN MASS ESTIMATES OF GROUPS OF GALAXIES

Gary A. Mamon
New York University

The knowledge of the masses of small groups of galaxies is important, because the timescales for collapse and virialization of groups depend on their mass-density. While the large inefficiency of the virial mass statistic is well known (Bahcall and Tremaine 1981, hereafter BT, and references therein), biases in the virial mass may produce wrong evolution timescales for small groups. These biases originate from group contamination by interlopers, and from the different galaxy and dark matter distributions inside the groups (caused by mass segregation). This second bias was first studied by Smith (1980, 1984) although already implicit in the work of Limber (1959). It formally arises because the ratio $2T/C$ where $C = \sum \mathbf{F}_\alpha \cdot \mathbf{R}_\alpha$ (the Clausius virial) is not the same for the luminous and global matter distributions. We illustrate here some quantitative aspects of this Limber bias from the output of N-body simulations of groups of 8 galaxies described elsewhere (Mamon 1986).

Table 1 below shows for the projected groups the median ratios of virial (luminosity weighted) and projected (BT) mass to total mass interior to the smallest sphere containing the centers of the galaxies.

Table 1

Initial Conditions	Group Density	Galaxy Type	Time (Gyr)	# in sample	M_{vir}/M	M_{pro}/M
Virialized	Dense	Halo	0.5	216	0.99	1.12
Virialized	Dense	Hubble	2	111	0.94	1.10
Virialized	Loose	Halo	20	267	0.89	1.01
Virialized	Loose	Hubble	20	264	0.81	0.91
Collapsing	Loose	Halo	20	195	0.98	0.87
Collapsing	Loose	Hubble	20	231	0.95	1.00

The numbers in Table 1 indicate that the median virial mass is a fair mass estimator, although it is generally too low by 10%. The projected mass also produces consistent values with a constant of $32/(\pi G N)$ in front of the sum of BT's equation (20). A similar result is reached by Heisler et al. (1986). The softening of the potential energy into a Clausius virial for dense groups of often overlapping galaxies has a negligible effect on the median mass estimate. Restricting the analysis to groups or subgroups of 4 galaxies produces similar results.

Bahcall, J. N. and Tremaine, S. 1981, Ap. J., **244**, 805.
Heisler, J., Tremaine, S., and Bahcall, J. N. 1986, preprint, and in this volume.
Limber, 1959, Ap. J., **130**, 414.
Mamon, G. A. 1986, to be submitted to Ap. J..
Smith, H. 1980, Ap. J., **241**, 63; 1984, in Clusters and Groups of Galaxies, p. 523.

J. Kormendy and G. R. Knapp (eds.), Dark Matter in the Universe, 114.
© 1987 by the IAU.

THE PECULIAR VELOCITY OF THE LOCAL GROUP IN THE DIRECTION OF THE VIRGO CLUSTER

L. Staveley-Smith & R.D. Davies
Nuffield Radio Astronomy Laboratories, Jodrell Bank

The measurement of the amplitude of the Local Group infall velocity towards the Virgo Cluster is a crucial test for the value of the universal density parameter Ω_0 and the ratio of the universal Hubble constant to its local value. However, a very large discrepancy exists between total infall velocities derived from peculiar velocity field observations and those derived from "scaling" methods using standard candles in the Virgo and Coma clusters. The former have tended to produce high Virgocentric peculiar velocities (350 to 500 km s^{-1}) whilst the latter give much lower values (-70 to 100 km s^{-1}).

To resolve this apparent discrepancy, we applied the standard infall model to our sample of over 200 spiral galaxies recently observed in HI at Jodrell Bank and Parkes. We have used the HI Tully-Fisher relation and the diameter-linewidth relation as distance indicators and in both cases find that the 68% upper confidence limit for the systematic infall amplitude at the position of the Local Group is 110 km s^{-1}. The total peculiar velocity (systematic plus thermal) in the direction of the Virgo Cluster is much less dependent on input model parameters and is equal to 83 ± 46 km s^{-1} when the sample redshift is restricted to 3000 km s^{-1}. However, this velocity increases with the redshift of our reference frame and reaches a value of 420 ± 140 km s^{-1} at a mean redshift of 3700 km s^{-1}. This leads us to the conclusion that our local reference frame (defined in part by the Local Supercluster) has a very large peculiar velocity of its own. The obvious implication is that substantial density inhomogeneities must exist on very large scales ($\gtrsim$ 40 Mpc) in the Universe.

We have also confirmed that there is substantial evidence for a rotation field in the inner Local Supercluster which is centred on the Virgo Cluster. The rotational velocity appears to peak at an angle of 35° from the Virgo Cluster where it reaches a value of 230 ± 70 km s^{-1}. The direction is such that the Virgo II complex is rotating away from us whilst the Canes Venatici/Ursa Major complex is rotating towards us.

We propose 2 reasons why peculiar velocity field observations have previously yielded unrealistically high infall velocities. The first is the neglect of the distance dependence of the Local Group peculiar velocity. The second is due to the confusion between infall and clockwise rotation that exists where there is an asymmetric north/south distribution of galaxies.

Our observations therefore indicate that the Local Supercluster is not responsible for the large values of the Local Group peculiar velocity that are deduced from measurements of the dipole anisotropy in the microwave background. The inhomogeneities that give rise to this anisotropy must be much more massive (10^{16-17} M$_\odot$) and occur at redshifts beyond 3000 km s^{-1}.

J. Kormendy and G. R. Knapp (eds.), Dark Matter in the Universe, 115.
© *1987 by the IAU.*

THE RING OF HI IN LEO:
PROBING THE MASS DISTRIBUTION IN A GALAXY GROUP

Stephen E. Schneider, Edwin E. Salpeter, Yervant Terzian
Department of Astronomy, Cornell University

The intergalactic neutral hydrogen in the M96 group (Schneider *et al.* 1983) provides an unusual probe of the detailed mass distribution in a group of galaxies. Previous observations (Schneider 1985) tentatively suggested the existence of a large ring structure to the gas. Sensitive new observations made at Arecibo confirm that the intergalactic gas forms a 200 kpc diameter* eccentric ring around the two central galaxies in the group, the giant elliptical galaxy NGC 3379 (M105) and the lenticular NGC 3384.

The pattern of measured radial velocities around the ring matches that of a non-circular Keplerian orbit with a uniquely defined focus. *Without any assumptions about the focus of the orbit,* the Keplerian model predicts a position and velocity for the center of mass closely coinciding with the two central galaxies. A mass distribution leading to significant deviations from a $1/r^2$ force law over the length scale of the ring (and likewise Milgrom's theory of gravitation) can neither account for a closed eccentric orbit nor the ability of the Keplerian model to predict the galaxies' position and velocity. The enclosed gravitational mass for the orbit is $\sim 6 \times 10^{11}$ M$_\odot$, leading to a modest mass-to-light ratio of ~ 25 for the total mass associated with the two galaxies out to ~ 100 kpc. This value for M/L exceeds internal estimates, based on the two galaxies' stellar velocity dispersions, by only a factor of ~ 2 (Tonry and Davis 1981).

There are some deviations from the Keplerian model at the ring's 70 kpc pericenter that might be explained by a halo distribution of matter about the central galaxies; however, these deviations may also result from M96's tidal influence, and the success of the Keplerian model implies that a massive halo, if present, must not extend much farther than the pericenter distance from the central galaxies. Hot massive neutrinos or any other form of dark matter spread evenly throughout the group are unlikely sources for any significant contribution to the group's mass; and a mass distribution not extending much beyond the luminous disks of the galaxies appears to be consistent with the data.

REFERENCES

Schneider, S. E. 1985, *Ap. J. (Letters)* **288**, L33.
Schneider, S. E., Helou, G., Salpeter, E. E., and Terzian, Y. 1983 *Ap. J. (Letters)* **273**, L1.
Tonry, J. L., and Davis, M. 1981 *Ap. J.* **246**, 666.

---

* Values quoted are for an assumed distance of 10 Mpc. Distance estimates for the M96 group range from 8 to 18 Mpc.

116

J. Kormendy and G. R. Knapp (eds.), Dark Matter in the Universe, 116.

CLUSTERING ON A SCALE OF 10^{18} M⊙

R. Brent Tully
Institute for Astronomy, University of Hawaii

This discussion represents an update of an article submitted for publication in The Astrophysical Journal that describes the concentration of Abell clusters to the supergalactic equator. Since that article was written, our sample of Abell-class clusters has been augmented by 44% to now total 309 clusters within a redshift of 0.1c. The distribution of those clusters with distance from the supergalatic equator is illustrated in the accompanying figure. The normalized histogram includes an attempt to correct for the diminished unobscured area in planes off the supergalactic equator. Within 4 bins on the equator (i.e., within $80/h_{75}$ Mpc) there are 105 clusters, rather than the 48 clusters that would be expected with a random distribution. The 57 excess clusters represents a 4.6 σ signal.

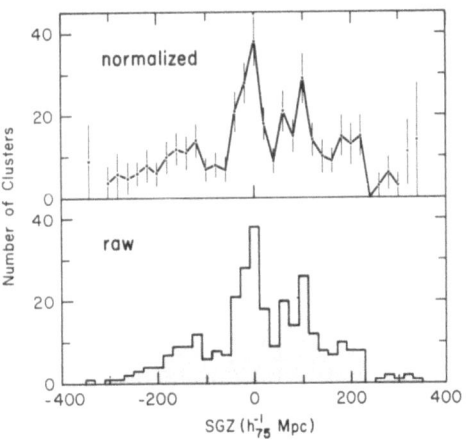

The bias toward greater completion locally can be crudely counteracted by the rejection of everything within $100/h_{75}$ Mpc radius. There are 22 nearby clusters that all contribute to the central signal. Without these clusters, there is still a significant 3.0 σ signal on the supergalactic equator.

It is concluded that 10^2 Abell-class clusters participate in superclustering on a very large scale. These clusters are contained within a region with a long dimension of about $600/h_{75}$ Mpc and a short dimension of less than $100/h_{75}$ Mpc. There is a coincidence between the plane of this structure and the plane defined by nearby galaxies that lead to the definition of the supergalactic coordinate system.

J. Kormendy and G. R. Knapp (eds.), Dark Matter in the Universe, 117.

PARAMETERS FOR DARK HALOS

K.C.Freeman
Mount Stromlo and Siding Spring Observatories
Research School of Physical Science
Australian National University

ABSTRACT. What are the characteristic scale lengths and densities for
the dark halos of galaxies, and the typical ratios of dark to luminous
mass? For elliptical galaxies, the best estimates come from X-ray data
which will be discussed in a later session. For spirals, the best
estimates come from rotation curves. I will concentrate on the halo
parameters for disk galaxies. At the end, there will be a few comments
on stellar dynamical data for ellipticals, and on the unique information
available for the dark halo of our Galaxy.

1. THE HALOS OF DISK GALAXIES

The disks of disk galaxies are supported in the radial direction mainly
by rotation. The equilibrium of disk matter is then given approximately
by

$$GM(r)/r^2 = V^2(r)/r$$

where $M(r)$ is the mass within radius r and $V(r)$ is the rotational
velocity at r. At large r, the integrated luminosity approaches the
total luminosity. If $M(r)$ also tends to a limit, then at large r we
would expect $V(r)$ to go like $r^{-1/2}$. This is not observed: usually $V(r)$
is flat at large r, which suggests that $M(r)$ increases like r, as for
the isothermal sphere, and this is the most direct evidence for the
presence of the massive dark corona component.

Does a flat rotation curve necessarily imply the presence of a dark
corona? There has been some argument about this: we will see that the
answer depends on how far out in radius the rotation data extend. Some
work by Kalnajs (1983) illustrates this clearly.

1.1 Flat Rotation Curves

Kalnajs considered disk galaxies with measured rotation curves $V(r)$ and
surface brightness distributions $I(r)$, and asked whether the
gravitational field of the luminous components (bulge and disk) alone

119

J. Kormendy and G. R. Knapp (eds.), Dark Matter in the Universe, 119–132.

could produce the observed flat rotation curves. He assumed that the
mass to light ratio M/L was constant, and devised an algorithm to
calculate the expected rotation curve from the observed surface
brightness distribution, assuming that the disk is in centrifugal
equilibrium.

For several examples shown by Kalnajs, the rotation curves
predicted from the surface brightness profiles were in excellent
agreement with the observed rotation curves: no dark halo was needed.
However, for these examples, as for most optically measured systems, the
rotation data extends only to about 3 disk scale lengths. The important
point that comes from this work is that, out to about 3 scale lengths,
the disk and bulge together can produce a fairly flat rotation curve.
This can also be seen from simple disk + bulge models. However, for a
disk + bulge potential to produce a flat rotation curve requires a
relationship between the scale lengths of the disk and the bulge, and
the characteristic density of the bulge and surface density of the disk
(see Freeman, 1985). This point is discussed further by Bahcall and
Casertano (1985).

Rotation curves out to 3 scale lengths do not go far enough to show
the gravitational effects of the dark halo unambiguously, so they tell
us little about the presence of a dark halo. However, beyond about 3
scale lengths, the rotation curve predicted from the disk + bulge alone
begins to fall, so a flat rotation curve extending beyond about 3 scale
lengths provides strong evidence for the presence of a dark halo. If the
rotation curve goes out far enough, into regions where the dark halo is
dominating the gravitational field, then the shape of the curve can be
used to estimate the scale length and density of the dark halo itself.

The procedure is to model the galactic rotation curve. The mass
distributions for the disk and bulge components are given by surface
photometry, and a simple model is used for the dark halo. The M/L ratios
for the luminous components and the parameters for the dark halo are
adjusted to fit the observed rotation curve. We will discuss this at
some length. However, we should first make some preliminary comments on
the shape and density distribution of the dark halo, and on the
contribution that the luminous components make to the total
gravitational field.

1.2 The shape of the dark halo

Here are some observations that are relevant to the question of the
shape of the halo.

Many galactic disks show evidence for warping of the gaseous
component. It is not yet understood how these warps survive. Some recent
dynamical theories of warps (Dekel and Shlosman, 1983; Toomre, 1983;
Sparke, 1985) require a somewhat flattened dark corona.

For several polar ring galaxies, rotation has been measured in both the disk of the parent galaxy and in the polar component, out to about $3R_{25}$ of the disk (see the paper by Whitmore et al. in this volume). The rotation curve in both components is flat, with V(polar)/V(disk) = 0.97 ± 0.08. If the dark component were in a disk, then this ratio would be in the range 0.6 to 0.8. Therefore, the dark matter is probably not disklike; the data suggests that it is more nearly spherical.

From the z-equilibrium of galactic disks, it is possible to make a direct estimate of the M/L ratio of the disk matter (Bahcall, 1984; van der Kruit and Shostak 1983; van der Kruit and Freeman, 1984). These M/L values are in the range 3 to 6, and include the contribution from the dark matter of the disk itself, but are not large enough to account for the amplitudes of the rotation curves in similar disk galaxies. It follows that about half of the galactic mass in typical spirals does not lie in the disk.

At this point, the shape of the dark halo is not well known, except that it is probably not disklike; it is usually taken to be spherical.

1.3 The density distribution of the dark halo

The main constraint is that the rotation curves are flat, in some cases to many disk scale lengths. This suggests that the density distribution $\rho(r) \propto r^{-2}$ at large r, although Bahcall et al. (1982) have shown that a steeper law may be appropriate. Some workers use an isothermal sphere to model the halo; others use a distribution of the form

$$\rho(r) = \rho_o [1 + (r/a)^\gamma]^{-1}$$

where $\gamma \approx 2$. Each model has a characteristic central density and scale length, and an associated velocity dispersion σ.

It would be better to use a selfconsistent bulge + disk + halo model, in which the halo is not simply imposed but is allowed to respond to the potential of the bulge + disk. Some work is already being done on such selfconsistent models, by Barnes and by van Albada.

1.4 Are the luminous components gravitationally significant ?

To estimate the halo parameters, the rotation curve for the disk + bulge + halo model is fitted to the observations by adjusting the model parameters. The shape of the rotation curve contribution from the luminous components (disk and bulge) is given by the surface photometry; the amplitude of this contribution is not known in advance, and depends on the adopted M/L ratios for the luminous components. The observed rotation curve puts an upper limit on these M/L ratios (ie when the luminous components provide the total potential gradient in the inner parts of the galaxy, as in Kalnajs's models).

Should the maximum values of M/L be used for the luminous components, or are smaller ones more appropriate ? Rubin and associates

(see Rubin's review in this volume) point out the similarities in shape
of rotation curves for different galaxies, and argue that the luminous
components may not dominate the potential gradient anywhere. However,
there are two further constraints on the M/L ratios for the luminous
components:

(i)   As mentioned above, it is possible to estimate M/L values for
      the disks of our Galaxy and other face-on systems, from the
      vertical velocity dispersions of gas and stars. This has been
      done now for several systems, and gives consistent values of
      M/L, between about 3 and 6. (Note that this includes the dark
      matter of the disk itself.)

(ii) Stellar population models give M/L ratios between about 1.2
      and 7 for old disk and bulge populations, depending on their
      color (eg Larson and Tinsley 1978).

It seems unlikely that the M/L ratios of the luminous components
(including the dark matter associated with the disk) could be much less
than 2.

2. PURE DISK GALAXIES

These are disk galaxies with very small bulge components. The parameters
of the dark halos can be measured fairly readily in these systems
because:

(i)   Their simple disk + dark halo structure makes it easier to
      identify the contribution of each component to the rotation
      curve.

(ii) They are usually late type galaxies with an extended HI
      distribution, so it is possible to measure their rotation
      curves out to many disk scale lengths. This is essential to
      tie down the halo parameters.

2.1 Carignan's work

Carignan estimated the halo parameters for three nearby pure disk
galaxies: NGC 247, 300 and 3109 (see Carignan and Freeman 1985). From
surface photometry he measured the I(r) distribution. Rotation curves
V(r) out to large radii were obtained from HI and Fabry-Perot data. He
used Kalnajs's procedure to calculate the rotation curve associated with
the I(r) distribution, and fitted this calculated V(r) to the observed
V(r) curve in the inner parts, where the disk probably dominates the
radial potential gradient. This procedure determines the M/L value for
the disk; it gives the maximum possible M/L for the disk (and therefore
the minimum halo), as discussed in section 1.4.

    After this fitting procedure was done, the observed V(r) curve lay well above the calculated V(r) in the outer parts of each galaxy. This shows the presence of a dark halo. The difference

$$\{V^2_{obs}(r) - V^2_{calc}(r)\}^{1/2}$$

was represented by the circular velocity curve for an isothermal sphere, which was chosen as a simple model for the dark halo. The isothermal sphere has a core radius r_c, central density ρ_o and velocity dispersion σ, where

$$4\pi G\rho_o r_c^2 = 9\sigma^2$$

At large radii, the circular velocity for the isothermal sphere tends to $\sqrt{2}\sigma$. This fitting procedure then gives estimates of r_c and σ for the dark halo. In some cases, it gives only the <u>ratio</u> σ/r_c, because the contribution of the isothermal sphere to the rotation curve is close to solid body rotation in the region where there is rotation data. However even this ratio is worth knowing, because it gives a direct estimate of the central density of the dark halo, from the above equation (subject to the lack of selfconsistency of the basic disk + halo model). Figure 1 shows how this procedure works for the galaxy NGC 3109.

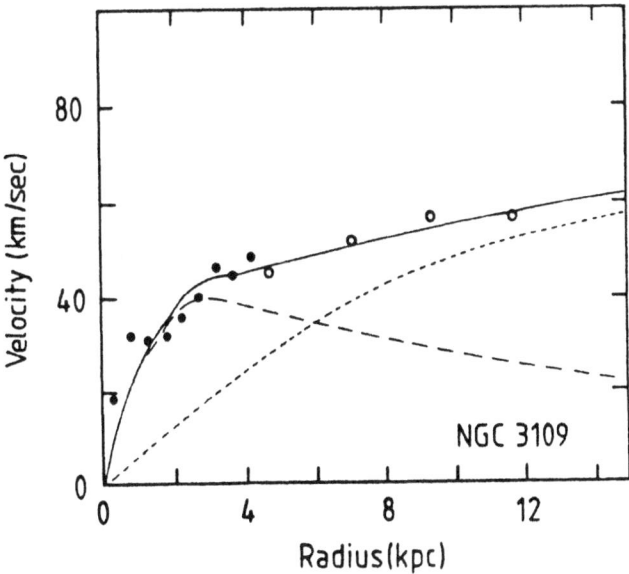

Figure 1.   Disk + halo model for the rotation curve of NGC 3109, for the parameters given in Table I. The filled and open points represent the observed rotation curve. The long and short dashed curves are the contributions from the disk and the halo respectively, and the full curve is their sum.

In a later study with the WSRT, Carignan et al. (1985) showed that the dwarf system UGC 2259 has a flat rotation curve and made a similar analysis to estimate the parameters of its dark halo. The results for these four pure disk galaxies are summarised in section 2.3.

2.2 NGC 3198: (van Albada et al. 1985)

The galaxies studied by Carignan are all fainter than $M_B = -18$. The Sc spiral NGC 3198 is brighter ($M_B = -19.4$), and is particularly interesting; for this galaxy, the WSRT rotation curve extends out to eleven disk scale lengths and the parameters for the dark halo can readily be estimated. van Albada et al. used a dark halo model of the form

$$\rho(r) = \rho_o [1 + (r/a)^{\gamma}]^{-1}$$

where $\gamma \approx 2$. The maximum disk/minimum halo solution has a M/L ratio of 3.6 for the disk, and the ratio of the halo mass to the disk mass is about 4, out to 30 kpc. However they point out that a model with less disk and more halo (say M/L = 1.2 for the disk) also gives an acceptable fit to the observed rotation curve. The fit is still adequate if the M/L ratio for the disk is taken as zero. (See however our arguments in section 1.4 against low values of M/L for the disk.)

2.3 Summary of results for pure disk galaxies

Table I gives parameters for the maximum disk/minimum halo solutions for the five pure disk galaxies described above. The columns give the name of the system, its absolute magnitude, the core radius (kpc), velocity dispersion (km s^{-1}) and central density ($M_\odot$ pc^{-3}) for the dark halo, the ratio of the halo mass to disk mass at the Holmberg radius and at the limit of the rotation data, and finally the maximum M/L ratio for the disk itself.

TABLE I

PARAMETERS FOR DARK HALOS OF PURE DISK GALAXIES

	M_B	r_c	σ	ρ_o	R_{Ho}	M(halo)/M(disk) Limit	$(M/L)_d$
NGC 247	−18.0	9	80	0.003	0.8	4	7.0
NGC 300	−18.0	12	60	0.004	0.7	2	2.2
NGC 3109	−16.8	10	45	0.002	1.5	7	3.0
UGC 2259	−16.4	7	57	0.013	1.0	2	7.0
NGC 3198	−19.4	12	100	0.008	1.5	4	3.6

For these pure disk systems, it is relatively straightforward to estimate the parameters of the dark halos: the disk and the halo are the only two components and, having the surface photometry, the M/L ratio is the only free parameter for the disk. Table I shows that the typical core radius for the dark halos is about 10 kpc and the typical central density is a few x 0.001 $M_{\odot}$ pc^{-3}. The mass of the halo out to the Holmberg radius is about the same as the disk mass. The last column shows that the maximum disk/minimum halo solutions give M/L values for the disk that are similar to those derived independently from the vertical equilibrium of the disks and from population models (section 1.4); there is no obvious reason at this stage to reject these solutions.

3. OTHER DISK GALAXIES

It is more difficult to study the dark halos of disk galaxies with significant bulges. The M/L ratio of the bulge comes in as a second free parameter for the luminous component, and in many cases the rotation curves do not extend much beyond the critical three disk scale lengths (see section 1.1). However there are many such galaxies with measured rotation curves, and recently Athanassoula, Bosma and Papaioannou (unpublished) have studied a sample of 60 systems with optical or 21-cm rotation curves.

Their procedure is similar in concept to that described earlier for the pure disk systems, with two extra features:

(i) for systems with bulges, the M/L value for the bulge appears as another free parameter;

(ii) swing amplifier theory (Toomre, 1981) was used provide another constraint on the M/L ratio of the disk. For example, if a particular galaxy shows no apparent m=1 mode but a clear m=2 mode, then its surface density was constrained to be low enough to inhibit the m=1 mode but high enough to allow the m=2 mode. In many cases, this constraint allowed the maximum disk permitted by the rotation curve, but in some cases it did not.

This is an interesting extra constraint. Is it an improvement on just assuming the maximum disk ? For systems in which the maximum disk is excluded by swing amplifier theory, the inferred dark halo will clearly have a core radius that is of the same order as the disk scale length, ie significantly shorter than the typical value of about 10 kpc found for the pure disk systems. The characteristic density for the dark halo will be correspondingly high. We would then need to know whether this high density and short core radius are just artifacts of the procedure, or are real features perhaps resulting from the dark halo responding to the potential of the disk and the bulge.

Preliminary results from this study include the following:

(i) For many of their systems, the parameters are fairly similar to those given in Table I for the pure disk galaxies. However for some systems, the halos do have significantly shorter core radii and characteristic densities that are higher by at least an order of magnitude, as discussed in the previous paragraph.

(ii) The ratio of halo mass to (disk + bulge) mass out to R_{25} is typically between 1 and 2, and shows a weak decreasing trend with increasing (B-V) color.

(iii) They also made minimum halo solutions for this sample. The M/L values for the disk derived in this way show an increasing trend with increasing (B-V) color. Quantitatively, this trend agrees well with the predictions from the Larson-Tinsley (1978) models (local IMF, monotonic decreasing star formation rates, ages 10 Gyr). Again, this favors the view that the minimum halo solutions are appropriate in most cases.

4. DARK MATTER IN ELLIPTICAL GALAXIES

The best estimates of dark matter parameters for elliptical galaxies at this time come from X-ray observations, which are discussed elsewhere. Here I will just mention briefly some stellar dynamical items.

The run of velocity dispersion with radius in ellipticals is typically flat or rising: see for example Illingworth (1983), Efstathiou et al. (1982), Dressler (1979). However, this does not necessarily mean that these systems have a dark halo: Tonry (1983) and Richstone (unpublished) have pointed out that such dispersion profiles can be consistent with the observed light distribution and a radially uniform M/L ratio, if the velocity ellipsoid becomes more tangential with increasing radius. For the dark matter problem, it would therefore be very useful to find an observational way to estimate the anisotropy of the velocity ellipsoid in ellipticals, well outside their cores.

One (difficult) possibility comes from the shape of the line of sight velocity distribution of stars in the outer parts of ellipticals; this shape will depend in some way on the distribution of the stellar orbital eccentricities and therefore on the degree of anisotropy of the velocity ellipsoid. As a simple example, take a spherical galaxy with the logarithmic potential $\Phi = V^2 \ln(r)$. Consider the extreme case in which the stellar orbits in the outer parts of this galaxy are circular and randomly orientated (ie the radial component of the velocity ellipsoid is zero). Then the distribution of observed line of sight velocities for these stars would be uniform between -V and +V, and zero elsewhere. On the other hand, for an isotropic velocity ellipsoid, the distribution of observed line of sight velocities would be peaked at zero, with wings going out beyond ±V. It would probably be difficult to

measure the shape of the broadening function accurately enough, from
integrated spectra and Fourier techniques, to estimate the anisotropy of
the velocity dispersion in the outer parts of ellipticals. A more
hopeful procedure would be to acquire a large sample of radial
velocities for the globular clusters around M87, say; this would give
the shape of the observed line of sight velocity distribution directly,
for the outer parts of the galaxy.

An independent dynamical approach to dark matter in ellipticals
comes from the shells observed around many elliptical galaxies. In
Quinn's (1984) dynamical theory for the origin of the shells, the
relationship between the radii of successive shells depends on the
potential field of the galaxy. The radii of shells in some well-observed
systems suggests the presence of dark matter, but not out to very large
radii.

We should mention here the question of dark matter in globular
clusters, which has been discussed by Peebles (1984). To test for dark
matter, we would need to know the radial behaviour of the velocity
ellipsoid in a globular cluster. This information is not yet available.
Probably the best data for this purpose comes from ω Cen, for which
Seitzer and Freeman (to be published) have measured the line of sight
velocity dispersion over the entire range of radius, from the cluster
center to the tidal radius. A simple isotropic King model, derived from
star counts, reproduces this velocity dispersion profile very well. Its
M/L ratio is about 2.5, which is well within the range of acceptable
values for a globular cluster stellar population. If there is a
significant amount of dark matter in the outer parts of this cluster,
then the velocity ellipsoid in the outer parts must be highly
anisotropic.

5. THE GALACTIC DARK HALO

From the galactic rotation curve, it is possible to estimate the
parameters of the galactic dark halo, just as for other spirals. We can
also use the kinematics of high velocity stars in the solar
neighborhood, the M31/Galaxy timing arguments, and the properties of the
outer spheroidal component to put additional constraints on the
parameters of the galactic dark halo.

5.1 The Rotation Curve

The rotation curve of the Galaxy can be used to estimate the parameters
of its dark halo, using procedures similar to those described in
Sections 3 and 3. However, there are some extra difficulties. (i) The
galactic rotation curve is not as well determined in the outer regions
as it is for some other spirals. (ii) The structure of the luminous
component (eg the scale length of the disk) is also not as well
determined. Several groups have recently constructed models to estimate
the dark matter content. Schmidt's (1985) model is simple and

illustrative. He uses three components to model the galactic rotation
curve: a bulge, an exponential disk with a scale length of 3.5 kpc, and
a dark halo with density distribution $\rho = \rho_0 (1 + r^2/a^2)^{-1}$. The density
scale of the disk (which for other galaxies was given by the M/L ratio)
comes in here through the local surface density Σ_0 at the sun. The model
parameters are not tightly constrained. For example, for $\Sigma_0 = 50\ M_\odot$
pc^{-2}, the halo scale length a = 4.6 kpc and the local density of the
halo is 0.010 $M_\odot\ pc^{-3}$; for $\Sigma_0 = 65\ M_\odot\ pc^{-2}$, a = 6.5 kpc and the local
density of the halo is 0.004 $M_\odot\ pc^{-3}$.

To reproduce the observed rotation curve with the potential field
of the luminous components alone would require unrealistic parameters,
such as a disk scale length of 8 kpc and $\Sigma_0 = 200\ M_\odot\ pc^{-3}$. From the
galactic rotation curve, it seems very likely that the Galaxy has a dark
halo, but it is probably too early to expect accurate values of the
parameters for the galactic dark halo from this approach.

5.2 The Escape Velocity

High velocity stars passing through the solar neighborhood have
velocities of up to at least 500 km s^{-1} relative to a nonrotating frame.
If these stars are bound to the Galaxy, they give some information about
the properties of the dark halo. For example, assume the galactic
rotation curve is flat (with velocity V) out to some radius R_m, and
Keplerian beyond R_m. If $v_\odot$ is the velocity of a star in the solar
neighborhood, relative to a nonrotating frame, and $R_\odot$ is the
galactocentric distance of the sun, then a star that just escapes from
the Galaxy has

$$v_\odot^2 = 2V^2 [1 + \ln(R_m/R_\odot)].$$

Therefore R_m is at least 40 kpc and the total mass of the Galaxy is then
at least $5.10^{11}\ M_\odot$. See Carney's paper in this volume for a more
extensive discussion of this work.

Such estimates are probably lower limits: we do not yet fully
understand the dynamics of the population of high velocity stars, and it
may be that all the high velocity stars in the solar neighborhood are
firmly bound to the Galaxy.

5.3 Timing

The M31/Galaxy timing arguments (eg. Gunn, 1974) give a galactic mass of
about $10^{12}\ M_\odot$, if the M/L ratios of the Galaxy and of M31 are similar.
If the mass is so large, then the flat rotation curve of the Galaxy must
continue out to radii of 80 to 100 kpc.

5.4 The Isothermal Spheroidal Component

Velocity dispersions have now been measured for many classes of objects
belonging to the spheroidal component (RR Lyrae stars, metal weak

giants, globular clusters, M stars), with galactocentric distances ranging from near zero up to about 60 kpc. The velocity dispersion of the spheroidal component is remarkably constant over this entire range in radius, at about 120 km s^{-1} (see Freeman 1985). The density distribution for the spheroidal component is observed to follow an $r^{-3.5}$ law (see for example Zinn (1985). Then, if the velocity ellipsoid is isotropic, the galactic mass within 60 kpc is about 7.10^{11} $M_{\odot}$. This agrees well with the value that would be estimated from the flat rotation curve out to this distance (see section 5.3).

5.5 Conclusion

It seems very likely that the galactic dark halo extends out to about 50-100 kpc and that its mass is $(5-10).10^{11}$ $M_{\odot}$. We can estimate the ratio of dark to luminous mass out to the Holmberg radius, for comparison with the other systems given in Table I. If we take the disk scale length as 4 kpc and the local column density as 65 $M_{\odot}$ pc^{-2}, then the luminous mass (including the bulge and the dark matter associated with the disk) is about 7.10^{10} $M_{\odot}$. Out to the Holmberg radius, the (dark + luminous) mass is about 18.10^{10} $M_{\odot}$, so the ratio of dark to luminous mass is about 1.5, which is similar to the values given in Table I. The ratio of the total dark mass to luminous mass could be as large as 15. However this is only a factor of 2 greater than the largest value of dark to luminous mass, out to the limit of the rotation data, for the galaxies given in Table I.

REFERENCES

Bahcall, J.N. 1984. Astrophys.J., **276**, 169.
Bahcall, J.N. and Casertano, S. 1985. Astrophys.J., **293**, L7.
Bahcall, J.N., Schmidt, M. and Soneira, R.M. 1982. Astrophys.J., **258**, L23.
Carignan, C. and Freeman, K.C. 1985. Astrophys.J., in press.
Carignan, C., Sancisi, R. and van Albada, T.S. 1985. Preprint.
Dekel, A. and Shlosman, I. 1983. In Internal Kinematics and Dynamics of Galaxies (IAU Symposium 100), ed Athanassoula, E., (Reidel, Dordrecht), p 187.
Dressler, A. 1979. Astrophys.J., **231**, 659.
Efstathiou, G., Ellis, R.S. and Carter, D. 1982. Mon.Not.R.astr.Soc., **201**, 975.
Freeman, K.C. 1985. Third Asian-Pacific Regional Meeting of the IAU, in press.
Freeman, K.C. 1985. In The Milky Way Galaxy (IAU Symposium 106), eds van Woerden, H. et al, (Reidel, Dordrecht), p 113.
Gunn, J. 1974. Comments Astrophys. Space Sci., **6**, 7.
Illingworth, G.D. 1983. In Internal Kinematics and Dynamics of Galaxies (IAU Symposium 100), ed Athanassoula, E., (Reidel, Dordrecht), p 257.
Kalnajs, A.J. 1983. In Internal Kinematics and Dynamics of Galaxies (IAU Symposium 100), ed Athanassoula, E., (Reidel, Dordrecht), p 87.

Larson, R.B. and Tinsley, B. 1978. Astrophys.J., **219**, 46.

Peebles, P.J.E. 1984. Astrophys.J., **277**, 470.

Quinn, P.J. 1984. Astrophys.J., **279**, 596.

Schmidt, M. 1985. In The Milky Way Galaxy (IAU Symposium 106), eds van
    Woerden, H. et al, (Reidel, Dordrecht), p 75.

Sparke, L. 1985. Astrophys.J., **280**, 117.

Tonry, J.L. 1983. Astrophys.J., **266**, 58.

Toomre, A. 1981. In The Structure and Evolution of Normal Galaxies, eds
    Fall, S.M. and Lynden-Bell, D., Cambridge University Press, p 111.

Toomre, A. 1983. In Internal Kinematics and Dynamics of Galaxies (IAU
    Symposium 100), ed Athanassoula, E., (Reidel, Dordrecht), p 177.

van Albada, T.S., Bahcall, J.N., Begeman, K. and Sancisi, R. 1985.
    Preprint.

van der Kruit, P. and Freeman, K.C. 1984. Astrophys.J., **278**, 81.

van der Kruit, P. and Shostak, S. 1983. In Internal Kinematics and
    Dynamics of Galaxies (IAU Symposium 100), ed Athanassoula, E.,
    (Reidel, Dordrecht), p 69.

Zinn, R. 1985. Astrophys.J., **293**, 424.

DISCUSSION

SCHECHTER: Could you elaborate on your assumption that the disk
mass-to-light ratio is constant as a function of radius? Isn't it
curious that disks are supposed to provide all of the observed mass
over the radius range in which they have roughly constant angular
velocity, while halos are needed only where disks cease to have
constant angular velocity?

FREEMAN: You know from the vertical equilibrium constraints that disk
M/L ratios are unlikely to be much less than 1, even in very blue
systems. Sometimes this forces you to make a maximum-disk model.
Otherwise, I can't say more than I've already said. The rotation-curve
decomposition procedure is not yet robust.

FABER: How would it affect the swing amplifier criterion if the spiral
structure were driven by a bar?

ATHANASSOULA: The swing amplifier criterion tells us only whether the
disk can respond to forcing, it doesn't distinguish between different
kinds of forcing. The point we are making is that the properties of a
galaxy (disk surface density, halo mass, velocity dispersion, etc.)
must be such as to allow the spiral structure that we see. This is
what the swing amplifier criterion tells us.

FREEMAN: In particular, you won't get spiral arms if Q is too large.

RICHSTONE: Your maximum-disk models show that Vera Rubin's statement
that dark matter contributes to g at all radii need not be correct.
They do not show that it is not correct. How can one do better?

FREEMAN: One can look at the stellar dynamics of bulges in early-type
systems. Velocity dispersions and stellar rotation measurements give
M/L ≈ 7 for bulges, which suggests that they are self-gravitating.
Also, recall that the vertical equilibrium of disks (Bahcall, van der
Kruit and associates) implies that disk M/L values are in the range 3 -
6. This puts a lower limit on the contribution of the disk to g.

E. TURNER: It would appear easier to understand the "conspiracy" of
the disk, bulge, and halo to produce flat and relatively featureless
rotation curves if these three components formed more or less at once
and from the same type of material (i.e., baryons).

FREEMAN: Maybe it would not make too much difference if everything
happens adiabatically, in whatever order.

RUBIN: I would like to urge you to tabulate values not only for the
model including the maximum disk, but also for the model including the
maximum halo, whenever both can be determined. As long as there is a
question as to which fit is most plausible, it will be valuable to know
the range of permitted halo parameters.

GUNN: I think that it is probably impossible to make a maximum halo
model, because all of these models assume some ad hoc form for the halo
density distribution. N-body simulations indicate that the central
concentration you get in a halo is critically dependent on how much
substructure there was initially. The amount of substructure was
certainly stochastically variable. So I don't think one knows a priori
what to expect for the form of the halo density distribution.

FABER: Yes. And if the dissipative infall of baryons further perturbs
the halo, then until we have a good theory for that process, we don't
know the shape of the halo density distribution today.

FREEMAN: If the halo is adiabatic, we can get a long way without
needing to know exactly how the baryons fell in. If we then make
self-consistent bulge/disk/halo models, I think we will be one step
closer to reality than we are now.

GUNN: We can't really make such models until we know more about halos.

OSTRIKER: In your decompositions, you stress the galaxies with small
bulges, and you find that things fit nicely with normal M/L ratios.
Did you look at the larger-bulge systems? Can you still model them
with normal M/L ratios? Or do you need very low values, maybe even
halos with negative masses.

FREEMAN: The cases of large bulges that I have looked at are from the
work of Athanassoula and Bosma. There the bulge M/L ratios had very
normal values, M/L ~ 4 - 5. They certainly found no negative values.

SELLWOOD: Are the disks resulting from these decompositions everywhere
locally gravitationally stable?

ATHANASSOULA: Yes.

SCHECHTER: Was it a matter of choice that your bulge-free spirals all
had very low luminosities or did you have difficulty finding high-
luminosity systems?

FREEMAN: The choice was to avoid systems with bulges. In the early
work that Carignan and I did, we had to choose nearby galaxies so that
we could make single-dish HI observations. That meant that we worked
on galaxies in the Sculptor group, and these have low luminosities.

LAKE: Jacqueline van Gorkom, Bob Schommer and I have measured an HI
rotation curve for an elliptical galaxy. The galaxy is about 0.5 mag
fainter than L*, has B-V = 0.9 and a de Vaucouleurs-law profile. In
other words, it is a perfectly normal elliptical. The rotation curve
is flat out to four or five times the effective radius. The value of
M/L thus changes by a factor of four over the radius range of the
observations.

HALO PARAMETERS OF SPIRAL GALAXIES

E. Athanassoula, A. Bosma, S. Papaioannou
Observatoire de Marseille

We have made a rotation curve analysis of a sample of spiral galaxies
for which both photometric and kinematical data of reasonable quality
are available in the literature. From the photometric radial luminosity
profile, assuming constant mass-to-light ratios for bulge and disk
separately, we calculate a rotation curve due to the luminous mass in a
galaxy. Comparison with the observed rotation velocities allows us to
derive a halo rotation curve, which can be used to derive characteristic
halo parameters. The decomposition into luminous and dark matter is not
unique, with as extremes a "minimum" disk (M/L = 0) and a "maximum"
disk (M/L as high as possible while requiring a realistic halo mass
distribution without a hollow core).

We have narrowed the range of M/L-values by introducing spiral structure
constraints. If the ratio of disk mass to halo mass is low the disk
may not be responsive enough to allow two-armed spiral structure. Thus
the requirement that the swing amplifier be active for m = 2 yields a
lower limit for the M/L ratio in the majority of spiral galaxies. On
the other hand, if the disk is too massive, a sizeable m = 1 component
is allowed, which in nearly all of our galaxies is incompatible with
the observed morphology. These considerations restrict the allowed
M/L-values for the disk to a range of 0.3 dex.

Comparison of our results with models of galactic evolution indicates
that our solutions for the maximum disk, with the m = 1 component
inhibited, are in agreement with current knowledge of present and past
integrated star formation rates. Their M/L-ratios as function of B-V
colour are in reasonable agreement with stellar population models. On
the other hand, the solutions involving the lower limit to the M/L
ratio lead to unreasonably high gas fractions in Sc galaxies of 50% or
more, depending on the adopted conversion for CO to H_2. Hence we consider
the maximum disk solutions, with the m = 1 component inhibited, as the
most adequate for further study of halo parameters.

In general, the ratio of halo core radius to optical disk radius is
smaller for Sa galaxies than for Sc galaxies.

J. Kormendy and G. R. Knapp (eds.), Dark Matter in the Universe, 133.

DYNAMICAL INTERACTION BETWEEN THE DIFFERENT COMPONENTS OF SO-Sd GALAXIES.

BACON R., MARTINET L.
Observatoire de Genève

Using both photometric and gas kinematic data of three galaxies of different type (NGC 7814 SO, NGC 2841 Sb, NGC 300 Sd), we estimate the density of the dark halo and visible components which gives the best fit with the observed rotation curves. We then compute the ratio of each component force to the total force as function of radius. Finally, we derive from the potential energy tensor the quantitative effect of each component on bulge, disk and halo dynamics.

1. MASS MODELS. Bulge and disk are oblate spheroid with spatial density inferred from the deprojected $r^{1/4}$ law and the exponential law, respectively. The dark halo is assumed spherical with a spatial density $\nu(r) = \nu_o(1+r/r_h)^{-3.05}$.

2. FITTING PROCEDURE. There are 4 free parameters to fit the observed rotation curves: the blue mass-to-light ratios of the bulge and the disk and the two parameters of the halo density. For NGC 7814 and NGC 2841 we use the stellar velocity rotation and dispersion for the computation of the bulge's mass-to-light ratio via the tensor virial formulae.

3. POTENTIAL ENERGY TENSOR. For each potential component U^i (i.e. bulge, disk and halo potential) we compute

$$(W_z)^i_j = \int \nu_j \, z \frac{\partial U^i}{\partial z} \, dV \quad \text{and} \quad (W_{\tilde\omega})^i_j = \int \nu_j \, \tilde\omega \, \frac{\partial U^i}{\partial \tilde\omega} \, dV$$

We define

$$(Q_z)^i_j = \frac{(W_z)^i_j}{\sum\limits_i (W_z)^i_j} \quad \text{and} \quad (Q_{\tilde\omega})^i_j = \frac{(W_{\tilde\omega})^i_j}{\sum\limits_i (W_{\tilde\omega})^i_j}$$

which measure the relative effect of the potential due to the i population on the j population.

4. CONCLUSION. These quantitative estimations of the dynamical interactions between dark halo, bulge and disc in three galaxies of types SO, Sb and Sd suggest:
1. Bulges of early-type spirals are probably self-gravitating.
2. SO-Sb disk dynamics are greatly affected by the bulge potential, whereas Sd disks are self-gravitating. The halo potential does not seem to play a key role in the stability of disk against bar-like perturbation.
3. Halos are very massive and almost self-gravitating.

J. Kormendy and G. R. Knapp (eds.), Dark Matter in the Universe, 134.

IS THE RATIO OF DARK-TO-LUMINOUS MATTER A FUNCTION OF GALAXY MASS AND/OR LUMINOSITY?

C. Carignan
Département de physique and
Observatoire du Mont Mégantic,
Université de Montréal,
C.P. 6128, succ. "A",
Montréal, Qué., CANADA H3C 3J7

ABSTRACT. The study of the ratio of dark-to-luminous matter in spiral galaxies has been the subject of several recent studies (Bahcall 1983; Casertano 1983; Carignan and Freeman 1985; van Albada et al. 1985; Carignan 1985). These studies have been possible because of the large number of high sensitivity HI observations which became available in the last few years, allowing to probe the halo potential to very large galactocentric distances. In the case of NGC 3198 (van Albada et al. 1985), it was even possible to derive the rotation curve out to 11 disc scale lengths. One of the important questions, forming the motivation for this type of work, is whether the ratio of dark-to-luminous matter is a function of galaxy mass and/or luminosity.

Mass models for late-type spirals (NGC 6946, 3198, 300, 3109, UGC 2259) covering the luminosity range $- 20 \leqslant M_B \leqslant - 16$ are presented.

The luminous disk is calculated using the luminosity profile with constant (M/L_B), except for UGC 2259 where an exponential disk was used since no surface photometry is yet available. The dark halo is represented by an isothermal sphere potential.

No real trend can be seen with ratios of dark-to-luminous matter varying from 0.6 to 1.8 ($M_H/M_D = 1.1 \pm 0.5$) at the Holmberg radius.

This strengthens earlier suggestions (Carignan and Freeman 1985; Carignan 1985; Bahcall and Casertano 1985; Carignan et al. 1985) that the ratio of dark-to-luminous matter appears to be independent of galaxy mass and/or luminosity.

References
Albada, T.S. van, Bahcall, J.N., Begeman, K. and Sancisi, R. 1985, submitted to Ap. J.
Bahcall, J.N. 1983, Ap. J., _267_, 52.
Bahcall, J.N. and Casertano,S. 1985, Preprint.
Carignan, C. 1985, Ap. J., _298_, in press.
Carignan, C. and Freeman, K.C. 1985, Ap. J., in press.
Carignan, S., Sancisi, R. and Albada, T.S. van 1985, In preparation.
Casertano, S. 1983, M.N.R.A.S.,. _203_, 735.

135

J. Kormendy and G. R. Knapp (eds.), Dark Matter in the Universe, 135.

SOME REGULARITIES IN THE MISSING MASS PROBLEM

S. Casertano and J. N. Bahcall
Institute for Advanced Study, Princeton, NJ 08540

We discuss available information on the distribution of luminous and dark matter in eight galaxies. The galaxies have been chosen according to the following criteria: 1) existence of a good rotation curve, extending well beyond the optical radius; 2) a mass model has been published; 3) valuable constraints can be put on the amount of dark matter *inside the optical radius*. A full description of the data and reduction procedures is in Bahcall and Casertano (1985).

For each galaxy, the mass models proposed by the original authors have been used. These models are based on a variety of methods. All assume $M \propto L$ for the disk. For some of the models the mass-to-light ratio was chosen in advance: for others it is determined by the "maximum disk" assumption. In two cases (the Milky Way and NGC 5907) independent dynamical evidence supports the value chosen for M/L. We feel that this is the best information available at present. However, the mass models are mostly based on plausibility arguments, and other models, with different disk and halo masses, are possible.

The properties of the luminous and dark matter in these galaxies, when compared to each other, exhibit suggestive regularities. Perhaps the most striking is in the value of the ratio $M_{\text{halo}}/M_{\text{disk}}$ within the optical radius, which is nearly constant ($\pm 30\%$) and very close to unity, although disk and halo masses separately vary by a factor of 100. This close correlation between the properties of luminous and dark matter is borne out by the fact that rotation curves do *not* show any major features at the transition between disk-dominated and halo-dominated regions.

One possible interpretation of the observed regularities may be the existence of some process that tends to equalize the amount of dissipational and dissipationless material inside some typical scale. If so, the observed regularities will provide valuable constraints on processes of galaxy formation.

An alternative explanation is that luminous and dark matter are actually different manifestations of the same component; in other words, that the nonluminous matter is baryonic. This is consistent with the absence of a break in the rotation curves and the constant disk-to-halo mass ratio. In addition, it is known that in the solar neighborhood at least some of the dark matter must be in a dissipative, disklike component.

For this explanation to work, the unobserved material must be in the form of small-mass compact objects (neutron stars, white dwarfs, etc.), brown dwarfs, or very massive black holes.

REFERENCE

Bahcall, J. N. and Casertano, S. 1985, *Ap. J. Letters*, **293**, L7.

J. Kormendy and G. R. Knapp (eds.), Dark Matter in the Universe, 136.
© *1987 by the IAU.*

MEASUREMENT OF DARK MATTER IN SPIRAL GALAXIES

Stephen M. Kent
Harvard-Smithsonian Center for Astrophysics

Luminosity profiles and rotation curves for 37 Sb and Sc galaxies have been combined to derive mass/light ratios for the stellar component and scale parameters for a dark halo component.

Galaxies were selected to have optical rotation curves measured by Rubin et al. (1985). Surface photometry was obtained using a CCD on the Whipple Observatory 61-cm telescope. Major and minor axis profiles were derived and, where necessary, decomposed into bulge and disk components.

The rotation curves were modeled as the sum of contributions from the bulge, disk, and a dark halo with a density law $\rho = \sigma^2/2\pi G(a^2 + r^2)$. The parameter σ was set to match the observed asymptotic velocity. A least-squares fit then yielded 3 parameters: the bulge and disk mass/light ratios and the scale radius a. The solutions were not always well constrained; usually many different combinations of parameters gave equally good fits to the observed rotation curves.

Out of 37 galaxies, 10 could be fit with no halo, 12 had halo-to-stellar mass ratios $M(H)/M(B+D)<1$, 14 had $M(H)/M(B+D)>1$, and 1 could not be fit. The mean M/L's were 3.7 for the bulge, 1.8 for the disk, and 4.7 for the total galaxy ($H_o = 50$, photometry in a Gunn r bandpass with no extinction corrections). The individual parameters showed large scatter; e.g. the scale radius a ranges from 1 to 30 kpc.

A typical decomposition and fit are shown in the figure.

REFERENCE

Rubin, V. C., Burstein, D., Ford, W. K., and Thonnard, N. 1985, Ap. J., **289**, 81.

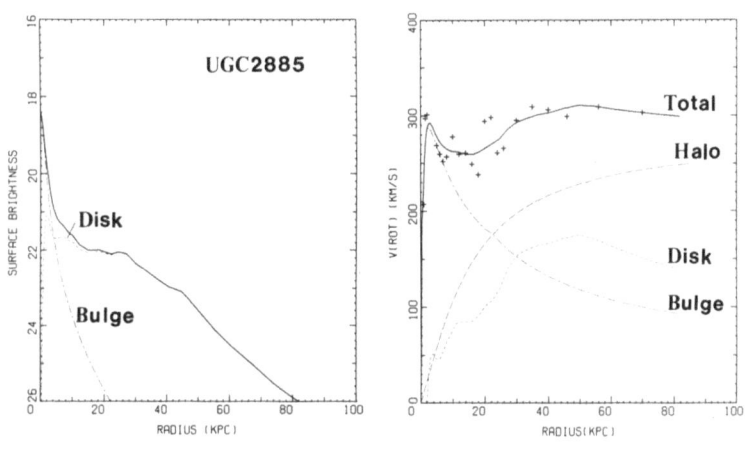

J. Kormendy and G. R. Knapp (eds.), Dark Matter in the Universe, 137.

MASS-TO-LIGHT RATIOS OF SPIRAL GALAXIES

J. Patricia Vader
Yale University Observatory

Mass-to-light ratios $M/L_{H-0.5}$ and M/L_B are plotted against color
$B-H_{-0.5}$ in Fig. 1 for 82 nearby spirals from the catalog of Aaronson et
al. (1982, Ap.J. Suppl. 50, 241), with B_T magnitudes from the RC2, $|b|$
> 20°, 45° $\leq$ i $\leq$ 80°, and HI mass estimates. Total masses and infrared
$H_{-0.5}$ magnitudes are measured within the blue isophotal radii R_{25} and
$R_{25}73$, respectively, which depend on galaxy color. This color bias is
corrected for by replacing R_{25} by R'_{25}, the radius a galaxy would have
at a standard color $B-H_{-0.5} = 2.17$. Stellar masses M_* and L_{HC}
luminosities within R'_{25} are obtained by substracting twice the HI
mass and by extrapolation, respectively. Corrected ratios M_*/L_{HC} and
M_*/L_B versus corrected color $B-HC$ are shown in Fig. 1 together with
theoretical model predictions. The corrected observed ratios are
systematically larger for bluer galaxies than predicted so that bluer
spirals seem to have relatively more massive halos, in agreement with
earlier results (Tinsley, B.M. 1981 M.N.R.A.S. 194, 63; Vader, J. P.
1984, in Formation and Evolution of Galaxies and Large Structures in
the Universe, eds. J. Audouze and J. T. Thanh Van, p. 227).

Fig. 1

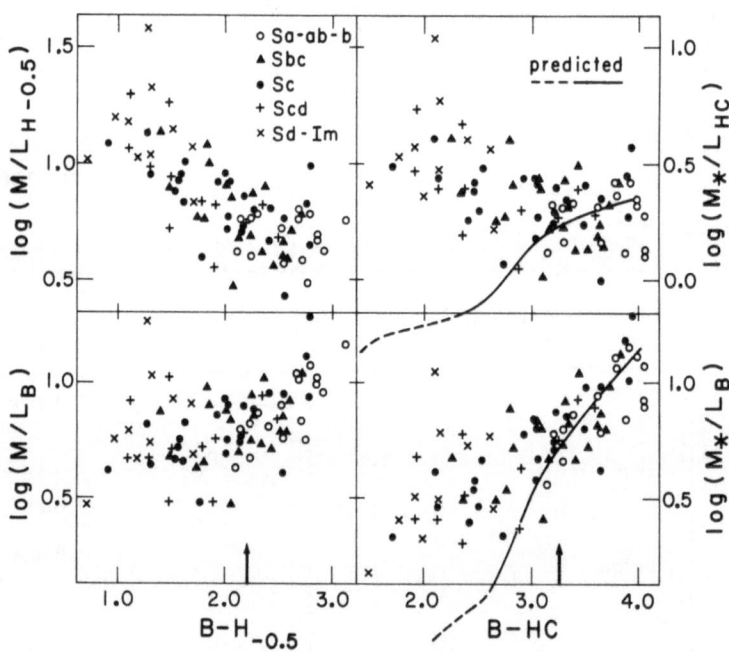

J. Kormendy and G. R. Knapp (eds.), Dark Matter in the Universe, 138.

DARK MATTER IN DWARF GALAXIES

John Kormendy
Dominion Astrophysical Observatory
5071 W. Saanich Road
Victoria, B. C. V8X 4M6
Canada

ABSTRACT

This paper reviews the observational evidence for dark matter (DM) in dwarf spiral (dS) and dwarf spheroidal (dE) galaxies. The most secure detection of DM in dwarf galaxies is given by HI rotation curves. They provide estimates of DM halo parameters, i. e., isothermal core radii r_c, central densities ρ_0 and one-dimensional velocity dispersions σ. The smallest DM halo measured so far is in DDO 127 ($M_B = -14.5$ for $H_0 = 75$ km s^{-1} Mpc^{-1}, $r_c \simeq 2.3$ kpc, $\sigma \simeq 27$ km s^{-1}). If this halo is made of neutrinos of mass m_ν, then phase-space constraints imply that $m_\nu > 110$ eV. This is difficult to reconcile with cosmological upper limits giving $\Omega \leq 1$. Ultimately, dE galaxies will provide the strongest constraints on DM in dwarf galaxies; a detailed look at present results shows that they are not yet conclusive.

1. INTRODUCTION

An effective test of any theoretical paradigm is to confront it with observations of extreme objects (Kuhn 1970). It is therefore important to look for dark matter in the smallest galaxies possible. Such observations give leverage to correlations of DM properties with mass. The results have implications for a variety of problems of galaxy formation and evolution. I have time to discuss only a few questions:

(1) Is the ratio of luminous to dark matter a function of galaxy mass? In particular, is there a lower mass limit below which galaxies do not contain DM? If even the smallest visible galaxies have substantial halos, do there exist galaxies which are completely dark?

(2) What are the core radii r_c, central densities ρ_0 and one-dimensional velocity dispersions σ of DM halos? Are the mass distributions isothermal? How do r_c, ρ_0 and σ correlate with galaxy mass and luminosity?

(3) What are the parameters of the smallest DM halos? These provide constraints on the kinds of sub-atomic particles that could make up DM (§4).

(4) Is the visible matter in the smallest dwarfs not self-gravitating? Does this turn off star formation, creating a natural lower limit to the sizes of small galaxies?

J. Kormendy and G. R. Knapp (eds.), Dark Matter in the Universe, 139–152.

We cannot yet answer these questions. It is only now becoming possible to address them observationally. Some preliminary results are discussed below. This review concentrates on the observations, and stresses the large uncertainties that remain. The smallest galaxies are very difficult to measure and interpret. At present we have estimates of halo parameters only to $M_B \simeq -14.5$, while even the detection of DM in fainter galaxies is uncertain.

Sections 2 and 3 discuss rotation curves of dS galaxies and velocity dispersions of dE galaxies. These appear to be measuring the same kind of object. Observations show that there is a closer kinship between dS and dE galaxies than between dE galaxies and giant ellipticals (see Kormendy 1985 and Aaronson 1986 for reviews). This has led to the suggestion that dEs are dS+Im galaxies that have lost their gas or turned it all into stars. I will therefore compare DM distributions in dS and dE galaxies without regard to optical morphology.

2. HI ROTATION CURVES OF dS + Im GALAXIES

The most clearcut results on DM distributions in small galaxies come from HI rotation curves. The techniques are discussed in this volume by Sancisi and van Albada (1986) and by Freeman (1986). The rotation curve $V(r)$ is measured to the largest possible radius R_{max}. A disk rotation curve is then estimated from the surface brightness distribution assuming that the mass-to-light ratio M/L is constant. This is subtracted in quadrature from $V(r)$, and a halo model (e. g., an isothermal) is fitted to the remainder to get the halo parameters: the maximum or asymptotic rotation velocity V_{max}, $\sigma = V_{max}/\sqrt{2}$, r_c and ρ_0.

Small, late-type galaxies are especially suitable for this procedure. They are relatively easy to interpret: they contain only a (usually exponential) disk and perhaps a halo, but not a bulge. Rotation-curve decomposition is uncertain even in this simple case, and very poorly constrained if there are more than two components. It is also important that dwarf spirals are often rich in HI, even at large radii. Since $V(r)$ for an exponential reaches a maximum at $2.2\alpha^{-1}, \alpha^{-1}$ the scale length, a DM halo can definitely be detected and its r_c measured only if $R_{max}/\alpha^{-1} \gg 1$. *It is worth a great deal of effort to reach the $V = $ constant part of the rotation curve.* Another advantage of dS galaxies is that they are numerous, so that nearby examples with suitable inclination, regularity and large R_{max} can be found. And the assumption that the gas is in circular motion is probably better than the assumption that the velocity distribution is isothermal in dwarf spheroidals.

However, there are also disadvantages and problems. Many are intrinsic to the method. The values of M/L and the assumption that $M/L \neq M/L(r)$ are uncertain. Also, the mass distribution of the disk must be corrected for the contributions of HI and H_2. Internal absorption corrections are poorly known and largest in nearly edge-on galaxies, which are otherwise attractive beause of their small velocity projection corrections. HI warps make the derivation of rotation curves uncertain. And there is always the problem that we do not know *a priori* what to use for the density distribution of the halo. Any error in the assumption that $M/L = $

constant translates directly into an error in the derived halo density distribution. Also, published models are not self-consistent: they ignore the effects of the visible and dark matter on each other. However, this problem is least severe in the smallest galaxies, which contain relatively little visible matter to pull on the dark matter.

Other problems are specific to small galaxies. As L decreases, the corrections for M_{HI} and M_{H_2} grow larger. Also, the rotation velocity approaches zero as L approaches interestingly small values $(M_B \gtrsim -14)$. When $V \lesssim \sigma_{HI}$ (the velocity dispersion of the gas), it needs to be corrected for pressure support. Also, the visible galaxy is then not flat, by amounts described by the usual $V_{max}/\sigma_{HI} - \epsilon$ diagram (Illingworth 1977). And the HI distribution and velocity field can be irregular. Finally, as luminosity decreases, the ratio of galaxy size to halo r_c decreases, so it is more and more difficult to reach the $V = $ constant part of the rotation curve. As a result of these problems, rotation curve measurement and decomposition become very difficult at $M_B \gtrsim -15$, and impossible in principle at some $M_B \gtrsim -13$.

There are few galaxies with $M_B \gtrsim -18$ for which adequate rotation curves are available. Several with $-18 \lesssim M_B \lesssim -17$ have been measured by Carignan and Freeman (1985), and by Carignan, Sancisi and van Albada (1985). Ken Freeman (1986) has already discussed this work; parameters of the halos are included in Table 1, below. Published data on several smaller galaxies show that they contain DM halos, but they have not been modelled to derive halo parameters. DDO 154 (Krumm and Burstein 1984) and NGC 6822 (Gottesman and Weliachew 1977) have rotation curves which continue to rise, respectively, to 3.8 and 3.3 times the radius r_{25} of the 25 B mag arcsec^{-2} isophote. Since they are very extended in HI, they would be suitable for modelling. They have $M_B = -16.5$ and -15.6, respectively.

The faintest dwarf spiral measured so far is DDO 127 $(M_B = -14.5)$. Surface photometry has been obtained by Souviron, Kormendy and Bosma (1986), and a Westerbork HI velocity field has been measured and modelled by Bosma, Kormendy and Souviron (1986). The galaxy turns out to be exceptionally well suited for halo measurement. It is a regular SABm spiral with a well-defined inclination of 54°. The HI distribution and velocity field are also regular, and are measured out to 1.7 r_{25}. The rotation curve rises almost linearly past r_{25} and then begins to level off. The maximum rotation velocity is ~ 34 km s^{-1}. The disk is exponential, with a scale length of only $\alpha^{-1} = 17'' = 0.4$ kpc. Since $R_{max} = 7.5\alpha^{-1}$, and since $V(r)$ is still rising slightly at R_{max}, it is clear that a substantial DM halo is present. DDO 127 is particularly interesting because a *"maximum disk" model does not fit the rotation curve*. The disk scale length is so short that the disk alone cannot account for most of the rising part of the rotation curve no matter what the M/L. Most of the rotation curve is apparently due to the DM; in particular, the minimum and maximum allowed amounts of disk are not very different. Therefore the halo parameters are relatively well determined despite the usual large uncertainties. Preliminary modelling gives the DM parameters listed in Table 1; these results include corrections for HI mass but not yet for σ_{HI}. (Beam smearing effects are small.) DDO 127 is the smallest galaxy in which DM is securely detected and in which its parameters have been estimated.

Table 1 summarizes DM parameters measured by assuming that halos are isothermal (van Albada *et al.* 1985; Carignan and Freeman 1985; Carignan, Sancisi and van Albada 1985; Bosma, Kormendy and Souviron 1986). For each galaxy I give both minimum- and maximum-halo results. Minimum-halo models use the disk to explain as much of the inner rotation curve as possible; usually they fit out to or slightly beyond the radius where $V(r)$ becomes nearly constant. Maximum-halo models use the halo to fit essentially all of the rotation curve (i. e., $M/L \sim 0$ for the disk). The true halo parameters are bracketed by these results.

Table 1
Parameters of DM Halos in Dwarf Spiral Galaxies

Galaxy	Type	M_B	σ	Min. Halo		Max. Halo	
				r_c	ρ_0	r_c	ρ_0
			$\frac{km}{s}$	kpc	$\frac{M_\odot}{pc^3}$	kpc	$\frac{M_\odot}{pc^3}$
NGC 3198	SBc	-19.8	105	12.	0.008	2.8	0.21
NGC 247	SAd	-18.7	90	22.	0.003	5.6	0.043
NGC 300	SAd	-18.1	60	12.	0.0042	3.2	0.060
NGC 3109	SBm	-17.4	40	10.5	0.0024	5.0	0.011
UGC 2259	SBdm	-17.1	57	8.7	0.0073	2.1	0.12
Mean					0.0050		0.089
Dispersion					0.0025		0.078
DDO 127	SABm	-14.5	27	2.3	0.023	1.8	0.044

The correlation of σ with M_B (corrected for internal absorption) is the Tully-Fisher (1977) relation. Otherwise, the sample in Table 1 is too small to show any new regularities. The best-constrained halo parameter is the central density. There is surprisingly little variation in ρ_0 from galaxy to galaxy, despite a range of 100 in L (cf. Bahcall and Casertano 1985). The values of ρ_0 are ~ 18 times larger for the maximum- than the minimum-halo case; the means and dispersions are given in the table. The truth must lie between these cases. Minimum-halo disks are often bar-unstable; not all of the galaxies are barred. Maximum-halo models are implausible because density waves and other self-gravitating structure are then impossible. And, indeed, the better-constrained halo parameters for DDO 127 are intermediate between the minimum- and maximum-disk averages for the other galaxies. The results suggest that there is a fairly canonical central density for DM halos in dwarf spiral galaxies, $\rho_0 = 0.01 - 0.05$ M$_\odot$ pc^{-3}. This value is compared with results for dE galaxies in the next section.

3. THE SEARCH FOR DARK MATTER IN DWARF SPHEROIDAL GALAXIES

For galaxies much fainter than $M_B \simeq -13$, the rotation velocity is less than the velocity dispersion of the gas (e. g., M81 dwA, with $M_B = -11.0$; Sargent, Sancisi and Lo 1983). Then the mass distribution cannot be derived from the rotation curve. A number of other techniques have been tried, including measurements of the velocity differences in binary pairs of dS+Im galaxies (Lake and Schommer 1984), and tidal radii of dE galaxies (Faber and Lin 1983). These have two advantages – they measure almost total masses, and they can easily be applied to large samples of galaxies. But they are plagued by technical problems, so they are not definitive. At present, the best way to look for DM in the smallest galaxies is to measure the dispersion of individual stellar velocities in dwarf spheroidals, and use the virial theorem. This was pioneered in an important series of measurements by Aaronson (1983) and Aaronson and Olszewski (1986).

It is convenient and to some extent plausible to adopt isothermal or King (1966) models for the light and mass distributions. This amounts to a particular choice of the geometric constants omitted in the simplest form of the virial theorem, $M \sim (3\sigma^2)r/G$. In the following, I estimate *central volume densities*; this minimizes the dependence on the assumption of a King model. Central light densities I_0, mass densities ρ_0, and mass-to-light ratios ρ_0/I_0 are given by

$$I_0 = \Sigma_0/pr_c; \tag{1}$$

$$\rho_0 = 166\,\frac{\sigma^2}{r_c^2}; \tag{2}$$

$$\frac{\rho_0}{I_0} = 166\,p\,\frac{\sigma^2}{\Sigma_0 r_c}, \tag{3}$$

where σ is in km s^{-1}, r_c is in pc, the central surface brightness Σ_0 is in $L_\odot$ pc^{-2}, ρ_0 is in $M_\odot$ pc^{-3}, and the geometric factor p is tabulated in Peterson and King (1975) as a function of the usual concentration index $\log r_t/r_c$.

The above procedure suffers from many complications and problems. (1) Like rotation curve decompositions, the models are not self-consistent. (2) King models may not apply to DM halos. The shapes of rotation curves in slightly larger galaxies could partially test this assumption, and King models do fit the central parts of halos, but the small observable radius range gives little leverage, and a good fit does not in any case guarantee that the velocity distribution is isothermal. (3) We think we know r_c for the visible matter, but we do not know it for the DM. This criticism has been emphasized by Madsen and Epstein (1984) and by Cowsik (1985). Certainly the fact that in spiral galaxies r_c^{DM} is larger than the characteristic radius of the visible matter should make us doubt that the two are equal in dwarf spheroidals. If we underestimate r_c, we overestimate ρ_0 and underestimate the total amount of DM. (4) Dwarf spheroidals have remarkably small central densities ρ_0^{vis} of visible matter. Nevertheless, if $\rho_0^{vis} > \rho_0^{DM}$, a central velocity dispersion tells us nothing about DM. We would then need to measure $\sigma(r)$ out to a radius where the visible and DM densities become comparable. This is beyond present capabilities. The visible matter density will in fact turn out to be too high in Fornax and possibly in other dwarf spheroidals.

The most obvious practical problem is the difficulty of measuring sufficiently accurate radial velocities of faint stars (B $\gtrsim$ 18). In the absence of DM, dwarf spheroidals would have $\sigma \lesssim$ 5 km s^{-1}. The required precision of $\sim$ 1 km s^{-1} takes $\sim$ 2 h per measurement with present techniques. Reducing systematic errors to comparable levels is more difficult. Carbon stars were the first to be observed (Aaronson 1983; Seitzer and Frogel 1985). They are easy to measure, because they are bright and have strong features. But most turn out to be velocity variables because of atmospheric or binary-star motions (Aaronson and Olszewski 1986). An extensive series of measurements by Aaronson and Olszewski (1986) seeks to determine velocities for ordinary K giant stars, and to eliminate binary stars and other velocity variables using multiple observations. Recent observations by McClure *et al.* (1986) provide an independent check of these results. The Dominion Astrophysical Observatory Radial Velocity Scanner (McClure *et al.* 1985) was used at the Coudé spectrograph of the Canada-France-Hawaii Telescope. Table 2 compares our results with those of Aaronson and Olszewski (1986). Here Δt is the difference in epoch of the two measurements; if it is not listed, Aaronson measured the velocity several times. All velocities are in km s^{-1}. The comparison is reassuring. Three of the stars show agreement as good as we can expect; star 536 may be a velocity variable. Also, it is reassuring that Pryor *et al.* (1986) find velocity dispersions of 2 − 3 km s^{-1} in four metal-poor globular clusters. The stars are like those measured in dEs, both in apparent magnitude and in spectral type. Therefore we *can* measure small dispersions when nature chooses to make them. Apparently the accuracy of the measurements is good.

Table 2

Comparison of Aaronson+ and McClure+ Radial Velocities in Draco

Star	McClure+ V_M	Aaronson+ V_A	$V_M - V_A$	Δt (yr)
249	-292.4 ± 1.0	-295.8 ± 1.0	$+3.4 \pm 1.4$	1.9
267	-291.6 ± 1.4	-290.0 ± 0.9	-1.6 ± 1.7	-
536	-306.0 ± 0.7	-301.9 ± 0.7	-4.1 ± 1.0	-
562	-298.3 ± 1.3	-297.2 ± 1.5	-1.1 ± 2.0	1.0

The available velocity dispersion measurements for five dwarf spheroidal galaxies are given in Table 3. Dispersions are not corrected for measuring uncertainties; the estimated error is only that due to the small number N of stars. The Seitzer and Frogel (1985) measurements of C stars have velocity accuracies of $\sim$ 1.3 km s^{-1}. There is no discrimination against binaries. The Aaronson and Olszewski (1986) stars are ordinary giants except for two C stars in Draco. Repeat measurements have allowed the elimination of 5 of 21 stars observed as binaries.

Table 3

Velocity Dispersions in Dwarf Spheroidal Galaxies

Galaxy	σ (km s^{-1})	N	Reference
Fornax	6.4 ± 2.0	5	Seitzer+ 1985
Sculptor	5.8 ± 2.4	3	Seitzer+ 1985
Carina	5.6 ± 1.6	6	Seitzer+ 1985
UMi	$11. \pm 3.$	7	Aaronson+ 1986
Draco	$9. \pm 2.$	9	Aaronson+ 1986

Table 4 lists structural parameters, central mass densities and mass-to-light ratios derived using equations 1 – 3. Sources of photometry are given in Kormendy (1985). The core radii of the light and dark matter are taken to be equal; this assumption is relaxed later. A self-consistent application of the King models requires several corrections to published data; otherwise M/L is systematically underestimated. Published core radii are corrected from the King (1962) definition in terms of a fitting function to the 1966 definition in terms of the dynamical models. Mean rather than major-axis radii are used; Schechter (1980) shows that the inclination correction factor to M/L then averages to 1 for an ensemble of randomly-oriented galaxies. In principle, the velocity dispersion should be corrected (i) for the mean radius of the stars measured in the galaxy, since $\sigma(r)$ decreases with increasing r in a King model, (ii) for projection, and (iii) for the lowering of the Maxwellian distribution assumed in the model. (The measured dispersion is smaller than the parameter σ in equations (2) and (3) by amounts that are tabulated in King 1966.) For a typical concentration $\log r_t/r_c = 0.6$, the cumulative correction to σ is a factor of 1.8. The corrections to σ *have not* been applied in Tables 3 and 4, because it is not clear (see below) that the physical assumptions are valid. If they are applied, ρ_0 and ρ_0/I_0 increase by as much as a factor of 3.

Table 4

Structural Parameters and Mass-To-Light Ratios of Dwarf Spheroidal Galaxies

Galaxy	M_B	$\log \frac{r_t}{r_c}$	μ_0	r_c	Σ_0	I_0	ρ_0	ρ_0/I_0
			$V\mu$	kpc	$\frac{L_\odot}{pc^2}$	$\frac{L_\odot}{pc^3}$	$\frac{M_\odot}{pc^3}$	
Fornax	-12.3	0.77	23.3	0.50	16.2	0.020	0.028 ± 0.018	1.4 ± 0.9
Sculptor	-10.2	0.76	23.9	0.17	9.7	0.035	0.19 ± 0.16	5.5 ± 4.9
Carina	-8.6	0.50	24.9	0.22	4.0	0.014	0.11 ± 0.07	7.8 ± 6.0
UMi	-7.4	0.75	26.1	0.15	1.2	0.005	0.91 ± 0.51	$175. \pm 131.$
Draco	-7.4	0.52	25.4	0.15	2.5	0.013	0.64 ± 0.34	$48. \pm 35.$

Taken at face value, Table 4 implies the following. There is no evidence for DM in Fornax, Sculptor or Carina (Seitzer and Frogel 1985). However, the central mass-to-light ratios in UMi and Draco are very large, suggesting that substantial amounts of DM are present (Aaronson and Olszewski 1986). But the estimated errors in ρ_0 and ρ_0/I_0 are very large. This is mainly due to the small number of velocity measurements and the large uncertainties in Σ_0. I have followed the usual practice of estimating the error in σ^2 using the variance of its distribution function: $\epsilon(\sigma^2) = \sigma^2\sqrt{2/N}$, $\epsilon(\sigma) = \epsilon(\sigma^2)/2\sigma$. Tremaine (1986) points out that this neglects the asymmetry of the χ^2 distribution of $\Sigma(v_i - \bar{v})^2$. A 68% confidence interval for σ^2 is much wider than $\epsilon(\sigma^2)$ on the high-σ side, but no narrower on the low-σ side. I. e., I have underestimated the sampling errors. Also, the formal errors do not take into account the corrections to σ discussed above, the possible inapplicability of King models, the fact that the models are not self-consistent, and any errors resulting from the assumption that r_c is the same for luminous and dark matter. (If dwarf spheroidals have exponential rather than King-model brightness profiles, then the above corrections to σ are not valid, but the geometric factors are reasonably realistic.) I therefore believe that the results are far from definitive.

An additional worry, or else a reason to wonder whether something very interesting is being discovered, is illustrated in Table 5. This compares central mass densities in dwarf spiral and spheroidal galaxies. Also shown are luminous matter densities in the latter. These are given by I_0 and by $M/L = 2$ as in globular clusters or $M/L = 7$ as in old disks (which assumes the existence of DM like that in the local galactic disk). There are a number of implications. First, there is no contradiction: $\rho_0^{vis} \lesssim \rho_0^{dynamical}$ in all the galaxies. Also, all could have DM halos like those in dwarf spirals (if σ has been overestimated in UMi and Draco). But in Fornax and maybe in Sculptor and Carina, $\rho_0^{DM} \lesssim \rho_0^{vis}$. Then we cannot find DM by measuring *central* velocity dispersions. No one tries to look for DM in galaxies like M31 by measuring the nuclear velocity dispersion. We were so impressed by the low surface brightnesses in dwarf spheroidals that we thought their densities would be "low enough". Evidently this is not the case. Then it is difficult to find DM in Fornax, Sculptor and Carina.

The most important implication of Table 5 involves UMi and Draco. Their dynamical central densities are shockingly high. They are higher than in dwarf spirals, from which we suspect they have evolved. Indeed, these are the highest central DM densities seen in any galaxy so far. In fact, $\rho_0^{dynamical} \sim 10^2\rho_0^{vis}$, so the luminous matter could not be self-gravitating. It is difficult to understand how stars could form without some self-gravity. (Of course, ρ_0^{vis} could have been larger in the past, when – if – dEs were still dSs.) Now, ρ_0^{DM} can be reduced by making $r_c^{DM} \gg r_c^{vis}$. But it is easy to show that as long as the DM remains gravitationally dominant, ρ_0^{DM} can be reduced by no more than a factor of 2 without affecting the visible structure. Then the above discussion remains valid. Only if $\rho_0^{DM} \lesssim \rho_0^{vis}$ (and r_c^{DM} is very large) do these problems disappear. Then we would conclude that DM halos are similar in dE and dS galaxies. If it is appropriate to correct σ upward, these conclusions become much stronger. I therefore suspect that σ has been over-

Table 5

Comparison of ρ_o in dE and dS Galaxies

Galaxy	ρ_o^{vis} $\frac{M_\odot}{pc^3}$		ρ_o $\frac{M_\odot}{pc^3}$	$\rho_o^{dS,min}$ $\frac{M_\odot}{pc^3}$	$\rho_o^{dS,max}$ $\frac{M_\odot}{pc^3}$
	$\frac{M}{L}=2$	$\frac{M}{L}=7$			
Fornax	0.040	0.14	0.028 ± 0.018	↑	↑
Sculptor	0.070	0.24	$0.19\ \pm 0.16$	↑	↑
Carina	0.028	0.10	$0.11\ \pm 0.07$	0.005	0.09
UMi	0.010	0.04	$0.91\ \pm 0.51$	↓	↓
Draco	0.026	0.09	$0.64\ \pm 0.34$	↓	↓
DDO 127				0.02	0.04

NOTES – Central DM densities for dS galaxies are averages for NGC 247, NGC 300, NGC 3109 and UGC 2259. Minimum- and maximum-halo results are given; the dispersions in ρ_0 are 0.003 and 0.08 $M_\odot$ pc^{-3}, respectively. DDO 127 is listed separately because the minimum- and maximum-halo solutions are similar.

estimated or that $r_c^{DM} \gg r_c^{vis}$ or both. In other words, even the detection of DM in dwarf spheroidal galaxies is presently very insecure.

A plausible guess, which requires testing, is that dwarf spheroidal galaxies contain DM halos like those in dwarf spirals. If so, then $\rho_0^{vis} \simeq \rho_0^{DM}$ in the smallest galaxies. Then self-gravity would only just be important even at the center. Are smaller galaxies prevented from being visible because they cannot make stars? Is the faint end of the luminosity function of galaxies created by the disappearance of "baryonic" self-gravity and star formation? No observation excludes the possibility that there exist many galaxies which are completely dark.

4. CONSTRAINTS ON NEUTRINO DARK MATTER

Tremaine and Gunn (1979) and Gunn (1982) have pointed out that the existence of DM in dwarf galaxies has important implications for the question of whether that DM could consist of massive neutrinos. Currently this seems unlikely, because it would then be difficult to explain galaxy clustering (White 1986). Nevertheless, it is important to examine all constraints on DM composition.

The Tremaine and Gunn argument is as follows. The phase-space density of non-interacting particles can only decrease during violent relaxation from values determined by thermal equilibrium in the early universe. Since neutrinos cannot be packed too closely together, they can make DM halos only if they are sufficiently massive. This provides a lower limit on the neutrino mass m_ν. The Pauli exclusion principle also provides a lower limit, which is less severe by a factor of $2^{1/4}$. Eventually we hope to compare these limits with actual measurements of neutrino masses; at present these are too uncertain (Turner 1986). But we can compare the lower limits with upper limits at which the neutrinos by themselves account for Ω. If $H_0 = 75$ km s^{-1} Mpc^{-1}, the closure density is 5930 eV cm^{-3} (Davis *et al.* 1981). Thermodynamic equilibrium in the early universe implies that the present number density of neutrinos plus antineutrinos is 109 g_ν cm^{-3} (Davis *et al.* 1981), where g_ν = (number of neutrino species)(number of spin states each). Then,

$$\text{if } \Omega_\nu \leq 1, \qquad\qquad m_\nu \leq 55\, g_\nu^{-1}\ \text{eV};$$
$$\text{if } \Omega_\nu \leq \Omega_{observed} \simeq 0.2, \quad m_\nu \leq 11\, g_\nu^{-1}\ \text{eV}. \tag{4}$$

The upper and lower limits are presently on the borderline of being inconsistent.

For an isothermal halo, the Tremaine and Gunn phase-space constraint is:

$$m_\nu^4 > \frac{9\, h^3}{2\,(2\pi)^{5/2}\, g_\nu\, G\, \sigma\, r_c^2}, \tag{5}$$

i.e.,

$$m_\nu > \left(120\ eV\right) \left(\frac{100\ km\ s^{-1}}{\sigma}\right)^{1/4} \left(\frac{1\ kpc}{r_c}\right)^{1/2} g_\nu^{-1/4}, \tag{6}$$

or, since

$$r_c^2 = 9\sigma^2/4\pi G\rho_0, \tag{7}$$

$$m_\nu > \left(106\ eV\right) \left(\frac{100\ km\ s^{-1}}{\sigma}\right)^{3/4} \left(\frac{\rho_0}{1\ M_\odot\ pc^{-3}}\right)^{1/4} g_\nu^{-1/4}. \tag{8}$$

Here h is Planck's constant and G is the gravitational constant. All neutrino species are assumed to have equal masses; otherwise the constraint is more severe. I prefer the form (8) in terms of ρ_0 and σ for several reasons. First, ρ_0 appears to vary relatively little from galaxy to galaxy. It is also better determined from the observations than r_c and σ (see Carignan and Freeman 1985). Then the uncertainties are concentrated in one parameter, σ. Finally, σ is immediately recognizeable from any rotation curve that reaches $V = \text{constant} = \sqrt{2}\ \sigma$.

The smallest galaxies for which σ and ρ_0 are sufficiently well determined give:

UGC 2259: $m_\nu > 47\, g_\nu^{-1/4}$ eV (min. halo); $\quad m_\nu > 95\, g_\nu^{-1/4}$ eV (max. halo);
DDO 127: $\;m_\nu > 110\, g_\nu^{-1/4}$ eV (best fit); $\qquad m_\nu > 130\, g_\nu^{-1/4}$ eV (max. halo). $\tag{9}$

These values are marginally inconsistent with the upper limits (4). The inconsistency is stronger for larger g_ν or smaller H_0 (the upper limit scales as H_0^2; the lower limit scales as $H_0^{1/2}$). However, since the rotation curves do not unambiguously reach $V = $ constant, σ may have been underestimated. Also, if we relax the assumption of isothermality and construct detailed models of neutrino halos, then the above limits are weakened (Madsen and Epstein 1984). At present, then, neutrino DM is on the borderline of being ruled out by these arguments. Of course, if Aaronson and Olszewski's dispersion measurements in UMi and Draco are taken at face value, they imply a more stringent limit $m_\nu > 560\ g_\nu^{-1/4}$ eV. But this is derived by assuming that $r_c^{DM} = r_c^{vis}$. It is not obvious how r_c^{DM} can be determined so that the σ measurements can be used to constrain m_ν. The best way to improve the ν DM limits appears to be to measure HI rotation curves in smaller dS galaxies.

5. CONCLUSION

The existence and properties of DM in the smallest galaxies have fundamental implications for the DM problem and for our understanding of galaxy evolution. To the extent that measuring techniques are adequate, DM halos have been detected in the smallest galaxies observed. However, we have only recently begun to measure halo parameters and still know little about their systematic behavior. For example, we need to look for relationships between halo parameters analogous to those for elliptical galaxies (Faber and Jackson 1976; Kormendy 1985). This presents us with an obvious opportunity, since HI rotation curves and central velocity dispersions have been measured for only a small fraction of the galaxies that could be observed. It is worth spending considerable effort to push the HI rotation curves to the largest possible radii, and to measure as many accurate radial velocities in dEs as possible. It is also possible that star formation turns off when the "baryonic" matter becomes non-self-gravitating, so that the smallest galaxies are optically invisible. It would therefore be important to extend searches for intergalactic HI clouds (e. g., Lo and Sargent 1979) to more groups and to fainter limits.

ACKNOWLEDGEMENTS

This paper is based in part on observations obtained at the Canada-France-Hawaii Telescope (CFHT) and at the Westerbork Synthesis Radio Telescope (WSRT). CFHT is operated by the National Research Council of Canada, the Centre National de la Recherche Scientifique of France, and the University of Hawaii. WSRT is operated by the Netherlands Foundation for Radio Astronomy with financial support from the Netherlands Organization for the Advancement of Pure Research. I thank A. Bosma, M. Fletcher, D. Hartwick, R. McClure and J. Souviron for permission to talk about our results before publication. I am also grateful to M. Aaronson, E. Olszewski, C. Pryor and J. Hesser for communicating their results before publication. Finally, I thank C. Pryor for several very helpful discussions, and for calculating King models to evaluate the projection corrections to σ and the effects of having $r_c^{DM} > r_c^{vis}$.

REFERENCES

Aaronson, M. 1983, *Ap. J. (Letters)*, **266**, L11.

Aaronson, M. 1986, in *Star Forming Dwarf Galaxies and Related Objects*, ed. D. Kunth, T. X. Thuan and J. T. T. Van (Paris: Editions Frontières), in press.

Aaronson, M., and Olszewski, E. 1986, in *IAU Symposium 117, Dark Matter in the Universe*, ed. J. Kormendy and G. R. Knapp (Dordrecht: Reidel).

Bahcall, J. N., and Casertano, S. 1985, *Ap. J. (Letters)*, **293**, L7.

Bosma, A., Kormendy, J., and Souviron, J. 1986, in preparation.

Carignan, C., and Freeman, K. C. 1985, *Ap. J.*, **294**, 494.

Carignan, C., Sancisi, R., and van Albada, T. S. 1985, preprint.

Cowsik, R. 1985, preprint.

Davis, M., Lecar, M., Pryor, C., and Witten, E. 1981, *Ap. J.*, **250**, 423.

Faber, S. M., and Jackson, R. E. 1976, *Ap. J.*, **204**, 668.

Faber, S. M., and Lin, D. N. C. 1983, *Ap. J. (Letters)*, **266**, L17.

Freeman, K. C. 1970, *Ap. J.*, **160**, 811.

Freeman, K. C. 1986, in *IAU Symposium 117, Dark Matter in the Universe*, ed. J. Kormendy and G. R. Knapp (Dordrecht: Reidel).

Gottesman, S. T., and Weliachew, L. 1977, *Astr. Ap.*, **61**, 523.

Gunn, J. E. 1982, in *Astrophysical Cosmology, Proceedings of the Study Week on Cosmology and Fundamental Physics*, ed. H. A. Brück, G. V. Coyne and M. S. Longair (Vatican City: Pontifical Academy of Sciences), p. 557.

Illingworth, G. 1977, *Ap. J. (Letters)*, **218**, L43.

King, I. 1962, *A. J.*, **67**, 471.

King, I. R. 1966, *A. J.*, **71**, 64.

Kormendy, J. 1985, *Ap. J.*, **295**, 73.

Krumm, N., and Burstein, D. 1984, *A. J.*, **89**, 1319.

Kuhn, T. S. 1970, *The Structure of Scientific Revolutions*, (Chicago: University of Chicago Press).

Lake, G., and Schommer, R. A. 1984, *Ap. J. (Letters)*, **279**, L19.

Lo, K. Y., and Sargent, W. L. W. 1979, *Ap. J.*, **227**, 756.

Madsen, J., and Epstein, R. I. 1984, *Ap. J.*, **282**, 11.

McClure, R. D., Fletcher, J. M., Grundmann, W. A., and Richardson, E. H. 1985, in *IAU Colloquium 88, Stellar Radial Velocities*, ed. A. G. Davis Philip and D. W. Latham (Schenectady: L. Davis Press), p. 49.

McClure, R. D., Fletcher, J. M., Hartwick, F. D. A., and Kormendy, J. 1986, in preparation.

Peterson, C. J., and King, I. R. 1975, *A. J.*, **80**, 427.

Pryor, C., McClure, R. D., and Hesser, J. E. 1986, *A. J.*, in preparation.

Sancisi, R., and van Albada, T. S. 1986, in *IAU Symposium 117, Dark Matter in the Universe*, ed. J. Kormendy and G. R. Knapp (Dordrecht: Reidel).

Sargent, W. L. W., Sancisi, R., and Lo, K. Y. 1983, *Ap. J.*, **265**, 711.

Schechter, P. L. 1980, *A. J.*, **85**, 801.

Seitzer, P., and Frogel, J. A. 1985, *A. J.*, **90**, 1796.

Souviron, J., Kormendy, J., and Bosma, A. 1986, in preparation.

Tremaine, S. 1986, discussion comment following this paper.

Tremaine, S., and Gunn, J. E. 1979, *Phys. Rev. Letters*, **42**, 407.

Turner, M. S. 1986, in *IAU Symposium 117, Dark Matter in the Universe*, ed. J. Kormendy and G. R. Knapp (Dordrecht: Reidel).

Tully, R. B., and Fisher, J. R. 1977, *Astr. Ap.*, **54**, 661.

van Albada, T. S., Bahcall, J. N., Begeman, K., and Sancisi, R. 1985, *Ap. J.*, **295**, 305.

White, S. D. M. 1986, in *IAU Symposium 117, Dark Matter in the Universe*, ed. J. Kormendy and G. R. Knapp (Dordrecht: Reidel).

DISCUSSION

TREMAINE: When you are dealing with these small number statistics the distribution of velocities follows a χ^2 distribution. This is quite asymmetric. Then it is not really appropriate to quote a ±σ error. I think that it is relatively easy to get a spuriously low value of M/L and more difficult to get a spuriously high value. Therefore, the errors may not be as bad as you have stated.

LAKE: Your rotation curve for DDO 127 has little curvature. You get constraints on a halo only because the maximum disk is so minimal. If you add the effect of the dispersion of the HI gas, this would raise the central few velocity points and allow you to add much more disk mass. Beam smearing has a similar effect.

KORMENDY: Of course, we will correct for both of these effects in the final analysis. It is true that the amount of disk mass could be somewhat larger than we derive without these corrections. But the disk scale length is so short that the derived halo rotation curve is not very much affected. Also, at large radii, where the rotation curve measures the halo, we observe $V \simeq 35$ km s^{-1}, while the velocity dispersion of the gas is < 10 km s^{-1}. Since the velocities add in quadrature, the dispersion of the gas does not have a large effect.

OSTRIKER: A comment on the apparently high central densities in dwarf spheroidals. If dark halos are made of massive black holes (a perhaps unlikely possibility), then Lacey and I find that a halo in a dwarf spheroidal will contract due to dynamical friction. The distribution of visible stars will expand, drastically reducing the stellar density. Then you expect a high central density of dark matter.

MADSEN: I am somewhat worried about the assumed isothermality of neutrino halos in the discussion of neutrino mass limits derived from dwarf galaxies. The lower mass limits may actually be reduced by a factor of a few if one uses more general assumptions for the neutrino velocity distribution.

KORMENDY: One of the things I did not have time to discuss was the important improvements made by you and others to isothermal assumptions for the halo.

SANDERS: Both you and the previous speaker have shown a decomposition of the rotation curve of UGC 2259 into disk and halo contributions. Is this really fair, since there is no surface photometry for this galaxy?

KORMENDY: You can't make an accurate decomposition without the surface photometry. The best you can do – and this is still more accurate than what I did for the dwarf ellipticals – is to derive an approximate scale length from the size of the object and the central surface brightness estimated by eye from a plate. However, the strongest constraints which I discussed are given by DDO 127, for which there does exist surface photometry.

The Search for Dark Matter in Draco and Ursa Minor: A Three Year
Progress Report

M. Aaronson and E. Olszewski
Steward Observatory, University of Arizona
Tucson, Arizona 85721 U.S.A.

ABSTRACT. We report the cumulative results of an on-going effort
to measure the stellar velocity dispersion in two nearby dwarf
spheroidal galaxies. Radial velocities having an accuracy $\lesssim 2$ km s^{-1}
have now been secured for ten stars in Ursa Minor and eleven stars in
Draco (including 16 K giants and 5 C types). Most objects have been
observed at two or more epochs. Stars having non-variable velocities
yield in both dwarfs a large (~ 10 km s^{-1}) dispersion. These results
cannot be explained by atmospheric motions, and circumstantial
evidence suggests that the effects of undetected binaries are also not
likely to be important. Instead, it seems that both spheroidals
contain a substantial dark matter component, which therefore must be
"cold" in form.

1. INTRODUCTION AND OBSERVATIONS

The question of whether or not small galaxies contain dark matter
has far reaching implications, both from the standpoint of the origin
and evolution of these systems, and for the nature of the nonluminous
material itself (e.g. Lin and Faber 1983, Aaronson 1983, Kormendy
1986). For dwarf irregulars, the issue is best attacked via H I
observations. However, the lowest luminosity systems we know of are
the halo dwarf spheroidals, and in particular Draco and Ursa Minor.
These galaxies contain no gas or other signs of recent star formation,
and the test for dark matter in them can only be done by measuring the
velocity dispersion of stars at the tip of an old giant branch on a
one-by-one basis. This necessitates velocities of ~ 1 km s^{-1}
accuracy for objects having V > 17 mag, measurements which until
recently would have been thought to be totally unfeasible.
Prompted by the developing technology (both hardware and
software), several groups have within the last few years initiated
attacks on the halo spheroidals. A summary of the current situation,
which has been rather controversial, was given by Kormendy (1985) at
this conference. Here we report the latest results of our continuing
efforts on Draco and Ursa Minor, where the observations have now
become considerably more extensive than those for any other
spheroidal.

J. Kormendy and G. R. Knapp (eds.), Dark Matter in the Universe, 153–160.

    The measurements we present were taken over a period from April
1982 to the most recent observing run in May 1985, and were all
obtained using the Multiple Mirror Telescope and echelle spectrograph
with a photon counting Reticon detector, and reduced using
cross-correlation techniques. This work initially began by
concentrating on the carbon stars, because of the ease with which
velocities could be obtained from the λ5636 C_2 Swan bandhead. Perhaps
fortuitously, however, the presence of only two C stars in Ursa Minor
and three in Draco quickly necessitated moving on to the K giants.
For this purpose we have chosen to work in the area of the Mg triplet
at λ5180, a region also densely populated by sharp features mostly due
to Fe and Cr.

    As a point of interest, the net photon counting rate with the MMT
echelle is 2 per second at V ~ 17 mag, which is sufficient, though, to
achieve ~ 1 km s^{-1} accuracy (10 km s^{-1} resolution) in 80 min when the
dark count is comparably low. However, by V = 17.5 mag, the situation
rapidly worsens, not so much from the smaller amount of flux, but
because the giant branch stars become hotter, and the absorption
features, already weak due to the low abundance, become barely
discernable. It then takes some two hours of integration to obtain ~
2 km s^{-1} precision.

    In both Draco and Ursa Minor, roughly ten proper motion members are
known between V = 17 and 17.5, and velocities for all of these have
now been obtained. The results are summarized in Figure 1. We note
that the majority of stars have been observed at two or more epochs,
separated by periods ranging from three months to two years, but most
typically by a year. The nine stars in Draco which do not show any

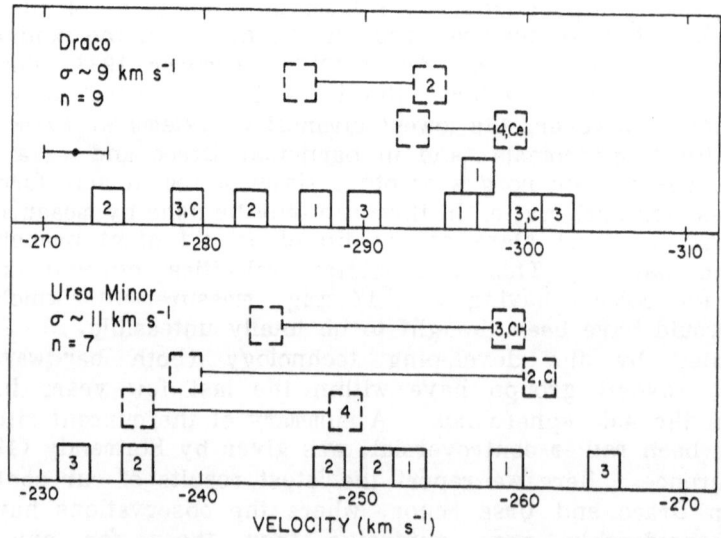

Figure 1. Velocity distributions in two dwarf spheroidals.
Quantities inside boxes are the number of epochs a star has been
measured. Single boxes represent mean speeds of stars showing no
variation larger than the estimated errors, while variables are shown
at the extreme range of their measured velocities. C stars are marked.

radial velocity variations give a dispersion (corrected for measuring error) of 9 ± 2 km s^{-1}, a value which remains unchanged if the two C stars are dropped; while the seven non-variable Ursa Minor stars yield a dispersion of 11 ± 3 km s^{-1}. A Kolmogorov-Smirnov test indicates that in both instances the velocity distribution does not differ from a gaussian at any level of statistical significance, but this is more a statement about small number statistics than anything else.

It is interesting that four of the five stars which show velocity changes might have been predicted to do so a priori. First, McClure (1984) has demonstrated that a high fraction of galactic CH stars are in binaries, as is apparently the CH star in Ursa Minor. Furthermore, the other Ursa Minor C star, which also appears to be binary, sits in a peculiar position in the HR diagram well below the tip of the giant branch. The third Ursa Minor binary candidate is star M, which is a somewhat anomalously bright K giant that may be a long period photometric variable, although no such stars have ever been conclusively detected in either Draco or Ursa Minor (see Aaronson and Mould 1985).

In Draco, the C star velocity variable has a peculiar emission line spectrum which almost certainly implies the presence of a degenerate companion (Aaronson, Liebert, and Stocke 1982). We note further that unlike the more luminous carbon stars found in the other halo spheroidals, the ones in Draco and Ursa Minor are of too small a bolometric magnitude to be accounted for by Iben's (1975) third dredge-up mechanism. On the other hand, the two remaining Draco C stars have yet to show any indication of duplicity; it was in fact the wide velocity difference between these two stars which led to the original suggestion of dark matter in this system (Aaronson 1983).

Finally, we point out that the velocity variations in the only two stars with observations at 4 epochs would have been apparent after the first 3 epochs, if the data had been reduced; i.e., these stars were not identified as velocity variables simply because we "tried harder" with them.

2. DISCUSSION

There are four possible contributors to the observed dispersion in Figure 1: atmospheric motion, tidal disruption, binarity, and self-gravitation. Now the multiple epoch observations appear to completely rule out atmospheric instability effects, as these stars simply do not show the jitter seen, for instance, by Mayor et al. (1984) in 47 Tuc. However, there is nothing inconsistent here, as 47 Tuc stars of the same color and magnitude as in our sample are stable, while the 47 Tuc stars that vary are considerably redder and more luminous.

Concerning tidal effects, there are two possibilities. First, if the systems are virialized and contain no dark matter, the expected dispersion is no greater than the escape velocity of ~ 1 km s^{-1}. The other alternative is that both systems are unbound. The present phase must then be short lived, since the crossing time is $\sim 10^8$ years. On the other hand, the stars are $\sim 10^{10}$ years old, which suggests only a

1% probability of viewing such an event once, let alone twice.

The binary question is the most problematic, but we can make the following points: First, Carney and Latham (1985) in an on-going survey of halo K giants are finding a binary fraction of merely ~ 15%. Second, during our most recent run in May 1985, ten stars were remeasured, and all but one were found to remain constant, which at face value does not suggest a high binary frequency. Third, there do exist systems for which very low dispersions have been obtained over a limited time base. For instance, Mathieu (1983) required just two years of observation to weed out large amplitude binaries and obtain an observed dispersion < 1 km s^{-1} in M67 (for a sample that included the ten brightest M67 K giants, with only one being eliminated as a binary). Fourth, Monte Carlo simulations indicate that if the dwarfs contain binaries with the characteristics of a "normal" galactic distribution, dispersions measured from the present set of multiple epoch observations are very unlikely to be much affected by undetected binaries, a point illustrated in Figure 2. Of course, all of this evidence is circumstantial. The possibility remains that binary

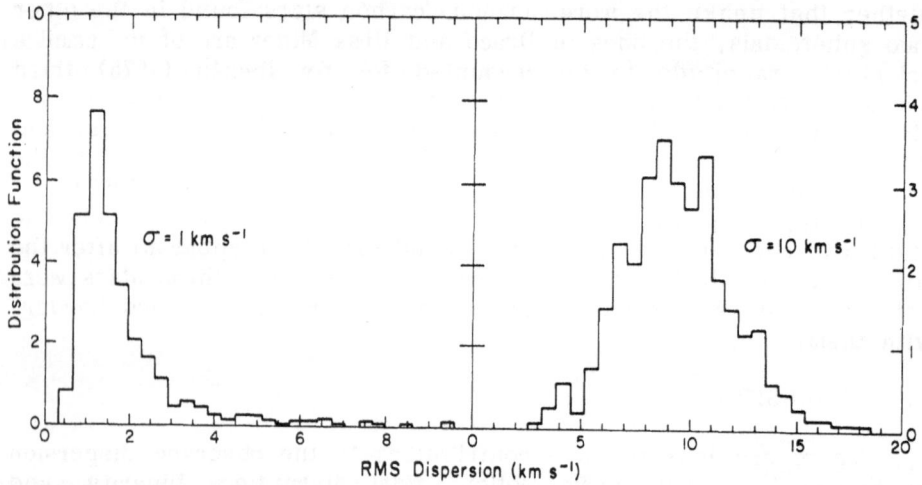

Figure 2. Monte Carlo simulations (500 trials each) of binary effects on observed dispersion in systems having a true dispersion of 1 and 10 km s^{-1}. Two observations of 10 stars taken one year apart are assumed, and cases with a velocity variation > 4 km s^{-1} are removed from the sample. The primary has a mass of 0.8 $M_\odot$, the phase, inclination angle, and eccentricity are chosen at random, and the period and secondary mass distributions are based on galactic field studies (see Mathieu 1983 for details). The binary fraction is set at 0.5. The distribution functions are arbitrarily normalized.

formation processes in the low density environment of the dwarfs differ radically from that in the galactic disk, and furthermore, that by very bad luck a high fraction of our observed stars are in "worst case" orbits. Continued monitoring of the stars is essential, along with increasing the sample, work which is clearly going to take some time.

We come then to the final possibility that the observed dispersions are gravitationally supported. Kormendy (1985) has discussed some of the uncertainties involved in deriving masses. Plowing ahead, we give in Table 1 the total mass-to-light ratios obtained by assuming a King model [and employing eq. (3) from Illingworth 1976]. The results yield rather large values of $M/L_V \sim 40$ for Draco and ~ 100 for Ursa Minor. The uncertainties in Table 1 involve the formal errors in the dispersions only, and it could be argued that with additional leverage in the luminosities and tidal radii, the results are not significant at much more than the one sigma level. On the other hand, because the χ^2 is assymetric, the dispersion errors quoted may be too conservative; e.g., the minimum Draco dispersion at the 10% confidence level given by a χ^2 test is 7.5 km s^{-1}, while the maximum dispersion is 15 km s^{-1}.

Table 1
Estimates of Dwarf Mass-to-Light Ratios

Name	M_V	$M_\odot$	M/L_V	$\langle V^2 \rangle^{1/2}$	M/L_V
Draco	-8.5	$9.8 \times 10^4 \langle V^2 \rangle$	$0.46 \langle V^2 \rangle$	9 ± 2	37 ± 11
UMin	-8.7	$2.2 \times 10^5 \langle V^2 \rangle$	$0.86 \langle V^2 \rangle$	11 ± 3	104 ± 40

Are such great M/L ratios tenable? To begin with, Tinsley (1981) pointed out that the observed M/L values for larger galaxies suggested an increase in the percentage amount of dark matter in later-type systems. Lin and Faber (1983) demonstrated that the fading of a dwarf irregular, with an initial $M/L_V = 6$, would yield a final spheroidal having $M/L_V = 28$, and in this regard the evidence that the spheroidals share a heredity closer to dwarf irregulars than to ellipticals continues to accumulate (see Aaronson 1985). Perhaps most significantly, Sargent (1985) has recently reported H I observations by himself, Lo, and Young, of very low luminosity dI systems which yield indicative masses that do result in a rough increase in M/L with decreasing magnitude, such that the faintest systems (with $M_V \sim -12$ to -10) have M/L values in the range 20 – 30. Additional fading by factors of 2-3 in such objects would give results in the range observed here. Indeed, recent theoretical models of biased galaxy formation with cold dark matter proposed by Dekel and Silk (1985) nicely predict the change in M/L with luminosity that may now be established, with the smallest systems having M/L values of 10 – 100 and velocity dispersions of 5 – 10 km s^{-1}.

A possible monkey wrench in all this is the fact that efforts made in three other spheroidals (Carina, Sculptor, and Fornax -- see Kormendy 1985 for a summary) have not yielded a clear indication of substantial dark matter. Three points can be made in this regard. First, unlike the present data set, all of these observations (mostly of small numbers of C stars) are resolution and not signal-to-noise limited, and higher quality measurements of large K giant samples are desperately needed. Second, Kormendy (1986) has argued that the observed central surface densities are so large in the other systems that the effect of a dark halo may not be apparent in the central velocity dispersion. Finally, the M/L values for dI's reported by Sargent (1985) do exhibit considerable variation, and it may be simply that the ratio of dark to luminous material varies from one system to the next.

In summary, while not yet totally compelling, we believe the evidence for dark matter in the Draco and Ursa Minor dwarfs spheroidals has grown considerably stronger. The possibility that the nonluminous material consists of neutrinos can then be ruled out by phase-space density arguments (Tremaine and Gunn 1979). Rather, free streaming effects must be quite small, and the dark matter is therefore thermally cold in nature.

We are indebted to Bob Mathieu for stimulating discussion and for providing Figure 2. M. A. also thanks the Center for Astrophysics, where the written version of this talk was prepared, for an SAO Summer Fellowship. Our research used the MMT Observatory, a facility operated jointly by the University of Arizona and the Smithsonian Institution, it was also partially supported by NSF grant AST 83-16629.

REFERENCES

Aaronson, M. 1983, Ap. J. (Letters), **266**, L11.
_____. 1985, in Star Forming Dwarf Galaxies and Related Objects, ed. D. Kunth and T. X. Thuan, in press.
Aaronson, M., Liebert, J., and Stocke, J. 1982, Ap. J., **254**, 507.
Aaronson, M., and Mould, J. 1985, Ap. J., **290**, 191.
Carney, B., and Latham, D. 1985, private communication.
Dekel, A., and Silk, J. 1985, preprint.
Iben, I. 1975, Ap. J., **196**, 525.
Illingworth, G. 1976, Ap. J., **204**, 73.
Kormendy, J. 1986, in IAU Symposium No. 117, Dark Matter in the Universe, ed. J. Kormendy and J. Knapp, (this volume).
Lin, D. N. C., and Faber, S. M. 1983, Ap. J. (Letters), **266**, L21.
Mathieu, R. D. 1983, Ph.D. Thesis, University of California, Berkeley.
Mayor, M., et al. 1984, Astr. Ap., **134**, 118.
McClure, R. D., 1984, Ap. J. (Letters), **280**, L31.
Sargent, W. L. W. 1985, in Star Forming Dwarf Galaxies and Related Objects, ed. D. Kunth and T. X. Thuan, in press.
Tinsley, B. M. 1981, M.N.R.A.S., **194**, 63.
Tremaine, S., and Gunn, J. E. 1979, Phys. Rev. Letters, **42**, 407.

DISCUSSION

GUNN: What is the mean time interval between different measurements of the same star?

AARONSON: The observations are separated by intervals of three months to two years. The typical spread is about a year. In fact, for many of the stars there is a problem with aliassing of one-year periods, which are typical for K giants that have measured periods.

GUNN: I think there is a very strong observational selection in favor of finding only the short-period systems. Griffin and I have had a lot of trouble finding radial velocity standards because all of the velocities change when we watch them long enough. We have been at this for ten years now. So I think that it is very difficult to find all the binaries with a short baseline. Of course, it is true that as you go to long periods, you necessarily get small velocity excursions. But they are not small compared to your dispersion values. We should be very cautious.

LYNDEN-BELL: If you believe that there is cold dark matter in UMi and Draco, why is there none in Fornax?

KORMENDY: The total density in visible matter of assumed M/L = 2 is high enough in Fornax so that a dark halo like those in dwarf spirals will not affect the central velocity dispersion. This statement is true for Fornax, and marginally for Sculptor and Carina. Ursa Minor and Draco are the easiest objects in which to look for dark matter.

DEKEL: You might expect the value of M/L to change as you go to fainter objects. Draco may have the highest value of M/L because it is the smallest dwarf spheroidal. We have a theory about the formation of dwarfs, which Silk will talk about tomorrow, in which we predict a trend like this. Unfortunately, Seitzer and Frogel's observations of carbon stars give very low values of the velocity dispersion, lower even than we expect. So my question is: What is the hope of repeating the observations of Sculptor and Fornax in the near future?

AARONSON: I have no such plans. I am waiting until we get an 8 m telescope in the southern hemisphere. I think you need to measure K giants; carbon stars are too uncertain. As you saw for my observations of carbon stars, two of the five look as though they are binaries and a third has a lot of atmospheric jitter.

DEKEL: What is the problem with measuring K giants?

AARONSON: No proper motion studies have been made, so these fields have a lot of contamination. This is particularly a problem for Fornax and Sculptor, because their velocities are not large. At the moment it takes 2 hours to measure one star, and one doesn't want to waste that sort of time on foreground objects. This project is going to become feasible when you can observe a star in five minutes.

FREEMAN: How do you get luminosities to derive M/L ratios?

AARONSON: I use Zinn's values. He has made star counts and assumed that the luminosity function is the same as in M3.

OSTRIKER: Would Milgrom tell us what he would predict for the velocity dispersions in these systems? It would seem as though they are ideal test cases.

MILGROM: These objects have low surface densities and therefore low accelerations, so the prediction is that they should show large mass discrepancies. There are complications due to the fact that their internal dynamics are affected by the external field of the parent galaxy, so that there is a prediction of some galactocentric distance dependence. Typically, mass discrepancies (not M/L) should be 10 - 15 at distances of ~ 100 kpc.

HI AND MASS DISTRIBUTION IN THE DWARF "REGULAR" GALAXY UGC 2259

C. Carignan, R. Sancisi, and
T.S. van Albada
Kapteyn Astronomical Institute
P.O. Box 800
9700 AV Groningen
Holland

ABSTRACT. HI synthesis observations of the dwarf "regular" galaxy UGC 2259 are presented. This system turns out to be most suitable for our purpose of studying the mass distribution in late-type spirals. It has a symmetrical HI distribution extending over 2 D_{25} and a regular velocity field which shows no sign of departure from axial symmetry. Despite the fact that UGC 2259 is fainter than most other objects of the same morphological type, its HI properties are typical of what is expected from an Scd galaxy with a $M_H/L_B = 0.43$.

While other low-luminosity galaxies, studied so far, exhibit solid-body type rotation curves, the most remarkable result for this dwarf system is that it has a flat rotation curve similar to what is seen in more massive spirals. The main difference between UGC 2259 and other low-luminosity galaxies is its regular appearance.

Despite the uncertainties in the mass distribution due to the lack of surface photometry, the most probable model implies that the dark halo and the luminous disk contribute about equally to the circular velocity at the Holmberg radius.

This result, by extending the luminosity range of well studied galaxies, strengthens earlier suggestions (Carignan and Freeman, 1985; Carignan, 1985) that the ratio of dark-to-luminous matter appears to be independent of galaxy mass and/or luminosity.

References

Carignan, C. 1985, Ap. J., 298, in press.
Carignan, C., and Freeman, K.C., 1985, Ap. J., in press.

J. Kormendy and G. R. Knapp (eds.), Dark Matter in the Universe, 161.

Neutral Hydrogen Observations of dI Galaxies in the Virgo Cluster

G. L. Hoffman.   E.E. Salpeter   G. Helou
Lafayette College CRSR, Cornell Univ. JPL-Caltech

We are in the midst of obtaining 21 cm observations of late-type
(Sdm through ImV, including BCD) dwarf galaxies in the Virgo Cluster,
selected from the Binggeli, Sandage and Tammann catalog (1985, in press).
To date, we have observed 174 of these objects; the sample is nearly
complete to B_T = 17.0. 114 have been detected, and 37 of the strongest
emitters have been mapped at 1!9 spacing to determine HI diameters and
rotation curves.

For every detected galaxy we have two bits of information relevant
to massive halos: the profile width ΔV and an indication of whether or
not the object is in disk-like rotation with a component of the rotation
velocity parallel to the line of sight (i.e., whether or not the profile
shows the classic "double-horned" shape at 8 km/s resolution). 37 of
the 114 detected dwarfs do indeed show double-horned profiles; this
fraction is largest for Sdm and Sm and decreases toward later types. It
is evident that the ratio of turbulent to rotational velocities increases toward the later types. On Fisher-Tully diagrams (with ΔV corrected for inclination as in Hoffman, Helou, Salpeter and Sandage 1985,
Ap. J. Lett. 289, L15), the dwarfs continue the ridge-line of the
brighter Virgo spirals irrespective of whether or not the profiles are
double-horned. Mass-to-light ratios computed from the inclination-
corrected ΔV and optical diameters are not very different those for
bright spirals.

For 16 of the 37 mapped dwarfs we have been able to determine
"isophotal" HI radii -- the outermost positions from which we think we
can reliably detect emission. This was possible only for dwarfs which
were clearly rotating. The mean ratio of HI to optical radii for these
resolved dwarfs is if anything larger than for a representative sample
of faint spirals (similar B_T and diameters), mapped in identical fashion.
These resoved objects allow a determination of indicative M/L using the
HI diameter (assuming the gas at that point is in an inclined circular
orbit); for the 16 dwarfs we get M/L = 12$\pm$2 (solar units) while for the
representative spirals 7$\pm$1.

A comparison of rotation curves for mapped dwarfs with those for
faint spirals similarly mapped shows that, for those that exhibit rotation, there is if anything a more pronounced tendency toward rising rotation curves for the dwarfs; however, a substantial fraction of the
dwarfs exhibit turbulent motions (i.e., Gaussian profiles) rather than
organized rotation.

Admittedly there may be selection effects; the resolved dwarfs are
among the brightest dwarfs in HI (yet spanning the full range of morphological types) whereas the spirals are a representative sample. But it
is clear that at least some of the dwarfs (of all types) have massive
halos very much like scaled-down versions of larger spirals.

162

J. Kormendy and G. R. Knapp (eds.), Dark Matter in the Universe, 162.
© 1987 by the IAU.

NEUTRINO MASS LIMITS AND DARK HALOS

Jes Madsen, Institute of Astronomy, Aarhus, Denmark
Richard Epstein, Los Alamos National Laboratory, USA

Lower mass limits for particles constituting the dark matter in galaxy halos can be derived from considerations of the initial and final phase space distribution. In the case of massive neutrinos, Tremaine and Gunn (1979) pointed out that the initial fine-grained occupation number of cosmological neutrinos and therefore the final coarse-grained phase space occupation, is less than 0.5. From this they were able to show that if the final neutrino distribution is an isothermal sphere, one can put lower limits on the neutrino mass from assumptions about the core radius of the neutrino sphere.

There is, however, no reason to believe that the final neutrino distribution should be an isothermal sphere. If more efficient packing in phase space takes place, halos may be made of neutrinos <u>less</u> massive than would be predicted from the isothermality assumption. A firm lower mass limit for an isotropic velocity distribution was derived by Madsen and Epstein (1984) based on the existence of a maximally compact sphere constructed from a cloud of primordial neutrinos.

But even limits based on isotropy may be misleading. The effects of velocity anisotropies were investigated by Madsen and Epstein (1985). If transverse velocity dispersion dominates in the outer parts of the halo, neutrino mass limits are increased relative to results in the isotropic case, but if radial velocity dispersion dominates, limits are weakened.

Velocity anisotropies can be described by the parameter α, where $\langle v_\theta^2 \rangle = \langle v_\phi^2 \rangle = (1-\alpha)\langle v_r^2 \rangle$. From a sample of well observed galaxies we conclude that $m_\nu > 35\ h^{\frac{1}{2}}eV$ if $\alpha \leq 0$, $m_\nu > 29\ h^{\frac{1}{2}}eV$ if $\alpha \leq 0.4$, and $m_\nu > 25\ h^{\frac{1}{2}}eV$ if $\alpha \leq 0.8$ (h is the Hubble parameter in units of $100\ km\ s^{-1}Mpc^{-1}$).

This research was performed in part under the auspices of the U.S. Department of Energy, and with support from the Danish Natural Science Research Council.

REFERENCES

J. Madsen, and R.I. Epstein, *Ap.J.* 282, 11 (1984).
J. Madsen, and R.I. Epstein, *Phys.Rev.Lett.* 54, 2720 (1985).
S. Tremaine, and J.E. Gunn, *Phys.Rev.Lett.* 42, 407 (1979).

J. Kormendy and G. R. Knapp (eds.), Dark Matter in the Universe, 163.
© *1987 by the IAU.*

X-RAYS FROM GALAXIES AND CLUSTERS OF GALAXIES: OBSERVATIONS AND PHENOMENOLOGY

C. R. Canizares
Massachusetts Institute of Technology
Deparment of Physics and Center for Space Research
37-501
Cambridge, MA 02139

ABSTRACT. X-Ray observations of galaxies and clusters can, in principle, trace the binding mass in these systems. I review some of the relevant work. The mass of hot gas in rich clusters is comparable to or exceeds the mass in visible stars. This proportion of gas to stellar material could be universal, although there is no direct evidence that it must be. Studies of the distribution of the gas indicate the presence of dark matter in the envelopes of some dominant cluster galaxies, most notably M87. The M/L_B values increase with radius to values of $\sim$ 400-600 $M_\odot/L_\odot$. Uncertainties in the temperature distribution of the gas have hampered these analyses and have made it difficult to draw definitive conclusions about the binding mass in clusters. Recent work on Coma suggests that M/L is falling with radius and the total M/L for the cluster may be as low as $\sim$ 120. Studies of early type galaxies show that many contain hot gas with temperatures $\sim 10^7$ K. There is evidence for the existence of cooling flows, and gravity rather than supernovae may be the dominant source of energy that heats the gas. The deduced binding masses for several bright galaxies are uncertain because of the unknown temperature profiles. Values of $M/L_B \approx$ 20 - 30 within $\sim$ 30 - 40 kpc are indicated if one assumes isothermality, but values as low as 5 and as high as 100 are allowed. With better models one may be able to reduce these uncertainties.

I. INTRODUCTION

X-ray observations have a very direct and clear relevance to the study of dark matter. X-rays from clusters and many galaxies are emitted by hot gas, and the X-ray surface brightness is directly related to the density of the gas (for some recent reviews see Forman and Jones 1982, Fabian, Nulsen and Canizares 1984, Sarazin 1985). The gas is a collisional fluid in hydrostatic equilibrium (or nearly so), so it traces the gravitational potential. Thus X-ray observations give us, in principle, an ideal method for tracing the binding matter in galaxies and clusters.

165

J. Kormendy and G. R. Knapp (eds.), Dark Matter in the Universe, 165–181.
© *1987 by the IAU.*

II. CLUSTERS OF GALAXIES

a. Morphology

Qualitatively, the X-ray images of rich clusters show a variety of
morphologies reflecting the variety of underlying potentials. The
extensive work of Jones and Forman (1984; also Forman and Jones 1982)
shows that some potentials are rather irregular whereas others are very
regular, symmetric and well developed. An important class of systems
has nearly circular X-ray surface brightness contours that are sharply
peaked and centered on a dominant (often a cD) galaxy. This is evidence
for a concentrated, spherically symmetric potential well.

b. Quantity of Hot Gas

Before looking at these potentials more quantitatively to deduce what
we can about dark matter, I want to consider the matter that is not
dark, namely the X-ray emitting gas, and explore its contribution to
the luminous mass of the universe. Jones and Forman (1984) have derived
the mass in hot gas within a radius of 3 Mpc for a large number of
clusters (I use H_o = 50 km s^{-1} Mpc^{-1} throughout). They obtain values of
$0.3-5 \times 10^{14}$ M_θ. (see also Abramopoulos and Ku 1983, who make somewhat
different assumptions). An important caveat is that, although some
clusters are observed to such large radii, most clusters are not. More
typically the measured surface brightness profiles extend only to $\sim$ 1 Mpc
or less (recall that the X-ray emissivity of a gas is proportional to the
square of the density, which makes the densest regions in the centers
much brighter than the outer parts), and so the estimates of total
masses to larger radii require extrapolations. Recently Henriksen and
Mushotzky (1985a) have noted the model dependence of the mass estimates
and the internal inconsistencies of the isothermal model usually assumed
(see also earlier work by Cavaliere and Fusco-Femiano 1976, 1978 and
Cavaliere 1979). An additional uncertainty comes from the possible
clumping of gas in the more irregular clusters.
    Bearing in mind these uncertainties, which could easily amount to
factors of 2-4, one can take as a rule of thumb that within rich
clusters the mass in X-ray emitting gas is roughly comparable to or
even exceeds the mass in stars. For fifteen clusters I obtained values
of $M_{gas}/L_v \sim 3 - 40$ M_θ/L_θ. Here I used values of M_{gas} within 3 Mpc from
Jones and Forman (1984) and from Abramopoulos and Ku (1983), as corrected
to 3 Mpc by Rothenflug and Arnaud 1985, and values of total L_v from
Oemler (1974) and Dressler (1978) ignoring differences in their
extrapolation techniques. Rothenflug et al. (1984) have looked at two
dozen clusters, and they conclude that if one takes only L_v within 3 Mpc
then M_{gas}/L_v increases with the luminosity of the cluster as $\sim L_v^{0.6}$.
Because of the uncertainties in both M_{gas} and L_v and in their
distribution within clusters, which could depend systematically on
cluster luminosity and type, one must treat this intriguing suggestion
with caution, although it bears further attention (it is related to the
question of whether the total mass to luminosity ratio increases with
the scale of the system [Blumenthal et al. 1984]).

Mushotzky (1984) and Rothenflug and Arnaud (1985) have shown that for 22 clusters the X-ray emitting gas has an iron abundance that is approximately half solar. This is important for questions of enrichment and evolution, which I will not address. Rothenflug and Arnaud (1985) and Rothenflug et al. (1984) argue that the approximate proportionality between gas mass and virial mass, but not luminosity, together with the apparently constant iron abundance is consistent with the suggestion that the virial mass is composed of stars or star remnants which produced the iron in the past. Again, the observations are subject to systematic uncertainties, but the relations deserve further study.

One can now ask whether the rough correspondence between gas and stellar mass is a property only of rich clusters or is it more general? Could there be comparable quantities of gas in poor groups or even around field galaxies (there is much less gas inside field galaxies as discussed below)? If there were, it could more than double the inventory of baryonic matter in the universe over that estimated from stars alone (e.g. see Blumenthal et al. 1984; Faber this meeting).

The answer is that although a universal relation between stellar and gas masses is certainly not demonstrated, neither can it be ruled out. In fact, X-ray emitting gas has been detected in poor clusters, but only those with reasonable high central galaxy densities and (not coincidentally) a central dominant galaxy (see Kriss, Cioffi and Canizares 1983, Bahcall 1982). For example the poor cluster AWM4 has richness -1 and contains $\sim$ 5 x 10^{12} $M_\odot$ of gas within 0.5 Mpc, which is comparable to its stellar mass.

Presumably, less dense poor clusters and groups could have comparable amounts of gas but have escaped detection because they have very low X-ray surface brightness. We have recently completed a study of the Pegasus I cluster, which is a loose, spiral rich system (Canizares et al. 1985). Although the X-ray image is dominated by the two central elliptical galaxies, there is evidence for diffuse intracluster gas with a mass of $\sim$ 3.5 x 10^{11} $M_\odot$ within a radius of $\sim$ 250 kpc. This is only $\sim$ 20% of the estimated stellar mass in the cluster. But the central regions of the cluster are just barely detected above the instrumental and diffuse background. The deduced mean gas density is $\sim$ 2 x 10^{-4} cm^{-3}, which is a factor of $\sim$ 3-5 lower than that of the least dense detected rich cluster. Therefore it is very likely that the gas is still more extensive but too diffuse to be seen. If the gas fills the $\sim$ 1-2 Mpc region occupied by the galaxies its total mass could easily be comparable to the stellar mass.

In summary, the evidence suggests that at least some poor clusters contain roughly the same proportion of hot gas relative to stars as do rich clusters (see also Rothenflug and Arnaud 1985). One can say still less about more isolated galaxies, although there is some recent evidence for possible circum-galactic material in a few cases (see below). That stars constitute one-half or less of the luminous matter in the universe remains an intriguing possibility.

c.   Total Binding Mass

I will now turn to the more central question of the determination of
binding masses from X-ray data. The operative equation is that of
hydrostatic equilibrium,

$$\frac{dP}{dr} = - \mu m G \, M(<r) n(r)/r^2 \tag{1}$$

which can be combined with the perfect gas law to give

$$M(<r) = (-k/\mu m G)[dln(n)/dln(r) + dln(T)/dln(r)]T(r)r. \tag{2}$$

Here P is gas pressure, r is radius, μm is the mean mass per
particle, n is the gas density, $M(<r)$ is the binding mass within r, T is
the temperature. The value of $n(r)$ can be deduced with reasonably few
assumptions from X-ray measurements with imaging detectors, like the IPC
on the Einstein Observatory (e.g. see Fabricant, Lecar and Gorenstein
1980). There is generally much less information about the radial
dependence of the temperature, which clearly complicates the application
of Equation 2. The standard procedure has been to use plausible models
for T vs. r; typical assumptions are that the gas is isothermal,
adiabatic or polytropic (see Sarazin's contribution to this meeting).
These models can be constrained by measurements of mean temperatures
and, in a few cases, by measurements of temperatures at several radii
or of temperature sensitive emission lines.
       By far the most successful application of Equation 2 has been to
the M87 in the Virgo cluster (Fabricant, Lecar and Gorenstein 1980,
Fabricant and Gorenstein 1983, Stewart et al. 1984), The X-ray image
and an extensive set of spectral measurements constrain both the total
binding mass and the distribution of mass with radius. The mass within
$\sim$ 200 kpc is $\sim$ 2 x 10^{13} $M_\odot$. The implied average value of M/L_B is $\sim$ 200,
and the very sharp concentration of light requires that M/L increase
with radius. Furthermore, the binding mass must have a core radius of
$\sim$ 10-30 kpc (if it has one at all), so it is clearly associated with
M87 rather than with the cluster as a whole (a model of Binney and Cowie
[1981] that attributed most of the mass to the cluster does not fit the
data [Fabricant and Gorenstein 1983, Stewart et al. 1984]).
       There is, of course, reason to believe that the approximately
central location of M87 in Virgo and its low relative velocity helped
it to acquire this large mass. The more isolated or less centrally
located galaxies discussed below do not show extensive X-ray halos.
       Data on other dominant cluster galaxies suggest that they too have
dark, massive halos, although in no other case is the temperature data
as complete as it is for M87. Matilsky, Jones and Forman (1985) derive
a mass of $\sim$ 2 x 10^{13} $M_\odot$ within 200 kpc for NGC4696 in the Centaurus
cluster. The dominant galaxies in poor clusters, which I mentioned
earlier, are similar (Kriss, Cioffi and Canizares 1983). Using the
assumption of isothermality of the X-ray gas, one derives mean values
of M/L_B of 70-100. In these cases the central galaxy contains typically
$\sim$ 30% of the light in the cluster and the local M/L_B increases to

400-600 at a few hundred kpc (see Fig. 1; I assumed B-V = 0.7 to convert L_V to L_B). The uncertainties in the temperature distributions cause uncertainties in the deduced mass distributions (see below), but the results are probably good to within a factor of 2-3, as are the trends in M/L vs. radius, (which cover an order of magnitude).

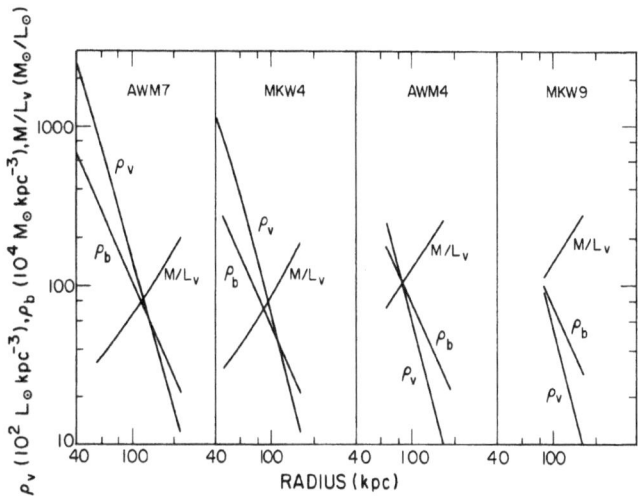

Figure 1. Density of light and binding mass deduced from the X-ray observations of four poor clusters with dominant galaxies, and the resulting M/L_V (from Kriss, Cioffi and Canizares 1983).

Several rich clusters have been studied in more detail in attempts to probe triaxial potentials (e.g. see Channan and Abramopoulos 1984, Fabricant, Rybicki and Gorenstein 1984, Fabricant et al. this meeting). Again the lack of temperature measurements is always a difficulty. For example, Figure 2 shows mass determinations for A2256 assuming either adiabatic or isothermal gas -- the values differ by factors of 2-3 (Fabricant, Rybicki and Gorenstein 1984). Nevertheless, the ellipticity of the potentials is demonstrated in these analyses.

Very recently, Henriksen and Mushotzky (1985b) and Cowie (1985) have tried to examine more carefully the model assumptions that go into analyses of this sort in the hopes of improving the accuracy of the mass determinations (there will be no improvement in the quality of the data for several years). Cowie has made a self-consistent fit to the X-ray data (assuming a polytropic gas model) and the optical velocity dispersions in the Coma cluster. Out to ~ 1 Mpc, where the X-ray surface brightness profile is still measureable, he obtains a total virial mass of $\sim 10^{14}$ $M_\odot$ and a total M/L_B of ~ 200. Two interesting results are first, that within this radius the gas mass is an increasing fraction of the virial mass and second, that M/L falls monotonically with radius, in marked contrast to the cases described above involving dominant galaxies. These results become even more extreme if one extrapolates to larger radii. At 4 Mpc the virial mass has increased by less than a factor of 2, and 30% of it is in the form of hot gas.

The mean M/L_B for the non-luminous component is only $\sim$ 120. The real question is how good is the extrapolation? The fits to the data require polytropic models with indices that differ from 1 (isothermal) and 5/3 (adiabatic), but these have no clear physical interpretation. So although these results are very provoking, they may not yet be the last words on the subject.

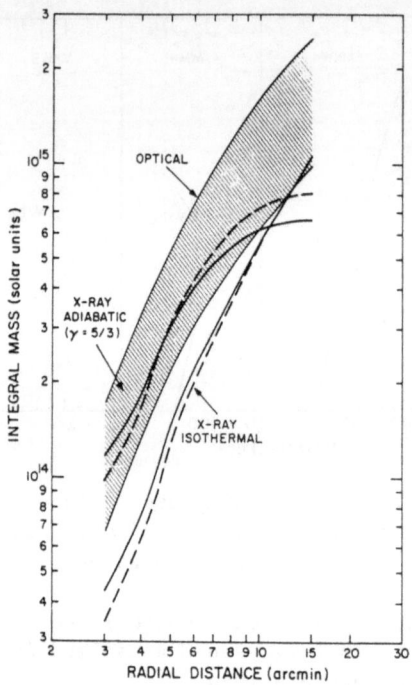

Figure 2. Integrated binding mass of cluster A2256 deduced from the optical galaxy counts and from the X-ray surface brightness distribution. The X-ray estimates are for oblate (dashed line) or prolate (solid line) geometries and for isothermal or adiabatic gas models, as indicated. For H_O = 50 kms^{-1} Mpc^{-1}, the scale is 100 kpc arc min^{-1}. (From Fabricant, Rybicki and Gorenstein 1984).

d.   Cooling Flows

I will conclude my discussion of observations of clusters with a few words about cooling flows, which involve pressure driven accretion flows of X-ray emitting gas that cools in the dense central regions of many clusters. This topic will be dealt with in greater detail by Fabian and Sarazin (this meeting) and I will return to it in the context of elliptical galaxies (see also reviews by Fabian, Nulsen and Canizares 1984, Sarazin 1985).

Here I just want to emphasize that there is ample evidence for the existence of cooling flows in clusters of galaxies. This evidence

includes X-ray images with high central surface brightnesses that
directly imply short cooling times and, less directly, low central
temperatures. It includes X-ray spectra, both broad band and in a few
cases X-ray emission line measurements, that indicate the presence of
cool gas over a range of temperatures. And it includes optical
studies of Hα emission from filaments that are condensing out of the
flow (see recent work by Hu, Cowie and Wang, 1985). To whatever extent
the details of the cooling flows are open to question, I think their
existence in a great many clusters is not.

III. Early Type Galaxies

a.   Global Properties
An increasing body of evidence accumulated over the past several years
has now established that many early type galaxies contain hot, X-ray
emitting gas (Forman et al. 1979, Bierman, Kronberg and Madore 1982,
Bierman and Kronberg 1983, Nulsen, Stewart and Fabian 1984, Dressel
and Wilson 1985, Forman, Jones and Tucker 1985, Trinchieri and Fabbiano
1985, Stanger and Schwarz 1985). In contrast, spiral galaxy X-ray
emission at ∿ 1 keV is all attributable to discrete sources (Fabbiano
and Trinchieri 1985).
        Figure 3 shows the X-ray and optical luminosities of ∿ 60 early
type (E and S0) galaxies. Notice that none of the upper limits is
restrictive; the data suggest that all elliptical galaxies brighter
than $L_B = 10^{10}$ $L_\odot$ (M_B = -19.5) have X-ray luminosities above 10^{39} erg s^{-1}.
Trinchieri and Fabbiano (1985) have argued that the X-ray emission of
the lower luminosity galaxies, and possibly of all those along the lower
envelope of the distribution, may be dominated by discrete sources, but
Forman, Jones and Tucker (1985) state that the discrete source
contribution will be important only for the least luminous systems.

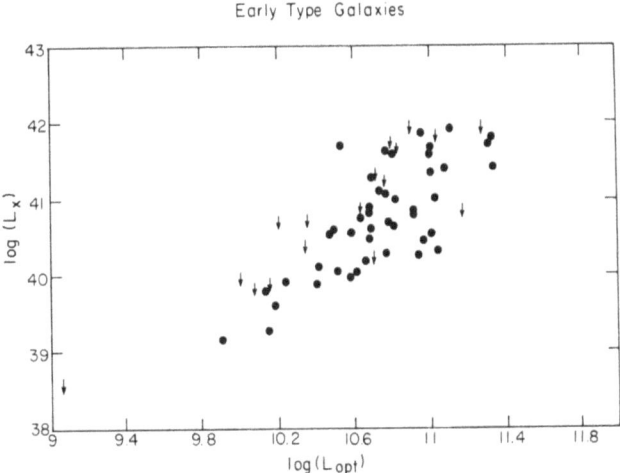

Figure 3. X-ray vs. optical luminosity for a sample of E and S0
galaxies, (From Canizares, Fabbiano and Trinchieri 1986).

For the X-ray luminous galaxies, there is ample evidence that the emission is from hot gas (Forman, Jones and Tucker 1985, Trinchieri and Fabbiano 1985). This is based on spectra of about a dozen galaxies, which appear thermal with temperatures of $\sim 10^7$ K, on a few cases of distorted surface brightness contours indicating that the galactic gas is being pushed or stripped by an external medium, and on a series of other arguments including the consistency of the hot gas picture and the inability of a simple discrete source model to give the observed non-linear L_x vs. L_{opt} distribution.

b.   Evidence for Cooling Flows

As pointed out by Nulsen, Stewart and Fabian (1984), the mere detection of the ellipticals in X-rays rules out a hot galactic wind that blows throughout the galaxy, as suggested by Mathews and Baker 1971 to explain what used to be the absence of gas in ellipticals (see also McDonald and Bailey 1981 and White and Chevalier 1984). This is because a sonic wind would flow away too quickly to maintain the required density of gas within the galaxy, and replenishment by stellar mass loss would be too small by one to two orders of magnitude. Nulsen et al. argued in favor of cooling flows in these galaxies (in agreement with an earlier speculation of mine; Canizares 1981), and there is good evidence that these do exist.

All the early type galaxies with sufficient data show the very high central X-ray surface brightness that implies short cooling times, as required for a cooling flow. For example, Figure 4 shows the deduced gas densitites and cooling times for NGC4636 and NGC4649 (Trinchieri, Fabbiano and Canizares 1986). The cooling times are less than a Hubble time throughout the galaxy, which indicates that a steady state cooling flow should have been established (see Fabian, Nulsen and Canizares 1984). A second piece of evidence is the existence of Hα filaments in many elliptical galaxies (c.f. Caldwell 1984, Demoulin-Ulrich, Butcher and Boksenburg 1984). These could be the galactic counterparts of the filaments seen in cluster cooling flows, as mentioned earlier. One direct piece of evidence that exists for some clusters but not for early type galaxies is X-ray spectral evidence of cooler gas. Such data are simply not available.

One thing to note is that a steady-state cooling flow powered by supernovae at the estimated rate of 0.22 per 100 years per 10^{10} L_B (Tammann 1982) overproduces the X-ray luminosity, and it does not give the correct dependence of L_x on L_{opt} (see Figure 2, White and Chevalier 1984 and Sarazin, this meeting). On the other hand, if the supernova energy input were lower, then a gravity-dominated cooling flow could roughly reproduce the observations (see Canizares, Fabbiano and Trinchieri 1986). In that case one would expect $L_x \sim \dot{M}(<\Delta\phi>_m + \varepsilon)$, where $\dot{M}$ is the mass loss rate of stars in the galaxy, $<\Delta\phi>_m$ is the mean gravitational potential properly weighted according to fraction of mass injected and ε is the specific energy due to thermalization of the gas by the stars. If one takes $\dot{M} \sim 1.5$ $M_\odot$ yr^{-1} $(10^{10}$ $L_\odot)^{-1}$ (Faber and Gallagher 1976) and computes the potential in terms of the galaxy velocity dispersion σ, one obtains a relation between L_x and L_{opt} σ^2.

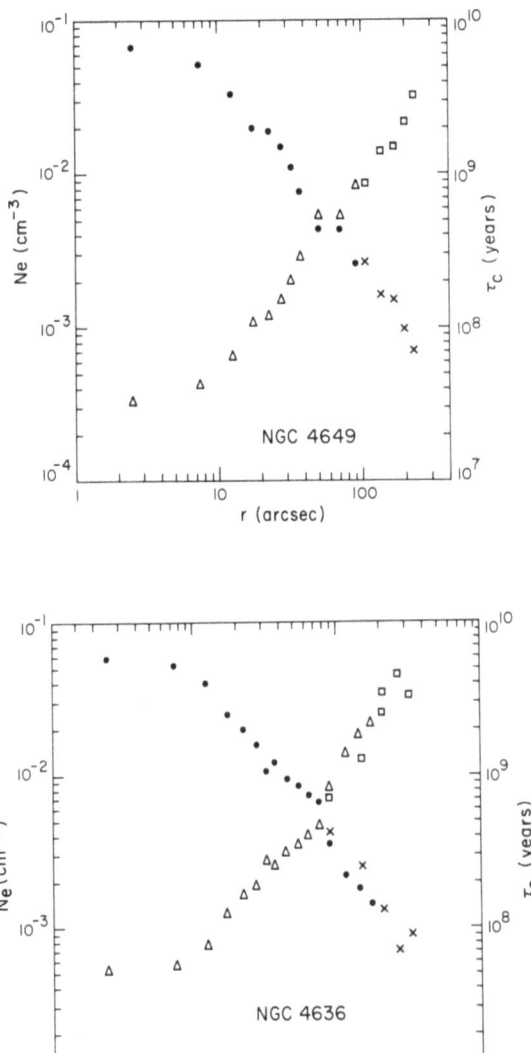

Figure 4. The electron densities and cooling times estimated from the X-ray surface brightness profiles of two early type galaxies. (From Trinchieri, Fabbiano and Canizares 1986).

This is shown in Figure 5 for two assumptions about the mass distribution
in the galaxy (see Canizares et al. 1986 for details). At high values
of $L_{opt}\sigma^2$ most of the luminosity will be emitted as X-rays, whereas at
lower values, where the curve is shown dashed, it could emerge in the UV.
There is a striking correspondence between the data and the simple model,
which has assumptions but no free parameters. A more detailed cooling
flow model is discussed by Fabian (this meeting).

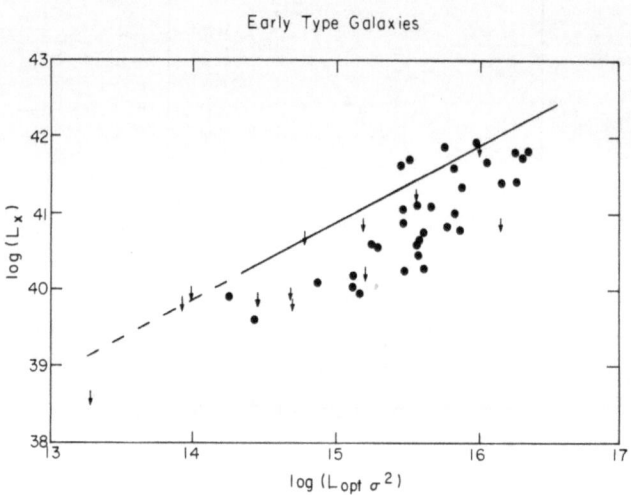

Figure 5. X-ray luminosity (in erg s^{-1}) vs. the product of optical
luminosity and the square of the line-of-sight velocity dispersion (in
units 10^4 $L_{\odot}(kms^{-1})^2$) for a sample of galaxies. The curve is the
expectation for the simple gravity dominated cooling flow model discussed
in the text for a typical galaxy potential. (From Canizares, Fabbiano
and Trinchieri 1986). The dashed portion of the curve corresponds to
cooling flows with temperatures below the X-ray band.

c.   Dark Matter in Early Type Galaxies

As in the clusters, the X-ray emitting gas in early type galaxies can
be used to probe their gravitational potentials and could reveal the
presence of dark matter in the form of galactic halos (Forman, Jones
and Tucker 1985). Let me first note however, that the observations
that galactic winds have been suppressed is not in itself a sufficient
argument in favor of massive halos. It is true that the observed mean
gas temperatures of $\sim 10^7$ K exceed the critical temperature for wind
formation (Bregman 1985, Forman, Jones and Tucker 1985, also McDonald
and Bailey 1981), which is $\sim 10^6$ K for a reasonable galaxy without halo.
But the addition of a plausible halo ten times more massive (but with a
density that falls as the square of the radius) only raises the critical
temperature by a factor of three, which in itself is not sufficient to
suppress the wind (see Canizares, Fabbiano and Trinchieri 1986). The
observed fact that the wind is suppressed is therefore not due to

energetics but is instead probably a dynamical effect: the very short
cooling time in the core of the galaxy is sufficiently small relative
to the flow time that a wind is not established (see Mathews and Baker
1971). At large radii, external pressure may play a role.

The evidence for massive halos must come from Equation 1, which in
units relevant for galaxies can be written as

$$M(<r) = 10^{12}(r/30 \text{ kpc})(T/10^7 \text{ K})[-d\ln(n)/d\ln(r) - d\ln(T)/d\ln(r)]M_{\odot}. \quad (3)$$

As was true for clusters, the first logarithmic derivative is the
quantity that is easiest to determine from the imaging observations.
For most galaxies studied in sufficient detail, the X-ray surface
brightness roughly follows the optical surface brightness out to the
faintest optical isophotes (Trinchieri and Fabbiano and Canizares 1986;
·Forman, Jones and Tucker 1985). This implies a gas density that falls
roughly as $r^{-1.5}$ over much of the galaxy. However, the most interesting
application of Equation 3 is at the largest radii where there are clear
departures from this relation for some galaxies. For example, in the
case of NGC4472 which lies in a subgroup of the Virgo Cluster, the X-ray
isophotes (see Figure 6) suggest strongly that the outer part of the
galactic gas is being pushed by an ambient medium (compare to the still
more extreme case of M86 in Virgo, where the ambient medium is
apparently stripping off much of the galactic gas [Forman et al. 1979]).
This gives different density profiles in different directions, and it
also suggests that dynamical effects might disturb the assumption of
hydrostatic equilibrium at the largest radii. Another example is
NGC4636 in the outskirts of Virgo. The X-ray image of this galaxy shows
a faint ring of excess emission at the largest radii (Stanger and Schwarz
1985, Trinchieri and Fabbiano and Canizares 1986). Unfortunately,
instrumental effects cannot be completely ruled out as the cause of this
feature. If it is real it suggests some interaction of galaxy gas with
circum-galactic material (and may indicate that this galaxy has a partial
wind in addition to its cooling flow). In any event the conservative
approach is to evaluate Equation 3 at radii inside these disturbances.
NGC4649 exhibits yet another behavior: its X-ray surface brightness
roughly follows the optical out to a radius of $\sim 2'$ and then falls
considerably more steeply.

Also as for clusters, the biggest difficulty in determining the
binding mass from Equation 3 is the uncertainty in the temperature and
its profile at the largest radii. Mean temperatures have been measured
for about ten galaxies (Forman, Jones and Tucker 1985, Trinchieri,
Fabbiano and Canizares 1986). These can be used directly if one
assumes that the gas is isothermal out to the radius in question. Such
an assumption is valid if conduction operates efficiently throughout
the galaxy, although there is as yet no detailed model for such a
quasi-static isothermal atmosphere (one difficulty is maintaining the
pressure at the outer boundary). Cooling flow models also give roughly
isothermal temperature profiles (White and Chevalier 1984).

Figure 7 shows the application of Equation 3 to the data of three
galaxies with the full range of possible assumptions (Trinchieri,
Fabbiano and Canizares 1986). The radii have been chosen conservatively,

as noted above, and several choices of possible temperature gradient are
indicated. We can be quite sure of the low temperature limit because
cooler gas could not easily be detected by the Einstein instruments.
The very high temperatures simply become implausible given the indicated
mean temperatures of $\sim 10^7$ K and a desire to limit M/L to values smaller
than the mean for rich clusters. Figure 7 also shows the values deduced
by Forman, Jones and Tucker 1985, who do assume isothermality at a
characteristic temperature of 1.2×10^7 K. Note that they generally
chose to apply Equation 3 at larger radii and used a mean value of
$d\ln(n)/d\ln(r)$ derived from a fit over the whole galaxy. For comparison,
the horizontal lines in the figures mark the range of masses deduced
from the central stellar velocity dispersions by Davies (1981), Tonry
and Davis (1982) and Katz and Richstone (1985).

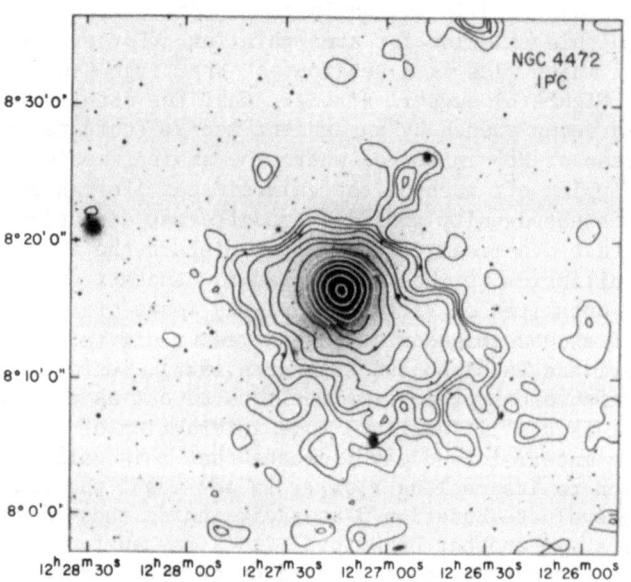

Figure 6. Isophotes of constant X-ray surface brightness superimposed
on an optical image of NGC4472, showing the asymmetry at large radii.
(From Trinchieri, Fabbiano and Canizares 1986).

One can see that conservative assumptions give an order of
magnitude uncertainty in the deduced values of binding mass (or M/L).
This situation could improve considerably if we had good physical models
of the gas which could be fit to the surface brightness profiles and
constrained by the measured mean temperatures (eg. see Fabian's
contribution to this meeting). The lowest values of M/L_B allowed by all
the data range from the value ~ 5 one might attributed to the stars
(Faber and Gallagher 1979) to ~ 20. The values of 10-20 suggested by

the optical data are quite acceptable. It is worth noting that the difficulty of interpreting stellar velocity dispersions is at least as great as those described here for the X-ray data (e.g. see the range of M/L values deduced by Katz and Richstone [1985] for NGC4636 in Figure 7). If the temperatures at the chosen radii are near to the mean value for each galaxy then the M/L_B would be $\gtrsim$ 20-30, and with the large uncertainties, values of $\gtrsim$ 100 are allowed (Forman, Jones and Tucker 1985). With the right theoretical tools and, eventually, better data one can expect X-ray observations to provide very detailed maps of the galactic gravitational potentials.

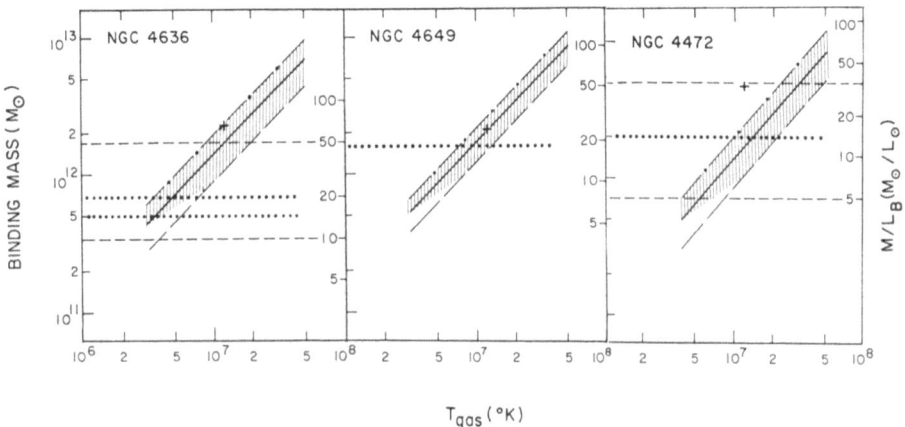

Figure 7. The range of binding masses within R_{max} deduced from the X-ray surface brightness profiles of three early type galaxies as functions of the temperature and temperature gradient at R_{max}. The solid line corresponds to $\alpha = dlnT/dlnr = 0$, while the upper and lower lines correspond to $\alpha = -0.5$ and $\alpha = 0.5$, respectively. The values of R_{max} are 30 kpc for NGC4636 and NGC4649 and 40 kpc for NGC4472. The crosses represent the values from Forman, Jones and Tucker (1985) for $T = 1.2 \times 10^7$ K and slightly different radii. The horizontal lines mark optical results from Davies (1981), Tonry and Davis (1981) and Katz and Richstone (1985). (From Trinchieri, Fabbiano and Canizares 1986).

AKNOWLEDGEMENTS

I thank Len Cowie for communicating his results prior to publication and for permission to quote them here. Much of the work on early type galaxies was performed in collaboration with Ginevra Trinchieri and Pepi Fabbiano. I have benefitted from conversations with Andy Fabian, Bill Forman, Christine Jones, Richard Mushotzky, Paul Nulsen and Wallace Tucker. I thank Elaine Aufiero for her expert preparation of the manuscript.

178   C. R. CANIZARES

REFERENCES

Abramopoulos, F. and Ku, W. 1983, Ap. J., 271, 446.
Bahcall, N. 1982 Ap. J. (Letters) 258, L17.
Bierman, P. and Kronberg, P.P. 1983, Ap. J. (Letters), 268, L69.
Bierman, P., Kronberg, P.P. and Madore, B.F. 1982, Ap. J. (Letters), 256, L37.
Binney, J. and Cowie, L. 1981, Ap. J., 247, 464.
Blumenthal, G.R., Faber, S.M., Primack, J.R. and Rees, M.J. 1984, Nature, 311, 517.
Bregman, J. 1985 private communication.
Caldwell, N., 1984, Pub. Astron. Soc. Pacific, 96, 287.
Canizares, C.R. 1981, in R. Giacconi (ed.), X-Ray Astronomy with the Einstein Satellite, (D. Reidel), 215.
Canizares, C.R., Donahue, M., Trinchieri, G., Stewart, G. and McGlynn, T. 1985 (submitted to Ap. J.).
Canizares, C.R., Fabbiano, G. and Trinchieri, G. 1986 (in preparation).
Cavaliere, A. 1979 in R. Giacconi and G. Setti (eds.), X-Ray Astronomy, (D. Reidel), 217.
Cavaliere, A. and Fusco-Femiano, R. 1976, Astron. Ap. 49, 137.
_____., 1978, Astron. Ap. 70, 677.
Channan, G. and Abramopoulos, F. 1984, Ap. J., 287, 89.
Cowie, L. 1985 private communication.
Davies, R. L. 1981, M.N.R.A.S., 194, 879
Demoulin-Ulrich, M.-H., Butcher, H. and Boksenberg, A. 1984, Ap. J. 285, 527.
Dressel, L. and Wilson, A. 1985, Ap. J., 291, 668.
Dressler, A. 1978, Ap. J., 226, 55.
Fabian, A.C., Nulsen, P.E.J. and Canizares, C.R. 1984, Nature, 310, 733.
Fabbiano, G. and Trinchieri, G. 1985, Ap. J. (in press).
Faber, S. and Gallagher, J. 1976 Ap. J. 204, 365.
Faber, S. and Gallagher, J. 1979, Ann. Rev. Astron. Ap. 17, 135.
Fabricant, D., Lecar, M. and Gorenstein, P. 1980, Ap. J., 241, 552.
Fabricant, D. and Gorenstein, P. 1983, Ap. J., 267, 535.
Fabricant, D. Rybicki, G. and Gorenstein, P. 1984, Ap. J. 286, 186.
Forman, W. and Jones, C. 1982 Ann. Rev. Astron. Ap. 20, 547.
Forman, W., Schwarz, J., Jones, C., Liller, W. and Fabian, A. 1979, Ap. J. (Letters), 234, L27.
Henriksen, M. and Mushotzky, R. 1985a Ap. J., 292, 441.
Henriksen, M. and Mushotzky, R. 1985b, (preprint).
Hu, E., Cowie, L., and Wang, Z. 1985, (preprint).
Jones, C. and Forman, W. 1984, 276, 38.
Katz, N. and Richstone, D. 1985 (preprint).
Kriss, G.A., Cioffi, D. and Canizares, C.R. 1983, Ap. J., 272, 439.
Matilsky, T., Jones, C. and Forman, W. 1985, Ap. J., 291, 621.
McDonald, J. and Bailey, M. 1981, M.N.R.A.S., 197, 995.
Mushotzky, R. 1984, in Proc. Symp. Hot Astrophys. Plasmas, Nice, 1982, Physica Scripta, T7, 157.
Nulsen, P.E.J., Stewart, G.C. and Fabian, A.C. 1984, M.N.R.A.S., 208, 185.
Oemler, A., 1974, Ap. J., 194, 1.
Rothenflug, R. and Arnaud, M. 1985, Astron. Ap., 144, 431.

Rothenflug, R., Arnaud, M., Boulade, O., and Vigroux, L. 1984, in M. Oda and R. Giacconi, (eds.), X-Ray Astronomy '84, (Tokyo: Inst. Space Astronautical Sci.), 391.

Sarazin, C. 1985, Rev. Mod. Phys. (in press).

Stanger, V.J., and Schwarz, J. 1986 submitted to Ap. J.

Stewart, G.C., Canizares, C.R., Fabian, A.C. and Nulsen, P.E.J. 1984, Ap. J., 278, 536.

Tammann, G. 1982, in M. Rees and R. Stoneham, (eds.), Supernovae: A Survey of Current Research, (D. Reidel), 371.

Tonry, J. and Davis, M. 1981, Ap. J., 246, 666.

Trinchieri, G. and Fabbiano, G. 1986, Ap. J. (in press).

Trinchieri, G., Fabbiano, G. and Canizares, C. 1986 Ap. J. (submitted).

White, R.E. and Chevalier, R. A., 1984, Ap. J., 280, 561.

DISCUSSION

DAVIS: Most of us would agree that the limitations on these
observations are the lack of information on temperatures and temperature
gradients. Could you tell us whether AXAF or other satellites are
likely to improve this situation?

CANIZARES: There will probably be some improvement already from ROSAT.
We could have done better with EINSTEIN if we'd known what to look for
during the lifetime of the satellite. Certainly, AXAF will help. It
will give us the ability to make spatially resolved spectral
observations with x-ray CCDs. But I don't know how many systems we
will be able to study. The observations are difficult, because the
surface brightnesses are very low.

P. QUINN: In the mass-radius curves you showed you were finding
a few x 10^{12} $M_\odot$ inside $r \simeq 30$ kpc. I'd like to ask Vera Rubin whether
these values are high compared to masses of spirals.

RUBIN: They are upper limits on the range of masses for spirals.

REES: Can you say something about the total amount of heavy elements?
Is it still true that the heavy element abundance can only be measured
in the cores of clusters, or can one now state a higher lower limit to
the total amount of heavy elements present?

CANIZARES: All of the information on heavy element abundances comes
from integrated spectra observed with instruments with large fields of
view; such observations are heavily weighted toward cluster cores. In
all systems which have been observed, the data are consistent with heavy
element abundances of about half the solar value. But there is no
information about abundances in individual elliptical galaxies, and no
information about the outskirts of even very rich clusters.

REES: But if one is having a hard time making the heavy elements, can
one just fit the core of a cluster, so that the total mass of heavy
elements is not too large?

CANIZARES: I think so. If you can find a good way of segregating the
heavy elements in the core, your assumption is consistent with the
observations.

GUNN: I find it very striking that the x-ray luminosity so nicely
follows the optical luminosity, in view of the fact that the x-rays come
from gas. The power-law dependences of the gas distribution and the
x-ray emission are entirely different. But there is evidence now that
elliptical galaxies have a higher proportion of globular clusters than
do later-type galaxies, and that the globular cluster distributions in
ellipticals have larger core radii than the galaxies. Could the x-rays
from elliptical galaxies entirely be due to emission from globular
cluster x-ray sources?

CANIZARES: The evidence for thermal emission is that the spectrum is thermal, rather than the harder spectrum you see from compact sources. Another problem with your suggestion is that the number of globular clusters in a galaxy could not be linearly proportional to the luminosity L, since we find $L_x \propto L^{1.6}$.

BINNEY: You said that many ellipticals have been detected. Does that include ellipticals in places like the Coma Cluster which have an appreciable intergalactic medium?

CANIZARES: No. The surface brightness of the cluster itself makes the measurement of individual galaxies difficult. The existing limits on the x-ray emission from galaxies in clusters don't really tell us much about whether the galaxies have their own gas or not. There have been reports of detections of galaxies in A1367 by Bechtold et al., but we have reexamined the data and are not convinced that you can detect galaxies against the cluster background. And in no case has anyone set a limit on the x-ray emission from a galaxy in a cluster that is lower than the luminosities observed for more isolated galaxies.

BINNEY: I think this is crucial for the question asked by Jim Gunn. If the x-ray emission were really from globular cluster sources, it would also be present in galaxies in clusters.

SCHECHTER: In all the images you've looked at, have you seen any evidence for halos without galaxies? I.e., have you seen any isolated diffuse sources?

CANIZARES: Not that I know of.

SCHECHTER: But, for example, Ed Turner needed offset mass distributions. And one might imagine that there are halos floating around that failed to form galaxies within them. Moreover, in some scenarios one could have failed galaxies, i.e., halos that didn't accumulate galaxies. These things might be in the vicinity of galaxies that we do see, so it makes some sense to look at existing exposures.

CANIZARES: That is being done. But in every case out of the 500 or so that have been looked at where there is an extended source, there is a rich cluster of galaxies with $z \lesssim 0.3$ which perfectly well explains the emission. So I don't see any evidence for empty halos at this point.

LAKE: I think it would be hard to find an empty halo. The cooling times you see involve flows of $1 - 10$ $M_\odot$ per year. If you don't have any supernovae to heat the gas, then over a Hubble time you are surely going to make a galaxy of mass $10^9 - 10^{10}$ $M_\odot$ at the center.

X-RAY HALOS IN GALAXIES AND CLUSTERS OF GALAXIES: THEORY

Craig L. Sarazin[*]
Department of Astronomy, University of Virginia,
Charlottesville, VA 22903 and
Joint Institute for Laboratory Astrophysics, University of
Colorado and National Bureau of Standards, Boulder, CO 80309

ABSTRACT. X-ray measurements provide an excellent method to determine
the amount and distribution of the dark matter in clusters. Unfortunately, accurate temperature profiles, necessary to this method, are
currently not available. However, if the intracluster gas is assumed
to have a monotonically decreasing temperature, one finds that the
dark matter is strongly concentrated to the cluster center, and has a
mass which only exceeds the known baryonic mass by a factor of about
three. On a second topic, cooling flows are shown to be a very common
feature of cluster central and normal elliptical galaxies. The cooling gas is probably ultimately converted into low mass stars.

1. INTRODUCTION

In this paper, two topics will be reviewed. First (Sec. 2), I will
discuss the use of x-ray measurements to derive mass profiles for
clusters of galaxies and individual galaxies, and I will report on
some recent applications of this method to clusters. Second (Secs. 3
and 4), I will discuss the cooling flows which appear to be a common
feature of ellipticals at the centers of clusters as well as relatively isolated elliptical galaxies. I will argue that the cooling
gas is consumed by low mass star formation. If very low mass stars
are formed, this could contribute to "dark matter" halos around galaxies.

---

[*]1985-86 JILA Visiting Fellow; permanent address, University of
Virginia.

J. Kormendy and G. R. Knapp (eds.), Dark Matter in the Universe, 183-199.

2. MASS DETERMINATIONS AND HYDROSTATIC EQUILIBRIUM

2.1. Method

Masses for individual galaxies and clusters can be derived by assuming
that the x-ray emitting gas is in hydrostatic equilibrium with the
gravitation field. This is a reasonable assumption as long as the
cluster is stationary (the gravitational potential does not change on
a sound-crossing time), other forces (magnetic fields, etc.) are not
important, and gas motions are significantly subsonic.

Under these circumstances, the gas obeys the hydrostatic equation
and the total mass can be determined from the variation of pressure
and density. This method has a number of advantages over the use of
stars (in galaxies) or galaxies (in clusters) as test particles to determine the gravitational potential. First, the gas is a collisional
fluid, and the particle velocities are isotropically distributed. On
the other hand, stars in galaxies or galaxies in clusters are collisionless, and uncertainties in the distribution of particle orbits
can significantly influence the derived mass distribution (see, for
example, Tonry 1983). Second, the statistical errors associated with
mass determinations for clusters from the gas distributions are much
smaller than those based on galaxy distributions, as there are only
~10^2 bright galaxies in a rich cluster. Third, better statistics in
the x-ray measurements means that it is considerably easier to avoid
background contamination, and to resolve possible uncertainties due to
subclustering (Geller and Beers 1982). Fourth, x-ray mass determinations are not very sensitive to the shape of the galaxy or cluster
(Strimpel and Binney 1979; Fabricant, Rybicki, and Gorenstein 1984).

The first applications of x-ray distributions to derive mass
distributions were by Bahcall and Sarazin (1977) and Mathews (1978)
in M87. The method has been developed extensively by Fabricant,
Gorenstein, and collaborators (Fabricant, Lecar, and Gorenstein 1980;
Fabricant and Gorenstein 1983; Fabricant, Rybicki, and Gorenstein
1984). Ideally, one would measure the spatially and spectrally resolved x-ray surface brightness of the cluster or galaxy $I_\nu(\vec{b})$, where
$h\nu$ is the x-ray photon energy and $\vec{b}$ is the projected position relative
to the center. This would be inverted to give the local x-ray emis-
sivity $\varepsilon_\nu(\vec{r})$, where $\vec{r}$ is the position relative to the cluster center.
This deconvolution is stable because the observed x-ray images of
regular clusters or galaxies are quite smooth. To deconvolve the projected surface brightness one must assume that the actual gas distribution is spherical or spheroidal (Strimpel and Binney 1979) or, more
generally, has an axis of symmetry in the plane of the sky (Fabricant,
Rybicki, and Gorenstein 1984). However, the resulting mass distributions are not affected strongly by the shape. For a spherical cluster, the Abel integral inversion for $\varepsilon_\nu(\vec{r})$ is:

$$\varepsilon_\nu(\vec{r}) = -\frac{1}{\pi}\frac{d}{dr^2}\int_{r^2}^{\infty}\frac{I_\nu(b)\,db^2}{(b^2-r^2)^{1/2}} \quad . \tag{1}$$

The x-ray emissivity of a hot plasma depends on its electron density n_e, its temperature T, and its abundances (Sarazin and Bahcall 1977)

$$\varepsilon_\nu = n_e^2 \, \Lambda_\nu \, (T, \text{ abundances}) \quad . \tag{2}$$

Heavy elements mainly produce discrete line features in x-rays; the strength of these features determines the heavy element abundances. The x-ray continuum is exponential $\varepsilon_\nu \propto \exp(-h\nu/kT)$, and thus the spectral shape of the emissivity $\varepsilon_\nu(r)$ determines $T(r)$, while its normalization gives $n_e(r)$. Then the hydrostatic equation gives the total mass $M(r)$ interior to r as

$$M(r) = - \frac{k \, T(r) r}{\mu \, m_p \, G} \left\{ \frac{d \ln n_e}{d \ln r} + \frac{d \ln T}{d \ln r} \right\} \quad , \tag{3}$$

where μ is the mean molecular weight. It is important to note that the mass depends only weakly on $n_e(r)$ (only on its logarithmic derivative), but depends strongly on the temperature $T(r)$.

2.2. Applications to Galaxy Clusters

The Einstein x-ray observatory produced x-ray images of clusters with excellent spatial resolution, but the imaging detectors had rather poor spectral resolution. Moreover, the telescope was sensitive only to x-ray photons with energies $h\nu \lesssim 4$ keV, while the typical x-ray cluster has a temperature $kT \sim 7$ keV. Since the exponential thermal bremsstrahlung spectrum is flat for $h\nu \lesssim kT$, the Einstein x-ray images do not give much information on the run of temperatures in the intracluster gas. On the other hand, since the instrumental response of Einstein was very insensitive to the temperature, excellent gas density distributions $n_e(r)$ were determined (Jones and Forman 1984). Unfortunately, the mass in equation (3) is mainly affected by the temperature distribution.

2.2.1. Isothermal Models. In order to resolve this uncertainty in the temperature distribution, one approach has been to assume a simple "polytropic" equation of state connecting the temperature and density $T \propto n_e^{\gamma-1}$, where $\gamma = 1$ means the gas is isothermal, while $\gamma = 5/3$ means the gas is adiabatic (isentropic). Some examples of these mass estimates include Fabricant, Rybicki, and Gorenstein (1984) and Vallée (1981).
     The assumption that the gas is isothermal leads to a particularly simple density distribution for the gas. If the total mass has a density distribution given by an "analytic King" model $\rho_{tot} \propto \left(1+(r/a)^2\right)^{-3/2}$ (King 1962) where a is the cluster core radius, and has an isotropic velocity distribution with a one-dimensional velocity dispersion σ, then the isothermal gas distribution is

$$n_e(r) = n_e(0) \left[1 + \left(\frac{r}{a}\right)^2 \right]^{-3/2 \, \beta} \tag{4}$$

where

$$\beta \equiv \frac{\mu\, m_p\, \sigma^2}{kT} \tag{5}$$

(Cavaliere and Fusco-Femiano 1976; Bahcall and Sarazin 1978).

Equation (4) provides an excellent fit to the observed surface brightnesses of most clusters, with a typical value of $3\beta/2 \approx 1$ or $\beta \approx 2/3$ (Jones and Forman 1984). The quality of these fits suggests that the gas may indeed be isothermal; in that case, the temperature can be determined by a global (not spatially resolved) x-ray spectrum of the cluster. Then, the total and gas masses can easily be found. These determinations give somewhat smaller total cluster masses than optical (virial) analyses, and give somewhat higher gas masses than most pre-Einstein analyses.

On the other hand, there are several arguments which suggest that the isothermal model cannot be consistently applied to gas distributions in clusters (Henriksen and Mushotzky 1985a; Henriksen 1985; Mushotzky 1985). The first problem is that when β is derived from x-ray temperatures (derived from global spectra) and optical galaxy velocity dispersions, larger values are generally found, with $\beta \approx 1.2$ being typical (Mushotzky 1985). A second problem with the isothermal model is that, for $\beta \approx 2/3$, the gas is more extended than the assumed total mass. In the outer parts of the cluster, the gas density actually dominates the total density, and thus it is inconsistent for the gas density and total density to have different distributions. Third, the HEAO-1 A-2 global x-ray spectra of some of the best studied clusters (for example, Coma; Henriksen and Mushotzky 1985b) cannot be fitted by emission from gas at a single temperature.

2.2.2. Global X-ray Spectra and Monotonic Temperature Gradients. Excellent global (not spatially resolved) x-ray spectra exist for bright clusters from the HEAO-1 A-2 detectors. These can generally not be fitted by emission at a single temperature (Henriksen 1985). These spectra can be used to determine how much gas (or, more precisely, how much $n_e^2 V$, where V is volume) is present at each gas temperature T, but they cannot tell us where this gas is located in the cluster. On the other hand, the Einstein x-ray images give us $n_e(r)$ (which can be integrated to give $n_e^2 V$), but give no information on temperatures. However, if we assume that T decreases monotonically with radius, we can match these two determinations of $n_e^2 V$ and derive $T(r)$.

Such analyses have been done for the Coma and Perseus clusters by Henriksen (1985), Henriksen and Mushotzky (1985b), and Cowie (1985). While the method described above is nonparametric (it involves direct determinations on $n_e^2 V$ from the x-ray spectra and images), these authors parameterized the distributions by assuming a polytropic equation of state $T \propto n_e^{\gamma-1}$.

Since the observed gas densities vary at large distances like an isothermal sphere ($n_e \sim r^{-2}$), while these determinations require that the gas temperature decreases, the total density will always decrease with radius more rapidly than the gas density. Since the total density is the sum of the gas density, the galaxy density, and the dark

matter density, these mass determinations indicate that the dark matter is concentrated to the cluster center. For Coma, this method
gives a dark matter distribution which is more centrally concentrated
than that of the galaxies, which is in turn more centrally condensed
than that of the x-ray emitting gas. Values of the mass-to-light
ratio for the entire cluster of $M_{tot}/L_V \approx 100$ ($H_o/50$ km/s/Mpc) are
found, with the hot gas contributing about 30% of the total mass.
    This leads to a picture for the dark matter in clusters which
is considerably different than has previously been given. First,
the ratio of dark mass to visible baryonic mass would only be two or
three, not the ten or twenty sometimes assumed. Moreover, if the dark
matter is more centrally concentrated than the visible baryonic matter, it would seem reasonable to assume that it has undergone dissipation. These properties of the cluster dark matter, taken alone, would
suggest a baryonic origin, perhaps in substellar condensations or
black holes.
    The major uncertainty in these analyses of cluster mass profiles
is the assumption that the temperature decreases monotonically outwards. While this seems quite plausible, and simple infall or galaxy
ejection models for the origin of the intracluster gas usually produce
such monotonic temperature gradients, there is no compelling physical
argument requiring a monotonic temperature gradient. Since cooling
and thermal condition timescales are long in the outer parts of clusters, the gas temperature profiles are probably determined by uncertain conditions at the time of cluster formation.
    With AXAF, it will be possible to simultaneously determine the
density and temperature profiles in clusters, without assuming a monotonic temperature variation. Until that is possible, these results on
the dark matter distribution in clusters must be viewed as tentative.

3. COOLING FLOWS ONTO CENTRAL CLUSTER GALAXIES

3.1. Evidence for Cooling Flows and Rates

At the centers of many rich and poor clusters of galaxies, the gas
density is high enough ($n_e \gtrsim 3 \times 10^{-3}$ cm^{-3}) that the intracluster gas
can cool over a Hubble time. Thus, we expect gas to cool and flow
into the centers of many clusters. Evidence that this is indeed
occurring includes the detection of peaks in the soft x-ray surface
brightness at the cluster center, central densities and temperatures
implying cooling times shorter than the Hubble time, and central temperature inversions ($dT/dr > 0$). The strongest evidence comes from
the detection of soft x-ray line emission from low ionization stages
produced at temperatures of 10^6-10^7 K coming from the cluster center.
An excellent review of cooling flows has been given by Fabian, Nulsen,
and Canizares (1984). A survey of cooling flows by Stewart et al.
(1984b) shows that the cooling rates range from 10-1000 M$_{\odot}$/yr. About
30 clusters are known with cooling rates on this order.
    The cooling rate is inferred directly from the x-ray luminosity
of the central surface brightness peak in x-ray emission from the

cluster, where the cooling time t_{cool} is less than a Hubble time.
The flow velocity, which is expected to be ~r/t_{cool}, is very subsonic
except possibly at the very center of the flow, and the thermal energy
of the gas is considerably larger than the change in the gravitational
potential. Under these circumstances the gas cools isobarically, and
the luminosity is

$$L_{cool} \approx \frac{5}{2} \frac{\dot{M}}{\mu m_p} kT$$

$$\approx 1.7 \times 10^{44} \text{ ergs/s} \left(\frac{\dot{M}}{100 \ M_\odot/yr}\right) \left(\frac{T}{8 \times 10^7 K}\right) \ . \quad (6)$$

In many cases, extended optical line emitting filaments (emitting
Balmer and Lyman lines, and [O II], etc.) are seen near the centers of
these cooling flows (Cowie et al. 1983). The line luminosities are
consistent with roughly the same cooling rates through the temperature
range T ~ 10^4 K as at x-ray temperatures, and the inferred gas pressure in the line emitting filaments are similar to those expected in
the cooling flows.

3.2. Accretion by Central Galaxies

In essentially every case of a cluster with a cooling flow, there is a
central dominant galaxy located at the center of the flow (Jones and
Forman 1984). In most cases, the central galaxy is a radio source,
and there is a correlation between $\dot{M}$ and the radio luminosity. Now,
these central galaxies cannot cause cooling flows. Since the velocity
dispersion of the galaxy is small compared to the cluster velocity dispersion or the sound speed in the intracluster gas, a central galaxy
has only a very small influence on the density of the gas and cannot
initiate cooling. However, once the gas has cooled so that its sound
speed is comparable to the galaxy velocity dispersion, the cooling
flow will be focused onto the central galaxy if it moves slowly enough
so the gas can cool before the galaxy moves away.
      Thus, the fact that cooling flows and central dominant galaxies
are strongly correlated means either that they both result from a
common cluster property (e.g., high central density) or that cooling
flows produce central dominant galaxies.

3.3. Fate of the Cooling Gas

If the cooling rates derived from the x-ray observations are correct,
and the cooling flows are long-lived, and gas is not expelled from the
central galaxies, a considerable mass M_{acc} will be accreted

$$M_{acc} = 10^{12} \ M_\odot \left(\frac{\dot{M}}{100 \ M_\odot/yr}\right) \left(\frac{t}{10^{10} \ yr}\right) \quad (7)$$

where t is the age of the cooling flow. This mass is comparable to
the mass of the central galaxy. What has happened to all this accreted gas?

Observations indicate that M_{acc} is not present in the form of H I or H II (Cowie et al. 1983; Burns, White, and Hayes 1981). It might be present in H_2, because CO observations have not been done for many cD galaxies. This much mass could not have been accreted by a central black hole, because the galaxy central velocity dispersion would be dramatically elevated and because the energy released by the accretion would vastly exceed the luminosities of these galactic nuclei. This much gas cannot be going into high mass star formation, because the galaxies would be bluer and have higher supernova rates than are observed.

3.4. Low Mass Star Formation

The cooling gas could be forming low mass stars $M_* \lesssim 1\ M_\odot$. While I cannot give a convincing argument as to why low mass star formation should be favored in cooling flows, the following argument is, at least, suggestive (Jura 1977; Fabian, Nulsen, and Canizares 1982; Sarazin and O'Connell 1983). As the gas cools, it becomes thermally instable, and slightly denser, cooler clumps will cool more rapidly. Presumably, these clumps form the observed optical emission line filaments. Now, as they cool further they will eventually become gravitationally unstable, if their mass exceeds the Jeans mass M_J. For static, nonmagnetic, isothermal spherical gas cloud, this is (Spitzer 1978)

$$M_J = 1.2\ \left[\left(\frac{kT}{\mu m_p}\right)^4 \frac{1}{G^3 P}\right]^{1/2}$$

$$\approx 0.54\ M_\odot\ \mu^{-2}\ \left(\frac{T}{10\ \mathrm{K}}\right)^2\ \left(\frac{P}{10^{-9}\ \mathrm{dynes\ cm^{-2}}}\right)^{-1/2} . \tag{8}$$

Now, the temperatures in molecular clouds in the cooling flows are probably somewhat lower than those in clouds in the disk of our galaxy. On the other hand, the ambient pressure in the cooling flows is $\sim 10^{-9}$ dynes/cm^{-2}, which is 10^3–10^4 times larger than that in the disk of our galaxy. Thus, in cooling flows $M_J \sim 1\ M_\odot$, whereas it is $\sim 100\ M_\odot$ in our galaxy. Since it is difficult to assemble a protostellar cloud with a mass exceeding M_J (it would collapse first), M_J may form an upper limit to the mass of star formation.

Is there any evidence for ongoing low mass star formation in cooling flow galaxies? The best known example is NGC 1275 in the Perseus cluster. It looks, from its surface photometry, like a typical giant elliptical galaxy, but over most of the optical extent its spectrum is dominated by A stars (Minkowski 1968; Rubin et al. 1977). Similarly, the cD in A1795 and PKS 0745-191, the central dominant galaxy in a southern x-ray cluster, have A-F star spectra (McNamara, O'Connell, and Sarazin 1985; Fabian et al. 1985). Thus these galaxies, which have several of the highest cooling rates observed, contain young stars with masses of 1–3 $M_\odot$.

3.4. Distribution of Stars Formed From Cooling Flows

If one takes the total accreted mass in a Hubble time for clusters
with large cooling rates, and divides it by the total optical luminosity of the central galaxy, mass-to-light ratios of ~10 $M_\odot/L_\odot$ are
found. Since the Jeans mass in the cooling flow is ~1 $M_\odot$, which is
about the mass of the luminous stars in ellipticals, it is possible
that all of the observed stars in these central galaxies were formed
from the cooling flows. Ignoring for the moment the evolution of the
cooling rate and the galaxy stellar distribution, one would predict
that stars are now forming with a distribution which is similar to the
observed luminosity of the galaxy (de Vaucouleurs' or Hubble law).

Alternatively, it might be the very low mass star formation is
favored in cooling flows, with masses <0.1 $M_\odot$. These stars would be
invisible optically. If the cooling rates were larger in the past,
these "black dwarfs" might form the "dark matter" halos which have
been observed around several central cluster galaxies (see the papers
by Canizares and by Fabian in this volume). In that case, the distribution of newly formed stars ought to mimic the distribution of
mass in the dark halo, with $\dot{M}(r) \sim r$.

It is very important to determine the distribution of the newly
formed stars within the cooling flow, or equivalently, the variation of
$\dot{M}(r)$ in the flow. As gas cools, forms clumps, and is converted into
stars, the rate at which hot, diffuse gas is flowing into the galaxy
center will decrease. Such a reduction in $\dot{M}(r)$ with decreasing r will
appear as a flattening in the x-ray surface brightness peak associated
with the cooling flow. Thus, the x-ray surface brightness profiles
can be used to derive the variation of $\dot{M}(r)$ with r, if the gas is
assumed to be in steady-state inflow (Fabian, Nulsen, and Canizares
1984; Stewart <u>et al</u>. 1984a; White and Sarazin 1985). These analyses
show that the x-ray surface brightness profiles require that $\dot{M}(r)$ decrease with decreasing r; hot gas is being removed from the flow and
presumably converted into stars even at large radii ~50 kpc from the
galaxy center. Unfortunately, White and Sarazin found that the uncertainties in temperature profiles prevented a unique determination of
the profiles of $\dot{M}(r)$. AXAF, with its better spectral response, should
allow a direct and accurate determination of $\dot{M}(r)$.

3.5. Cooling Flow Models with Star Formation

As an alternative to the direct deconvolution of $\dot{M}(r)$, White and
Sarazin (1985) have calculated cooling flow models including star
formation. The star formation rate was assumed to be proportional
either to the cooling rate, or to the rate of growth of linear thermal
instabilities.

These models give the equilibrium distribution of the newly
formed stars. These distributions generally are fairly close to the
de Vaucouleurs' profile assumed for the background galaxy, suggesting
that cooling flows form the optical portions of galaxies.

4. COOLING FLOWS ONTO NORMAL ELLIPTICAL GALAXIES

4.1. Observations

Recent x-ray observations indicate that many early-type galaxies which
are not in the cores of rich compact clusters have extended x-ray
emission (Forman, Jones, and Tucker 1985; Trinchieri and Fabbiano
1985; Nulsen, Stewart, and Fabian 1984). These galaxies have x-ray
luminosities of $L_x \sim 10^{39}-10^{42}$ ergs/s, and sizes of typically $R_x \sim$
50 kpc. There is a strong correlation between the x-ray and optical
luminosities of the galaxies $L_x \propto L_B^{1.6-2.0}$, where L_B is the blue luminosity. It seems most likely that the x-ray emission is thermal
with typical gas temperatures $T \sim 10^7$ K. The gas densities in the
inner parts of the coronae exceed 0.01 cm^{-3}, and vary roughly as
$r^{-3/2}$. The total gas mass is typically $M_g \sim 10^9-10^{10}$ $M_\odot$, and the
ratio of gas mass to luminous stellar mass M_* is typically $M_g/M_* \sim$
0.02. (See the paper by Canizares in this volume.)

4.2. Source of the Gas

These galaxies are sufficiently far from the cores of clusters that it
is unlikely that the gas is accreted intracluster gas. It seems most
likely that this gas is simply the result of normal stellar mass loss.
The present rate of stellar mass loss per unit volume $\dot{\rho}$ can be written
as $\dot{\rho} = \alpha_* \rho_*$, where ρ_* is the mass density of stars, and α_* is the inverse of the stellar mass loss timescale. Stellar evolution studies
suggest that $\alpha_* \approx 1-2 \times 10^{-12}$ yr^{-1}. Thus, if the lifetime of the
galaxy is t, the total accumulated gas mass M_g is

$$M_g/M_* \approx 0.01-0.02 \left(\frac{t}{10^{10} \text{ yr}}\right) \frac{\langle\alpha_*\rangle}{\alpha_*} \qquad (9)$$

where $\langle\alpha_*\rangle$ is the average value of α_* over the lifetime of the galaxy.
Given that the rate of stellar mass loss was considerably higher in
the past, the observed mass of gas can easily be produced in this way.

4.3. Dynamics

What is the dynamical state of the gas? Assuming that the gas can be
treated as spherically symmetric and homogeneous over scales comparable to the radius from the galaxy center, there would appear to be
four possibilities, determined by the rates of heating and cooling in
the gas. First, if the gas is heated sufficiently, it could form a
wind and blow out of the galaxy (Mathews and Baker 1971). Prior to
the discovery of diffuse x-ray emission from ellipticals, galactic
winds were generally invoked as a mechanism to remove the gas produced
by stellar mass loss, and to explain the absence of significant amounts
of H I in these galaxies. This argument no longer appears tenable.
The amount of hot gas in ellipticals agrees with the amount expected
from stellar mass loss. Thus, the absence of cool gas in ellipticals
is due to the fact that most of the gas is hot, and not to the lack

of any sort of gas. In fact, x-ray emission from ellipticals is completely incompatible with the presence of a global galactic wind. In
a wind, the gas would be removed on the sound crossing time of the
galaxy ($\lesssim 10^8$ yr). Unless the stellar mass loss rate was $\sim 10^2$ times
larger than expected, the gas density would be $\sim 10^2$ times smaller and
the x-ray luminosity $\sim 10^4$ times smaller than is observed.

If neither heating or cooling were important (or if they were
in a stable equilibrium), the gas might just accumulate in the galaxy
potential well (Forman, Jones, and Tucker 1985). Using the density
profiles of Forman et al. and assuming a temperature of 10^7 K, I have
calculated the radius r_c at which the cooling time is equal to 10^{10} yr;
this lies in the range $r_c \approx (\frac{1}{2} - 1) \times R_x$. Thus, over nearly all the
observed gas extent, cooling is very effective. Unless the heating
rate balances cooling, the gas would cool and flow into the galaxy
center. Even if there were a heating mechanism which balanced cooling, this thermal equilibrium would be very unstable and the gas would
form dense, cool clumps which fall into the galaxy center. Given that
the cooling time is short, it is unlikely that the gas would just accumulate in the galaxy potential.

Third, the gas might form a cooling flow, in which gas was
ejected by stars, cooled, and flowed into the galaxy center (White
and Chevalier 1984). Given the short cooling time in the gas and the
problems with wind or static models, I believe this must be the correct dynamical model for the gas.

A fourth, hybrid possibility is a "partial wind" (White and
Chevalier 1984). Here, gas forms a cooling flow in the inner parts
of the galaxy where the gas density is large, but forms a wind in the
outer parts of the galaxy where heating can overcome both cooling and
the gravitational binding of the gas. Unless the gas in elliptical
galaxies is pressure-confined by "intergalactic" gas, it seems likely
that a wind would form in the outermost parts of a galaxy. However,
the gas density in such a wind would probably be too low to be observed in x-rays (as argued above). Since the observed x-ray halos
in ellipticals extend to essentially the full optical extent of the
galaxies ($R_x \sim 50$ kpc), any possible partial wind must start very far
out in the galaxy.

I conclude that most of the x-ray emitting gas in elliptical
galaxies forms a cooling flow.

4.4. Heating of the Gas

4.4.1. Gravitational Heating.
Gas ejected from a star is initially
moving with the orbital velocity of the star. The distribution of
these orbital velocities is determined by the stellar velocity dispersion σ_* (one-dimensional). Thus, the ejected gas is given an energy
per unit mass of $3\sigma_*^2/2$. If the gas forms a cooling flow, then subsequent infall through the galaxy gravitation potential will give it
a similar amount of energy. Based on numerical calculations, I find
that the total gravitational heating is about $3\sigma_*^2$. Let us define an
injection temperature T_{inj} such that $kT_{inj}/\mu m_p$ is the energy per unit
mass given the gas. Then, for gravitational heating T_{inj} is

$$T_{inj} \approx 1.4 \times 10^7 \text{ K } \left(\frac{\sigma_*}{250 \text{ km/s}}\right)^2 \quad . \tag{10}$$

4.4.2. Type I Supernova Heating. The other major source of heating in the gas is likely to be Type I supernovae. If the rate of mass ejection in supernovae is $\alpha_{SN}\rho_*$ and gas is ejected at a velocity V_{SN}, then the heating per unit mass by supernovae is $(\alpha_{SN}/\alpha_*) V_{SN}^2/2$. For the Type I supernova rate of Tammann (1974), I find

$$T_{inj} \approx 6 \times 10^7 \text{ K} \tag{11}$$

although this is very uncertain, in part because of the small number of supernovae used in computing the rate.

Comparing the heating of supernovae and gravitational heating, it seems supernova heating should dominate. However, I will now argue that gravitational heating must dominate in elliptical galaxies, and that the Type I supernova rate must be lower than previously thought.

4.5. X-ray Luminosities

Balancing the heating of the gas with its cooling, the x-ray luminosity must be $L_x = \dot{M}(kT_{inj}/\mu m_p)(\Lambda_x/\Lambda)$. Here, (Λ_x/Λ) is the fraction of the cooling radiation which occurs in the observed x-ray band, which is expected to be very nearly unity if $T \sim 10^7$ K. $\dot{M}$ is the total rate of gas injection by stellar mass loss, which is $\dot{M} = \alpha_* M_* = \alpha_* (M/L_B)_* L_B$, where M_* is total stellar mass, $(M/L_B)_*$ is the mass-to-light ratio of the stellar population, and L_B is the blue luminosity of the galaxy. Combining this with equations (10) and (11) gives

$$L_x = L_B \alpha_* \left(\frac{M}{L_B}\right)_* \left(\frac{\Lambda_x}{\Lambda}\right) \left[3\sigma_*^2 + \left(\frac{\alpha_{SN}}{\alpha_*}\right)\frac{V_{SN}^2}{2}\right] \quad . \tag{12}$$

If supernovae dominate the heating, then the second term in brackets is larger. Now, all of the terms in equation (12) except L_B and σ_*^2 are either unity $[(\Lambda_x/\Lambda)]$ or determined by the stellar population of the galaxy $[\alpha_*, (M/L_B)_*, \alpha_{SN}, V_{SN}]$. Thus, if elliptical galaxies of different luminosities have essentially the same stellar population, then we expect $L_x \propto L_B$ if supernova heating dominates. However, the observed relationship between L_x and L_B is steeper, $L_x \propto L_B^{1.6-2.0}$. Moreover, if we take typical values of $\alpha_* \approx 1.5 \times 10^{-12}$ yr^{-1}, $(M/L_B)_* \approx 8$ $M_\odot/L_\odot$, and T_{inj} from equation (11), we find $L_x \approx 6 \times 10^{41}$ ergs/s $(L_B/10^{11} L_\odot)$. For most galaxies, this greatly exceeds the observed x-ray luminosity. Thus, if supernova heating dominates, neither the observed x-ray-optical correlation nor the observed x-ray luminosities can be explained.

If the heating of the gas is gravitational, then the energy per unit mass is determined by the stellar velocity dispersion σ_*. Now, the velocity dispersion is known to correlate strongly with the optical luminosity of the galaxy, with $L_B \propto \sigma_*^{3-4}$ (Tonry 1981; Faber and Jackson 1976). If gravitational heating dominates, then equation (12) gives $L_x \propto L_B^{1.5-1.67}$, where the higher exponent in the L_x-L_B

relationship corresponds to the lower exponent in the $L_B-\sigma_*$ relationship (Nulsen, Stewart, and Fabian 1984). If the $L_B-\sigma_*$ relationship is written as $L_B \approx 10^{11}\, L_\odot\, (\sigma_*/350\ km/s)^{3-4}$, then the x-ray luminosity is

$$L_x \approx 2.8 \times 10^{41}\, [\frac{L_B}{10^{11}\, L_\odot}]^{1.5-1.67}\ \mathrm{ergs\ s}^{-1} \tag{13}$$

which fits the observed x-ray luminosities reasonably well. The corresponding inflow rate of gas due to stellar mass loss is

$$\dot{M} \approx 1.2\, [\frac{L_B}{10^{11}\, L_\odot}]\ M_\odot/yr \quad . \tag{14}$$

I conclude that the gas in elliptical galaxies is heated primarily by gravitation and not by supernovae, and that the rate of Type I supernovae in ellipticals probably has been overestimated by a factor of at least three.

4.6. Distribution of the Hot Gas

The form of the gas distribution in the cooling flows can be derived from the energy equation in the flow. It is easy to show that these flows are subsonic except possibly very near the center, and thus kinetic energy can be ignored. The change in the enthalpy flux is balanced by the rates of heating and cooling in the gas. The rate of heating is associated with stellar mass loss, and thus is proportional to the stellar density. For an elliptical galaxy, the stellar density increases rapidly toward the galaxy center, roughly as $\rho_* \propto r^{-3}$. This means that heating due to mass loss dominates over convection of enthalpy as the primary means of delivering energy to each volume of the gas. Moreover, subsonic flows are nearly hydrostatic, and the gas must cool in order to flow inward. The result is that the convective energy transport is proportional to the cooling. Under these conditions, the flow nearly satisfies a local energy balance, with heating due to mass loss balancing cooling:

$$\alpha_* \rho_* \frac{kT_{inj}}{\mu m_p} \approx n_e^2\, \Lambda(T) \tag{15}$$

where the right-hand side is the cooling rate.

Now, if T_{inj} and T do not vary rapidly, we will have $n_e \propto \rho_*^{1/2}$. Since $\rho_* \sim r^{-3}$ outside the core of an elliptical galaxy, this gives $n_e \sim r^{-3/2}$, basically as is observed. This statement can be made more directly in terms of the optical and x-ray surface brightnesses, I_B and I_x respectively. As long as $T \gg 10^6$ K, the cooling radiation appears in the x-ray band and $n_e^2\, \Lambda$ is the x-ray emissivity. Let us integrate equation (15) along a line-of-sight through the galaxy. The right-hand side gives I_x, and the left-hand side is proportional to I_B:

$$\alpha_* \ (\frac{M}{L_B})_* \ I_B \ (\frac{kT_{inj}}{\mu m_p}) \approx I_x \ . \tag{16}$$

Thus, simple cooling flow models predict that the x-ray and optical surface brightnesses vary in proportion to one another, essentially as is observed (Trinchieri 1985).

4.7. Fate of the Cooling Gas

About 1 $M_\odot$ per year of gas will flow into the center of a large elliptical galaxy. We expect this gas will be thermally unstable, and will form optical line emitting filaments, as have been seen in many ellipticals (Caldwell 1982). I believe that this gas must ultimately be consumed through star formation; this may explain the blue population seen in the ultraviolet light of some ellipticals. Finally, some of this gas may flow into the galactic nucleus and power activity there.

This work was supported in part by NSF Grant AST81-20260, and in part by NASA Astrophysical Theory Grant NAGW-764.

REFERENCES

Bahcall, J. N., and Sarazin, C. L.: 1977, Astrophys. J. Lett. **199**, L89.
Bahcall, J. N., and Sarazin, C. L.: 1978, Astrophys. J. **219**, 781.
Burns, J. O., White, R. A., and Hayes, M. P.: 1981, Astron. J. **86**, 1120.
Caldwell, N.: 1982, Yale Ph.D. Thesis.
Cavaliere, A., and Fusco-Femiano, R.: 1976, Astron. Astrophys. **49**, 137.
Cowie, L. L.: 1985, preprint.
Cowie, L., Hu, E., Jenkins, E., and York, D.: 1983, Astrophys. J. **272**, 29.
Faber, S. M., and Jackson, R. E.: 1976, Astrophys. J. **204**, 668.
Fabian, A. C., Nulsen, P. E., and Canizares, C. R.: 1982, Mon. Not. Roy. Astron. Soc. **201**, 933.
Fabian, A. C., Nulsen, P. E., and Canizares, C. R.: 1984, Nature **310**, 733.
Fabian, A. C., Arnaud, K. A., Nulsen, P. E., Watson, M. G., Stewart, G. C., McHardy, I., Smith, A., Cook, B., Elvis, M., and Mushotzky, R. F.: 1985, preprint.
Fabricant, D., and Gorenstein, P.: 1983, Astrophys. J. **267**, 535.
Fabricant, D., Lecar, M., and Gorenstein, P.: 1980, Astrophys. J. **241**, 552.
Fabricant, D., Rybicki, G., and Gorenstein, P.: 1984, Astrophys. J. **286**, 186.
Forman, W., Jones, C., and Tucker, W.: 1985, Astrophys. J. **293**, 102.
Geller, M. J., and Beers, T. C.: 1982, Publ. Astron. Soc. Pac. **94**, 421.

Henriksen, M. J.: 1985, Univ. of Maryland Ph.D. Thesis.

Henriksen, M. J., and Mushotzky, R. F.: 1985a, Astrophys. J. 292, 441.

Henriksen, M. J., and Mushotzky, R. F.: 1985b, preprint.

Jones, C., and Forman, W.: 1984, Astrophys. J. 276, 38.

Jura, M.: 1977, Astrophys. J. 212, 634.

King, I. R.: 1962, Astron. J. 67, 471.

Mathews, W. G.: 1978, Astrophys. J. 219, 413.

Mathews, W. G., and Baker, J. C.: 1971, Astrophys. J. 170, 241.

McNamara, B., O'Connell, R. W., and Sarazin, C. L.: 1985, preprint.

Minkowski, R.: 1968, Astron. J. 73, 842.

Mushotzky, R. F.: 1985, in Proc. Conf. on Non-thermal and Very High Temperature Phenomena in X-ray Astronomy, edited by G. Perola and M. Salvati, in press.

Nulsen, P. E., Stewart, G. C., and Fabian, A. C.: 1984, Mon. Not. Roy. Astron. Soc. 208, 185.

Rubin, V., Ford, W., Peterson, C., and Oort, J.: 1977, Astrophys. J. 211, 693.

Sarazin, C. L.: 1985, Rev. Mod. Phys., in press.

Sarazin, C. L., and Bahcall, J. N.: 1977, Astrophys. J. Suppl. 34, 451.

Sarazin, C. L., and O'Connell, R. W.: 1983, Astrophys. J. 268, 552.

Spitzer, L. Jr.: 1978, Physical Processes in the Interstellar Medium (New York: Wiley), p. 283.

Stewart, G. C., Canizares, C. R., Fabian, A. C., and Nulsen, P. E.: 1984a, Astrophys. J. 278, 536.

Stewart, G. C., Fabian, A. C., Jones, C., and Forman, W.: 1984b, Astrophys. J. 285, 1.

Strimpel, O., and Binney, J.: 1979, Mon. Not. R. Astron. Soc. 188, 883.

Tammann, G. A.: 1974, in Supernovae and Supernova Remnants, edited by C. B. Cosmovici (Dordrecht: Reidel), p, 155.

Tonry, J. L.: 1981, Astrophys. J. Lett. 251, 1.

Tonry, J. L.: 1983, Astrophys. J. 266, 58.

Trinchieri, G.: 1985, preprint.

Trinchieri, G., and Fabbiano, G.: 1985, Astrophys. J. 296, in press.

Vallée, J. P.: 1981, Astrophys. Lett. 22, 193.

White, R. E., and Chevalier, R. A.: 1984, Astrophys. J. 280, 561.

White, R. E., and Sarazin, C. L.: 1985, preprint.

DISCUSSION

KORMENDY: Regarding the discussion of cooling flows: In a high-resolution photometric survey that I have been making with the Canada-France-Hawaii Telescope, I find that elliptical galaxies very often have dust near their centers. There is a strong correlation between the presence of dust and the detection of x-ray emission.

SARAZIN: I'm very interested to hear of such a correlation, because there's another new result that ties into this. Mike Jura has recently examined the IRAS data base, and found that x-ray emitting galaxies have infrared emission from their centers. He argues that they must contain 10^9 $M_\odot$ of molecular hydrogen to produce the amount of infrared emission he sees.

YAHIL (to KNAPP): Wouldn't we see CO from these 10^9 $M_\odot$ of H_2?

KNAPP: If CO is present, and if the ratio of CO emissivity to H_2 mass is the same as in the Galaxy, then observations of the necessary sensitivity are possible. To date, less than half-a-dozen ellipticals have been observed carefully, and none of them shows CO emission. But this number is too small to restrict the argument Craig just presented.
        If I may ask a related question: Can dust possibly survive in a cooling flow?

SARAZIN: If Jura is right about the amount of infrared emission from these ellipticals, then I think the grains must come out in planetary nebulae and never get mixed into the hot gas. Otherwise they would certainly be destroyed.

KNAPP: But John Kormendy just said that dust is observed and is well correlated with the presence of x-rays.

SILK: I do not find the argument for exclusively low-mass star formation in cooling flows to be very compelling, for the following reason. The infall rate at a given radius is effectively taken to be the mass of a shell of gas divided by the local cooling time scale. However, if the cooling gas forms any massive stars, the resulting supernovae will heat the gas. This could reduce the rate of infall to the extent that adoption of even a solar neighborhood initial mass function might be consistent with the observed colors of galaxies at the centers of cooling flows.

SARAZIN: It seems to me that the required supernova rate would be much too high. The cooling luminosity of the cooling flow in the Perseus Cluster is more than 3×10^{44} erg s^{-1}. For supernovae to provide the majority of this energy and seriously lower the cooling rate, the supernova rate would have to be ten per year. NGC 1275 has been sufficiently observed that I very much doubt that it could have one supernova per month.

OSTRIKER: My question is a variant on the last one. If I consider isolated galaxies, where the x-ray luminosity is low, I find an approximate upper bound on the supernova rate of less than one every 300 years if a cooling flow is to be produced. I'm assuming an energy of 10^{51} ergs per supernova. This supernova rate is less than the rate people actually observe.

SARAZIN: That's right, it's about 1/3 of the rate found by Tammann. But remember that these rates for E and S0 galaxies are based on a total of only nine observed supernovae.

GUNN: Craig, if the cooling time for the gas in galaxies is short compared to the Hubble time, then you have to look for some equilibrium situation. Why then would you expect the amount of gas present to be the total amount output by the stars over a Hubble time?

SARAZIN: The cooling time becomes equal to the Hubble time just at the outer edge of the flow. The mass is mainly coming from the outer edge, because the gas density is dropping off as $1/r^{3/2}$. This argument is uncertain by a factor of two or three. If you do the integral properly, then you find that you can have a steady-state distribution with a reasonable current gas inflow rate.

GUNN: But then you run into the problem that most of the gas is not where most of the stars are. The stellar density drops as $1/r^3$. Are the distributions of gas and stars consistent?

SARAZIN: Yes. If gas is coming out of stars, then you measure the x-ray luminosity due to the cooling of the gas, while the amount of heat input is just proportional to the mass being shed by the stars. In a steady state, these two ought to agree. So you expect that the x-ray surface brightness is proportional to the optical surface brightness. But because the x-ray surface brightness depends on the square of the density, the gas density will go as the square root of the stellar density, as observed.

TUCKER: Two points. (1) In addition to the supernova heating mentioned by Joe Silk, there is another source of heating, namely the relativistic electrons associated with the radio sources known to be present in most galaxies with cooling flows. If this energy source is included, the accretion rates can be reduced by two orders of magnitude or more. (2) I would like to point out that a steady state model cannot explain the large amount of gas observed to be present in x-ray halos. The observations require that the halos store up the mass lost by stars over the lifetime of the galaxy, not just over a cooling time. You must use a non-steady model, in which case you can get $L_x \propto L_B^2$ for supernova heating.

SARAZIN: All we know about the gas in cooling flows is that it is cold. We have never seen it actually flow into the center. It is possible to have enough heating to make up for the extra cooling near the center, so that the gas is in thermal equilibrium. But this situation is violently thermally unstable. The gas would clump and fall into the center on a time scale which is at least as short as the cooling time. So you end up with a cooling flow anyway.

WHITE: When you analyze the cooling flows for structure, you just assign a unique temperature and density to each point. At the same time you say that there is a thermal instability occurring which is forming stars. I wonder whether the clumpiness in the gas can seriously affect the solution.

SARAZIN: That's the point I was making when I talked about consistently determining the cooling rates. Both Ray White and I, and Andy Fabian and colleagues, do the calculation by putting in gas at a range of temperatures at each radius, in order to include the effect of cooling.

SANDERS: If a substantial fraction of stars in cD galaxies form from gas in cooling flows, these stars should lie on essentially radial orbits. One might expect the stellar velocity distribution to be extremely anisotropic. Might this not be apparent in the observed dependence of velocity dispersion on projected radius?

SARAZIN: Yes, if the stars formed from comoving lumps in a radial cooling flow, the orbits would indeed be radial. However, it is possible that star-forming clouds might have a significant velocity dispersion. For example, the cooling rate is very high between 10^6 K and 10^4 K. Clouds of gas should cool isochorically through this range, and then be repressurized by shocks. These shocks, which probably produce the observed optical emission-line filaments, could give the star-forming lumps a significant amount of transverse velocity, and result in less radial stellar orbits.

COOLING FLOWS AND THE FORMATION OF DARK MATTER

A.C. Fabian, K.A. Arnaud & P.A. Thomas
Institute of Astronomy,
Madingley Road,
Cambridge CB3 0HA,
United Kingdom

ABSTRACT. The distribution of matter condensing out of cooling flows in clusters of galaxies and individual elliptical galaxies has been studied using X-ray data and is found to resemble the expected mass profiles of the underlying galaxies. Most of the cooled gas must create objects of high mass-to-light ratio, although some more normal stars are produced. Cooling flows provide an observable mechanism for the continual formation of dark matter around galaxies. Since the conditions at galaxy formation are similar to those in cooling flows if the gas reaches the virial temperature, we suggest that they are local models of galaxy formation.

1. INTRODUCTION

The evidence for cooling flows in clusters of galaxies has been reviewed at this symposium by Canizares (1986) and Sarazin (1986) and elsewhere by Fabian, Nulsen & Canizares (1984) and Sarazin (1985). Briefly, the process involves the cooling of hot gas in the cores of clusters and galaxies by X-ray emission. The cooling gas is pushed inward by the pressure of outer, less dense, gas. Thermal instability causes small density perturbations to grow and drop out of the flow over a wide range of radii within the cooling region. Some of the blobs are visible through optical line radiation. Here we discuss new results on cooling flows in individual elliptical galaxies and on the variation of mass flow rate with radius in galaxies and in central cluster galaxies. There are strong indications that the envelopes of the central galaxies are still being formed at the present epoch by this processs and that much of the cooled gas condenses into objects of high mass to (visible) light ratio. At least some 'dark matter' is baryonic. The profile of mass deposition in a cooling flow resembles that of the mass profile of the underlying galaxy.

The nature of the condensed objects remains a problem, (as does that of most dark matter). There is clear evidence that a solar neighbourhood initial-mass-function is **not** involved. What we do infer from the X-ray and other data of the large flows is that matter

201

J. Kormendy and G. R. Knapp (eds.), Dark Matter in the Universe, 201–213.

condenses from hot gas at a rate and in a manner that builds a large extended galaxy. Some stars of solar mass and more are also formed. Cooling flows may be a model for the formation of galaxies. Distant cooling flows (z ∿ 1) are probably already observed around some of the 3C radio galaxies.

2.   INTERPRETATION OF THE X-RAY DATA

Imaging and spectroscopic X-ray data, primarily from the Einstein Observatory, are the major source of information on cooling flows. We have made extensive use of the X-ray surface brightness profiles from the Imaging Proportional Counter (IPC) and High Resolution Imager (HRI); much of these data are discussed by Jones & Forman (1984). In order to obtain gas density and temperature profiles, we have employed three separate methods.

2.1. Deprojection of surface brightness profiles:

The X-ray surface brightness from most circularly symmetric clusters rises so steeply that a direct deprojection of the counts gives a stable solution (Fabian et al. 1981). X-ray counts accumulated in concentric annular bins are thereby converted into count emissivities, ε, in nested shells at the cluster. The cluster is assumed to be spherically symmetric. (Small departures from symmetry are not important, Fabian et al. 1981). ε is related to the gas density and temperature, n and T, through the emission process (thermal bremsstrahlung and line emission), the detector response and the intervening photoelectric absorption, i.e. ε = f(n, T) where f is a known , but complicated function. We can then separate n and T with the equation of hydrostatic equilibrium,

$$\frac{dP}{dr} = - \rho \frac{d\phi}{dr}$$

where P and ρ are the gas pressure and mass density and ϕ is the gravitational potential. This potential is estimated from an assumed mass profile (say, a King law) and galaxy velocity dispersion, both obtained from optical data. Gas density profiles are thereby obtained which, due to the limited response of the detectors to temperature variations, are consistent and accurate to about 10 percent. The temperature profiles are much less certain since they depend strongly on ϕ. Some limits can be placed by ensuring that the re-projected solution does not conflict with the overall spectrum of the cluster or any spectroscopic components (e.g. lower temperatures or lines, see e.g. Mushotzky et al. 1981; Canizares 1981). Where there are strong constraints on the temperature profile, we can of course determine ϕ directly from the X-ray data (Fabricant & Gorenstein 1983; Stewart et al. 1984a).
    Where the radiative cooling time of the gas is less than the Hubble time we can solve for the rate at which matter is cooling out of the flow at each radius, $\delta\dot{M}$, from the local luminosity,

$$\delta L = \dot{M} \left[\frac{5}{2} \frac{k\Delta T}{\mu m_H} + \Delta\phi \right] + \delta \dot{M} \left[\frac{5}{2} \frac{kT}{\mu m_H} + f\Delta\phi \right]$$

where $\dot{M}$ is the rate at which matter flows completely through that
annulus and f is a weighting factor (~ 0.5) for the gravitational work
done on $\delta \dot{M}$ (see Fabian et al. 1985a). $\dot{M}$ is found to increase
with radius to substantial rates (20 $M_\odot yr^{-1}$ in M87, Stewart et al.
1984a; 300 $M_\odot yr^{-1}$ in NGC 1275, Fabian, Nulsen & Canizares 1984). A
selection of results for $\dot{M}(r)$ is shown in Fig. 1.

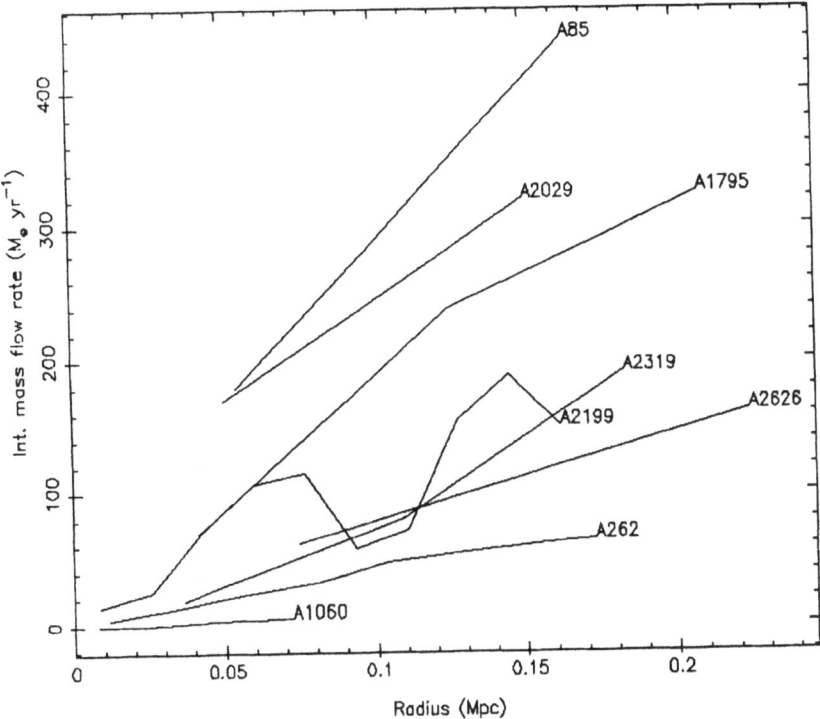

Fig. 1. Integrated mass flow rates, $\dot{M}$, in a number of cooling flow
clusters. They increase in a similar manner, although there is a wide
spread in the distances and in the quality of the data.

2.2. Model fitting:

The equations of continuity, force and energy for gas in a spherical
potential can be directly integrated to obtain surface brightness
profiles which can then be compared with the data (Fabian et al
1981; Thomas et al. 1985). Once again, ϕ is estimated from optical
work. As discussed above, the X-ray data definitely require that $\dot{M}$ is
not constant, and some theoretical prejudice must then go into the
selection of the form of its radial dependence. In this respect, the
deprojection method is much simpler and freer from pre-conceived ideas.
        The results for detailed fits to the X-ray profile of the Virgo

elliptical galaxy, NGC 4472 (Thomas et al. 1985) are shown in fig. 2. In this case we include the possibility of distributed mass and energy sources (from stellar mass loss and supernovae) as well as sinks for cooled matter. The profile beyond ∿ 20 kpc requires a dark halo, as reported by Forman, Jones & Tucker (1985). Cooling is important within ∿ 60 kpc and the data require that the matter drops out of the flow in a distributed manner since the gravitational energy liberated in falling to the luminous core radius is too large. The X-ray emission cannot be explained by any near- or super-sonic flow such as would be expected if a wind were operating. The inflow velocities are highly subsonic ($1 - 10\,km\,s^{-1}$) and the mass flow rate falls steadily inward from ∿ $0.8\,M_\odot yr^{-1}$ at   60 kpc (fig. 3).

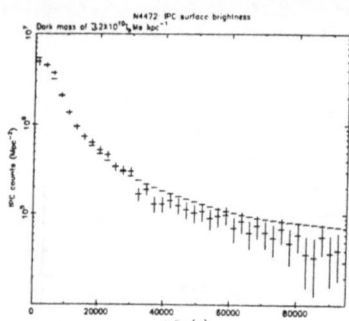 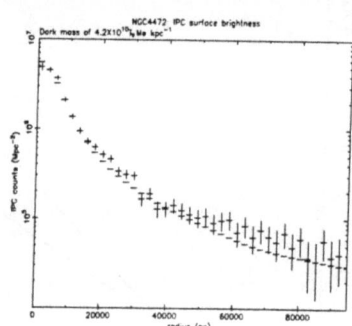

Fig. 2.   Comparisons of the IPC surface brightness profile of NGC 4472, at a distance of 15 Mpc, (crosses) with the results of a detailed cooling flow model including an isothermal dark halo of core radius 10 kpc and the spatial response of the detector. The total dark mass out to 100 kpc is $3.2 \times 10^{12}T_7 M_\odot$ (left-hand panel) and $4.2 \times 10^{12}T_7 M_\odot$ (right-hand panel), where the outer gas temperature is $10^7 T_7$ K. The influence of the dark mass is seen even at 20 kpc. Further systematic uncertainties in the data beyond ∿ 50 kpc   due to background subtraction and departures from spherical symmetry are not included.

The heating rate due to supernovae, assumed by White & Chevalier (1983) in a study of flows in elliptical galaxies to be that due to ∿ 1 event per 30 y, must be reduced to ∿ 1 per 100 y. Also it seems that some of the stellar mass loss - red giant winds and planetaries - condenses back into stars, or low-mass objects before it has a chance to shock on other cold gas (see also White & Chevalier 1983).
The X-ray data are clearly opening a new era in the study of gas in galaxies. The amount of gas and star formation in a large elliptical galaxy is only a little less than that in a spiral galaxy such as our own.

2.3. Comoving inflow of cooling blobs:

The necessity that matter drops out of the flow at all radii suggests

to us and Paul Nulsen that the gas within a cooling flow consists of
many phases which are intimately mixed.   To make any progress, we
assume that these phases are comoving, as expected for small blobs at
X-ray emitting temperatures (Cowie, Fabian & Nulsen 1981).   It is
possible that the hottest phase does not comove and it may even flow
outward (Nulsen, private communication).

    Beyond the outermost radius where the cooling time equals the age
of the cluster atmosphere we assume that there is a small range of
density variation.   At smaller radii, they develop according to the
thermal instability.   The cooling gas is pushed subsonically inward by
the pressure of the outer gas and the initially densest blobs cool
first and drop out from the flow at the largest radii.   The remaining
gas flows further inward with blobs of decreasing initial density
finally cooling and collapsing.   We can solve for this process using a
modification of the deprojection method where there are as many phases
as annular bins and we (numerically) follow each phase outward from the
centre.   The count emissivities are now expressed as $\varepsilon(n_1, n_2, T_1, T_2)$
as the gas cools from n_1, T_1 to n_2, T_2, over an annulus.   To obtain
this function we have integrated the spectra from Raymond & Smith
(1977), folding in the detector response and absorption as before.

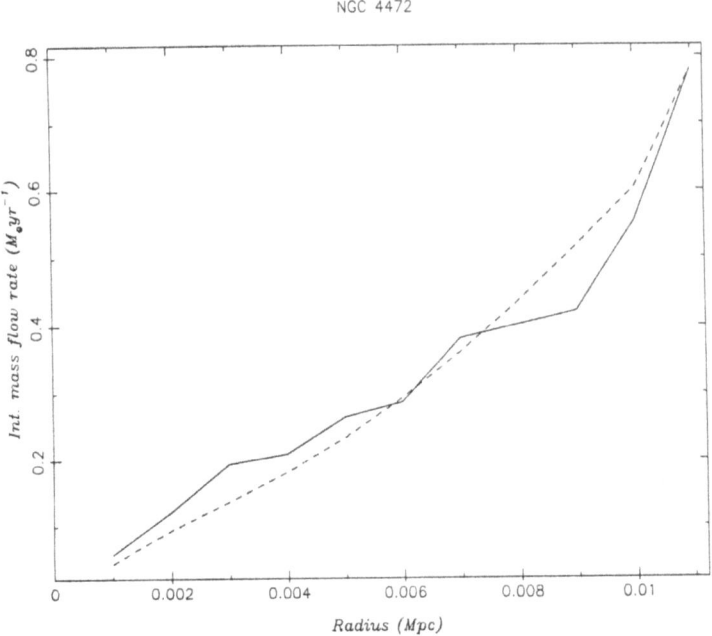

Fig. 3.   The integral mass flow rate within NGC 4472 from the
deprojection method (solid line). The model fit (not shown) is in good
agreement.   The dashed line indicates the gravitational mass profile
scaled to the outer radius shown.

We have applied this method to a few clusters for which there is excellent imaging data (e.g. $> 10^5$ counts) and find good agreement with the other methods. The results for the region around NGC 1275 in the Perseus cluster are shown in fig. 4. $\delta \dot{M}$ is approximately constant with radius beyond $\sim$ 20 kpc (there are central-point-source subtraction uncertainties at smaller radii). Matter is dropping out of the flow in a form consistent with the mass profile of a large elliptical galaxy (fig. 5). The total rate of $\sim$ 300 $M_\odot yr^{-1}$ can supply the total mass of a large central galaxy over a Hubble time.

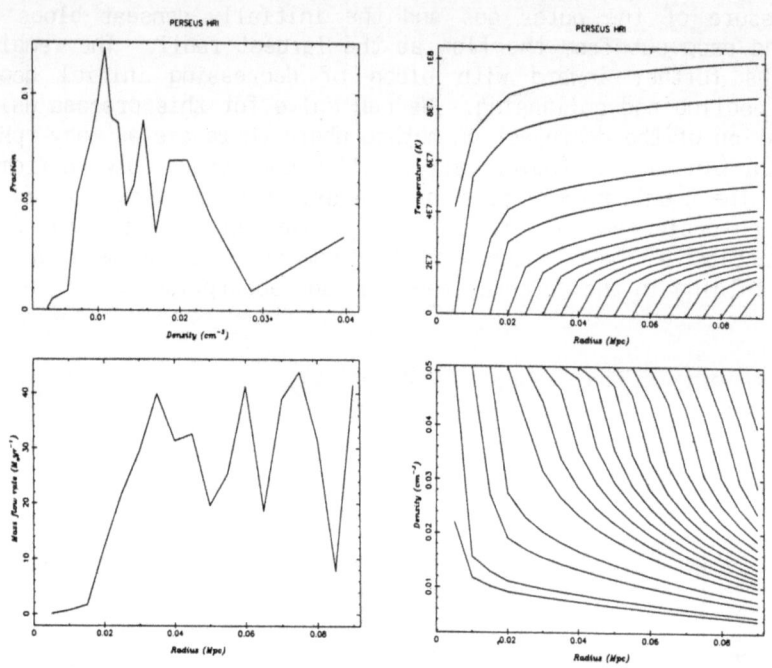

Fig. 4. Results from a multiphase analysis of HRI data from the region around NGC 1275. The right-hand panels show temperature and density profiles for each phase. The upper left-hand panel contains the relative distribution of mass flow with density at the outer cooling radius. The lower left-hand panel shows the rate at which mass drops out of the flow at each radius, $\delta \dot{M}$. Note that $\delta \dot{M}$ is almost constant with radius beyond 2 kpc. Other clusters for which there are good data give similar results.

We conclude this section by stressing that cooling is relatively common in rich clusters (Stewart et al. 1984b), poor clusters (Canizares, Stewart & Fabian 1983) and even isolated elliptical galaxies (Nulsen, Stewart & Fabian 1984). In some cases, very large quantities of matter are condensing out, at a rate that can build a complete galaxy over a Hubble time. The mass of the underlying galaxy is compatible with this possibility. However, there is a problem with the light if the initial-mass-function of the condensed objects is similar to that inferred for the stars in our Galaxy.

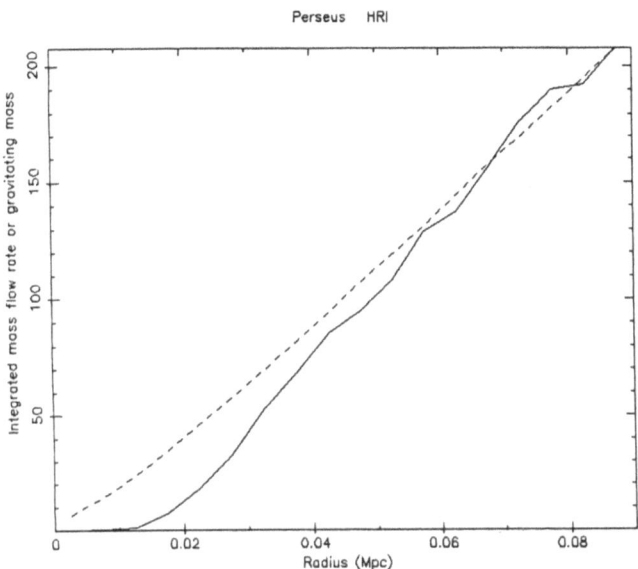

Fig. 5. The integrated mass flow rate, $\dot{M}$, for NGC 1275 (solid line) compared with the assumed mass profile producing ϕ, $\rho_* \propto (1 + (r/a)^2)^{-1}$. The mass profile is scaled to agree with $\dot{M}$ at the outer radius. They would agree quantitatively if the dark matter has a velocity dispersion of ~ 240 km s^{-1} and the flow continues for H_o^{-1}.

3. THE FORMATION OF STARS AND CONDENSED OBJECTS IN COOLING FLOWS

The X-ray data on the Perseus cluster around the galaxy NGC 1275 indicate a cooling rate of ~ 300 M$_\odot$ yr^{-1} (Fabian et al. 1981). The gas is also observed at $\sim 10^7$ K (Mushotzky et al. 1981), by which time it has lost ~ 80 percent of its initial thermal energy (initially, $T \sim 8.10^7$ K), and then as optical and ultraviolet emitting filaments (see Lynds 1970; Kent & Sargent 1979; Cowie et al. 1983 for Hα and Fabian, Nulsen & Arnaud (1984) for Lyα). Absorption by still cooler gas ($T \ll 10^4$ K) has been detected at 21 cm and a velocity of 5300 km s^{-1} by Crane, van der Hulst & Haschick (1982). We have little doubt that gas is condensing out of the flow. The nature of what it condenses into is highly uncertain.

   The implied star formation rate is ~ 100 times that in our galaxy, yet NGC 1275 is not 5 magnitudes brighter than our galaxy. The continuum spectrum of parts of NGC 1275 does resemble that of an A star

(Rubin et al. 1977), so at least some massive stars (M $\gtrsim$ 3 M$_\odot$) are formed. The IUE data on it show no evidence for an extra-nuclear continuum so we have concluded that less than $\sim$ 2 percent of the mass flow produces OB stars (Fabian, Nulsen & Arnaud 1984). An uncertainty in this limit due to reddening is eliminated by recent infrared observations of the galaxy at wavelengths of 10 - 400 μm. Gear et al. (1985) show that thermal re-radiation by dust does dominate the spectrum, but can be explained by $\sim$ 2 percent of the 300 M$_\odot$yr^{-1} cooling flow producing OB and A stars. This is considerably less than would be expected from a solar-neighbourhood initial-mass-function and, together with the total magnitude limits, requires that the star formation is biased towards low mass stars. The high pressure and lack of dust may be contributing factors (Fabian, Nulsen & Canizares 1982; Sarazin & O'Connell 1983).

Arnaud & Gilmore (1985) have recently observed several central galaxies in cooling flows at the 2.3μm CO feature. This is a sensitive indicator of the relative giant to dwarf ratio of late-type stars. There is no evidence for a dominant population of low-mass dwarfs.

The accumulated mass of condensed matter around NGC 1275 is

$$\dot{M} H_o^{-1} = 6 \times 10^{12} \left(\frac{H_o}{50 \text{ km s}^{-1}\text{Mpc}^{-1}} \right)^{-3} \left(\frac{\dot{M}}{300 \text{ M}_\odot\text{y}^{-1}} \right) \text{M}_\odot$$

comparable to the total expected mass. A simple estimate of the mass-to-(visual) light ratio for the condensed matter is M/L$_v$ $\approx$ 10. For PKS 0745-191 where $\dot{M}$ $\approx$ 1000 M$_\odot$y^{-1}, Fabian et al. (1985) obtain M/L$_v$ > 25. This is, of course, reduced if the flow has not operated at such a high rate for H$_o^{-1}$ or if H$_o$ > 50 km s^{-1}Mpc^{-1} (M/L$_v$ varies as H$_o^{-1}$) and is increased if some of the observed light is due to a pre-existing galaxy. The A and earlier stars may form within transient low pressure regions in large isochorically cooled blobs . These blobs may also produce most of the observed line emission (Cowie, Fabian & Nulsen 1980). Most of the gas may cool via smaller blobs which emit much less light and form very low-mass stars. (This could accentuate a problem in explaining the high optical line luminosities.) Whether these are nuclear-burning stars, brown dwarfs, Jupiters or even smaller objects is impossible to tell at the present time. Perhaps fragmentation (Low & Lynden-Bell 1976) really works. It does appear that the conditions which produce massive stars may be special to spiral and irregular galaxies (e.g. relatively low pressures, low relative velocities, Giant Molecular Clouds etc.). Cooling flows provide one mechanism for the formation of baryonic 'dark' matter.

4. COOLING FLOWS AT EARLIER EPOCHS AND GALAXY FORMATION

Cooling flows contain both radio-loud (e.g. Cygnus A, $\dot{M}$ $\sim$ 100 M$_\odot$y^{-1}; Arnaud et al. 1984) and radio-quiet (e.g. AWM 7, $\dot{M}$ $\sim$ 40 M$_\odot$y^{-1}; Canizares et al. 1983) galaxies. The existence of flows in poor clusters suggests that they are common in sub-clusters, and are subseqeuntly disrupted when they merge to form present-day rich clusters (Stewart et al. 1984b; McGlynn & Fabian 1984). Cooling

flows may then be more common in the past. A small fraction of the
flow reaching the centre of the galaxy can easily power an active
nucleus and so the relatively recent decline of luminous active
galaxies, quasars and radio sources may follow the disruption of
subclusters. Some traces remain of extended radio sources in the form
of wide-angle tails and cluster halo sources (Arnaud et al. 1984).
    Although cooling flows with $\dot{M} \gtrsim 100$ $M_\odot y^{-1}$ were detectable beyond
redshifts of ~ 0.7 by the Einstein Observatory, a more X-ray luminous
active nucleus could easily mask its appearance. Consequently, they
need not be immediately apparent. The strong optical and ultraviolet
emission lines from the cooled filaments and the formation of
massive stars are much more readily found out to greater redshifts. The
3CR and other distant galaxies studied by Spinrad & Djorgovski
(1984a,b) Butcher & Oemler (1984) and by Lilly & Longair (1984) are
good examples (Fabian et al. 1985b). The luminosity of their [OII]
emission is similar to that from the 1000 $M_\odot y^{-1}$, z = 0.1 cooling flow
around PKS 0745-191 (fig. 6; Fabian et al. 1985a). Searches for
similar objects which are radio-quiet have not often been carried out,
although Hazard & McMahon (private communication) are finding similar
spectra on objective-prism Schmidt plates.

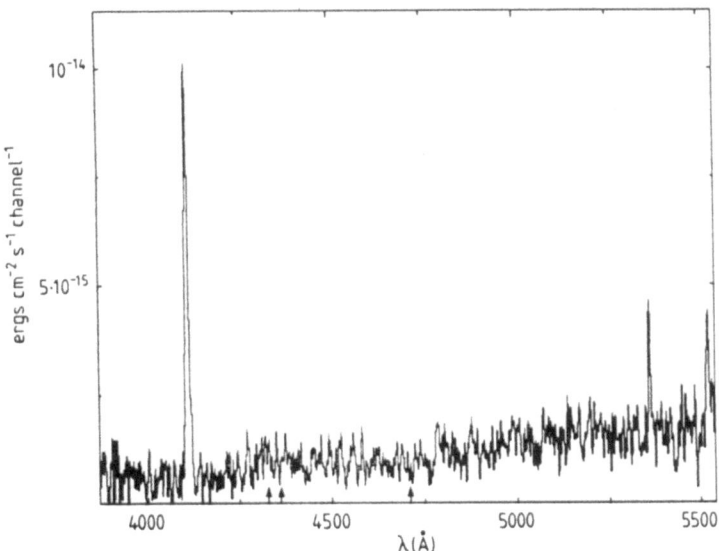

Fig. 6. The strong [OII] line and blue continuum of the 1000 $M_\odot y^{-1}$
cooling flow galaxy, PKS 0745-191 (Fabian et al. 1985a).

    The cooled gas probably collapses into sheets which can have a
significant covering fraction. Detectable absorption in lines due to
CII etc. may then appear in the spectra of background quasars (Crehan
private communication).

In summary, we note that the formation of cold, dark baryonic matter, together with a small fraction of 'ordinary' stars , out of X-ray hot gas is continuing around and within many of the largest galaxies. Some dark matter is then relatively young. If galaxies formed from gas which was heated by collapse to temperatures close to the virial temperature (as suggested by, for example, Rees & Ostriker 1977 and Silk 1977) then the pressure and lack of dust resemble a cooling flow (Fabian et al. 1985b). Only a small fraction of the mass needs to condense into high-mass stars which subsequently explode to seed the rest of the gas with metals. Massive galaxies and their envelopes form over a comparatively long time stretching up to the present epoch, and without many bright stars.

REFERENCES

Arnaud, K.A. & Gilmore, G.: 1985, preprint.
Arnaud, K.A., Fabian, A.C., Eales, S.A., Jones, C. & Forman, W.: 1984, M.N.R.A.S. **211**, 981.
Butcher, A.R. & Oemler, A.: 1984, Nature, **310**, 31.
Canizares, C.R.: 1981, X-ray Astronomy with the Einstein Satellite, ed. R. Giacconi, Reidel, 215.
Canizares, C.R., Stewart, G.C. & Fabian, A.C.: 1983, Ap.J., **272**, 449.
Canizares, C.R.: 1986, this volume.
Cowie, L.L., Fabian, A.C. & Nulsen, P.E.J.: 1980, M.N.R.A.S., **191**, 399.
Cowie, L.L., Hu, E.M., Jenkins, E.B. & York, D.G.: 1983, Ap.J., **272**, 29.
Crane, P.C., van der Hulst, J.M. & Haschick, A.D.: 1982, IAU Symposium 97, eds. D.S. Heeschen & C.M. Wade, 307.
Fabian, A.C., Hu, E.M., Cowie, L.L., Grindlay, J.: 1981, Ap.J., **248**, 47.
Fabian, A.C., Nulsen, P.E.J. & Canizares, C.R.: 1982, M.N.R.A.S., **201**, 933.
Fabian, A.C., Nulsen, P.E.J. & Arnaud, K.A.: 1984, M.N.R.A.S., **208**, 179.
Fabian, A.C., Nulsen, P.E.J. & Canizares, C.R.: 1984, Nature, **310**, 733.
Fabian, A.C., Arnaud, K.A., Nulsen, P.E.J., Watson, M.G., Stewart,G.C., McHardy, I., Smith, A., Cooke, B., Elvis, M. & Mushotzky, R.F.: 1985, M.N.R.A.S., in press.
Fabian, A.C., Arnaud, K.A., Nulsen, P.E.J. & Mushotzky, R.F.: 1985b, Ap.J., submitted.
Fabricant, D. & Gorenstein, P.: 1983, Ap.J., **267**, 535.
Forman, W., Jones, C. & Tucker,W.: 1985, Ap.J., **293**, 102.
Gear, W.K., Gee, G., Robson, E.I. & Nolt, I.G.: M.N.R.A.S., in press.
Jones, C. & Forman, W.: 1984, Ap.J., **276**, 38.
Kent, S.M. & Sargent, W.L.W.: 1979, Ap.J., **230**, 667.
Lilly, S. & Longair, M.S.: 1984, M.N.R.A.S., **211**, 833.

Low, C. & Lynden-Bell, D.: 1976, M.N.R.A.S., **176**, 367.
Lynds, R.: 1970, Ap.J., **159**, L151.
McGlynn, T.A. & Fabian, A.C.: 1984, M.N.R.A.S., **208**, 709.
Mushotzky, R.F., Holt, S.S., Smith, B.W., Boldt, E.A. & Serlemitsos,
    P.J.: 1981, Ap.J., **244**, L47.
Nulsen, P.E.J., Stewart, G.C. & Fabian, A.C.: 1984, M.N.R.A.S.,
    208, 185.
Raymond, J. & Smith, B.W.: 1977, Ap.J.Supp.,
Rees, M.J. & Ostriker, J.: 1977, M.N.R.A.S., **179**, 541.
Rubin, V.C., Ford, W.K., Peterson, C.J & Oort, J.H.: 1977, Ap.J.,
    211, 693.
Sarazin, C.L. & O'Connell. R.W.: 1983, Ap.J., **268**, 552.
Sarazin, C.L.: 1985, Rev.Mod.Phys., in press.
Sarazin, C.L.: 1986, this volume.
Silk, J.: 1977, Ap.J., **211**, 638.
Spinrad, H. & Djorgovski, G.: 1984a, Ap.J., **280**, L9.
Spinrad, H. & Djorgovski, G.: 1984b, Ap.J., **285**, L49.
Stewart, G.C., Canizares, C.R., Fabian, A.C. & Nulsen, P.E.J.: 1984a,
    Ap.J., **278**, 536.
Stewart, G.C., Fabian, A.C., Jones, C. & Forman, W.: 1984b, Ap.J.,
    285, 1.
Thomas, et al.: 1985, preprint.
White, R.E. & Chevalier, R.A.: 1983, Ap.J., **275**, 1983.

DISCUSSION

OSTRIKER: This is related to Jim Gunn's question to Sarazin and has to do with the continuity equation. As I understand it, the mass is dropping out of cooling flows in a way that would make mass distributions that are like dark halos, not like visible ellipticals. That is, the mass is making things that are less centrally concentrated than the stars we see now. Then the gas cannot have come out of the stars in the galaxy. Do you agree?

FABIAN: Yes, I agree.

OSTRIKER: So we don't have continual recycling and inflow. Then how did the metals get into the gas? Either they are primordial, or they came out of other galaxies. So either the central galaxy or other galaxies had previously to have had a wind.

FABIAN: It is possible that the initial mass function is pressure-dependent. If so, then low pressures produce high-mass stars. Then to start with you produce lots of high-mass stars, which eject high-metallicity matter into the surrounding envelope. This gas could then make a cooling flow and form the rest of the galaxy. All this requires a wind sufficient to put the metals into the halo. If that wind were very metal-rich, and if there were already some gas at large radii, then the total mass outflow could have been less than the total present mass inflow.

SHAPIRO: In view of the anomalously large number of globular clusters in M87, I'd like to ask whether you can form globulars in these radiative cooling flows?

FABIAN: We have thought about that, and so have Martin Rees and Mike Fall. Conditions in the cooling flows are almost ideal for forming globular clusters. The problem is to keep the gas at 10^4 K long enough to allow it to go Jeans unstable. So maybe the globular clusters in M87 are related to the cooling flow, and maybe they're not.

E. TURNER: If you take seriously the idea that giant ellipticals form continuously, this could have an effect on the Hubble diagram of brightest cluster galaxies. There are at least two possible consequences. First, you might form stars which could become giants in a Hubble time. Second, if you deposit new material within the existing stellar population, you could contract or expand the orbits and therefore affect the number of stars inside some particular observational aperture. In a normal stellar-evolutionary model of elliptical galaxies, luminosity evolution is rapid shortly after the stars form, and then slow thereafter. But your scenario might imply steep luminosity evolution at recent times.

STEIGMAN: Do you require extremely efficient star formation?

FABIAN: 100%.

STEIGMAN: Does that bother you?

FABIAN: Perhaps it should. The big problem is that we see evidence for the gas at 7 keV, 1 keV, in the UV and in the optical. There is evidence for HI absorption in the low-velocity system seen against NGC 1275. But after that, we don't see the gas.

SCHECHTER: Do you expect angular momentum transport in the cooling flow? If not, is the star formation fast enough to keep disks from forming?

FABIAN: This is discussed in a paper by Paul Nulsen. He points out that if the gas is hot, it may be possible for turbulence to transport angular momentum outwards. You could then get HI rings forming in some of these systems.

LARSON: I am very uncomfortable with the suggestion that a hundred $M_\odot$ per year, or even a few $M_\odot$ per year, could disappear inconspicuously into low-mass stars. Even if nobody understands anything about star formation, there are, after all, a few facts known just from looking up at the sky. One is that we are familiar with star formation in giant molecular clouds. It has been suggested that because the pressure in cooling flows is orders of magnitude higher than that in the interstellar medium in our Galaxy, star formation can be very different. But the pressures in the cores of giant molecular clouds are quite comparable to the pressures in cooling flows. Star formation in the cores of molecular clouds is vigorous and anything but inconspicuous. Massive stars form and large amounts of radiation are produced. I would expect even more extreme things to happen if condensations were to occur in cooling flows. A second fact which has been ignored is that, whenever a mass spectrum has been observed in any kind of stellar system, it has always had a long power-law tail toward high masses which is not too different from Salpeter's original power law. This includes solar neighborhood stars, star clusters in our galaxy, and star clusters in other galaxies. You would require a kind of star formation in which the high-mass tail is completely absent. So the fate of the gas in your cooling flows has little or nothing to do with star formation as we know it.

FABIAN: I agree. We are worrying about all this. It's not just the pressure that is different. If you have shocks, then it is likely that the material is in very thin sheets, and whatever forms at the other end of the shock is sprayed out at shock velocities of hundreds of km s^{-1}. There are many other differences from molecular clouds. For example, there are no molecules, or certainly there are no molecules except H_2 distributed across the face of the cooling flows, because the dust couldn't have survived.

HOT CORONAE AROUND EARLY-TYPE GALAXIES: EVIDENCE FOR DARK HALOS

W. Forman, C. Jones, and W. Tucker
Harvard-Smithsonian Center for Astrophysics
60 Garden Street
Cambridge, Massachusetts 02138 USA

ABSTRACT. The analysis of the X-ray emission from a sample of 55 bright early-type galaxies shows that hot gaseous coronae are a common and perhaps ubiquitous feature of such systems. The X-ray emission can be explained most naturally as thermal bremsstrahlung from hot gas ($kT \approx$ 0.5-1.5 keV) which may be accumulated from mass loss during normal stellar evolution. The presence of these coronae shows that matter (10^9-10^{10} $M_\odot$) previously thought to be expelled in a galactic wind is instead stored in a hot galactic corona which may be heated and powered by supernova explosions. Perhaps the single most important feature of these coronae is that they provide a unique tracer of the gravitational potential in the outer regions of bright early-type galaxies. In this paper we describe the X-ray properties of these coronae (gas mass, temperature, and extent) and discuss their implications for the presence of massive dark halos around individual early-type galaxies. We find total masses of early-type galaxies up to 5×10^{12} $M_\odot$. We estimate mass-to-light ratios for early-type galaxies and find values up to ~100 (in solar units), similar to those found for the larger dynamical systems of groups and clusters.
Reference: Astrophys. J. 293, 102 (1985).

J. Kormendy and G. R. Knapp (eds.), Dark Matter in the Universe, 214.

Extended X-ray emission from early-type Galaxies:
Comparison with optical

G. Trinchieri   G. Fabbiano
Harvard-Smithsonian Center for Astrophysics

In this poster we display the results from a detailed analysis of
the distribution of the X-ray emission in early type galaxies. Two
major results have come out of the analysis so far:
  a) The surface brightness radial profiles of isolated elliptical
and SO galaxies are smoothly decreasing functions of radius out to a
R_{max} (similar to the optical radius). Outside R_{max} a flattening in
the slope is observed, although the exact shape of the profiles at
large radii, where the data are poorest, cannot be determined at
present.
  b) For $R \ll R_{max}$, the X-ray and optical surface brightness profiles
are similar (the nuclear region could be an exception). At larger
radii, the X-ray profile could be flatter (NGC 4649 and NGC 4472
northern sector) or steeper (NGC 4472 southern sector) than the
optical profile, or have a similar shape (NGC 4636).
  It is likely that the different profiles reflect the action of
slightly different environments and/or a different ambient density
around each galaxy, combined with the "history" of each galaxy. The
tail in NGC 4472 could be the result of the motion of this galaxy in a
dense intracluster medium. The X-ray deficiency in NGC 4649 could be
due to either ram pressure stripping or the action of wind in the
outer regions of the galaxy.
  The flat profiles observed at large radii suggest the interaction
with an external medium that pressure confines the interstellar gas.
This complicates the simple assumption that the hot gas is a tracer of
the binding mass at large radii. Moreover, the lack of direct
measurements of the temperatures and temperatures gradients (or of a
unique model that can be applied to the X-ray data) introduces a large
uncertainty in the determination of the binding mass even at radii
where the hydrostatic equilibrium approximation should hold. However,
the X-ray data can be used to obtain an independent measure of the
total mass of elliptical galaxies. As already suggested by the
optical data, the mass-to-light ratios are higher than expected from
the stellar component of these objects, thus suggesting the presence
of non-luminous matter also in early type galaxies.

215

J. Kormendy and G. R. Knapp (eds.), Dark Matter in the Universe, 215.

FILAMENTARY NEBULOSITY SURROUNDING M87

S. van den Bergh
Dominion Astrophysical Observatory

C.J. Pritchet
University of Victoria

Recently we have obtained both Hα + [NII] and broad-band red exposures of a number of galaxies with an RCA 320 x 512 CCD at the prime-focus of the 3.6 m CFH Telescope. Figure 1 shows the difference between Hα and red exposures (each with a total integration time of 60 min) of M87 = NGC4486.

1. We confirm the observations of Ford and Butcher (1979), albeit at much fainter limiting surface brightness and at higher resolution, that M87 is embedded in a system of filamentary Hα emission.

2. No major filaments are seen south of a line joining the jet, the nucleus and the Arp (1967) "counterjet". De Young, Condon and Butcher (1980) have interpreted this, together with the swept-back appearance of the M87 radio lobes, as evidence for subsonic motion of M87 towards the north.

3. The figure shows no obvious relationship between the Hα filaments and the M87 jet.

4. Arp's counterjet has a knotty structure that seems quite different from that of the other M87 Hα features.

5. For an assumed distance of 15 Mpc the filaments in M87 have lengths of up to 3 kpc and widths of ~ 0.2 kpc.

6. Comparison with Lynds' (1970) Hα photograph of NGC1275 shows that the emission nebulosity associated with Per A is much patchier than that surrounding Virgo A.

7. M49 = NGC4472 in the Virgo cluster does not contain a system of Hα filaments like that seen near M87.

REFERENCES

Arp, H.C. 1967. Ap.Letters 1, 1.
De Young, D.S., Condon, J.J. and Butcher, H. 1980. Ap.J. 242, 511.
Ford, H.C., and Butcher, H. 1979. Ap.J.Suppl. 41, 147.
Lynds, R. 1970. Ap.J.(Letters) 159, L151.

J. Kormendy and G. R. Knapp (eds.), Dark Matter in the Universe, 216–217.

N

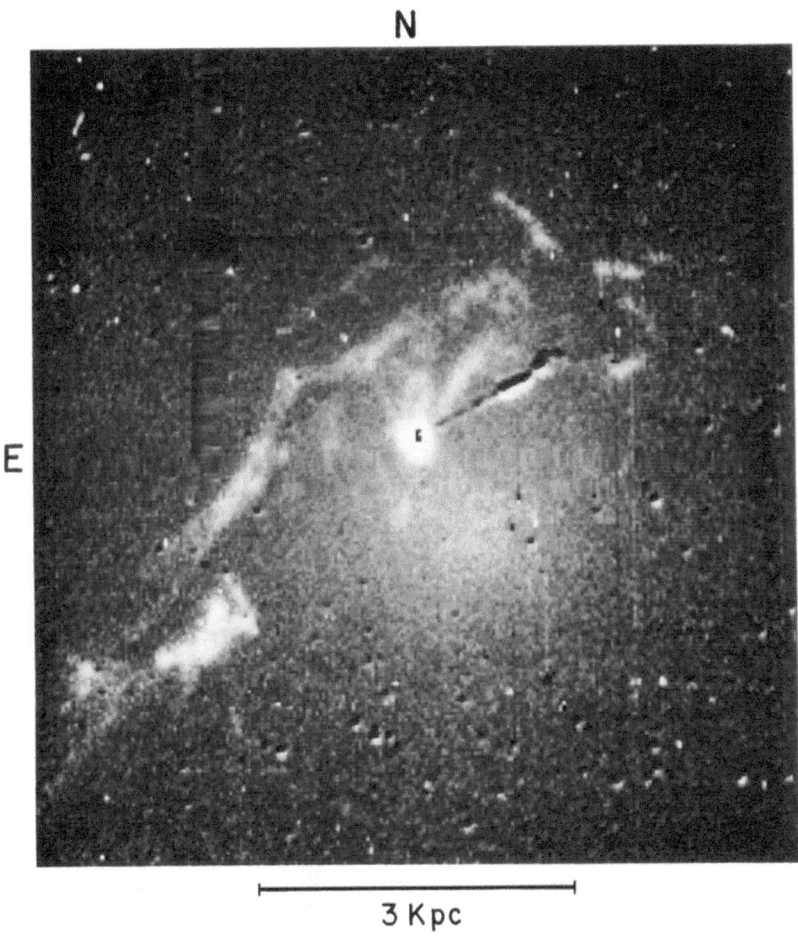

3 Kpc

Figure 1. Hα emission filaments surrounding M87. The bright knots SE
of the nucleus are Arp's "counterjet". The jet and some globular
clusters are visible because of the imperfect alignment of the Hα and R
images.

GRAVITATIONAL LENSES AND DARK MATTER: THEORY

J. Richard Gott, III
Princeton University Observatory
Princeton, NJ 08544 USA

ABSTRACT: In principal, gravitational lenses can be used to study dark
matter in a variety of ways. They can provide information on masses of
lensing galaxies, circular velocities in halos, the value of Ω in
condensed objects, and the constitution of heavy halos (whether they
are made of low mass stars or not). They can tell us whether mass and
light are distributed equally, provide insights on mass in groups of
galaxies and on exotic dark matter such as strings and non zero Λ. As
an example, the existance of the normal lens case QSO 2016 with z_Q =
3.27 allows one to set the limit $q_0 > -2.3$.

1. INTRODUCTION

Gravitational lenses are an important new tool for studying dark matter
in the Universe. I will discuss a number of ways in which these
systems may be used in principal.

2. MASSES OF LENSING GALAXIES

Image separations may be used to put lower limits on the mass
associated with lensing galaxies. Suppose the lensing galaxy lies at a
distance d from us and is about halfway between us and the lensed QSO.
The galaxy produces a bend angle of $\alpha \sim 4GM/rc^2$ where r is the impact
parameter. Typically two relatively bright QSO images will be seen on
opposite sides of the lensing galaxy with a third faint image of the
QSO lying near the center of the galaxy. The separation of the two
bright images is approximately $\Delta\theta \sim (2r/d) \sim \alpha \sim [8GM(1+z_L)/dc^2]^{1/2}$.
With $\Omega_0 \sim 0$, $H_0 \sim 50$ km s^{-1} Mpc^{-1}, z_Q large, $z_L \sim 0.4$, d ~ 1500 Mpc one
finds

$$\Delta\theta \sim 4" \ (M/10^{12} \ M_\odot)^{1/2}$$

Since the observed cases have separations of 2" to 7" this shows
immediately that the masses of lensing galaxies (the mass contained
within the two QSO images) is surprisingly large, implying some dark

219

J. Kormendy and G. R. Knapp (eds.), Dark Matter in the Universe, 219–226.

matter. E.L. Turner, in the following companion paper, presents
detailed lower mass limits of this type for each case.

3. CIRCULAR VELOCITIES IN HALOS

If we model galaxies as singular isothermal spheres then we find that
the two bright images should have

$$\overline{\Delta\theta} \sim 4GM/rc^2 \sim 2\pi(V_{CIR}/c)^2 \sim 0".6 \ (V_{CIR}/200 \ km \ s^{-1})^2$$

where V_{CIR} is the circular velocity in the lensing galaxy halo [cf.
Turner, Ostriker, Gott (1)]. This shows that the observed lensing
galaxies either have larger circular velocities than average or are
being helped by additional matter in the beam most likely from the
group or cluster of which the galaxy is a member (1). This also
prompts us to search for lensing cases with small angular separation.
Tyson et al. (2) have shown that measurements of the lensing distortion
of background galaxies by foreground galaxies can be used to set limits
on V_{CIR} for typical galaxies. They find $V_{CIR} \leqslant 190 \ km \ s^{-1}$ out to ~ 120
Kpc which compares with values of $V_{CIR} \sim 130$ to $220 \ km \ s^{-1}$ at these
separations deduced from binary galaxies [Turner (3), Yahil (4), White
(5)].

4. Ω IN COMPACT DARK OBJECTS

Press and Gunn (6) showed that out to a redshift of $z_Q \sim 2$ the optical
depth to lensing is of order

$$\tau \sim 0.3 \ \Omega_{LENS} \ .$$

Separations of order $\Delta\theta \sim 0".001 \ (M/10^5 \ M_\odot)^{1/2}$ would be seen. Since
VLBI studies show that most QSO's are not multiply lensed on milli-
arcsecond scales there can not be a closure density of compact objects
with masses of $\geqslant 10^5 \ M_\odot$. Using the mini-lensing effect [twinkling of
the compact QSO continuum region produced by the lensing action of
small masses (7,8)] Canizares (9) has shown that $\Omega < 0.1$ in objects of
Jupiter mass or larger or otherwise QSO's would show more line versus
continuum variations than they do.

5. ARE MASS AND LIGHT DISTRIBUTED EQUALLY?

Lenses provide a particularly clear test of this question. As E. L.
Turner will discuss in the next paper the observed cases rather
strongly suggest that the mass and light are not distributed equally.

6. ARE HEAVY HALOS MADE OF LOW MASS STARS?

Gott (7) and Chang and Refsdal (8) showed that for a double lensed QSO
like 0957 if the heavy halo of the lensing galaxy were made of low mass

stars of mass M_S then there would be an uncorrelated twinkling in the images with timescales of

$$\Delta t \sim 1.4 \text{ yrs } (M_S/M_{JUPITER})^{1/2} .$$

For 0957 the two images have optical depths to such mini-lensing of 0.25 and 3. Long term monotoring of lensed QSO's is needed to do this test for low mass stars. Further detailed studies of such mini-lensing effects have been carried out by Young (10), Narasimha et al. (11), Ostriker and Vietri (12), and Paczynski (13).

7. ARE THERE BLACK HOLES OR CUSPS IN GALACTIC NUCLEI?

Chang and Refsdal (14) showed that if a black hole is placed in a galactic nucleus it can block completely the central faint overfocused image. A similar effect occurs if there is a cusp in the center as in a singular isothermal sphere. If the mass distribution is smooth, there is a theorem by Burke (15) that there must be an odd number of images. Black holes or cusps in galactic nuclei swallowing the faint central image may be responsible for the fact that all the observed cases show an even number of images.

8. ARE HALOS NON-BARYONIC?

If $H_0 = 50$ km s^{-1} Mpc, $\Omega_{baryon} \sim 0.1 \sim \Omega_{HALOS}$ Gott, Gunn, Schramm, Tinsley (16) but if $H_0 = 100$ km s^{-1} Mpc, $\Omega_{baryon} \sim 0.025 < \Omega_{HALOS}$. Refsdal (17) has shown that in a double lensed QSO intrinsic variations in the QSO will appear in the two images with a time delay of order

$$\Delta t \sim \Delta\theta^2 d/c \sim 1 \text{ yr } (\Delta\theta/3'')^2 (50/H_0).$$

With good modeling we can get H_0 [Borgeest and Refsdal (18), Young et al. (19)]. We really only get upper limits on H_0 because smooth matter in the beam can artifically lower the time delay [Falco et al. (20), Borgeest and Refsdal (18), Alcock and Anderson (21)]. Long term monotoring is needed to measure Δt.

9. DARK MATTER IN GROUPS AND CLUSTERS

Dense groups and clusters can serve as gravitational lenses (1,22). An interesting possible example of this is VV172 which is being investigated by a group of us at Princeton and by Hammer and Nottale (23). The group VV172, at a redshift of $\sim 15,800$ km s^{-1}, may be lensing the background galaxy at a redshift of $\sim 36,900$ km s^{-1}. This could explain why the background galaxy has a high apparent luminosity for its color.

10. EXOTIC DARK MATTER-STRINGS

The exact exterior metric for a string is

$$ds^2 = -dt^2 + dz^2 + dr^2 + (1-4\mu)^2 r^2 d\phi^2$$

[Gott (24)] where μ is the mass per unit length in the string in Planck masses per Planck length. For string assisted galaxy formation $\mu \sim 10^{-5}$. This is a conical space with an angle deficit $D = 8\pi\mu$. A string can produce double lens images of distant QSO's with separations of

$$\Delta\theta < 8\pi\mu \sim 50"$$

which should be observable. For futher studies on strings see Vilenkin (25), Kaiser and Stebbins (26) and Hogan and Rees (27). Moving strings produce fluctuations in the cosmic microwave background of order $\Delta T/T \sim 8\pi\mu \sim 10^{-4}$ (24, 26).

11. MEASURING Ω

Several methods are available for measuring Ω. After many cases have been discovered we may infer a curve of optical depth versus z_Q. The optical depth due to galaxies in an $\Omega=0$ model rises more steeply with z_Q than in an $\Omega=1$ model (1) being a factor of 2 higher at at $z_Q \sim 3$. We can plot $\Delta\theta$ versus z_Q; in the $\Omega=1$ case this is flat but in the $\Omega=0$ case it falls by about 20% by a redshift of $z_Q \sim 3$. The ultimate method wuld be to measure H_0 using time delays from a low redshift set of QSO's and compare it with the value of H_0 deduced from a high redshift set of QSO's, thus measuring the decelleration of the universe directly.

12. DARKEST MATTER - Λ

Paczynski and Gorski (28) and Alcock and Anderson (29) have noted that a positive cosmological constant can produce large image separations. Gott and Park (30) have found an interesting effect that occurs in closed universes with $\Lambda > 0$. Here the cosmological metric is:

$$ds^2 = -dt^2 + a^2(t) [d\chi^2 + \sin\chi^2 (d\theta^2 + \sin^2\theta d\phi^2)]$$

They find that the displacement in the sky produced by a singular isothermal sphere galaxy is

$$\beta_{cr} = \alpha \sin (\chi_Q - \chi_L)/\sin(\chi_Q)$$

where $\alpha = 2\pi (V_{CIR}/c)^2 = $ const. and χ_L and χ_Q are the co-moving distances to the lens and QSO respectively. The antipode is at $\chi = \pi$. If $\chi_L < \chi_Q < \pi$ then $\beta_{cr} > 0$ and three images are formed if the QSO's true position is within β_{cr} of the center of the lensing galaxy. A bright double with separation $\Delta\theta = 2\beta_{cr}$ is formed plus a third extremely faint image between them which is located at the position of the lensing galaxy nucleus (see fig. 1). As $\chi_Q \rightarrow \pi, \beta_{cr}$ blows up; the lensing cross sections, image magnifications and separations also blow up. Thus if the antipodal redshift were within the observed range of QSO's we would expect to see a large number of bright double lensed

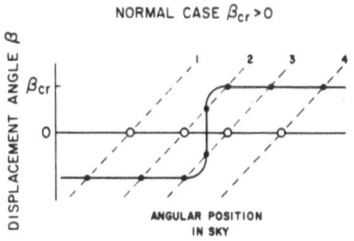

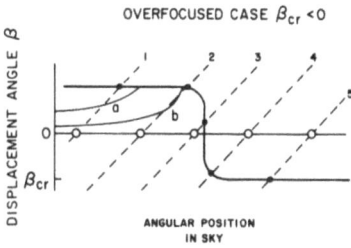

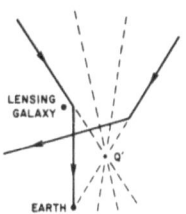

Figure 1. Normal and overfocused lensing cases for a galaxy with an
isothermal sphere heavy halo and small core radius (nearly a singular
isothermal sphere). Image displacement is plotted as a function of
angular position in the sky. The true unlensed position of the QSO is
shown as an open circle, the positions of the images are shown as
filled circles. In the normal case ($\beta_{cr} > 0$), if the QSO is at
positions 1 or 4 only one image will be formed, but if it is at
positions 2 or 3, three images will be formed. In the overfocused case
($\beta_{cr} < 0$) one image is formed at positions 1 through 5. If the
rotation curve falls slowly as in curve a still only one image will be
formed, but if the rotation curve falls rapidly enough as in curve b,
three images will be formed for a QSO at position 2. In the
overfocused case the universe as a whole forms a real image of the QSO
at Q' as illustrated. The bottom illustration shows how light beams
are bent in the overfocused case when QSO is at position 4.

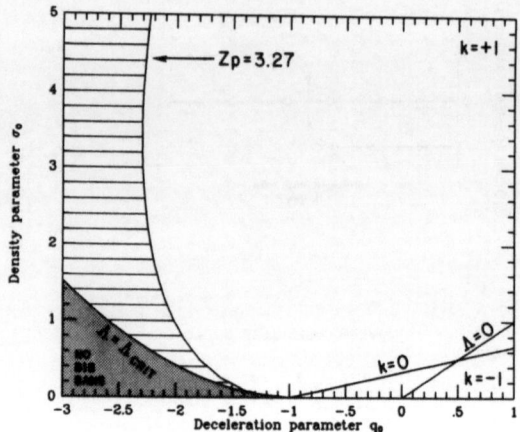

Figure 2. Constraints implied by gravitational lensing are plotted on
the σ_0, q_0 plane. The vertical coordinate is the density parameter
$\sigma_0 = \Omega_0/2$ and the horizontal coordinate is the deceleration parameter
q_0. The region below the line $\Lambda = \Lambda_{CRIT}$ is not allowed because the
models have no big bang. The observed lens cases including a normal
one with $z_Q = 3.27$ imply that in a closed model the antipodal redshift
z_p should be greater than 3.27. The curve showing models with $z_p =$
3.27 is plotted. The hatched region has $z_p < 3.27$ and is not allowed
by the lensing criterion. For all σ_0, the allowed value of $q_0 > -2.3$.
The graph shows the regions for which the models have positive,
negative or zero curvature k = +1, −1, 0 and the line showing models
with $\Lambda = 0$.

QSO's of wide separaton which could have hardly escaped notice.
For $\chi_Q > \pi$ and $\chi_Q - \chi_L < \pi$, $\beta_{cr} < 0$ and we have the over focused case
illustrated in fig. 1. The universe itself forms a real image of the
quasar at Q' just in front of us at a distance of $\chi_Q' = \chi_Q - \pi$. The
lensing galaxy lies behind this deflecting the light beams out of the
way. With a singular isothermal sphere galaxy only one image is
formed. And if the QSO's true position is within β_{cr} of the center of
the lensing galaxy this one image will be an extremely faint one going
through the galactic nucleus. If the galaxy has a rotation curve that
falls rapidly enough there is a possibliity of forming three images all
to one side of the lensing galaxy (see fig. 1) as Paczynski (31) has
noted. Even if galaxies have heavy halos that cut off sharply at 100
kpc, the rotation curves will not fall rapidly enough to allow the
offset triple image case unless the galaxy is close enough to the
antipode so that $\beta_{cr} < -14"$. As long as the rotation curves are either
flat or rising as discussed at this conference we will see only one
image in the overfocused case. Davis et. al's (32) data on velocity
diferences between pairs of galaxies suggests that this is the case at
least out to the correlation length $r \sim 10$ Mpc ($H_0=50$). In any case
for the overfocused case we never see two bright images with the
lensing galaxy in between. The lensed QSO 2016 has (33) $z_L \sim 0.8$, z_Q
$= 3.27$. It is a bright double with the lensing galaxy roughly between
the two images. It is thus an example of a normal lens case where β_{cr}
> 0 and not an example of an overfocused lens case where $\beta_{cr} < 0$.
Since the lensing galaxy is roughly halfway between us and the QSO Gott
and Park (30) show that the only way to reasonably get $\beta_{cr} > 0$ for this
case is if $\chi_Q < \pi$, and the antipodal redshift $z_p > 3.27$. The existence
rather simple normally lensed cases with $z_Q = 1.41$, $z_Q = 2.15$, $z_Q =$
3.27 and the observed lack of correlation of $\Delta\theta$ with z_Q strongly
implies that $z_p > 3.27$. The regions in the σ_0, q_0 plane ruled out by
this limit are shown by hatched lines in fig. 2. We see that the only
allowed models have $q_0 > -2.3$.

13. SUMMARY

We can see that gravitational lenses provide many exciting posibilities
for studying dark matter in the universe. A number of these tests are
difficult and making them work in practice will require a careful
observational treatment with due attention to selection effects and
modeling uncertainties. Some may have to await discovery of
particularly simple or auspicious lensing cases. E.L.Turner in the
next paper will discuss what we have learned so far from the six
observed cases and the prospects for the future.

This work supported by NASA grant NAGW-626

REFERENCES

1. Turner, E.L., Ostriker, J.P., Gott, J.R. 1984, Ap.J. 284, 1.
2. Tyson, J.A., Valdes, F., Jarvis, J.F. and Mills, A.P. 1984, Ap.J. 281, L59.
3. Turner, E.L. 1976, Ap.J. 208, 304.
4. Yahil, A. 1977, Ap.J. 217, 27.
5. White, S.D.M. 1981, M.N.R.A.S. 195, 1037.
6. Press, W.H. and Gunn, J.E. 1973, Ap.J. 185, 397.
7. Gott, J.R. 1981, Ap.J. 243, 140.
8. Chang, K. and Refsdal, S. 1979, Nature 282, 561.
9. Canizares, C.R. 1982, Ap.J. 263, 508.
10. Young, P. 1981, Ap.J. 244, 756.
11. Narasimha, D., Subramanian, K. and Chitre, S.M. 1984, M.N.R.A.S. 210, 79.
12. Ostriker, J.P., Vietri, M. 1985 (preprint).
13. Paczynski, B. 1985 (preprint).
14. Chang, K. and Refsdal, S. 1984, Astron. Astrophys, 132, 168.
15. Burke, W.L. 1981, Ap.J. 244, L1.
16. Gott, J.R., Gunn, J.E., Schramm, D.N., Tinsley, B. 1974, Ap.J. 194, 543.
17. Refsdal, S. 1964, M.N.R.A.S. 128, 307.
18. Borgeest, U. and Refsdal, S. 1984, Astron. Astrophysics 141, 318.
19. Young, P., Gunn, J.E., Kristian, J., Oke, J.B., Westphal, J.A. 1981, Ap.J. 244, 736.
20. Falco, E.E., Gorenstein, M.V., Shapiro, I.I. 1985, Ap.J. 289, L1.
21. Alcock, C., Anderson, N. 1985, Ap.J. 291, L29.
22. Narayan, R., Blandford, R., Nityananda, R. 1985 (preprint).
23. Hammer F. and Nottale, L. 1985 (preprint).
24. Gott, J.R. 1985, Ap.J. 288, 422.
25. Vilenkin, A. 1981, Phys. Rev. D. 23, 852.
26. Kaiser, N. and Stebbins, A. 1984, Nature, 310, 391.
27. Hogan, C., Rees, M.J. 1985 (preprint)
28. Paczynski, B. and Gorski, K. 1981, Ap.J. 248, L101.
29. Alcock, C., Anderson, N. 1985 (preprint).
30. Gott, J.R., Park, M-G. 1985 (preprint).
31. Paczynski, B. 1985 (comment later in this meeting).
32. Davis, M. and Peebles, P.J.E. 1983, Ap.J. 267, 465.
33. Lawrence, C.R., Schneider, D.P., Schmidt, M., Bennett, J.N., Hewitt, B., Burke, B.F., Turner, E.L. and Gunn, J.E. 1984 Science 223, 46.

GRAVITATIONAL LENSES AND DARK MATTER: OBSERVATIONS

Edwin L. Turner
Princeton University Observatory
Princeton, NJ 08544 USA

ABSTRACT. Following a few general comments on gravitational lenses
from an observer's perspective, the currently available observations of
the six known gravitational lenses are summarized. Attention is then
called to some regularities and peculiarities of the properties of the
known lenses and to how they might be interpreted. The most important
conclusions relevant to the dark matter problem which can be obtained
from the current observations are that the distributions of mass and
light appear to be quite different in at least some of the lensing
objects and that objects with projected M/R values about ten times
larger than those ordinarily associated with galaxies exist and are not
too rare (assuming $\Lambda = 0$).

1. INTRODUCTION

Before reviewing the currently available observations of the six known
gravitational lenses, I would like to make a few general remarks about
the advantages and disadvantages of lenses as tools for studying
cosmology and dark matter from an observer's perspective.

First, gravitational lenses are typically technically difficult
objects to locate and study. They are quite rare with only 5 found
among the roughly 3000 known quasars to date and only 1 found by
studying individual galaxies. They are usually fairly faint; the
quasar images in the known lenses vary from about 16th to about 23rd
magnitude. Their angular sizes are small (1/2 to 7 arc sec) and strain
the resolution of ground based observations (except for VLB). The time
scales over which they need to be monitored are unpleasantly long
(weeks to decades). None of these problems are insuperable,
particularly given the power of present and planned facilities such as
4m class telescopes equipped with CCD's, the VLA, ST, the VLBA, the
NNTT, etc. Nevertheless, it is clear that the potential of lenses
(described in the preceding paper by J. R. Gott) cannot be realized by
small programs or using modest facilities.

Second and probably more serious, it is not clear whether the
elegant but idealized lens experiments devised by theorists can

J. Kormendy and G. R. Knapp (eds.), Dark Matter in the Universe, 227–239.
© *1987 by the IAU.*

actually be carried out in the "dirty laboratory" provided to us by the real Universe. The complexities of the lensing mass distributions and the competition of various lensing and cosmological effects and quasar properties will undoubtedly result in many complications. The situation may well prove analogous to that for the "standard candle" Hubble diagram q_0 test, namely elegant and beautifully simple in principle but elusive and beset by systematic uncertainties in practice. Nevertheless, I do not believe we should become discouraged at this early stage of lens studies. We have no tools for studying cosmology against which similar objections cannot be raised. At least lenses offer us a new tool to attack problems against which we have worn the old ones dull. Moreover, lenses have the encouraging property that they can be cosmologically informative on an individual basis, not just in statistical samples. Thus, if we are lucky, we may find the gravitational lens counterpart of the binary pulsar and learn a great deal from the careful study of a single special object. This is a nearly unique possibility for cosmologists. Some of us have taken to referring to such putative, specially useful lens systems as "Rosetta Stone" lenses.

Leaving these larger issues for the future to resolve, I turn now to a review of what is currently (June 1985) known about the six lenses so far discovered and what conclusions or hints may be drawn from their properties.

2. OBSERVED PROPERTIES OF THE SIX KNOWN GRAVITATIONAL LENSES

Given the constraints on the length of this review, it is obviously impossible to consider each of the six known lenses individually in any detail. Thus, Figure 1 and Table I are intended to summarize the available information on 0957+561, 1115+080, 1635+267, 2016+112, 2237+031, and 2345+007.

Figure 1 displays a representation of the optical image of each system, all to the same scale. The quasar images are shown as filled circles while the positions of foreground galaxies, which must participate in and may be responsible for the lensing effect, are indicated by open circles. These plots are based on the best available optical images of each of the lens systems listed above (1 - 6, respectively). The most striking properties of these images are the absence of the third (fifth in 1115+080) quasar image required by the transparent lens theorem in all cases, the general lack of co-linearity between the images and the putative lensing object expected for spherically symmetric lenses, and the unexpectedly large (7) angular splittings of the images.

Table I gives the name, discovery reference, and discovery date for each lens system; the number of lens images detected; the status of attempts to detect the lensing object; the redshift of the lensed quasar; the redshift of the lensing galaxy, shown in parentheses if it is merely a photometric estimate; the maximum angular separation of the lensed images; the brightness ratio of the two brightest images; a brief statement of the evidence that the object is a lens system; the

TABLE I Lens Parameters and Status

Name Ref Date	No. of Images	Lensing Object	z_Q	z_L	$\Delta\theta_{MAX}$	R_{AB}	Lens Evidence	Model	Comments
0957+561 (8) 1979	2 of 3+	Yes	1.41	0.39	6"	1.4	spectra. radio. time delay?	Yes	brightest cluster galaxy
1115+080 (9) 1980	4 of 5+	Yes?	1.72	(0.3)	2"	<1.2	spectra. close bright pair.	No	lens outside image circle ?
1635+267 (10) 1983	2 of 3+	No	1.96	–	4"	4.4	spectra.	No	large $\Delta\theta$ and no lens
2016+112 (11) 1983	2 of 3+	Yes	3.27	(0.8)	3"	1.1 to 1.6	spectra. radio.	No	very faint third image
2237+031 (12) 1984	2 of 3+	Yes	1.70	0.04	2"	1.1	coincidence with galaxy.	No	very small z_L
2345+007 (13) 1981	2 of 3+	Yes?	2.15	(1.5)	7"	3.7	spectra.	No	large $\Delta\theta$ and very weak lens?

status of attempts to make a detailed model of the lensing process; and comments on any unusual or surprising properties of the system. Perhaps the most noteworthy features of this table are the general lack of success in attempts to construct detailed lensing models and the fact that all of the known lenses appear to be unusual or surprising in one way or another.

The presentation of the available data in summary form as given above obscures the fact that there is an enormous variation in how much effort has been devoted to studying the various lens systems. Only 0957+561 has been studied exhaustively so far, and some of the systems could be looked at far more closely than they have to date.

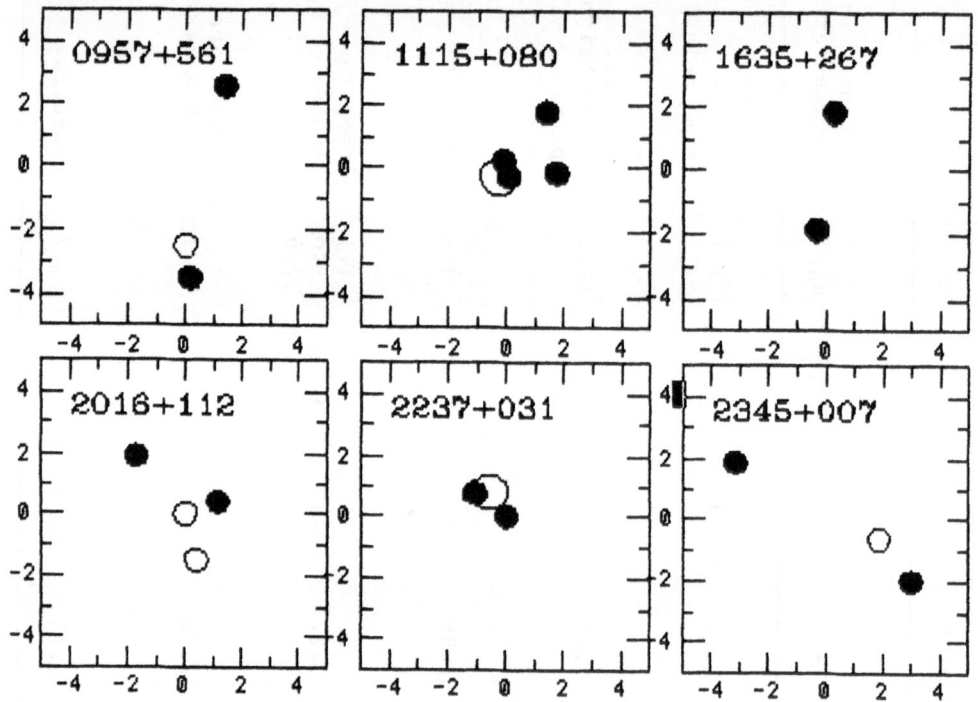

Figure 1. Optical images of the six known gravitational lens systems. Filled circles indicate quasar images, and open circles, probable lensing objects. Each box is 10" on a side.

3. IMPLICATIONS OF THE OBSERVATIONS

Although the six known gravitational lenses do not constitute a statistically valid sample in any sense and probably do not include an example of a "Rosetta Stone" lens, some physically interesting quantities may be calculated from their observed properties and some

intriguing hints may be discerned. Several of these are described
briefly below.

3.1 Lens Masses

Table II gives the mass of a point mass and the one dimensional
velocity dispersion of an isothermal sphere required to produce the
observed maximum image angular separation for each of the six known
lenses. It also gives the physical radius of the circle just enclosing
the two most widely separated images, an upper limit on the radius of
the mass producing the splitting via Ricci focusing. These numbers are
based on the lens redshifts given in Table I and are thus affected by
their substantial uncertainties. A lens redshift of 0.5 has been
assumed for 1635+267, and values have been calculated with that lens
redshift as well as 1.5 for 2345+007 since it is unclear that the
recently discovered (6) and apparently large redshift galaxy in that
system is the primary lensing object.

TABLE II Lens Masses and Sizes

Name	$M_p/10^{11}M_\odot$	σ (k/s)	r (kpc)
0957+561	13	427	9.5
1115+080	1.4	241	3.2
1635+267	20	505	8.1
2016+112	7.8	342	7.0
2237+031	0.054	154	0.3
2345+007(1.5)	120	878	14.9
2345+007(0.5)	21	459	12.4

H_o = 100 k/s/Mpc, q_o = 1/2

It should be emphasized of course that these numbers are not based
on real models of each lens system and do not attempt to account for
any of their observed properties beyond the maximum image separation.
These numbers also neglect the effects of shearing due to mass outside
the image circle which clearly plays a role in 0957+561 and quite
possibly other systems; of course, the mass required to account for the
separations by such shearing effects grows very rapidly with the
projected distance from the image circle. All things considered, the
tabulated numbers probably give a reasonably reliable rough estimate of

the projected masses involved in producing the observed lens systems. Their surprisingly large magnitude is of course the same surprise as that associated with the unexpectedly large (7) splittings.

3.2 Mass Distributions vs. Light Distributions

Well observed gravitational lenses offer the opportunity to directly check the classical but now out of favor hypothesis that the mass content of the Universe is traced by the distribution of stars. This is possible because the properties of a lens depend only upon the projected mass distribution in the plane of the sky which is equivalent to the observed lens surface brightness distribution if the hypothesis were correct. The only free parameter left to reproduce the positions and brightness ratios of the observed images is the unperturbed position of the background quasar.

Such calculations have been carried out for 0957+561 (14), 1115+080 (15), and 2016+112 (16). In no case does the observed light distribution account for the observed image properties. In fact no mass distribution in which the mass concentrations are even concentric with the observed light concentrations (i.e., galaxies) has been found to satisfactorily account for the data. The problems are accounting for the dog leg in the image-lens-image line in 0957+561, explaining the asymmetry of the four images with respect to the lens in 1115+080, and reproducing the acute image-lens-image angle and the very faint third image in 2016+112.

This negative result is probably the most important contribution of lens studies to our knowledge of the properties of dark matter to date. It strongly reinforces some earlier clues (17) that the distribution of the dark matter may be extremely poorly correlated with that of stars. If it is confirmed as a general property of lens systems, particularly when based on ST observations which will give us far more reliable and complete determinations of the lens surface brightness distributions, it will amount to a major discovery in my view.

3.3 Dark Objects

One possible exciting gravitational lens discovery would be a system in which the combination of the angular splitting of the images, the redshift of the lensed object, and very deep negative searches for the lensing object effectively ruled out the possibility of lensing by any known type of astronomical object and thus required the existence of a massive dark object. In such a system it would be possible to place a lower limit on the M/L of the lensing object and thus to detect in effect a dark object. Such a conclusion would have to contend with the possibility that the particular system was actually not a gravitational lens at all and the possible alternative explanation that $\Lambda \neq 0$, which allows less massive and higher redshift objects to produce larger splittings.

None of the known six lens systems are unambiguous examples of this situation, but two of them, 1635+267 and 2345+007, are promising.

Very deep CCD images of the former reveal no candidate lensing object
(18) while only a very faint and apparently distant object which is
probably incapable of producing more than a small fraction of the
observed 7" split has been found for the latter (6). The existing data
probably rule out all but the most extreme sorts of known astronomical
objects as the lenses in these systems; even these possibilities could
be ruled out by deep ST images.

3.4 Number of Images

A quite rigorous mathematical theorem (19) requires that all
transparent gravitational lenses produce an odd number of images. The
observations of the six known lens systems are unanimous in revealing
an even number of images (two in all but one case) in each system. It
is not clear whether this clear disagreement of theory and observation
is an important hint of some sort, a fluke, or merely a misleading
reflection of some bias or shortcoming of the observations.

The most straightforward explanation for the discrepancy is the
possibility that the lenses are not transparent; at optical wavelengths
dust could be the culprit, but in the radio (only relevant for 0957+561
and 2016+112) there is no such natural possibility. The other obvious
possibility is that the n+1 th images are generally much fainter than
the first n images. In cases with spherical symmetry, this can be
arranged by assuming very cuspy or even singular (i.e., a black hole)
central density distributions.

Of the known lenses, this problem is probably most serious for
2016+112, a system for which high quality VLA maps rule out a third
image as bright as 0.1% of the observed images over most of the field
(20).

3.5 Distribution of Image Angular Separations

The definitive comparison of observations of the distribution of image
angular separations to theoretical predictions (21) will have to await
better controlled and understood statistical samples of lenses;
however, if the current predominance of 5 ± 2 arc second splittings and
the absence of <1" cases is not reversed by such samples, it will
present a major puzzle (or clue?) concerning the nature of the lensing
masses. Possible explanations include a remarkably close coincidence
of the surface densities of large structures in the Universe with the
critical lensing surface density, and $\Lambda \neq 0$. It would be particularly
hard to understand the absence of a substantial number of <1"
separation systems since these should be produced by the inner regions
of ordinary galaxies whose mass distributions are thought to be well
known from rotation curve studies (22). Of course, this dilemma is not
yet upon us since it is based on a rather uncertain extrapolation of
the current observational situation.

3.6 Brightness Ratios

It is interesting to note the rather bimodal distribution of observed

brightness ratios (of the two brightest images) in Table I in which
four of the systems have $R_{AB} < 1.5$ while the other two have $R_{AB} \approx 4$.
It is also curious that the two systems with large brightness ratios
are 1635+267 and 2345+007, the same two which are possible candidates
for being lensed by dark objects (23). The distribution of brightness
ratios can be used as a diagnostic of the mass distribution in the
lensing objects in principle (24). Could this be another indication
that the lensing object in these systems is of an unusual nature?

3.7 Lensing By Low Mass Objects

No unambiguous, or even very suggestive, evidence for lensing by low
mass objects (minilensing) has been reported to date. Of the known
lenses, 0957+561 and 2237+031 appear to be particularly good candidates
for minilens searches (25). The absence of a third image in any
particular system could be blamed on a minilens event, but this would
not be a satisfactory statistical explanation since such events can
also brighten images.

4. SUMMARY AND DISCUSSION

The observations of gravitational lenses to date lead to two fairly
definite conclusions relevant to the dark matter problem. First, it is
clear that the distribution of mass in the Universe is not the same as
the distribution of light (i.e., stars); moreover, in some systems the
center of mass is not even coincident with the center of light.
Second, projected mass distributions with $M/R > 10^{12} M_\odot/10$ kpc exist
and are not too rare (or perhaps $\Lambda \neq 0$).

In addition, the existing lens observations hint at two further
conclusions without establishing them with any certainty. First, the
possibility that some of the lensing objects may be dark (very high M/L
compared to ordinary astronomical systems) must be taken seriously.
Second, there is some indication that the lensing objects have very
compact, or even singular, cores.

To date, the observations have not provided us with examples of
small angular separation (<1") lenses, minilensing events, or simple
"Rosetta Stone" lenses. Observers must continue to seek for examples
of these intriguing possible systems.

Four general types of observations are needed to advance
gravitational lens studies: lens surveys designed to produce
statistically useful samples and to reveal "Rosetta Stone" lenses,
detailed studies of known lens systems (not just 0957+561), flux
monitoring programs designed to detect minilens events and determine
time delays, and measurements of distortions in resolved background
objects produced by the lensing effect of foreground galaxies. Some
effort is being made in each of these areas. J. N. Hewitt describes a
major VLA -optical lens survey, and A. Tyson describes observations of
the fourth type listed above elsewhere in this volume.

Given that the first known gravitational lens was discovered only six years ago, it seems to me quite reasonable to hope that they yet have much to tell us about dark matter and other cosmological issues.

Preparation of this review was supported in part by NSF grant AST84-20352 and NASA grant NAGW-626. Several colleagues cited in the text for "private communications" provided results in advance of publication and gave permission for their citation here. Without their generosity, this review would not have been even nearly up to date.

REFERENCES

1. Stockton, A. 1980, Ap.J. Lett., 242, L141.
2. Henry, J.P. 1984, private communication.
3. Gunn, J.E., and Schneider, D.P. 1985, private communication.
4. Schneider, D.P., Lawrence, C.R., Schmidt, M., Gunn, J.E., Turner, E.L., Burke, B.F., and Dhawan, V. 1985, Ap.J., 294, 66.
5. Tyson, J.A. 1985, private communication.
6. Tyson, J.A. 1985, private communication.
7. Turner, E L. 1980, Ap.J. Lett., 242, L135.
9. Weymann, R.J., Latham, D., Angel, J.R.P., Green, R.F., Leibert, J.W., Turnshek, D.E., Tyson, J.A. 1980, Nature, 285, 641.
10. Djorgovski, S., and Spinrad, H. 1984, Ap.J. Lett., 282, L1.
11. Lawrence, C.R., Schneider, D.P., Schmidt, M., Bennett, C.L., Hewitt, J.N., Burke, B.F., Turner, E.L., and Gunn, J.E. 1984, Science, 223, 46.
12. Huchra, J., Gorenstein, M., Kent, S., Shapiro, I., Smith, G., Horne, E., and Perley, R. 1985, A.J., 90, 691.
13. Weedman, D., Weymann, R.J., Green, R.F., and Heckman, T.M. 1982, Ap.J. Lett., 255, L5.
14. Young, P.J., Gunn, J.E., Kristian, J., Oke, J.B., and Westphal, J.A. 1980, Ap.J., 241, 507; Greenfield, P.E., Roberts, D.H., and Burke, B.F. 1985, Ap.J., 293, 370; Falco, E.E., Gorenstein, M. V., and Shapiro, I.I. 1985, Ap.J. Lett., 289, L1.
15. Lee, H.M., and Gott, J.R. 1986, Ap. J., submitted.
16. Narashimha, D., Subramanian, K., and Chitre, S.M. 1984, Ap.J., 283, 512; and reference 4 above.
17. Yahil, A. 1977, Ap.J., 217, 27; Baldwin, J.E., Lynden-Bell, D., and Sancisi, R. 1980, MNRAS, 193, 313.
18. See references 3 and 10 above.
19. Burke, W.L. 1981, Ap.J. Lett., 244, L1.
20. Lawrence, C.R. 1985, private communication.
21. Turner, E.L., Ostriker, J.P., and Gott, J.R. 1984, Ap.J., 284, 1; and reference 7 above.
22. Rubin, V.C., Ford, W.K., and Thonnard, N. 1978, Ap.J. Lett., 225, L107; and see paper by V. C. Rubin in this volume.
23. Hewitt, J.N. 1985, private communication.
24. Hewitt, J.N., and Turner, E.L. 1986, in preparation.
25. Gott, J.R. 1981, Ap.J., 243, 140; Paczynski, B. 1985, private communication.

ADDENDUM

The editors have generously allowed me to update the summary of the
gravitational lens observational situation described in the preceding
paper (current for June 1985) as the proceedings go to press (currently
late December 1985). There have been several important developments and
changes:

A new probable gravitational lens 0023 + 171 (Hewitt et al. 1986)
has been discovered. Two images but no lensing object are seen. The
source redshift is 0.95 and the angular splitting of the images is 5".
Both components are radio sources and one shows radio jets/lobes. The
optical line brightness ratio between the two components is about 3 to 1.

The lensing galaxy in 1115 + 080 reported by Henry is, after
further image analysis, said to be located essentially directly in
between the A and A' images (Henry and Heasley 1986). In addition,
Shaklan and Hege (1986) dispute the existence of the galaxy reported by
Henry and Heasley and instead report a galaxy in a different position.
The small angular size of the 1115 + 080 system and the bright apparent
magnitude of the A - A' component make detection of a lensing galaxy
particularly difficult for this system.

Schneider et al. (1986) report new observations of 2016 + 112 which
reveal the third image located very near the center of the lensing
galaxy C, the redshift of the lensing galaxy D (1.01), and the existence
of two slightly resolved emission line regions to the northwest and west
of components A and B, respectively.

Further image analysis by Tyson of his 2345 + 007 data indicate
that the tentative detection of a very faint lensing galaxy in that
system was spurious. It is now reported that any lensing galaxy between
the two images must be fainter than J = 25.5 (Tyson 1986).

All references in this addendum are to preprints.

DISCUSSION

WHITMORE: It seems possible that a slightly asymmetric massive halo may
have its center of mass displaced from the center of light. This could
explain the discrepancy you report. Can you tell us how great the
displacement might be?

E. TURNER: The results are fairly model dependent, but 20 kpc might be
a typical value.

BALBUS: What is the timescale for twinkling due to minilensing in the
Huchra lens?

PACZYNSKI: If the minilenses are stars of $0.5 - 1 \, M_{\odot}$, the timescale is
something like two years.

MADSEN: Are diffraction effects negligible for minilensing?

GOTT: You are observing a distant object with a telescope that's a
minilens with a projected size of $\sim 10^{-6}$ arcsec. Diffraction becomes
important if you look at radio wavelengths longer than about 60 cm.

REES: Some of the things you said about M/L in the lens assume there is
no minilensing, don't they? Your evidence wouldn't be the same if you
took minilensing into account.

E. TURNER: You could use minilensing to explain our inability to get
the brightness ratios right. But you have the additional problem of not
having the right number of images in the right places. All of the
models fail for that reason. You can't get out of that with minilensing,
except to the extent that you could temporarily erase one of the images.

OSTRIKER: On the question of the absence of the central image:
Isothermal-sphere models certainly predict that there will be one, but
has anyone looked at what happens if you model galaxies using de
Vaucouleurs laws, which have singular centers? Are they singular enough
to prevent the formation of a central image?

GOTT: When you use a singular isothermal sphere, then as the core
radius shrinks to zero, the little core images shrink to zero intensity.
That is singular in the sense that we want. For example, our Galaxy has
a singular enough core to knock out the central image.

OSTRIKER: But I'm asking in particular about the de Vaucouleurs model,
which is not as singular as the isothermal sphere.

PHINNEY: The answer is no, because the model has a finite central
surface density. All that matters is the projected surface density.

E. TURNER: In any particular system you can always make the third image
faint by making the core radius small. But as long as it is finite,
there will be configurations where the central image will be fairly

bright. For example, in 2016+112 I think a de Vaucouleurs law is not centrally concentrated enough to eliminate the central image.

BURKE: Bonometti, Shapiro et al. have shown that for 0957+561 the compact VLBI source which had been suspected as the third image is in fact not the third image. It may be the core of the giant galaxy G1.

WHITE: Can the absence of a third image plausibly be ascribed to obscuration in the lens in any of the known systems?

E. TURNER: Yes, this may be important for 1115+080, 2345+007, 1635+267, and 2237+031. It cannot be the problem for 0957+561 and 2016+112, which are radio sources with excellent VLA images.

PACZYNSKI (to GOTT): I think that your limit $q_0 > -2.3$ is not quite as stringent as you say, because if the opposite side of the Universe is not exactly at the redshift of the source, but between us and the source, then the argument you presented doesn't work.

GOTT: If the object you were looking at were just beyond the antipodal point, it would not in general produce a double lens image. A QSO just beyond the antipodal redshift could be double lensed only by a very low-redshift galaxy, and the optical depth for this is small. Another thing that happens is that, as you approach the antipodal point from this side, all the lens cross-sections blow up, because the Universe is acting like a big magnifying glass. If this happened, say, at a redshift of 3, you would see a tremendous number of large-separation double quasars at redshift 3. Just beyond that there would be a wall: you would see no quasars. So the optical depth is not a linear function of distance. A third effect is that we would see a big dependence of the lens splitting on z. We don't see that; the splittings are more-or-less constant, as you would expect in a Friedmann model.

TYSON: Regarding the apparent paradox if 2345+007 is lensed by a single $M_\ast$ galaxy at $z \approx 1.5$: There is evidence for at least one foreground cluster.

E. TURNER: Yes, it would be much easier to understand 2345+007 if the dominant contribution to the total bending angle were at a $z < 1$.

CARR: It is worth stressing that VLBI observations already limit the number of lenses much smaller than galaxies. Of a sample of 50 sources, 5 show double structure on the milliarcsec scale with the components having comparable spectra. Even if this double structure does not result from lensing, one can infer that dark objects in the mass range around $10^6 M_\odot$ cannot have more than a tenth of the critical density.

E. TURNER: Yes. Given the absence of a statistically well defined sample of lenses, the strongest statistical results are presently based on the absence of the lenses predicted by some models of dark matter.

BURKE: The range of separations from 1" down to 0".1 or maybe 0".01 has

been investigated little or not at all. Some efforts are now under way
to remedy this lack.

E. TURNER: The VLA survey, which should cover $\Delta\theta > 0\overset{..}{}3$, is briefly
described in a poster paper by J. Hewitt, et al.

SHAPIRO: What range can we assume for the observational coverage of
angular splittings for lensed images? Can we, for example, assume that
all 1' splittings are known? Would we identify lensed images if their
splitting were even larger than 1' , say 1°? How about splittings which
are smaller than 2"?

E. TURNER: The current situation is very complex; clearly some QSO's
have been examined very carefully and others hardly at all. My rough
guess is that for splittings between several arcseconds and about 1
arcminute and for small to moderate brightness ratios we have not missed
many lenses among the known quasars.

SCHECHTER: Can you tell us a bit about VV172? Is it an interesting
lens system?

E. TURNER: Yes, it is interesting. Tod Lauer has taken a picture of
the chain and subtracted all the nearby galaxies to show the shape of
the background galaxy. You can then put enough mass into the group to
bind it, and ask what the intrinsic shape of the background galaxy must
be so that it looks the way it does after lensing. The result is
slightly banana-shaped. There is considerable freedom in these models,
which are by Hyung-Mok Lee and Gott. In principle, though, you could
use systems like this to put constraints on how the mass is distributed
by saying what you are willing to believe about the undistorted image of
the background object.

PACZYNSKI: In this model, what is the amplification factor? Does the
luminosity of the background galaxy agree with its observed color?

E. TURNER: The amplification is not very large, about a factor of two.
That helps, but does not entirely account for the color-luminosity
discrepancy. It's hard to produce enough amplification to explain the
color discrepancy without producing an unreasonable distortion which
then has to be exactly cancelled by some unreasonable shape.

PACZYNSKI: You get the distortion if you assume that mass is distributed
like light. You suggested that this is not the case in the known lenses.

E. TURNER: Yes. A smooth mass distribution gives magnification without
disortion.

WHITE: Has the velocity dispersion of the background galaxy in VV172
been measured to see if it is what you'd infer for the color and
distance you adopt?

E. TURNER: No, but that's a good idea which had crossed my mind.

A SEARCH FOR GRAVITATIONAL LENSING

J. N. Hewitt, G. I. Langston, J. H. Mahoney, B. F. Burke (MIT);
E. L. Turner (Princeton); C. R. Lawrence (Caltech);
C. L. Bennett (NASA/GSFC)

Gravitational interactions allow one to investigate the nature of matter in the universe independent of the properties that make it luminous. Much as studies of the dynamics of galaxies and clusters of galaxies have indicated the presence of dark matter, gravitational lensing provides an independent probe of the large scale distribution of dark matter in the universe.

There are six known cases of gravitationally lensed quasars, and they are being discovered at a rate of one per year. As the sample of gravitational lenses grows, we begin to be able to study the properties of these objects as a group. Already the known cases raise some questions:

(1) Why do we observe such large image angular separations?
(2) Why do we always see an even number of images?
(3) Why do we observe brightness ratios so close to unity?
(4) Why do we detect the lens in only three (four?) out of six cases?
(5) Does the correlation between image brightness ratio and lens detection mean anything?

The observed properties could be at least partly due to (poorly-understood) observational selection effects. We wish to study the characteristics of gravitational lenses drawn from a well-defined sample of radio sources.

Our strategy is to (1) observe many (thousands!) of radio sources with the Very Large Array, selecting sources that exhibit multiple point structure; (2) observe the lens candidates optically, selecting sources with radio-optical counterparts; and (3) for these best candidates, carry out optical spectroscopy. Identical spectra will be evidence of gravitational lensing. 3172 radio sources from the MIT-Green Bank 5 GHz survey have been observed with the VLA. To date, 1362 of these sources have been examined for multiple point structure, and 160 appear to be promising gravitational lens candidates. The observations can resolve image angular separations as small as 0".3 and 1".0 in the VLA's A and B arrays, respectively. The range of image brightness ratios detectable is limited by the dynamic range of the maps produced, which in these snapshot observations is, at worst, 17 to 1.

Of the 160 gravitational lens candidates, 80 have been observed optically to approximately 24th magnitude in the R band. Eight sources display radio-optical counterparts. Some of these, of course, will be chance superpositions of radio and optical sources in the sky, but these eight sources are prime candidates for spectroscopic observations. So far, one gravitational lens, MG2016+112, has been found, and another triple radio source shows spectra consistent with the lensing hypothesis.

J. Kormendy and G. R. Knapp (eds.), Dark Matter in the Universe, 240.
© 1987 by the IAU.

GALAXY MASS DISTRIBUTION FROM GALAXY-GALAXY GRAVITATIONAL LENSING

J. Anthony Tyson
AT&T Bell Laboratories

The average gravitational lens distortion of background galaxy images by foreground galaxies is an independent, non-kinematical measurement of galaxy mass distribution $M(r)/r$ (Tyson, et al. 1984). The upper limit we obtained for the equivalent circular velocity, while small compared with some heavy halo models, is consistent with dynamical estimates for samples of galaxies of all types (e.g. Turner's binary data and the Rubin, et al. rotation curves). For example, for a mean cutoff radius of 65 kpc/h, our 3σ upper limit for the equivalent circular velocity $(GM/r)^{\frac{1}{2}} = 190$ km/sec. For a mass cutoff at 190 kpc/h our 2σ upper limit is 175 km/sec. If I weight a sample of asymptotical rotation curve velocities by recent field luminosity functions, I get mean circular velocities less than 170 km/sec.

Is it possible to improve on the accuracy of this coherent lens distortion technique? Statistical gravitational lens image distortion of background galaxy images about the positions of foreground galaxies is complicated by several low-level systematics. Limits to many systematics were found by substituting either random positions or stars for the "foreground" or "background" galaxies. I have re-examined 46,954 images of galaxies which are behind and within 30 arcsec of 11,789 foreground galaxies, using new techniques for correcting some small systematics. The two largest offenders are: (1) contamination of the background galaxy sample by foreground dwarf galaxies, and (2) light contamination from the foreground galaxy.

(1) The average magnitude of galaxies in our foreground sample is 20.75 J mag, and the background sample is 23.06 J mag. Thus, any physical companions of foreground galaxies, masquerading as background galaxies, have absolute magnitudes about 2 mag fainter than the average foreground galaxy in our sample. The cross-correlation $w_x(\theta)$ between these samples clearly reveals clustering of foreground dwarf galaxies, with an amplitude slightly less than we found in our original simulation. Thus, this effect does not change our conclusions regarding the average galaxy $M(r)$.

(2) Fatal systematic errors occur when working within 4 arcsec of the foreground galaxy, due to its light. Up to now, we have avoided this region. I now have a program which fits the foreground galaxy's 2-d light profile, and subtracts it. This permits automated detection, moments and photometry of faint galaxies closer to the center of the foreground galaxy. Pending confirmation from simulated data, this may offer a much improved mass determination.

The limits for the mass distribution of an average field galaxy are consistent with our previous results, but are not consistent with the hypothesis that all galaxies have heavy halos corresponding to circular velocities over 180 km/sec extending beyond 80 kpc/h.

Tyson, A., Valdes, F., Jarvis, J., Mills, A., 1984, Ap.J. <u>281</u>, L59

J. Kormendy and G. R. Knapp (eds.), Dark Matter in the Universe, 241.

DARK MATTER: OBSERVATIONAL ASPECTS

Jaan Einasto, Mihkel Joeveer and Enn Saar
Tartu Astrophysical Observatory
202444 Toravere
Estonia, USSR

ABSTRACT. A review of observational work on dark matter in USSR is given. Dynamically the dark matter can be located (i) in the galactic disk and/or in dwarf galaxies, (ii) in coronas of galaxies and in clusters of galaxies, and (iii) distributed smoothly in voids. The possible amount of matter in all three forms is discussed. Physically dark matter can be baryonic or non-baryonic, in the latter case either hot, warm or cold. Available information on the nature of dark matter is indirect, coming from theories of the formation of structure in the Universe. Two constraints to the formation scenarios are discussed, the galaxian correlation function and their morphology.

1. INTRODUCTION

There are two kinds of matter, the visible or luminous matter, and the dark matter. If one believes in recent inflationary models of the Universe, the total density of matter equals the closure density of the Universe. From dynamical considerations dark matter can be divided into the local one in galactic disks and/or in dwarf galaxies, the halo (or coronal) dark matter around giant galaxies and in clusters of galaxies, and the smoothly distributed background in voids.

Direct dynamical data give us little information on the physical nature of the dark matter. However, indirect data on the distribution of galaxies in space, characterized by the galaxian correlation function and other statistics, as well as data on the microwave background, chemical composition of matter etc. can be used to draw conclusions on the nature of the dark matter. These conclusions are indirect and enter into the calculations through various scenarios of the formation of structure in the Universe. The scenarios depend on the nature of the dark matter, whether it is baryonic, hot (neutrinos), or cold (axions) or is it simulated by the λ-term. Observational data mentioned can be used to test the formation scenarios and respective dark matter models. The situation is illustrated in Fig. 1.

In the following we give a review of recent observational work in the USSR on both the dynamical and physical aspects of the problem.

243

J. Kormendy and G. R. Knapp (eds.), Dark Matter in the Universe, 243–261.
© *1987 by the IAU.*

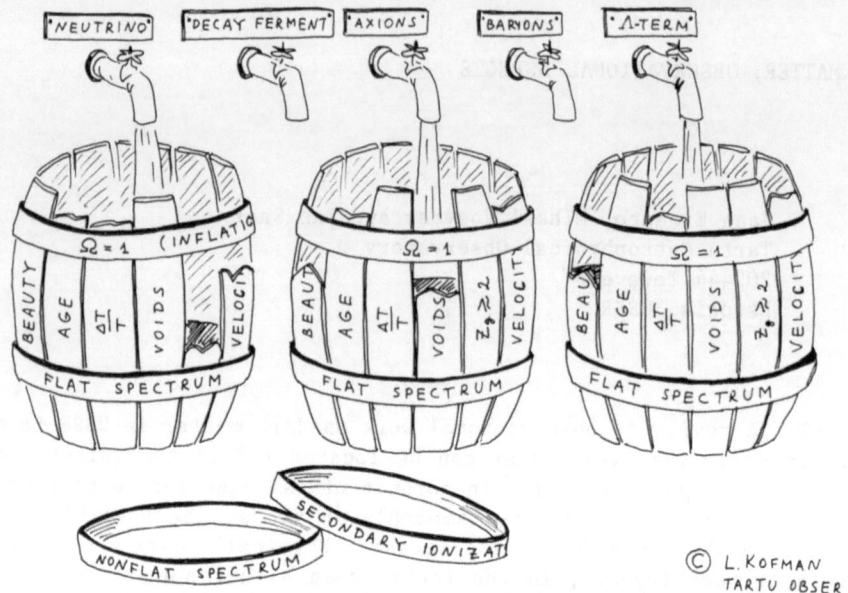

Figure 1. The barrel diagram - principal models of the formation of the structure in the Universe using different candidates of dark matter are shown as barrels. There are three main candidates of the dark matter: neutrino, axions (and other cold particles), and the cosmological constant. The barrels are hooped together by two principal assumptions, $\Omega = 1$, and a flat spectrum of initial perturbations. If these assumptions do not work, there are some hoops in reserve: a nonflat spectrum and secondary ionization. Various observational tests are expressed as staves. The height of a stave indicates the degree of accordance of the model with this particular test. One test is the beauty or internal harmony of the model. The level of the liquid in the barrel is equal to the height of the shortest stave, which determines the degree of acceptance of the model. If necessary, a coctail from several liquids can be made, or some ferment as neutrino decay added.

Idea and artwork by L. Kofman.

2. LOCAL DARK MATTER IN GALACTIC DISK

The local mass density problem was introduced by Oort (1960). He demonstrated that dynamical mass density in the solar neighborhood, ρ_{dyn}, exceeds the mass density of known objects, ρ_{vis}, and explained this discrepancy by the gravitational attraction of an unknown population of invisible objects.

The dynamical density can be determined from the Poisson equation which in cylindrical coordinates has the form

$$4\pi G\rho_{dyn} = -dK_z/dz-dK_R/dR-K_R/R, \tag{1}$$

where G is the gravitational constant, and K_R and K_z are the components of the gravitational acceleration in the radial and vertical direction, respectively. For the solar vicinity the radial gravitational accelera- tion can be expressed through the Oort dynamical constants, A and B. Similarily, the vertical acceleration can be expressed as follows (Kuzmin 1952):

$$-dK_z/dz = C^2. \tag{2}$$

Using the Oort-Kuzmin constants one has instead of (1)

$$4\pi G\rho_{dyn} = C^2 - 2(A^2 - B^2). \tag{3}$$

In this expression the vertical acceleration term C^2 is dominating, thus the dynamical density is determined essentially by the gradient of the vertical acceleration.

Oort (1960) derived the dynamical density by calculating the vert- ical gravitational acceleration K_z for a fairly large z interval and by determining its vertical gradient near the galactic plane. The dynamical density was also determined by Kuzmin (1952, 1955). Instead of calculat- ing the acceleration in a large z interval Kuzmin concentrated from the very beginning on its gradient. Near the galactic plane K_z is pro- portional to z. For populations located entirely in this small z inter- val, the vertical velocity dispersion $\sigma_{\dot{z}}$ is independent of z, and the constant C can be expressed through $\sigma_{\dot{z}}$ and the dispersion of z- coordinates of stars of the same population, σ_z (Kuzmin 1952):

$$C = \sigma_{\dot{z}}/\sigma_z. \tag{4}$$

This expression is valid for relatively young stars. Older popula- tions have higher z-velocities and their stars move outside the linear K_z regime. To determine C stars are to be chosen from a narrow belt a- round the galactic equator. In his pioneering study Kuzmin used A and gK stars. The z-coordinates were calculated from the galactic latitudes, the z-velocities from proper motions perpendicular to the galactic plane, and distances were estimated from apparent magnitudes. This approach has the advantage that systematic errors in adopted distances of stars change both dispersions in the same manner, thus distance errors cancel out. A similar method was used later by a number of authors from Tartu whose results are summarized in Table 1. For cepheids instead of proper motions radial velocities were used, and the vertical component of the velocity dispersion was calculated using the standard relations between the dispersions σ_R, σ_θ, and σ_z.

As seen from Table 1, all determinations made in the Tartu Observa- tory yield dynamical densities about 0.10 $M_\odot pc^{-3}$ in good agreement with direct determinations of the mass density, $\rho_{vis} = 0.092\ M_\odot pc^{-3}$ (Joeveer, Einasto 1976). Eelsalu (1961) and Joeveer and Einasto (1976) have stud- ied the reason for the discrepancy between the results obtained in these studies and in the Leiden studies (Oort 1960). The main reason lies in the inhomogeneity of statistical samples selected on the basis of the HD spectral classification. This classification does not separate the metal deficient old disk and halo stars. The inclusion of a small number of old stars may significantly increase the velocity dispersion whereas the spatial dispersion remains practically the same.

TABLE 1.   KUZMIN CONSTANT C AND LOCAL DYNAMICAL DENSITY

Stars	N	Method	C km/s/kpc	ρ_{dyn} M_o/pc^3	Reference	
Bright stars		z, μ_b, $D(z)$	68	0.081	Kuzmin[1]	1955
Bright stars		z, $\phi(M)$	66	0.076	Eelsalu	1958
B8-B9	17	τ, z, $\dot{z}$	70	0.086	Joeveer	1972
Cepheids	179	τ, z	65	0.074	Joeveer	1974a
Cepheids	100	v, $D(z)$	80	0.114	Joeveer	1974b
gK			75	0.10	Balakirev	1976
B8-A5	278	μ_b, $D(z)$	89	0.142	Joeveer	1985a
gK	253	μ_b, $D(z)$	66	0.076	Joeveer	1985a

[1])A rediscussion of earlier determinations of C by Oort 1932, Kuzmin 1952, Safronov 1952, and Parenago 1954

In such a confused case it is reasonable to use independent observational information to get a more reliable answer. Joeveer (1968, 1972, 1974a) noticed that very young populations are not in a stationary state, but oscillate in the z-direction:

$$z = z_0 \cos Ct, \tag{5}$$
$$V_z = -C z_0 \sin Ct. \tag{6}$$

Here z_0 is the maximum distance from the galactic plane for an individual star. The oscillation of a population as a whole can be easily explained if stars form outside the galactic plane in large complexes and "fall" together toward the plane (Dixon 1967a, b). Formula (5) was used by Joeveer (1974a) for a sample of cepheids, for which individual ages can be determined. Stars were grouped according to the age, t, and dispersions of the z-coordinates, σ_z were calculated. The σ_z versus t plot (Fig. 2) demonstrates the presence of a sinusoid as expected. The value of C can be estimated from two minima and one maximum.

If for a group of stars ages and the z-coordinates as well as the z-velocities are available, then each individual star yields an equation to determine C. C can be determined by minimizing the sums (Joeveer 1972)

$$f(C) = \sum [(z_i)_{obs}^2 - (z_i)_{calc}^2] \tag{7}$$
and
$$g(C) = \sum [(V_{zi})_{obs}^2 - (V_z)_{calc}^2]. \tag{8}$$

Accurate data are available for 17 B8-B9 stars. The plot of f and g versus C is given in Fig. 3. We see a pronounced minimum at C=70 km/s/kpc.

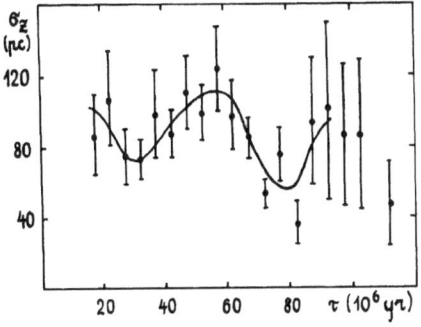 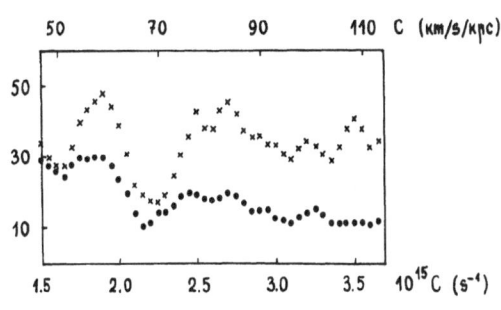

Figure 2.(left).   Relation between the coordinate dispersion in the z-direction σ_z and the age parameter $\tau = [\rho(R)/\rho(10)]^{-1/2}t$ in case of classical cepheids (Joeveer 1974).

Figure 3.(right).   Dependence of the sums $10^{-15}f(C_j)[km]$ and $g(C_j)[km/s]$, (dots and crosses, respectively) on the Kuzmin constant C, for nearby late B-type stars calculated on the basis of their evolutionary ages (Joeveer 1972).

This result as well as recent redeterminations of C using the classical method demonstrates that one can adopt a round value C = 70 km/s/kpc which leads to the dynamical density ρ_{dyn} = 0.09 $M_\odot pc^{-3}$ in good agreement with the direct density estimation. Thus there is no room left for the local dark matter. In a series of papers Bahcall (1984a, b) has obtained a considerably larger value. The reason for the discrepancy is not clear.

3. DARK MATTER IN DWARF SPHEROIDAL GALAXIES

According to Aaronson (1983) and Faber and Lin (1983) dwarf spheroidal galaxies may contain appreciable amounts of nonluminous matter. The preliminary evidence for this comes from unexpectedly large velocity dispersions in the Draco (Aaronson, 1983) and Carina (Cook, Schechter, Aaronson, 1984) dwarf spheroidal galaxies and from large tidal masses of the Sculptor, Draco, UMi and Carina galaxies (Faber, Lin, 1983). Both the virial as well as tidal mass-to-light ratios in these galaxies are an order of magnitude larger than in typical globular clusters.
    The basic assumption to derive the virial and tidal masses is the dynamical equilibrium of gravitational systems under study. An alternative explanation to the rather large velocity dispersions in the Draco and Carina galaxies could be that these galaxies are just now being tidally disrupted, as already noted by Aaronson (1983). To decide between the two possibilities, the presence of dark matter or tidal disruption, a comparative analysis of morphological properties of dwarf spheroidal galaxies was performed by Joeveer (1985b). Several arguments in favour of the tidal disruption scenario were found.

A strong evidence for this comes from the tidal radii r_t versus absolute blue luminosities L_B diagram (Fig. 4). Here and in the following the data about the Virgo cluster galaxies are from Binggeli et al. (1984) and about dwarf spheroidal satellites of our Galaxy from Hodge (1966), and Faber and Lin (1983). As one can see the fit of three distant satellites of our Galaxy to the regression line determined by the Virgo cluster galaxies is excellent, but all nearby satellites are 2-5 times larger than can be expected on the basis of their luminosities. The deviations from the regression line are correlated with the tidal mass-to-luminosity relations by Faber and Lin. The simplest interpretation of this is that the nearest dwarf spheroidal satellites of the Galaxy are actually not tidally limited but tidally expanded. If one treats the diameters of these expanded galaxies as normal, he gets the fictitious abnormally large tidal masses and mass-to-luminosity ratios.

The data about the satellite r_t/r_c ratios and ellipticities e. = 1-b/a are also in accordance with the tidal disruption scenario. The outer regions of satellite galaxies are more strongly influenced by tidal forces and must expand more intensively in comparison with inner regions, which define the core radius r_c. Indeed, on Fig. 5 such an effect is clearly seen. The probable expanded satellites of the Galaxy have larger r_t/r_c ratios in comparison with satellites outside the strong tidal field.

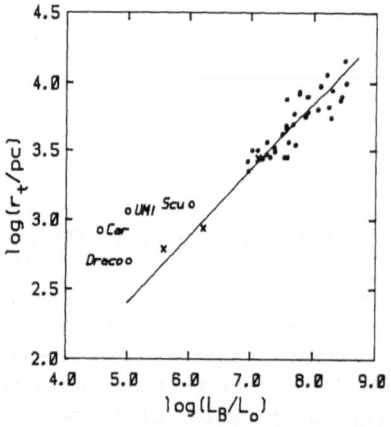 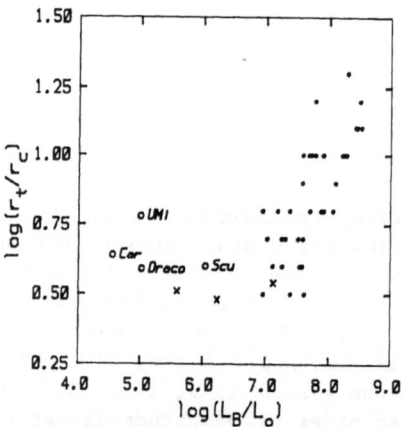

Figure 4.(left).   Correlation of the tidal radius r_t with the absolute blue luminosity L_B. Dots stand for the Virgo cluster dwarf spheroidal galaxies, crosses and open circles represent resolved dwarf spheroidal satellites of the Galaxy. Open circles are nearby (R<100 kpc) satellites with high tidal mass-to-luminosity ratios (M/L_V>6.8 M_O/L_O), crosses are more distant satellites with M/L_V<1.0 M_O/L_O from Faber and Lin (1983). The straight line is a mean regression for Virgo dwarfs.
Figure 5.(right).   The ratio of the tidal radius r_t to the core radius r_c as a function of the absolute blue luminosity L_B. Coding is the same as in Fig. 4.

As the tidal expansion takes place mainly along the orbital plane, it is expected that tidally expanded galaxies have obtained additional ellipticities. So another support to the tidal disruption scenario comes from the mean ellipticities, which are $\langle e \rangle$ = 0.22±0.09 in case of three distant probably undisturbed satellites and 0.38±0.05 in case of four nearby satellites (Joeveer 1985b).

Summarizing, it is highly probable that the four nearest dwarf spheroidal satellites of our Galaxy (Sculptor, Draco, UMi and Carina) are strongly disturbed by tidal forces and the estimation of virial and tidal masses of these galaxies is impossible. It seems that only outer satellites can give us more or less reliable information about the masses of dwarf spheroidal galaxies. Of course, the result for the Fornax galaxy by Cohen (1983) leaves little hope to find a significant amount of dark matter in galaxies with masses $<10^8$ M_o.

4. DECOMPOSITION OF GALAXIES INTO VISIBLE AND DARK POPULATIONS

In a series of papers Einasto (1974 and references therein) developed a method to decompose galaxies into visible and dark populations. The method is based on the assumptions that physically homogeneous galactic populations can be described by ellipsoidal models and that mass-to-luminosity ratios within the populations are constant. The decomposition can be made by comparing photometric and dynamical data (velocity dispersion, rotation). Photometric data describe visible components, whereas dynamical data depend on both components (Haud 1985).

Models are available for a number of nearby galaxies. Table 2 lists the galaxies and calculated mass-to-luminosity ratios for visible populations. We see that for all galaxies studied so far, this ratio is rather small, between 2 and 7 in blue light. As a mean value we can adopt $(M/L_B)_{gal}$ = 4.3.

TABLE 2. VISIBLE POPULATIONS IN GALAXIES

Name	Type	M_B	Distance (Mpc)	M/L_B	Reference
M 32	E2	-16.0	0.69	1.95	Einasto, Tenjes, Traat, 1980
M 87	E0-1	-22.2	20.	5.2	Tenjes, Einasto, Oleak, 1985
NGC 3115	S0	-20.0	10.	3.0	Tenjes, 1985
M 104	Sa	-22.3	20.	2.05	Tenjes, 1985
M 81	Sab	-20.2	3.3	7.0	Einasto, Tenjes, Zasov, Barabanov, 1980
M 31	Sb	-20.8	0.69	6.2	Einasto, Tenjes, Zasov, Barabanov, 1980
NGC 4565	Sb	-21.7	10.	5.1	Tenjes, 1985
NGC 4321	Sbc	-20.7	20.	2.0	Tenjes, 1985

5. LOCAL GROUP

The Local Group of galaxies presents an unique case where the total mass of the system can be derived from the internal dynamics of both subsystems as well as from the external dynamics by calculating the mutual orbit of its main concentration centres, M31 and the Galaxy.

The external method to calculate the mass of the Local system was first applied by Kahn and Woltjer (1959), followed by Gunn (1974), Lynden-Bell and Lin (1977) and others. A basic uncertainty of the method lies in the value of the circular velocity of the Galaxy at the Sun which must be used to correct the observed motion of Andromeda to the Galactic centre. Another uncertainty is the determination of the motion of our Galaxy in respect to the barycentre of the Local Group.

Einasto and Lynden-Bell (1982) redetermined the orbit of the Galaxy in the Local Group. They used recent determinations of the circular velocity of the Sun, which yield V_0 = 220 km/s. To calculate the motion of the Galaxy in the Group only independent members of the Local Group were used, i.e. galaxies not located in one of its main concentration centres around Galaxy and Andromeda. The total mass of the Local Group depends on the age adopted. Adopting on the basis of nucleocosmochronology and globular cluster ages an interval of 13-18 billion years, they found for the mass $M_{LG} = 5\pm1 \times 10^{12} M_0$.

The masses of both subgroups in the Local Group can be derived individually from their internal kinematics. The level of the constant circular velocity in the Galaxy and Andromeda is well determined. The extent of the dark corona can be estimated from the extent of the respective satellite system. For the Andromeda subgroup Einasto and Lynden-Bell found the mass $2.3 \times 10^{12} M_0$. For our subgroup much more data are available, in particular the high velocity clouds of neutral hydrogen and Magellanic Stream yield valuable information, as well as distant globular clusters. Using all these data Einasto with collaborators (Einasto et al. 1975, 1976, 1978) found values between $1.2-2.0 \times 10^{12} M_0$.

We come to the conclusion that both internal and external data yield for the Local Group and its both subgroups masses in good mutual agreement. This suggests that the most uncertain element in the chain of the mass calculations, the radius of the dark corona of the Galaxy and Andromeda, cannot be considerably in error.

Adopting the above total mass and the total luminosity from Vennik (1985) one has for the total mass-to-light ratio of the Local Group M/L_B = 90 and M_{tot}/M_{lum} = 20.

6. DOUBLE GALAXIES

Karachentsev (1980, 1981a, b, c, d) has observed and analyzed the majority of double galaxies from his list of relatively isolated pairs (Karachentsev 1972). At the time of the analysis redshifts were available for 440 pairs. Of this sample 17 galaxies were singles (a star was included as a galaxy in the catalog), and 59 were considered as optical pairs on the basis of the calculated mass-to-luminosity ratios. Karachentsev adopted a rather strict criterion M/L<100 in solar units

for physical pairs (photographic magnitudes and the Hubble constant
H = 75 km/s/Mpc were used). For mean mass-to-luminosity ratio Karachentsev obtained $M/L_B = 10.8$, and considered this low value as an indication against the presence of dark matter around galaxies.

This results seems to contradict other evidence and needs some comments. A weak point in calculating the mean mass-to-luminosity ratio
is the exclusion of all pairs with $M/L>100$. The velocity difference distribution shown in Fig. 6 demonstrates that pairs with $M/L>100$ and with
the velocity difference $\Delta V <1000$ km/s form a natural tail of the distribution. The study of spatial distribution of galaxies has shown
(Joeveer, Einasto, Tago 1978, Einasto et al. 1980, 1984, Tago et al.
1984) that practically all galaxies are located in systems which form
thin filaments separated by huge empty voids. Thus optical pairs are
usually separated by $\Delta V>>1000$ km/s. Only in cases when we look along a
filament, smaller velocity differences in optical pairs are expected.
In this case other galaxies at the same Hubble velocity should be observed, since galaxies are not isolated.

To have a rough estimate of this selection effect we calculated the
mean M/L for the sample with $M/L>100$ but $\Delta V <800$ km/s, there were 24
pairs in this additional sample. For the combined sample we have
$M/L_B = 23.7$ for H = 75 km/s/Mpc or $M/L_B = 16$ for H = 50 km/s/Mpc. The
mean separation of pairs is 60 kpc (H = 50) and the M/L_B quoted refers
to this distance.

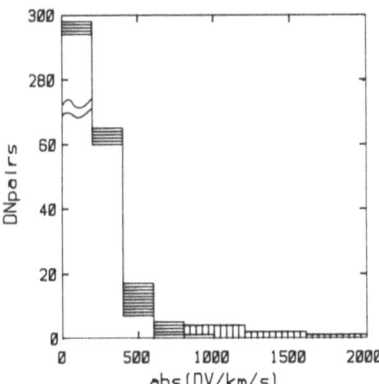

Figure 6. Distribution of velocity differences for the Karachentsev sample of double galaxies. Open areas - physical pairs by Karachentsev, horizontal shading - pairs added in present paper, vertical shading -pairs considered optical both by us and Karachentsev.

7. GROUPS OF GALAXIES

Recently Vennik (1984) compiled a new group catalog. Only galaxies
in the Northern Hemisphere far from the galactic belt (b>30°) were considered for group membership, and the velocity limit V<3200 km/s was
used, which corresponds to the limit of the Fisher-Tully (1981) HI survey of nearby dwarf galaxies. The majority of galaxies in the sample

not observed by Fisher and Tully have new accurate redshifts from the
CfA optical study. Groups were selected using a hierarchical clustering
algorithm (Materne 1978, Tully 1980). The density level at which a
group was selected was taken as a function of the mean spatial density
of galaxies in a particular field. Using this adaptive cut-off it was
possible to separate density enhancements in large clouds of galaxies
which in earlier studies were considered as single groups.

The catalog has been studied for dynamics of groups by Vennik
(1985). Groups were combined on the basis of similarity of morphological types of main galaxies. The dynamics of groups is determined by the
dark component, thus all galaxies were considered as test particles and
entered into dynamical calculations with equal weights. The extent of
synthesized groups has been estimated from the distribution of number
density of all member galaxies.

TABLE 3.   MASS-TO-LUMINOSITY RATIOS IN SYSTEMS OF GALAXIES

	R Mpc	M/L_B	M/M_{1um}	Reference
Galaxies	0.02	4	1	Table 2
Double gal.	0.06	16	4	Karachentsev 1980, 1981
Local Group	0.7	90	20	Einasto, Lynden-Bell 1982
Groups Sc-Irr	1.5	37	9	Vennik 1985
Sa-Sbc	1.5	61	14	Vennik 1985
E-So/a	1.5	95	20	Vennik 1985
Clusters	3	150	20:	Faber, Gallagher 1979
Closure		460	100	Felten 1985

The results obtained demonstrate that mass-to-luminosity ratio in
groups is slightly lower than thought earlier. The difference is caused
by a more strict definition of groups and more accurate velocities.
However, the principal conclusion is the same: M/L in groups exceeds
respective values in visible populations of galaxies by a factor of 10-
20. It is also important to note that the mass-to-luminosity ratio decreases considerably with increasing de Vaucouleurs morphological type
of main galaxies of groups.

8. LARGE-SCALE DISTRIBUTION OF DARK MATTER

In Fig.7 we plot mass-to-luminosity ratios of various systems as a
function of the extent of the system. Data on clusters of galaxies were
taken from Faber and Gallagher (1979), data corresponding to the critical cosmological density from Felten (1985). All data are given in the
B system for H = 50 km/s/Mpc.

Practically all galaxies are located in systems: groups, clouds,
clusters and superclusters. The percentage of galaxies situated in rich
clusters is rather small. The most common form of galaxy environment is
a group, a cloud or a loose cluster. If one adopts for these systems

mass-to-luminosity ratios given above one can argue that the majority of galaxies are located in systems which comprise about 20 per cent of the cosmological critical density. All visible systems of galaxies fill a volume of several per cent of the total volume of space, the rest being void of visible galaxies (Einasto et al. 1980). ·

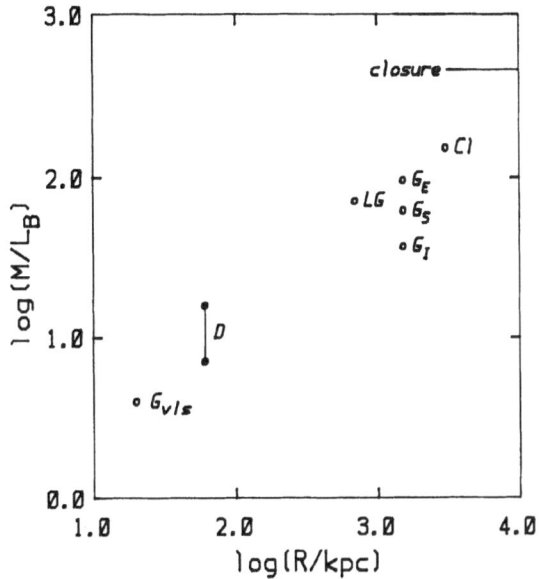

Figure 7. Mass-to-luminosity ratios for systems of different scale R. For double galaxies a range from the Karachentsev original determination to our present estimate is shown. G_{vis} denotes the normal mass-to-luminosity ratio of stars, D - double galaxies, LG - the Local Group, G_E, G_S and G_I - groups with the main galaxy of elliptical, spiral or irregular morphology, respectively, and Cl stands for rich clusters of galaxies.

Voids cannot be entirely empty. The smoothness of the microwave radiation background demonstrates that at the epoch of decoupling of matter from radiation the density distribution was rather uniform. The only force which can evacuate voids is gravitation. Calculations demonstrate that the available cosmological time is insufficient to evacuate voids completely, there should be primeval matter there, both the primordial gas and the dark matter (Zeldovich, Einasto, Shandarin 1982).

Astronomical determinations of the total density of matter in the Universe are rather uncertain. But if one accepts the inflationary scenario of the creation of the observed Universe, its total density should be equal to the critical density. In this case about 80 per cent of matter should be located in voids.

Physical properties of gas in cosmic voids have been studied by Ozernoy and his coworkers (Ozernoy and Chernomordik 1985, Chernomordik and Ozernoy 1983 and references therein), using UV spectral data by Brosh and Gondhalekar (1984).

9. STRUCTURE OF THE UNIVERSE: CORRELATION FUNCTION

In order to test various scenarios of the formation of the structure of the Universe a number of quantitative tests have been proposed, such as the correlation analysis, cluster analysis, etc. By far the most popular method used is the study of the correlation function.

The correlation function, $\xi(r)$, is defined as the excess of the observed number of pairs of galaxies at a given distance, $N_O(r)$, over the respective number of pairs in a random catalog, $N_P(r)$,

$$1 + \xi(r) = N_O(r)/N_P(r). \tag{9}$$

All studies carried out so far indicate that the correlation function at small distances can be represented by a power law

$$\xi(r) = (r/r_0)^{-\gamma}, \tag{10}$$

where r_0 is the correlation length. For galaxies $r_0 = 5$ h^{-1}Mpc (Peebles 1980 and references therein), whereas for clusters it is 25 h^{-1}Mpc (Klypin and Kopylov 1983, Bahcall and Soneira 1983).

To clarify the reason of this discrepancy Einasto, Klypin and Saar (1985) determined the correlation function for a number of observational samples of galaxies and clusters of galaxies of various depth using Huchra's (1983) compilation of redshifts which includes the CfA survey. To avoid incompleteness of data, conical samples were taken with boundaries used in the CfA redshift survey. Samples were cut off at certain V_O and an absolute magnitude M_O corresponding to the apparent magnitude limit of the CfA survey, 14.5, at the limiting redshift V_O. Redshift limits from 1000 to 10000 km/s were used. The use of the same absolute magnitude limit over the whole particular sample makes samples homogeneous, and no magnitude dependent weighting is necessary.

The correlation functions found for the northern galactic hemisphere are plotted in Fig. 8. One can see that with increasing sample depth the curves are shifted up and right. Thus respective correlation length also grows with the sample depth (Fig. 9). For samples containing only a part of a supercluster, the length r_0 is much smaller than the conventional value, 5 h^{-1}Mpc; for samples containing a whole supercluster, the length is approximately equal to the conventional value, and for samples containing several superclusters, the correlation length is about twice the conventional value. This result is not due to differences in absolute magnitude cutoff or mean spatial density of objects, as suggested by Szalay and Schramm (1985). Samples picked up from the same space with different absolute magnitude limit and mean density have practically identical correlation functions.

Einasto, Klypin and Saar determined also the correlation function for theoretical samples, calculated from N-body experiments. Adiabatic and isothermal scenarios were used, both either with all test particles included, or with particles only from high-density regions included (biased galaxy formation). In one case, the adiabatic scenario with biased galaxy formation, theoretical samples behave as observational ones: the correlation length increases with sample volume (Fig. 10).   This model shares one important feature with observations - it contains large empty regions between systems of test particles. In all other cases either the regions between clusters are filled with a rarefied population of particles (both scenarios, all test particles included) or empty regions are rather small, of size of systems of particles (I-scenario with biased galaxy formation).

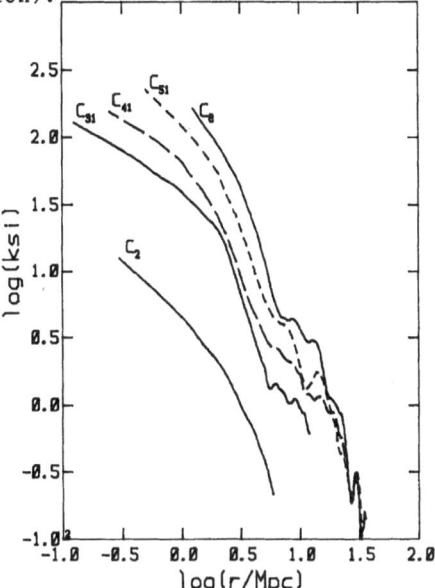

Figure 8. Correlation functions for samples in the Northern Galactic Hemisphere in the direction of the Coma supercluster. Samples C_2, C_{31}, C_{41}, C_{51} and C_6 have limiting redshifts 1200, 2000, 4000, 8000 and 10000 km/s, respectively.

This test indicates that the volume dependence of the correlation length is crucially dependent on the presence of large empty voids. Actually, the dependence is due to differences in the normalizing factor, $N_p(r)$, of formula (9). For small radii this factor is proportional to the relative volume of a spherical shell of radius r

$$N_p(r) = 2\pi r^2 dr N^2/V, \qquad (11)$$

where N is the total number of particles in the Poisson sample and V is the volume of the sample. Let us compare two identical samples of galaxies, surrounded by empty regions of different volume, V_1 and V_2. $N_0(r)$ in (9) is identical for both samples, but $N_p(r)$ is not, due to differences in volume V_i. Let V_0 be the volume filled with systems of galaxies, $C_i = V_0/V_i$ - the filling factor of the sample, and we get for correlation functions of our two samples

$$1+\xi_1(r) = C_2/C_1(1+\xi_2(r)). \qquad (12)$$

Within superclusters the filling factor is about 0.1 (Einasto et al. 1984), but in large volumes it is only about 0.01, thus the difference in filling factors explains the difference in correlation functions. Since the correlation length is defined as the value of the argument at which $\xi(r) = 1$, it also increases with decreasing filling factor.

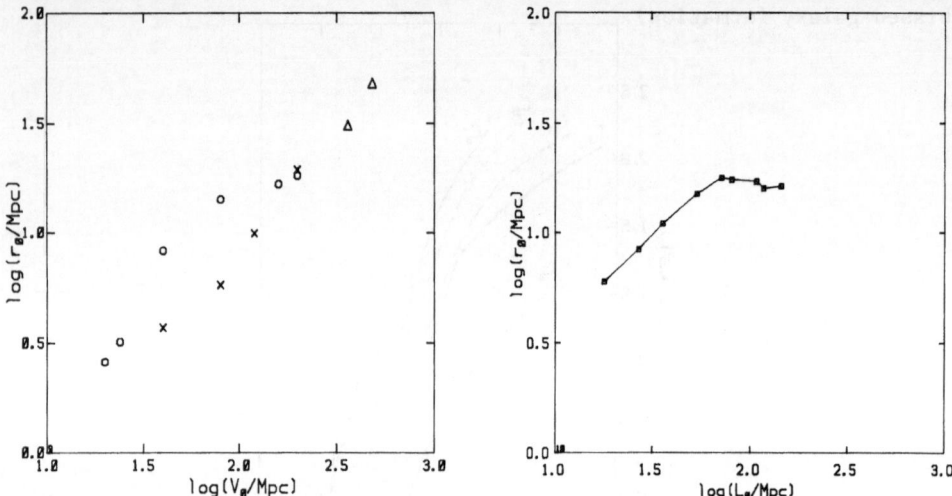

Figure 9.(left). Plot of the correlation length, r_0, versus limiting redshift, V_0, in Megaparsecs (H = 50 km/s/Mpc). Open circles and crosses designate samples in the Northern and Southern Galactic Hemisphere, respectively, triangles - Abell clusters of galaxies.

Figure 10.(right). The correlation length versus cube size plot for theoretical samples (A-scenario, biased galaxy formation: points from low density regions removed). For comparison with Fig. 9 r_0 and L are given in Megaparsecs.

Model calculations demonstrate that for samples exceeding in volume the characteristic volume of a void between superclusters, the correlation length reaches its global value (see Fig. 10). On the other hand, observed samples show no trend to converge at large sample sizes to a constant correlation length (see Fig. 9). Even clusters of galaxies sampling much larger distances than galaxies continue the same r_0 versus V_0 trend. Thus we come to the conclusion that presently available samples are not deep enough to derive the global value of the correlation length. The conventional value of the correlation length is definitely too low, available deepest galaxy samples indicate a value $r_0 = 10 \ h^{-1}$Mpc.

10. STRUCTURE OF THE UNIVERSE: MORPHOLOGY OF GALAXIES

The morphology of galaxies enters as an important parameter of the formation and evolution of galaxies. In earlier studies Dressler (1980) and Postman and Geller (1984), among others, studied the morphology of clusters and groups of galaxies of various mean densities. A pronounced

variation of the morphology with density was found. This dependence is
most probably due to evolutionary effects. Galaxies in low-density regions, in particular, isolated galaxies, do not change their morphology.
For this reason the study of isolated galaxies is of great interest for
theories of galaxy formation.

    Einasto and Einasto (1985) used CfA redshift survey, supplemented
with other available data, to study morphology of galaxies both in systems of galaxies as well as for isolated galaxies. Galaxies were attributed to systems of various richness (isolated galaxies, small systems,
large systems) using cluster analysis. The neighborhood radius at which
galaxies were included into various systems was varied by a factor of
ten, which corresponds to a variation of three orders of magnitude in
density enhancement. Three areas on the sky were studied which correpond
to the Local Supercluster, and to the Coma and Perseus superclusters.
Fig. 11 plots the number of galaxies in systems of different richness as
a function of the neighborhood radius for the Local supercluster,
Fig. 12 gives the relative distribution of morphological types. Other
superclusters have similar distributions.

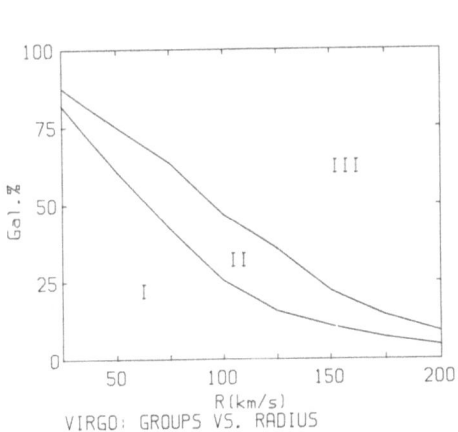

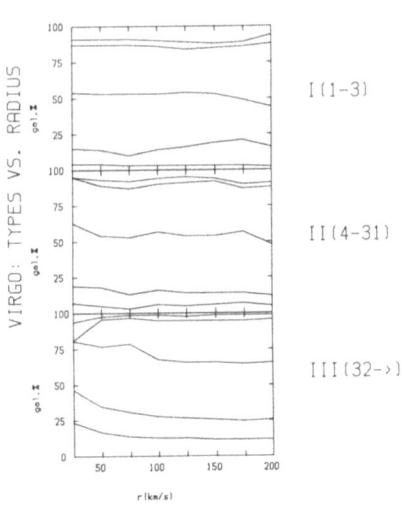

    Figure 11.(left). Plot of relative number of members of systems
of galaxies of different richness as a function of neighborhood radius
R in the Virgo supercluster. Isolated galaxies and small systems (number
of galaxies in the system n≤3) are designated as field I, intermediate
systems (4≤n≤31) as II, and large systems (n≥32) as III. Neighborhood
radius is expressed in Hubble expansion velocity units.
    Figure 12.(right). Plot of relative numbers of galaxies of various
morphological types in the Virgo supercluster as a function of neighborhood radius. Systems of various richness are plotted in panels I, II,
and III. In all panels six areas are given, representing (from bottom to
top) the following morphological types: E, S0, Sa - Sd, Irr, Pec, un-
classified.

As expected (Dressler 1980, Postman and Geller 1984) in systems of galaxies (groups and clusters) morphology changes with neighborhood radius (i.e. density). However for isolated galaxies the morphological distribution is remarkably stable and has little, if any, change with the degree of isolation. This distribution should reflect the formation conditions of galaxies.

11. CONCLUSIONS

The principal results of the studies reviewed can be summarized as follows:

1) There is yet no firm evidence for the presence of large amounts of invisible matter in the solar vicinity and in dwarf spheroidal galaxies.

2) The total amount of matter in visible galaxies, around galaxies and clusters, and smoothly distributed in voids relates as 1:20:80.

3) The correlation function of galaxies strongly depends on the presence of large voids in the sample. In largest observed samples the correlation length is 10 h^{-1}Mpc, about twice the conventional value.

Acknowledgements. We thank Dr. J. Huchra for supplying us with a copy of his redshift compilations which made the study of the large-scale properties of the Universe possible, and Drs. A. Dekel, P.J.E. Peebles and S.F. Shandarin for useful discussion.

REFERENCES

Aaronson, M., 1983. Astrophys. J. (Letters), **266**, L11.
Bahcall, J.N., 1984a. Astrophys. J., **276**, 156.
Bahcall, J.N., 1984b. Astrophys. J., **276**, 169.
Bahcall, N., and Soneira, R., 1983. Astrophys. J., **270**, 20.
Balakirev, A.N., 1976. Astr. Zh., **53**, 119.
Binggeli, B., Sandage, A. and Tarenghi, M., 1984. Astr. J., **89**, 64.
Brosch, N., and Gondhalekar, P.M., 1984. Astr. and Astrophys. **140**, L43.
Chernomordik, V.V., Ozernoy, L.M., 1983. Astrophys. Sp. Sci., **97**, 19.
Cohen, J.G., 1983. Astrophys. J. (Letters), **270**, L41.
Cook, K., Schechter, P., and Aaronson, M., 1984. BAAS, **15**, 907.
Dixon, M.E., 1967a. Astr. J., **72**, 429.
Dixon, M.E., 1967b. Mon. Not. R. astr. Soc., **137**, 337.
Dressler, A., 1980. Astrophys. J., **236**, 351.
Eelsalu, H., 1958. Tartu Astron. Obs. Publ. **33**, 153.
Eelsalu, H., 1961. Tartu Astron. Obs. Publ. **33**, 416.
Einasto, J., 1974. Proc. First European Astr. Meeting, **2**, 291.
Einasto, J., Haud, U., Joeveer, M., and Kaasik, A., 1976. Mon. Not. R. astr. Soc. **177**, 357.
Einasto, J., Haud, U., Joeveer, M., Kaasik, A., Traat, P. 1978. Astr. Circ. USSR No 1023, 2.
Einasto, J., Joeveer, M., Kaasik, A., and Haud, U., 1975. Astr. Circ. No 895, 1.

Einasto, J., Joeveer, M., Saar, E., 1980. Mon. Not. R. astr. Soc. **193**, 353.

Einasto, J., Kaasik, A., Kalamees, P., Vennik, J., 1975. Astr. Astrophys. **40**, 161.

Einasto, J., Klypin, A. and Saar, E., 1985. Mon. Not. R. astr. Soc. (in press).

Einasto, J., Klypin, A.A., Saar, E., Shandarin, S.F., 1984. Mon. Not. R. astr. Soc. **206**, 529.

Einasto, J. and Lynden-Bell, D., 1982. Mon. Not. R. astr. Soc. **199**, 67.

Einasto, J., Tenjes, P., Traat, P., 1980. Astr. Circ. No 1132, 5.

Einasto, J., Tenjes, P., Zasov, A.V., Barabanov, A.V., 1980. Ap. Space Sci. **67**, 31.

Einasto, M., and Einasto, J., 1985. In preparation.

Faber, S., Gallagher, J., 1979. Ann. Rev. Astr. Astrophys. **17**, 135.

Faber, S.M. and Lin, D.N.C., 1983. Astrophys. J. (Letters), **266**, L17.

Felten, J.E., 1985. Comments on Astrophys. (in press).

Fisher, J.R. and Tully, R.B., 1981. Astrophys. J. Suppl. Ser. **47**, 139.

Gunn, J.E., 1974. Comments Astrophys. and Space Phys. **6**, 7.

Haud, U., 1985. (in preparation).

Hodge, P.W., 1966. Astrophys. J. **144**, 869.

Huchra, J., 1983. Personal communication.

Joeveer, M., 1968. Tartu Astr. Obs. Publ. **36**, 84.

Joeveer, M., 1972. Tartu Astr. Obs. Teated, No 37, 3.

Joeveer, M., 1974a. Tartu Astr. Obs. Teated, No 46, 35.

Joeveer, M., 1974b. Tartu Astr. Obs. Teated, No 46, 18.

Joeveer, M., 1985a. Tartu Astr. Obs. Publ. (in press).

Joeveer, M., 1985b. Preprint.

Joeveer, M., Einasto, J., 1976. Tartu Astr. Obs. Teated, No 54.

Joeveer, M., Einasto, J., Tago, E., 1978. Mon. Not. R. astr. Soc., **185**, 357.

Kahn, F.D., Woltjer, L., 1959. Astrophys. J. **130**, 105.

Karachentsev, I.D., 1972. Soobsh. Spec. Astr. Obs., **7**, 3.

Karachentsev, I.D., 1980. Astrofizika **16**, 217.

Karachentsev, I.D., 1981a. Astrofizika **17**, 249.

Karachentsev, I.D., 1981b. Astrofizika **17**, 429.

Karachentsev, I.D., 1981c. Astrofizika **17**, 675.

Karachentsev, I.D., 1981d. Astrofizika **17**, 693.

Klypin, A.A. and Kopylov, A.I., 1983. Pis'ma v AZh. **9**, 75.

Kuzmin, G.G., 1952, Tartu Astr. Obs. Publ. **32**, 5.

Kuzmin, G.G., 1955, Tartu Astr. Obs. Publ. **33**, 3.

Lynden-Bell, D., and Lin, D.N.C., 1977. Mon. Not. R. astr. Soc. **181**, 37.

Materne, J., 1978. Astr. Astrophys. **63**, 401.

Oort, J.H., 1932. Bull. Astr. Inst. Neth. **6**, 249.

Oort, J.H., 1960. Bull. Astr. Inst. Neth. **15**, 45.

Ozernoy, L.M., Chernomordik, V.V., 1985. Pis'ma Astr. Zh. (in press).

Peebles, P.J.E., 1980. The large-Scale Structure of the Universe, Princeton Univ. Press.

Postman, M., and Geller, M.J., 1984. Astrophys. J. **281**, 95.

Szalay, A., and Schramm, D., 1985 (preprint).

Tago, E., Einasto, J., Saar, E., 1984. Mon. Not. R. astr. Soc. **206**, 559.

Tenjes, P., 1985. (in preparation).

Tenjes, P., Einasto, J., Oleak, H., 1985. Astr. Nachr. (in press).
Tully, R.B., 1980. Astrophys. J. **237**, 390.
Vennik, J., 1984. Tartu Astr. Obs. Teated No 73.
Vennik, J., 1985. Preprint.
Zeldovich, Ya.B., Einasto, J., Shandarin, S.F., 1982. Nature **300**, 407.

DISCUSSION

PEEBLES: The deep galaxy surveys in narrow fields, the Durham and
Edinburgh deep angular distributions, the Jagellonian field and Lick
sample, and the redshift sample of Kirshner, Oemler, Schechter and
Shectman all indicate $r_0 \sim 5h^{-1}$ Mpc, consistent with the shallow CfA
result. How are these results to be reconciled with your proposal that
the effective value of r_0 increases with increasing depth? Are you
proposing that the galaxy distribution is not a stationary random
process?

EINASTO: The correlation length determined from a particular sample is
inversely proportional to the filling factor of this sample. The deep
samples you mentioned cover small areas on the sky; thus voids in these
samples are essentially one-dimensional, and the filling factor is
relatively large. It is possible that this effect is responsible for
the disagreement quoted. We have used large areas on the sky; thus
voids enter as three-dimensional objects and suppress the filling
factor. In two-dimensional distributions (the Lick sample and
Jagellonian field counts), information on the presence of voids is
incomplete. To transform two-dimensional correlation functions into
three-dimensional ones, additional information on voids must be used.
    Our data indeed indicate that on scales smaller than the
characteristic diameter of voids, the galaxy distribution is not a
stationary random process. Available galaxy samples have depth smaller
than this characteristic diameter.

DEKEL: Like Peebles, I am worried about the growth of $\xi(r)$ with the
depth of the sample. This contradicts all previous results. In
particular, I believe that the sample you use is incomplete beyond
10,000 km s^{-1}, which may be the reason for the effect you find. Could
you elaborate on how you weight the galaxies at different redshifts, and
how you deal with the boundaries of the volume sampled?

EINASTO: The growth of $\xi(r)$ with sample depth is clearly seen in the
whole interval of depths of subsamples. Thus the result is not
dependent on the deepest samples. The absolute magnitude cutoff for
each subsample of different depth is taken equal to the absolute
magnitude corresponding to the CfA apparent magnitude limit at the
cutoff redshift of the subsample. Thus the absolute magnitude cutoff
within all subsamples is redshift independent and no weighting of

galaxies at different redshifts is needed. Sample volume boundaries
have been taken into account in generating the Monte-Carlo catalogue of
test particles used in the calculation of the respective number of
pairs.

DRESSLER: Although my study of galaxy morphology vs. environmental
density did not extend to regions as sparse as you discussed, Postman
and Geller used the CfA redshift survey to show that a correlation
persisted down to very weak enhancements. The work you described showed
no such effect, although the parameterization seemed somewhat different.
Is there a contradiction between these two studies and, if so, have you
any explanations?

CHINCARINI: In support of Dressler's question, I refer to work on
Perseus-Pisces by Giovanelli, Haynes and Chincarini (preprint). We
find, in agreement with Postman and Geller, that the autocorrelation
function depends on morphological type. We also find a strong
dependence between type percentage and density. The same density
dependence is detected in a catalogue complete to 14.5 mag by DeSouza,
Vettolani and Chincarini (preprint). We find that the relation between
type and density depends also on the luminosity function of each type.

EINASTO: Our results are consistant with the other studies mentioned
for high-density systems (clusters and groups). In low-density regions
our results seems to be in contradiction with others. One possible
reason may be the method of analysis used. The correlation technique
used in some of the quoted studies is volume dependent whereas the
clustering method is not. A more detailed comparison of both sets of
results is needed to clarify the situation.

J. BAHCALL: You have raised the interesting question of why the Soviet
studies of the local missing mass have given a different answer than
those made in the West. I believe that the primary reason for this
discrepency is the incorrect assumption by Kuzmin that the gravitational
potential in the z direction is quadratic. If one wants to limit the
error in the potential caused by this approximation to less than 10%,
then one must use stars at no more than 18 pc ($\sigma/4$ km s^{-1}) above the
plane. No population that I know about satisfies this condition. In
addition, some of the samples you mentioned are not sufficiently pure or
homogeneous to use in this context (e.g., bright stars or B stars).

EINASTO: The population used in Kuzmin's study and in later work has a
V_z dispersion of about 7 km s^{-1} and a z coordinate dispersion of ~ 100
pc. In this z interval the inaccuracy of the quadratic assumption can
hardly explain the factor of two difference in results. I agree that
Kuzmin's original sample was inhomogeneous, but in recent work modern
data have been used, so the inhomogeneity, if present, is small.
Finally, a completely independent method which uses the ages of stars
also supports Kuzmin's original results. Presently it is difficult to
see where the weak points of density determinations are. So I come to
the conclusion that further detailed work is needed to find a better
value of the local density.

N-BODY RESULTS ON DARK MATTER

Simon D. M. White
Steward Observatory
University of Arizona
Tucson, Arizona 85721
U.S.A.

ABSTRACT. The structure of the dominant "dark" component of the Universe may evolve primarily under the influence of gravity. A number of models for the evolution of the Universe make specific predictions for the statistical properties of density fluctuations at early times. N-body simulations can follow the nonlinear development of such fluctuations to the present day. A major difficulty arises because we cannot observe the present mass distribution directly. Recent N-body work has concentrated on models dominated by weakly interacting free elementary particles. Neutrino-dominated but otherwise conventional cosmologies pass rapidly from a smooth distribution to one dominated by lumps with masses greater than those of any known object. Cosmologies dominated by "cold dark matter" produce mass distributions which fit the observed galaxy distribution (i) if $\Omega = 0.1 - 0.2$ and galaxies follow the mass distribution, or (ii) if $\Omega = 1$, $H_0 < 50$ km/s/Mpc and galaxies form preferentially in high density regions. In the latter case, clumps form with flat rotation curves with about the amplitude and abundance expected for galaxy halos.

1. INTRODUCTION

Two major problems confront any attempt to make detailed evolutionary models for the distribution of dark matter in the Universe. We have no idea what the stuff is, and so cannot specify the physical interactions it is subject to. In addition, since we cannot see it, we have only indirect indications of where it is, and we do not know how much of it lies between galaxy groupings. The first difficulty may not be serious when considering recent evolution, because we have good evidence that the dark matter is presently in a non-gaseous, effectively collisionless form. These evolutionary phases can therefore be studied by N-body methods. However, although gravity may be the sole driver of evolution today, other interactions which depend on the nature of the dark matter condition the distribution of density fluctuations with which it emerged from the early universe. These fluctuations set the initial conditions from which later structure

J. Kormendy and G. R. Knapp (eds.), Dark Matter in the Universe, 263–278.
© *1987 by the IAU.*

must grow by gravitational processes. In addition, other processes might have influenced the dark matter before it condensed into its present form. The situation is clearly simpler in the absence of any other significant processes. This is a major motivation for concentrating considerable effort on models which the dark matter is composed of freely moving elementary particles -- neutrinos, photinos or axions, for example -- even though they may seem inherently less plausible than "jupiters" or stellar remnants. For such models the astrophysical uncertainties associated with star formation do not affect the large-scale distribution of dark matter.

The uncertainties associated with galaxy formation cannot, however, be avoided when comparing the mass distribution predicted by any model with what we see in the sky. Some assumption about how galaxies "light up" the mass is required before such a comparison can be attempted. Most early N-body studies of structure formation assumed that "galaxies trace the mass", meaning that the statistical properties of the mass distribution in the models were identified directly with the corresponding properties of the observed galaxy distribution. This is an assumption of convenience which has little theoretical justification. We shall see below that during certain evolutionary phases of a neutrino-dominated universe, much of the mass lies in very large low density regions which contain no nonlinear structures, and thus no galaxies. In general, while it seems likely that regions where the mass density is high will form more galaxies than low density regions, there are many situations in which the overdensity in galaxies will be a nonlinear function of the overdensity in mass. In this case the morphology of the galaxy distribution will reflect that of the underlying mass distribution, but quantitative statistical properties such as correlation functions will differ. When trying to test N-body models it is important to remember the uncertainties introduced by our ignorance of how galaxies form, and to focus on those aspects of the "observed" mass distribution which are least subject to such uncertainties.

Early N-body simulations of the growth of structure in the universe were more concerned with elucidating the nonlinear clustering process than with the nature of the material that was clustering (Press and Schechter 1974, Peebles and Groth 1976, Aarseth, Gott and Turner 1979, Efstathiou 1979, Efstathiou and Eastwood 1981). Distributions of particles were evolved from initial conditions selected for simplicity, rather than on the basis of any theory for prior evolution, and particles were rather loosely identified as galaxies when making comparisons with observation. The first simulations of evolution from initial conditions with a coherence length were motivated by to theories for the origin of structure from adiabatic initial fluctuations. However, they still represented the theoretical predictions in an extremely schematic way (Klypin and Shandarin 1983, Centrella and Melott 1983, Frenk, White and Davis 1983). In addition these studies still identified the distribution of mass with that of the galaxies. Since 1983 it has become clear that modern N-body techniques can follow the nonlinear development of structure from a precise representation of the fluctuations predicted by detailed theories for the linear phases of evolution, albeit over a limited range of mass scales (Efstathiou et al. 1985). Studies to date have considered evolution from the linear fluctuation spectra predicted for universes dominated by massive neutrinos (White, Frenk and Davis 1983, Fry and Melott 1985) and by cold dark matter (Davis et al. 1985). When comparing with observation, this more recent work has made some attempt to account for uncertainties in how galaxies form. The rest of this review is based mainly on results from my collaborative research program with Marc Davis, George Efstathiou and Carlos Frenk.

2. INITIAL CONDITIONS AND NUMBERICAL TECHNIQUES

The evolution of strucure in the early universe is usually discussed in terms of the linear gravitational instability of a Friedmann universe (see e.g. Peebles 1980). This is in accord with the intuitive expectations that gravitational effects will cause clumping to increase with time, and that protogalactic perturbations must therefore have had small amplitude when the mean density of the Universe exceeded present galactic densities. Further support comes from upper limits on fluctuations in the microwave background which show that the amplitude of large-scale density perturbations was $< 10^{-4}$ when the primeval plasma recombined at $z \approx 10^3$. Once the nature of the particle and radiation fields present in the early universe has been specified, linear theory can be used to calculate the amplitude of a plane wave perturbation of comoving spatial frequency $k = 2\pi/\lambda$ as a function of time --

$$\delta_k(t) = T(k,t) \, \delta_{k,P} \ . \tag{1}$$

In this expression δ_k is the amplitude of the relative density fluctuation in some particle or radiation field, $\delta_{k,P}$ is the primordial amplitude with which it was generated, and $T(k,t)$ is a transfer function. If fluctuations are a result of quantum effects during an early inflationary epoch, then under very general conditions they are expected to have the scale-free Harrison-Zel'dovich constant curvature spectrum,

$$\left| \delta_{k,P} \right|^2 \propto k \ , \tag{2}$$

with random phases. The inflationary model also predicts $\Omega = 1$, of course, unless some fine-tuning is invoked. (See Guth 1985 for a review of these questions.) Apart from an overall normalization, equations (1) and (2) determine the "initial" conditions for galaxy formation completely once the contents of the Universe are specified.

Figure 1 shows power spectra at late times in a universe now dominated by weakly interacting massive particles (from Bond and Szalay 1983, Bond and Efstathiou 1984). The quantity $k^3 \left| \delta_k \right|^2$ is the total power in waves of scale $\lambda \sim 2\pi/k$. Objects of this size will condense out of the general expansion when $k^3 \left| \delta_k \right|^2 \sim 1$. Such nonlinear behavior can in general be followed only by numerical simulation, but

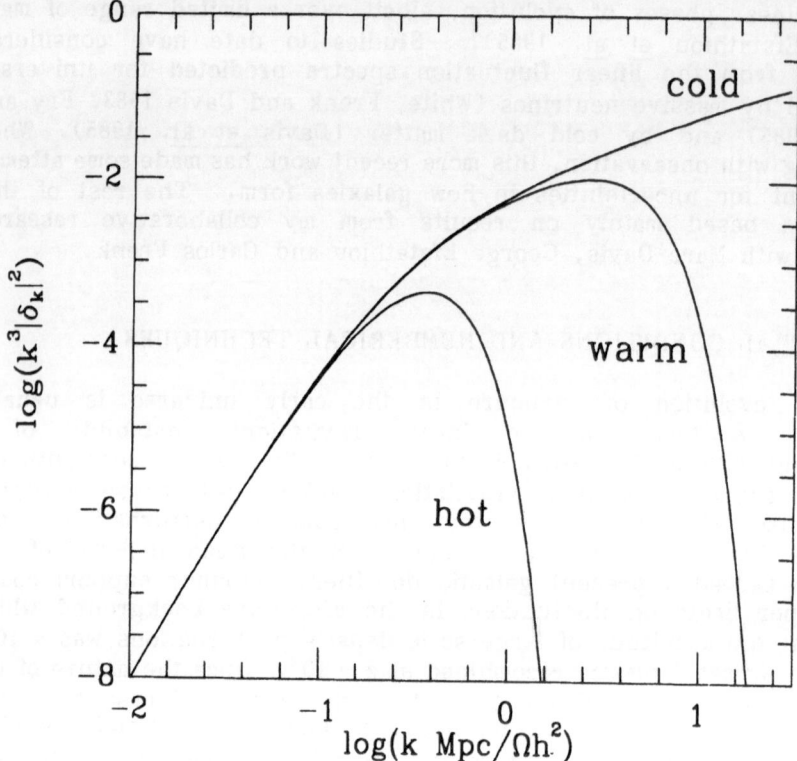

Figure 1. Power per decade as a function of spatial frequency in a universe dominated by collisionless elementary particles. These are the linear power spectra at late times in a universe which initially had the adiabatic constant curvature fluctuations predicted by inflationary models. The three cases are differentiated by the random velocities of the particles involved.

until nonlinear effects are important the spectra of Figure 1 evolve by increasing their amplitude while maintaining their shape. The three curves in the figure correspond to different types of weakly interacting particle. Neutrino-like particles can dominate the universe either if their mass is of the order of 30 eV, or if it is of order 3 GeV. In the former case the particles remain relativistic until $z \sim 10^5$ and all small-scale fluctuations in their density are wiped out as they move around. This effect is reflected in the high frequency cut-off of the curve labelled "hot"; note that the peak here corresponds to the rather large wavelength

$$\lambda \approx 17/\Omega h^2 \text{ Mpc} \approx 17 \ (100 \text{ eV}/m_x) \text{ Mpc}, \tag{3}$$

where Ω and h are the usual cosmological parameters and m_x is the mass of the particle. If the particles have a mass near 3 GeV, free streaming effects do not influence fluctuations on scales of interest; the particles are then a particular example of cold dark matter. Other cold dark matter candidates are axions and supersymmetric partners of certain known particles. These matters are reviewed by Turner in this volume. Turner also discusses the possible intermediate case of warm dark matter. Notice that in a hot dark matter universe the first objects to collapse will have a characteristic scale given by equation (3), whereas in a cold dark matter universe small objects will collapse first, and will then aggregate into larger systems. Thus structure in a neutrino-dominated universe will grow according to a variant of Zel'dovich's "pancake" scenario (Zel'dovich 1970) whereas an axion-dominated model will cluster hierarchically in the manner discussed by Peebles (1965, 1980).

When N-body methods are used to follow nonlinear evolution from the initial conditions summarized in Figure 1, a number of technical limitations arise. The mass of the particles used to represent the mass distribution in a "typical" region of the universe can be quite large. For example White et al. (1983) used 32768 particles to simulate the neutrino distribution in a comoving cubic region of present side $L = 65 \ h^{-2}$ Mpc. In these $\Omega = 1$ models each particle thus represented $2 \times 10^{12} \ h^{-4} \ M_\Theta$ or $10^{76} \ h^{-6}$ neutrinos. The smoothest possible way to put down N particles in a cubic region is to site them on a regular cubic lattice. Periodic boundary conditions can be used to embed this region in an infinite universe. When such a distribution is perturbed in order to obtain the desired linear power spectrum, the largest wavelength that can be represented is L, while the shortest is $2.0 \ N^{-1/3} \ L$. Thus the initial conditions can be modelled over a dynamic range of 16 in length or 4000 in mass for N = 32768. As a simulation evolves, bound clumps cease to follow the general expansion and so shrink relative to the size of the computational volume. At late times the mass distribution can therefore be studied over a wider range of length scales, provided that the numerical scheme gives an accurate representation of the Newtonian force law at small separations. An N-body model is a useful representation of the evolution of clustering from the time when bound clumps of a few particles first condense out of the expansion until times when waves of scale L begin to go nonlinear. When the fluctuation spectrum, as plotted in Figure 1, is relatively flat, these two times can be uncomfortably close. This is the case for the cold dark matter spectrum on small scales. Efstathiou et al. (1985) give a detailed discussion both of the technical limitations of N-body codes, and of our technique for obtaining a random phase, growing mode realization of any given power spectrum. I now discuss the results that our group has obtained for neutrino-dominated and cold dark matter universes.

3. NEUTRINO-DOMINATED UNIVERSES

White et al. (1983) used the hot dark matter spectrum of Figure 1 as an initial condition for simulations of a neutrino-dominated universe. This work superceded earlier studies which had used schematic and rather poor representations of the predictions of linear theory; furthermore it demonstrated that the major features of the

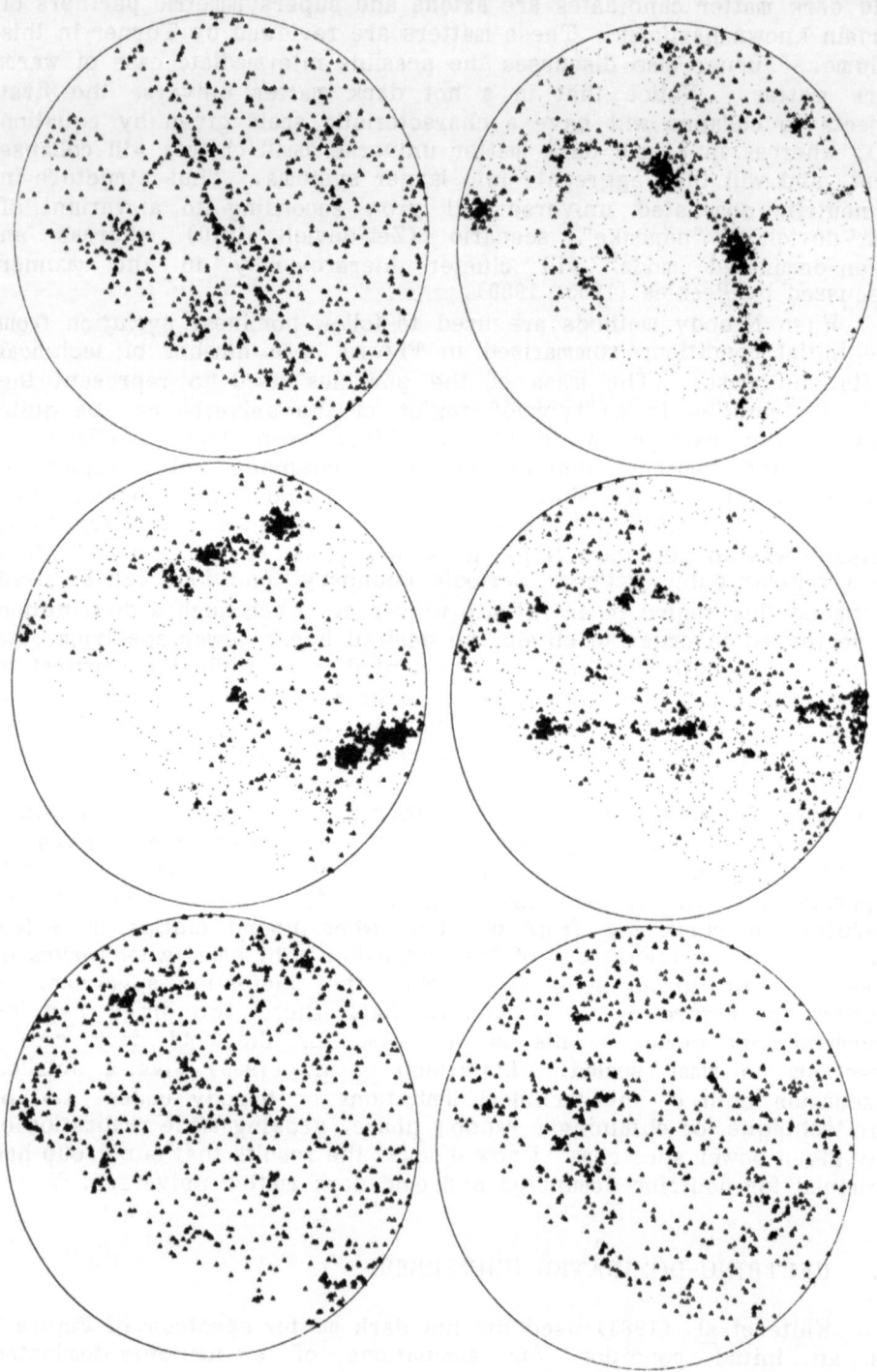

subsequent evolution could be reproduced using different numerical techniques. Structure forms in these models very much as expected on the basis of earlier theoretical work (e.g., Doroshkevich et al. 1980). The first things to collapse are sheet-like structures which rapidly link up across the computational volume. Filaments form at the intersections of sheets, and clusters at the intersections of filaments. Although there is a clear tendency for uncondensed matter to flow primarily into sheets, for sheets to flow into filaments, and filaments into clusters, all three kinds of structure are present simultaneously. Evolution is quite rapid; by the time the universe has expanded by a factor $a/a_f = 3$ since the first formation of collapsed structure, almost half the total mass is associated with dense, massive, roughly spherical clusters. Thereafter the distribution evolves mainly by merging of clusters in a hierarchical fashion. When $a/a_f \approx 2.5$ the autocorrelation function of the mass distribution in the models has the same slope as the observed autocorrelation function of galaxies. The amplitudes agree provided the cosmological parameters satisfy $\Omega h \approx 1.0$. However, in the real universe we see manifestly nonlinear objects at $z > 3$; for $a/a_f > 4$ the model correlation function is steeper than that of galaxies and has comparable amplitude only for $\Omega h > 1.5$.

The major disagreement between observation and these models surfaces when we realize that galaxies can only form in regions where the local matter distribution has collapsed. Thus the many simulation particles which lie in low density regions between pancakes can represent neutrinos and uncondensed gas, but they cannot represent galaxies. White et al. (1983) found that if they calculated correlation functions using only particles in regions which had undergone local collapse, then for any value of a/a_f the result agreed in amplitude with the observed galaxy correlations only for $\Omega h > 2$. This unacceptable constraint reflects the fact that for acceptable cosmological parameters the amplitude of the correlation function of "galaxies" in the model exceeds that observed by a factor of at least 15. This discrepancy is illustrated in Figure 2 which compares the distribution of galaxies in the Center for Astrophysics survey with that predicted in three realizations of a flat neutrino-dominated universe with $a/a_f = 3.5$ and an age of 12 Gyr. The disagreement here is manifest. Although it depends on the identification of "galaxies" in this diagram, White et al. (1984) point out that for $a/a_f > 3.5$ the neutrino clumps contain most of the mass of the universe, are too

Figure 2. Equal area projections of the galaxy distribution in the northern sky and in five artificial catalogs made from simulations. The picture at top left is the CfA survey volume limited to 4000 km/s. The next three pictures show neutrino-dominated flat universes with $h = 0.54$ in which galaxy formation began at $z = 2.5$. Triangles show particles which could represent galaxies while dots show particles in uncollapsed regions which cannot represent galaxies. The bottom two pictures show catalogs made from CDM models with $\Omega = 0.2$ and $h = 1.1$ in which galaxies trace the mass.

massive to be identified with any known object and would be very hard to hide. Thus, regardless of the details of galaxy formation, the conventional neutrino-dominated model appears to be in severe difficulty.

4. COLD DARK MATTER UNIVERSES

From a numerical point of view cold dark matter (CDM) universes are considerably more difficult to simulate than neutrino-dominated universes. This is because the spectrum shown in Figure 1 has no pronounced characteristic length and is rather flat at high frequencies. As a result, during the early evolution of a CDM universe nonlinear structure is expected to form simultaneously on a wide range of scales. Davis et al. (1985) carried out a large series of simulations of CDM universes to study the evolution of the mass distribution on scales of a few Mpc. They modelled both open and Einstein-de Sitter universes and used a numerical scheme which reproduced the interparticle force correctly at separations greater than $\varepsilon = L/600$. Their simulations followed 32768 particles within comoving cubic regions of present size $L = 32$ and 64 $(\Omega h^2)^{-1}$ Mpc, giving individual particle masses of 3 and 23 x 10^{11} $(\Omega h^2)^{-2}$ $M_\odot$ respectively. They were thus able to resolve structures with scales intermediate between those of galactic halos and of superclusters of galaxies. Large filamentary structures, superclusters of clumps and large low-density regions appear at certain time in all of these models. The autocorrelation of their mass distribution steepens gradually with time, reflecting the lack of self-similarity expected as a result of the curvature of the linear CDM spectrum. Their three-point mass correlations are stronger relative to the two-point correlations than is the case for the observed galaxy distribution. Finally, dense clumps form in these models with a wide range of masses.

In order to compare these results with observation it is necessary to make some assumption about the location of galaxies. The simplest and most convenient choice is to assume that bright galaxies "trace the mass" and can therefore be identified with a random subset of the simulation particles. With this choice one is forced to consider open models in order to get an approximate match to the dynamics of observed groups and clusters. The open models of Davis et al. (1985) give a reasonable fit to the observed galaxy distribution with the parameter choice $\Omega = 0.2$, $h = 1.1$. Although this value of h is rather high, they would have been able to get a equally good fit to the data for smaller h if they had considered open models with a lower initial fluctuation amplitude. The bottom two plots in Figure 2 show sky maps of the distribution of "galaxies" in two of these open simulations for comparison with the plot of the CfA data. The qualitative agreement between observation and the models is quite good. A similar level of agreement is seen in wedge diagrams which plot galaxy redshift against one of the angular coordinates on the sky. This is shown in Figure 3 which compares part of the northern CfA survey with similarly shaped regions in catalogs constructed from two further models with $\Omega = 0.2$.

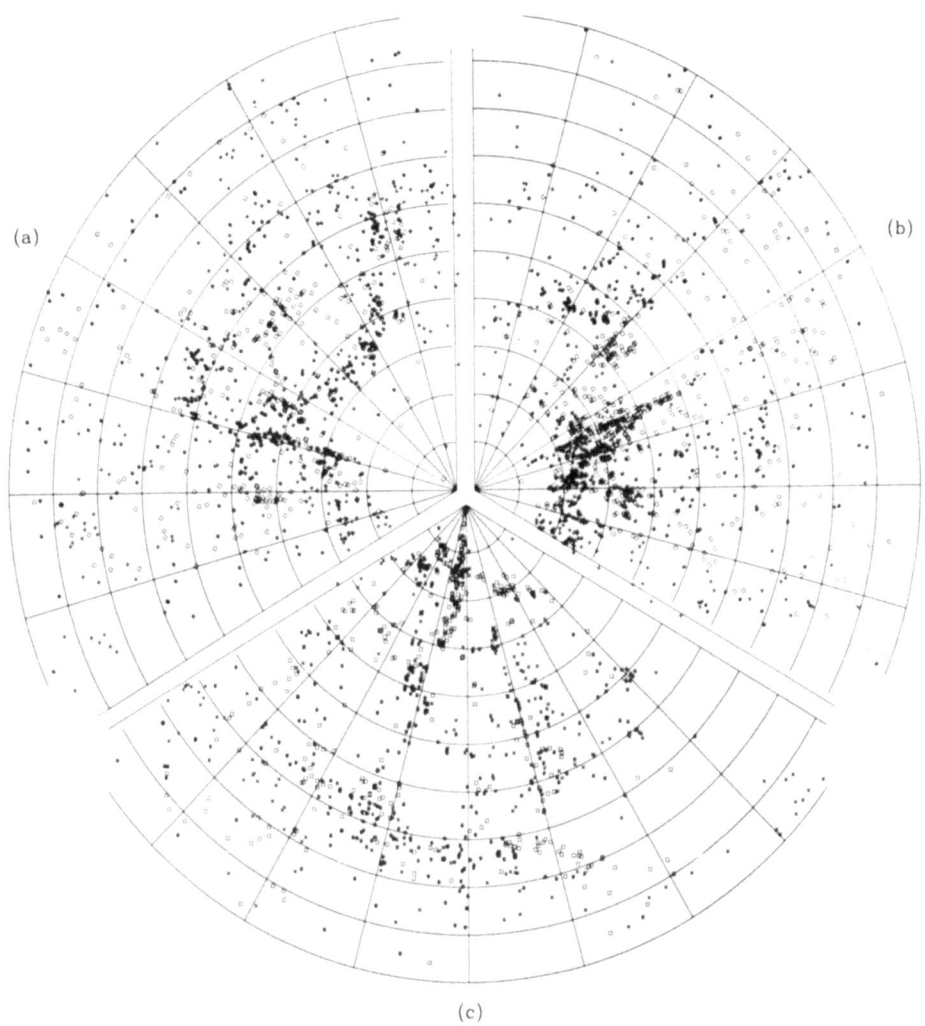

Figure 3. Wedge diagrams for the CfA survey (c), and for two catalogs from CDM models with $\Omega = 0.2$ and $h = 1.1$ in which galaxies trace the mass. "Galaxies" with $0° < \delta < 45°$ are plotted. Circles are placed at intervals of 1000 km/s and radial lines are at intervals of one hour in R.A. The three symbols refer to three $15°$ ranges in δ.

Notice the fingers-of-God effect both in the models and in the real data, as well as the large filament in model (a) which is transverse to the redshift coordinate and is similar to the Coma supercluster in the CfA catalog. These model catalogs differ from the real data mainly in that their clusters are somewhat tighter and the associated velocities are somewhat higher. This shows up quantitatively in a two-point correlation function which is too steep on small scales, in overly strong three-point correlations, in an rms relative velocity of pairs which is too large on small scales, and in apparent M/L ratios for groups which are about a factor of 2 too large (see Nolthenius and White, this volume). It is not clear how serious these discrepancies are, because physical effects related to the neglected internal structure of galaxies and their halos may become important on small scales (see, for example Barnes 1985). In addition some of the problems would be alleviated in open models with a lower initial fluctuation amplitude.

If Ω is indeed unity, as suggested by inflation, galaxies cannot trace the mass. Rather they must be over represented by a factor of about 5 in the dense regions from which dynamical mass estimates are obtained. Such a bias arises if galaxies form only near high peaks of the linear density field. This effect is discussed from a theoretical point of view by Kaiser (1985) and Bardeen (1985), and was demonstrated explicitly by Davis et al. (1985) in their simulations. It is intuitively attractive to assume that galaxies form preferentially at peaks of the dark matter density field, but it is far from obvious that a sufficiently sharp threshold will be imposed for the bias to be large (see for example, Rees 1985). The bias is illustrated in Figure 4 which compares the mass distribution in an Einstein-de Sitter CDM model with the distribution of "galaxies" which were initially placed at 2.5σ peaks of the smoothed linear density field. The same structure is evident in both pictures but it has much higher contrast in the "galaxies". Davis et al. found that for $h \approx 0.45$ the "galaxy" distribution in these $\Omega = 1$ models provided a somewhat better fit to observation than the unbiased $\Omega = 0.2$ models discussed above. In particular, the three-point correlations and the properties of groups fitted better in the biased models. Thus it would clearly be premature to conclude that the observed kinematics of galaxies on $0.5 - 5\ h^{-1}$ Mpc scales requires an open universe.

Any consistent model for the growth of structure must explain not only the properties of groups and clusters of galaxies, but also the unrelaxed filaments and voids seen on very large scales, as well as the structure of individual galaxy halos on small scales. Frenk et al. (1985) have begun a program to study the second of these questions by simulating the evolution of small regions of a CDM universe. Figure 5 shows results from a simulation which followed a comoving cubic region of side $L = 3.0\ h^{-2}$ Mpc from $z = 6$ to the present day in an Einstein-de Sitter universe. The fluctuation amplitude was chosen to agree with that needed to fit galaxy clustering (for $h = 0.45$) in the "biased" models; with this scaling each particle in the model weighs $6 \times 10^9\ M_\odot$. The three frames on the left show projections of the whole simulation at $z = 2.5$, $z = 1$, and $z = 0$, while those on the right show the

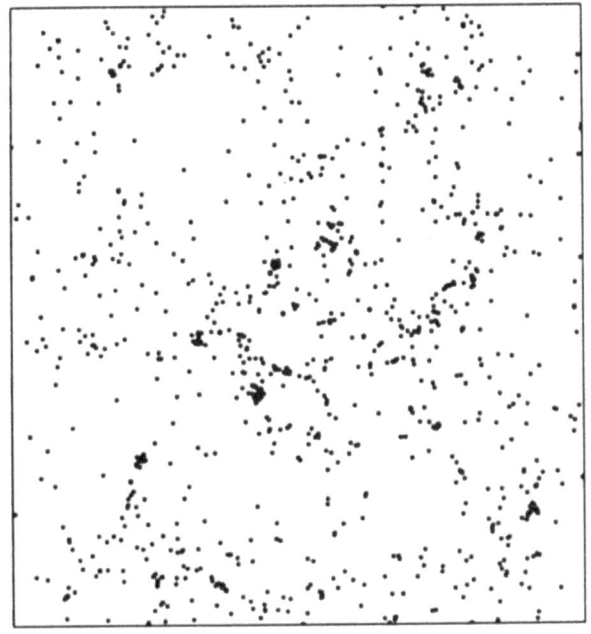

Figure 4. The projected distribution of all particles (left) and of "galaxies" (right) in a simulation of a flat CDM universe. The side of the box is 160 Mpc for h = 0.45. The "galaxies" correspond to 2.5σ peaks of the smoothed linear density field.

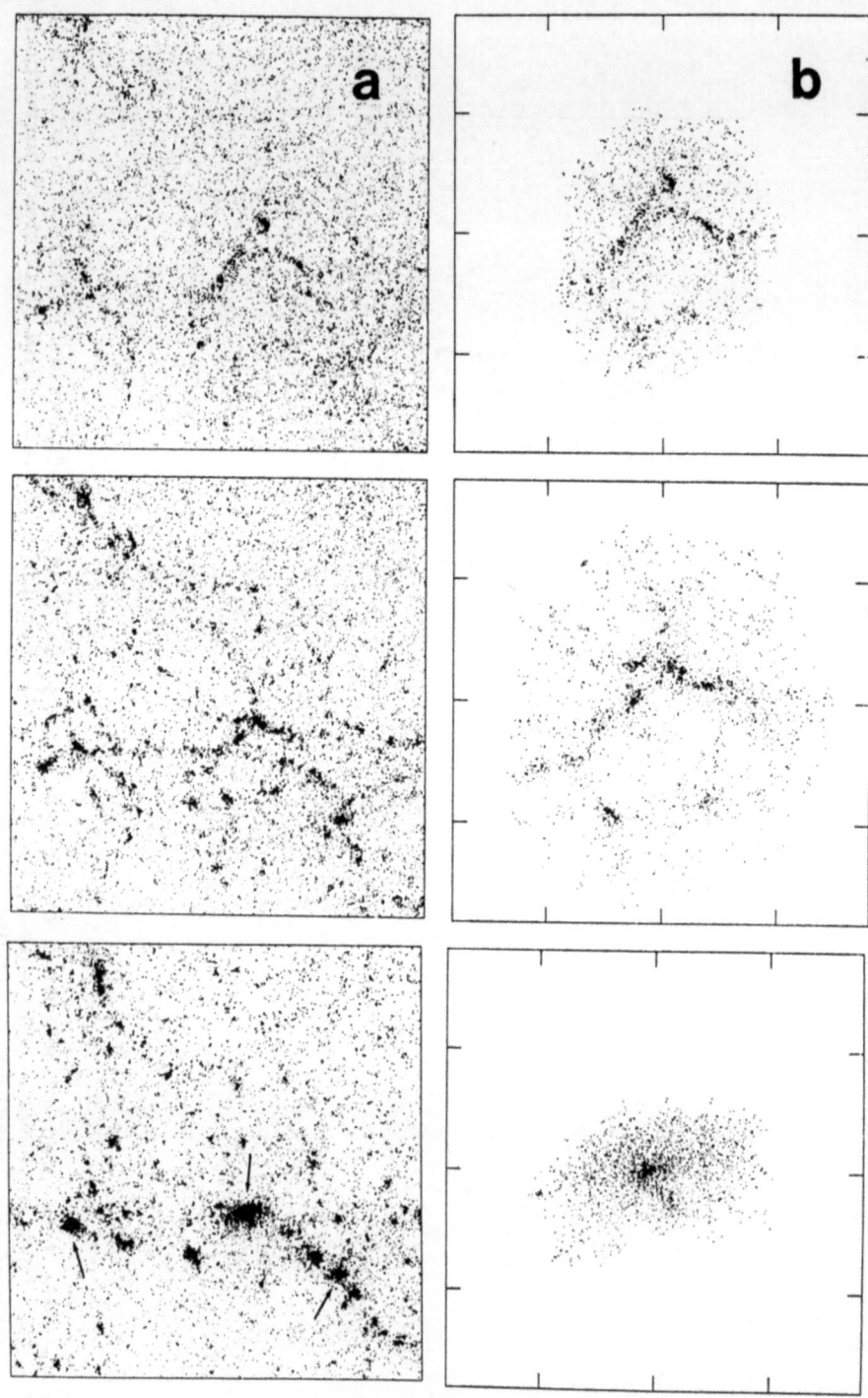

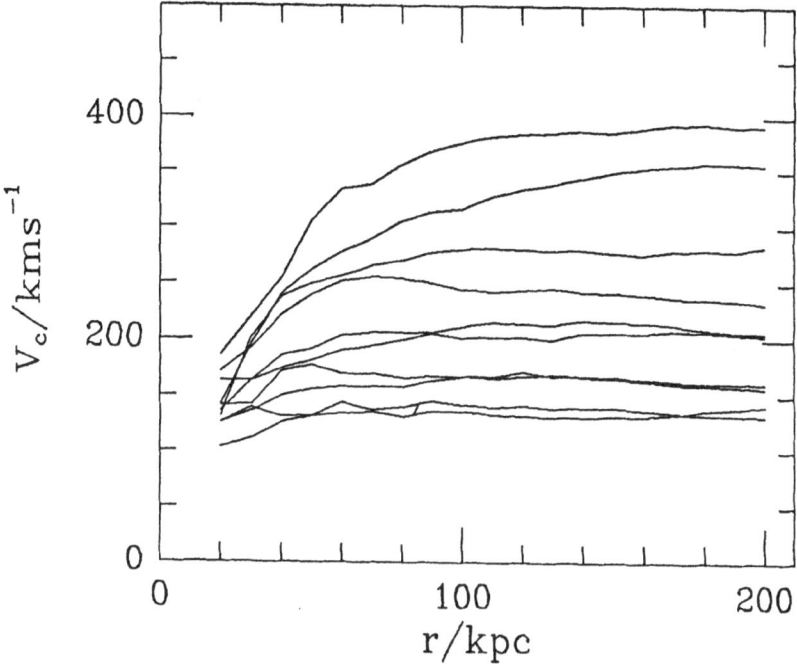

Figure 6. Rotation curves defined by $V_c^2 = GM(r)/r$ for the ten most massive lumps at $z = 0$ in the simulation shown in Figure 5.

positions in physical (non-comoving) coordinates of the particles which end up in the most massive clump. Clustering grows extremely rapidly in this model -- the clump shown in (b) contains 1/8 of the total mass. By $z = 2.5$ about 20 clumps have collapsed which have circular speeds in excess of 100 km/s at $\sim$ 30 kpc; two have circular speeds in excess of 200 km/s. About half of these accrete more material in a rather quiescent way during their later evolution. The rest are involved in a variety of merger events. The "rotation curves" of the ten most massive clumps at $z = 0$ are shown in Figure 6. The highest amplitude curve corresponds to the object of Figure 5(b) which formed by the merging of 4 or 5 large lumps. The second curve corresponds to a massive binary which is just about to merge near the bottom left of Figure 5(a); the individual components had flat rotation curves with $V_c = 220$ km/s. The other curves are all remarkably flat outside $r = 20$ kpc, the resolution limit of our model. A priori about 10 galaxies brighter than M33 ($V_c \approx 110$ km/s) and about 2 galaxies brighter than M31 ($V_c \approx 250$ km/s) are expected in a volume of this size. Thus it

Figure 5. Evolution of a 14 Mpc region of a universe like that shown in Figure 4. Column (a) shows the projected distribution of all the particles at $z = 2.5$, 1. and 0. Column (b) shows the evolution of the particles which end up in the biggest lump marked by an arrow. Physical coordinates are used in (b) where tick marks are at 1 Mpc intervals.

appears that galaxy halos form with flat rotation curves of about the right amplitude in a biased CDM model with $\Omega = 1$. Frenk et al. sketch a theory for the origin of the Hubble sequence based on the evolutionary properties of this model. If it is correct, it requires galaxy formation to be a long drawn-out process occurring predominantly at quite recent redshifts ($z = 1$-3).

Recent work on this project has been supported by NSF grant AST-8352062 and NATO travel grant # 689/84. I am grateful to Marc Davis, George Efstathiou and Carlos Frenk for making our collaboration both enjoyable and fruitful.

REFERENCES

Aarseth, S. J., Gott, J. R., and Turner, E. L. 1979, Ap. J., **228**, 664.
Bardeen, J. M. 1985, Inner Space/Outer Space, (eds. E. W. Kolb and M. S. Turner), University of Chicago Press, in press.
Barnes, J. 1985, M.N.R.A.S., **215**, 517.
Bond, J. R., and Szalay, A. S. 1983, Ap. J., **174**, 443.
Bond, J. R., and Efstathiou, G. 1984, Ap. J. (Letters), **285**, L45.
Centrella, J., and Melott, A. L. 1983, Nature, **305**, 196.
Davis, M., Efstathiou, G., Frenk, C. S., and White, S. D. M. 1985, Ap. J., **292**, 371.
Doroshkevich, A. G., Khlopov,M. Yu., Sunyaev, R. A., Szalay, A. S., and Zel'dovich, Ya. B. 1980, Ann. N.Y. Acad. Sci., **375**, 32.
Efstathiou, G. 1979, M.N.R.A.S., **187**, 117.
Efstathiou, G., and Eastwood, J. W. 1981, M.N.R.A.S., **194**, 503.
Efstathiou, G., Davis, M., Frenk, C. S., and White, S. D. M. 1985, Ap. J. Suppl., **57**, 241.
Frenk, C. S., White, S. D. M., and Davis, M. 1983, Ap. J., **271**, 417.
Frenk, C. S., White, S. D. M., Efstathiou, G., and Davis, M. 1985, Nature, in press.
Fry, J., and Melott, A. L. 1985, Ap. J., in press.
Guth, A. 1985, Inner Space/Outer Space, (eds. E. W. Kolb and M. S. Turner), University of Chicago Press, in press.
Kaiser, N. 1985, Inner Space/Outer Space (eds. E. W. Kolb and M. S. Turner), University of Chicago Press, in press.
Klypin, A. A., and Shandarin, S. F. 1983, M.N.R.A.S., **204**, 891.
Peebles, P. J. E. 1965, Ap. J., **142**, 1317.
Peebles, P. J. E., and Groth, E. J. 1976, Astron. Ap., **53**, 131.
Peebles, P. J. E. 1980, The Large-Scale Structure of the Universe, Princeton.
Press, W. H., and Schechter, P. L. 1974, Ap. J., **187**, 425.
Rees, M. J. 1985, M.N.R.A.S., **213**, 75P.
White, S. D. M., Frenk, C. S., and Davis, M. 1983, Ap. J. (Letters), **274**, L1.
White, S. D. M., Davis, M., and Frenk, C. S. 1984, M.N.R.A.S., **209**, 15p.
Zel'dovich, Ya, B. 1970, Astron. Ap., **5**, 84.

DISCUSSION

YAHIL: I have a question about the normalization of the power spectrum.
The nice thing about the cold dark matter scenario is that it predicts
the shape of the initial perturbations, but the normalization of those
perturbations is up to you. I am concerned because the calculations
were carried out over only a small range of the expansion factor. If I
understood correctly, you put the perturbation spectrum down at a
relatively low initial redshift. I understand that computer time
limitations make it difficult to do the calculations over longer periods
of time, but do you have any feeling about what would happen if you took
the same shape and started it earlier with a lower amplitude?

WHITE: It's not a question of computer time. It costs nothing to do
the early stages. All that would happen if we took a lower amplitude
and started earlier would be that the initial condition would look more
like a purely regular grid. In fact, the real limitation is the number
of particles we used, not the epoch at which we started.

ZUREK: I am very concerned about the range over which your calculations
reproduce the evolution of the cold dark matter universe. The problem is
not the smoothing of the potential, but rather the initial conditions.
At $z = 6$, much of the structure on the galactic ($\sim 10^{12}$ $M_\odot$) mass scale
is already nonlinear ($\delta M/M \sim 1$, and in the peaks of the density
distribution which are·presumably future galaxies, $\delta M/M > 1$). Therefore,
if one starts the simulation at $z \sim 5$ with an unevolved cold dark matter
spectrum, one does not give galactic and smaller scales enough time to
evolve. Consequently, I would expect that your calculation underestimates the richness of the structure on galactic and smaller scales.

WHITE: The fundamental resolution limit of our model is set by the mass
of the individual particles (7×10^9 $M_\odot$). At the start, structures made
up of ~ 10 particles are not expected to have collapsed, but they do so
soon thereafter. It is clear that we cannot model the full richness of
substructure expected on scales below 10^{10} $M_\odot$. However, the binding
energy invested in such substructure is very small, and I don't believe
it will significantly influence the structure of the much more strongly
bound objects which form later; these we identify with galaxy halos.

BARNES: I've been asked to comment on the following question: are
massive halos a massive liability for galaxies in groups? N-body
simulations of groups, with each galaxy modeled by many particles
(Barnes 1985, M.N.R.A.S., 215, 517), show that galaxies with extensive
dark halos merge very quickly. Low-velocity encounters in groups allow
the halos to stick together, trapping the luminous galaxies, which then
merge in a few local dynamical times. Thus, groups of galaxies with
massive halos are dynamically unstable.
    We observe plenty of groups with short crossing times (~ 0.1 H_0^{-1}),
so it is logical to ask if groups of halo galaxies last long enough. If
all the dark mass is initially in individual halos, the galaxies merge

too quickly, but models with 0.5 - 0.75 of the DM in a common group halo pass this test. Even so, the merging rates are rather high, and one might argue that groups of halo galaxies yield too many isolated merger remnants. But it is not clear to me (1) how many remnants are too many, (2) whether they would really be isolated, and (3) what to identify them with: ellipticals, cD galaxies, or things like V Zw 311. There is also some uncertainty in choosing the initial conditions. Finally, an old merger remnant may later accrete enough gas to form a new disk and disguise itself as a spiral galaxy.

While I have tried to argue that rapid merging in groups may not be unreasonable, other people have reached very different conclusions. Ishizawa (this conference) has assumed that multiple mergers within a group yield a cD, and concluded that individual galactic halos cannot extend beyond $\sim$ 40 kpc. Mamon (Ph. D. thesis, Princeton) has argued that most compact groups are really chance projections within loose groups. Coming back to my original question, I feel that there is as yet no convincing answer one way or the other, and I hope the rest of you are now as confused as I am.

B. JONES: When looking at the "galaxies" in your simulations, should one worry about two-body effects modifying the density profiles and the rotation curves? This could be quite serious for the smaller galaxies which show the flat rotation curves in your simulations.

WHITE: The very central parts of the final "halos" do indeed contain only a few tens of particles, as do the first clumps that form near the start of the simulation. However, structure evolves so rapidly to large mass scales in this model that I don't believe there is sufficient time for two-body effects to produce any substantial rearrangement of binding energy. The only way to demonstrate this conclusively would be to repeat the model with a much larger number of particles.

DEKEL: Cold dark matter seems successful on galactic scales, and one can even get the galaxy correlation function right if $\Omega \sim 0.2$ or if the galaxies are biased. But could you reproduce the large-scale structure with cold dark matter alone? In particular, I am afraid that you can't reproduce the observed cluster correlation function, and that you don't have the pronounced filamentary structure observed on scales of 20 - 200 h^{-1} Mpc. This seems a severe difficulty for the $\Omega = 1$ biased model, which is somewhat reduced if $\Omega = 0.2$, because the length scale is stretched accordingly.

WHITE: This is an important question which we have begun to address but have not yet fully investigated. We seem to get roughly the right abundances of rich clusters in our models, but they seem less correlated than the results of Bahcall and Soneira suggest. Most of our models concentrate on too small a volume of space to give any useful information about the existence or otherwise of very large-scale structures such as the void in Bootes or the Perseus-Pisces Supercluster chain.

ARE NUMERICAL SIMULATIONS RELIABLE? PROPOSED IMPROVEMENTS AND TESTS

F. R. BOUCHET
I.A.P. and CPT-Ecole Polytechnique

When one builds a code to simulate numerically a process, the first concern is the range of validity of the results. This can be accessed empirically, though the results can be misleading if the tests are too naïve. For particle-mesh codes simulating the gravitational clustering, an analytical theory has been proposed in Bouchet et al. 1985. It yields the numerical dispersion relation of the system in the linear regime,and thus describes how the linear growth rate is affected by the discretisation. The theoretical predictions are in agreement with the results of actual numerical experiments: both show that the results of standart particle-mesh codes should not be trusted at distances smaller than 6 to 8 grid-spacing Δx (depending on the detail of the algorithm).

Bouchet and Kandrup (1985) use this theory to first propose a prescription for building an "optimal" algorithm, which amounts to introduce such "errors" in the resolution of the Poisson equation that the inaccuracies arising from the other steps of the computation are compensated.

Second, we argue that to test a simulation, a natural quantity to consider is the fractional difference $\Delta = (F_T - F_C)/F_T$ between the true force F_T and the computed one F_C, for a given 1D density perturbation. In a true computation the force field is spatially rapidly varying (it's high frequency component is important), so it might be misleading to consider a two particles density perturbation, since the description of the high frequency part of the two-particle force is not important to achieve a good accuracy. It appears more appropriate to consider Δ for structures of typical length λ (e.g. a sinusoid) and ask: how large must be λ to achieve, say, a 5% accuracy. For plain particle-mesh (PM) codes, the answer is again $\lambda \gtrsim 6-8\Delta x$. To increase the dynamical range, it has been proposed (P^3M algorithm) to add to the mesh force a short-range correction computed by summing the two-body interactions of the closest particles (nearer than r_s). This of course improves the short-range description, but outside r_s, the code is a plain PM one, and $\Delta_{P^3M} \equiv \Delta_{PM}$. For the code of Efsthathiou et al. (1985), a 5% accuracy requires $\lambda \gtrsim 11\Delta x$ (and Δ can be as large as 19% for $\lambda = 2r_s$!) Since the errors are limited to some intermediate range, specific statistical properties as the 2-point correlation function might be unaffected, but this has to be empirically proved for any new studied property.

Bouchet F.R., Adam J.-J., Pellat R., 1985, A&A, 144, 413.
Bouchet F.R., Kandrup H., To appear in ApJ, dec 1, 1985.
Efstathiou G., Davis M., Frenk C., White S., 1985, ApJ Suppl, 57, 241.

J. Kormendy and G. R. Knapp (eds.), Dark Matter in the Universe, 279.

Cold Dark Matter: Galactic Halos

C. S. Frenk
University of Sussex

A flat universe dominated by cold dark matter (CDM) is an attractive arena for the formation of galaxies and large scale structure. Current upper limits on anisotropies of the cosmic microwave background and the standard theory of primordial nucleosynthesis are both compatible with such a universe. Furthermore a flat CDM model in which galaxy formation is biased towards high density regions provides a good match to the observed distribution of galaxies on Megaparsec scales. In collaboration with M. Davis, G. Efstathiou and S.D.M. White, we have carried out a high resolution N-body simulation which shows that this model can also account for the abundance and characteristic properties of galactic halos. The initial conditions for this simulation were based on the results of our previous work which gave both the scaling and overall normalisation of the initial CDM fluctuation spectrum appropriate to the biased galaxy formation model. We simulated a cubic region of present size 14 Mpc (H_0 = 50km/s/Mpc) from a redshift of 6 to the present day, with a resolution of 2kpc initially and 14 kpc at the end. We found that by a redshift of 2.5 about 20 clumps with circular speeds exceeding 100 km/s had collapsed near high peaks of the initial linear density field. Between Z = 2.5 and the present most of them remained isolated and accreted extensive outer halos, while others merged into larger systems. The rotation curves of the final smooth systems were impressively flat at large radii resembling the measured rotation curves of spiral galaxies. Furthermore, the abundance of clumps with circular velocities larger than 150 km/s was about the same as the abundance of galaxies brighter than M33 expected in a volume the size of our simulation. Significant transfer of angular momentum to surrounding material occurred as large subclumps merged. Most of this angular momentum was originally invested in the orbital motions of the subclumps. As a result, the central regions of merged objects showed little rotation.

Our simulation suggests a possible explanation for the Hubble sequence and for the observed correlation between galaxy morphology and environment. The first systems that can sustain significant star formation are likely to form on a dynamical timescale and to resemble spiral bulges. Disks may then form during extensive periods of quiescent evolution within relatively isolated halos. In high density regions, tidal interactions and mergers would prevent disk formation. The inner bulge-like stellar systems thus exposed may appear as faint ellipticals, or when merging has been significant, as bright ellipticals.

One important prediction of this scheme is that galaxy formation is a protracted process which begins late and continues until the present day.

J. Kormendy and G. R. Knapp (eds.), Dark Matter in the Universe, 280.
© 1987 by the IAU.

SIMULATIONS OF COMPACT GROUPS OF GALAXIES WITH HALOS

T. Ishizawa
Department of Astronomy, University of Kyoto

Self-consistent simulations of seven groups are performed from the maximum expansion to the present using Aarseth's N-body code. An initial galaxy consists of 100 stars. Its mass, half-mass radius, and central velocity dispersion are 1, 0.41, and 0.96. Units of mass, length, velocity, and time are $1.4 \times 10^{12} M\odot$, 100 kpc, 245 kms^{-1} and $4.0 \times 10^{8} y$. Table 1 gives the elapsed time from the Big Bang to the formation of a multiple merger $tm+Tc^*/2$. For $H_0 = 80$ $kms^{-1} Mpc^{-1}$, the Hubble time $H_0^{-1} = 30.6$ in our units. Dense groups except B form multiple mergers in a Hubble time.

Table 1. Initial Parameters of Simulated Groups

Model	No. of galaxies	radius	velocity dispersion	β^*	$-E_0^*$	$tm+Tc^*/2$	cD now expected?
A shell	10	10	0.12	0.01	6.21	18.6	Yes
B sphere	10	20	0.	0.	2.99	61.6	No
C disk	10	10	0.10	0.006	8.35	13.8	Yes
D sphere	10	10	0.10	0.009	5.88	22.5	Yes
E sphere	10	10	0.18	0.03	5.09	31.2	Yes
F sphere	10	10	0.58	0.34	4.99	31.6	Yes
I sphere	50	20	1.52	0.80	71.86	30.0	Yes

E_0^*=the initial total energy of a group when galaxies are point masses, β^*=the initial ratio of the random kinetic energy to $-E_0^*$ when galaxies are point masses, $Tc^*=2\pi(3/5)^{3/2}GM^{5/2}/(-2E_0^*)^{3/2}$, tm=the epoch of merger formation when at least four galaxy cores are merging.

Figure 1 shows the virial diagram for Geller-Huchra's(1983) groups. The loci of A, E, F and I are obtained using the projected positions and line-of-sight velocities of surviving cores in the merging phases. They fall along a line of constant density $3Tc^*/2 = H_0^{-1}$. Thus, almost compact groups with $3Tc^*/2 < H_0^{-1}$ are expected to have cD galaxies now. However the frequency of cD galaxies in Geller-Huchra groups is 7% in the range $3Tc^*/2 < H_0^{-1}$ and $M < 10^{14} M\odot$. The scarcity of cD's requires us to reduce such a high merging rate by decreasing the size of halos.

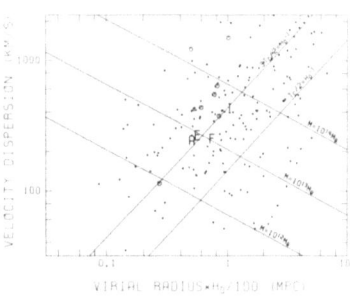

Figure 1

Thus, if galaxies in a group have been formed with not a large velocity dispersion by the phase of maximum expansion and if $H_0 < 80$ $kms^{-1} Mpc^{-1}$, the half-mass radii of the halos should be less than 41 kpc.

REFERENCE

Geller, M. J. and Huchra, J.P.: 1983, Astrophys. J. Suppl. **52**,61.

J. Kormendy and G. R. Knapp (eds.), Dark Matter in the Universe, 281.
© 1987 by the IAU.

VOIDS IN THE 2D CFA CATALOG

M. Lachièze-Rey and F. Bouchet
CEA, Service d'Astrophysique; Ecole Polytechnique

The void probability function was known from a long time to be an efficient tool in analysing the galaxy repartition (White, Sharp, Schaeffer). We developed a method to determine this quantity with a minimum noise, for any proposed distribution resulting from observations, simulations, or theory. We apply it here to the 2D-CFA catalog. This will be later compared to other 2D galaxy catalogs, to the 3D CFA catalog and to the results of numerical simulations.

The void probability $P(\Omega)$ is defined (in 2 dimensions) as the probability that a randomly chosen zone of the sky, of solid angle Ω contains no galaxy. We measure it in the sample as a function of Ω. The method and the error estimates will be presented in a subsequent paper (Bouchet and Lachièze-Rey 1985). The estimated noise appears however to be very low. The complete results will be presented in Bouchet and Lachièze-Rey (1985).

In order to compare them to theoretical predictions they were expressed as functions of the scaling variable $q = \sigma\Omega \cdot (\theta)\overline{W}(\theta)$ where $\overline{W}(\theta)$ is the averaged 2-point angular correlation function (see Schaeffer 1984).

We present in the figure the result of our measure, compared with the theoretical predictions of Schaeffer (1984). All models $\nu \neq 1$ are excluded. The difference with $\nu = 1$ may be due to a bad estimation of the scaling variable q. It remains that the good fit up to $\theta = 1.2°$ is significant and has to be explained.

In conclusion we point out that we are able to estimate $P(\Omega)$ with a very good precision; this provides a very efficient and powerful tool to compare observational, computational and theoretical work. The fact that we are able to exclude some models which reproduce the low order correlations functions is very encouraging. On the other hand, the fit, although presently imperfect, between the data and the prediction of $\nu = 1$ Schaeffer's model is stimulating and appeals for more theoretical work on interpreting $P(\Omega)$.

References

Bouchet F. and Lachièze-Rey M., 1985, in preparation for ApJ Letters
Schaeffer R., A&A, <u>134</u>, L15
Sharp N., 1981, MNRAS, <u>195</u>, 857
White S., 1979, MNRAS, <u>186</u>, 145

J. Kormendy and G. R. Knapp (eds.), Dark Matter in the Universe, 282.

AN UPPER LIMIT ON THE MASSES OF GALAXIES IN CLUSTERS

David Merritt
University of California, Berkeley

Simon D. M. White
Steward Observatory, University of Arizona

Clusters of galaxies are believed to be dominated by dark matter. Some of this matter is presumably bound to galaxies in the form of massive halos, while the rest moves freely in the cluster potential well. The exact fraction of dark matter bound to galaxies is an important datum for models of cluster evolution, since time scales for orbital decay, merging, stripping, etc. are sensitive functions of galaxy mass. In this study we attempt to put a firm upper limit on the amount of dark matter associated with galaxies in clusters, by calculating the response of a galaxy with an initially massive halo to the *mean* tidal field produced by the overall cluster potential well. If the velocity dispersions of galactic halos are roughly equal to those of luminous galaxies, σ_g, it is easy to show that the truncated mass of a spherical galaxy orbiting near the center of a cluster is roughly $m_g \approx G^{-1}\sigma_g^3\sigma_c^{-1}R_c \approx 4 \times 10^{11}M_\odot$, where σ_c and R_c are the cluster velocity dispersion and core radius. The precise value of m_g must depend on the orbital geometry, as well as the number of pericenter passages since cluster formation, among other factors.

We used a novel scheme for following the evolution of a galaxy in a cluster potential well. At every time step, the force field in the neighborhood of the galaxy was assumed to be the sum of two components: a fixed component from the cluster, and a time-varying component from the galaxy. In order to be able to evolve a large number of galaxies over a Hubble time, we assumed that each galaxy remained *spherical*, and computed its potential on a radial grid fixed on its center. The accuracy of the spherical code was verified in a few cases by comparison with a more realistic, quadrupole-order code. The cluster dark-matter density was assumed to follow an analytic King model, i.e. $\rho_c(R) = \rho_c(0)\left(1 + R^2/R_c^2\right)^{-3/2}$. The initial galaxy mass for the runs shown below was $m_i = 2\,G^{-1}\sigma_g^3\sigma_c^{-1}R_c$, roughly twice the predicted, tidally truncated value; the initial galaxy velocity dispersion was 0.3 times that of the cluster. All galaxies were evolved until a time of $T = 50R_c\sigma_c^{-1} \approx 10^{10}$ years for a typical rich cluster.

The Table gives final galaxy masses in units of $G^{-1}\sigma_g^3\sigma_c^{-1}R_c$, as a function of pericenter and apocenter distances in units of R_c. Mass loss continues throughout a Hubble time, but most takes place during the first few billion years. Galaxies on elongated, high-energy orbits appear to be the most strongly truncated, although the dependence of final mass on orbital parameters does not appear to be great. Scaling our results to galaxies of various luminosities gives an upper limit of $\sim 15\%$ for the fraction of the dark matter bound to galaxies in a cluster like Coma. This value is consistent with upper limits based on mass segregation arguments.

R_{peri}	0	1		0	1	2		0	1	2	3
R_{apo}	1	1		2	2	2		3	3	3	3
m_g	1.17	1.33		0.81	1.10	1.25		0.88	1.08	1.29	1.47

J. Kormendy and G. R. Knapp (eds.), Dark Matter in the Universe, 283.

CLUSTERING PROPERTIES OF COLD DARK MATTER UNIVERSES AND THE CfA REDSHIFT SURVEY: A UNIFORM COMPARISON

Richard Nolthenius, Simon White
Steward Observatory

We have compared the properties of galaxy groups in the CfA red-shift survey to those in the cold dark matter simulations of Davis, Efstathiou, Frenk and White (1985). Redshift catalogs (four realizations) were made from the open universe (unbiased; mass follows light) simulations at expansion factors of 1.8 (Ω=.3, labelled T1) and 3.2 (Ω=.2, labeled T2) relative to initial conditions, and from the biased formation version of the Ω=1 simulation (5 realizations) at expansion factor 1.4. The T1, T2 and biased catalog sets were volume limited at distances of 5000, 3000 and 6500 km s^{-1} respectively, and magnitude limited (m$\leq$14.5) beyond. They were then randomly culled by 25%, 9% and 10%, respectively, to match the density of comparison versions of the CfA Redshift Survey which were similarly volume limited. We have devised a percolation type grouping algorithm which lets the linking distance on the sky increase with distance in the magnitude limited regions, while keeping the radial velocity linking criteria fixed. We've verified that this produces groups with density contrast un-correlated with distance and (unlike the Geller-Huchra criteria) M/L uncorrelated with group size for the open models. M/L may, however, show a slight negative correlation with distance.

We have compared the properties of group catalogs made from these redshift catalogs using a range of linking cutoffs. The biased formation Ω=1 catalogs produce groups with properties in good agreement with corresponding CfA groups at all linking cutoffs. The T1 and T2 group catalogs are in significant conflict with the corresponding CfA groups in the following areas; too few galaxies are grouped. The characteristic rise and fall of group number is delayed to significantly larger cut-offs, and dispersions and group size are larger, producing M/L's a factor of two too high. There is a hint that M/L may be correlated with cluster size in the CfA data, possibly indicating dark matter less clustered than the galaxies. The biased and unbiased models do not differ significantly in their percolation properties - all percolate at cutoffs$\sim$75% of that for a Poisson distribution. The CfA survey percolates more easily yet, at $\sim$ 60% of Poisson.

We conclude that not only does biasing allow an Ω=1 universe to be compatible in the observations, it actually provides a good fit where the best fitting unbiased models do not.

References

Davis, M., Efstathiou, G., Frenk, C., and White, S. D. M., 1985, Ap.J 292, 371
Geller, M. J. and Huchra, J., 1983, Ap. J Supp. 22, 61

284

J. Kormendy and G. R. Knapp (eds.), Dark Matter in the Universe, 284.
© 1987 by the IAU.

Formation of Galactic Halos in the Cold Dark Matter Universe :
Computer Simulations

P.J.Quinn, J.K.Salmon and W.H.Zurek
California Institute of Technology &
Los Alamos National Laboratory

We simulate the formation of structure on the galactic scale in the cold dark matter, $\Omega=1$ universe.

Numerical calculations are initiated at z=24 in a cube with a volume of 10^3 Mpc^3. Such large redshifts are necessary to capture galactic scale perturbations in the linear regime. Initial conditions are imprinted by deforming a 64^3 cubic lattice of particles so that the Fourier transform of its density has the power spectrum :

$$P(k)=Ak(1+\alpha k+\beta k^{3/2}+\gamma k^2)^{-2}$$

where k is the wavenumber, $\alpha=1.71 h^2$ Mpc., $\beta=9.0(h^2 Mpc.)^{3/2}$, $\gamma=1.0(h^2 Mpc.)^{1/2}$ and $A=4.63\times10^3 h^4$ Mpc^4. (Peebles 1982, Blumenthal et al 1984, Davis et al 1985). Dynamical evolution is then followed with an FFT cloud-in-cell code on a 64^3 mesh.

N-body simulations of spheres(radius=5Mpc.)cut out from the FFT cube are used to follow the nonlinear development of structure on galactic scales. The change over from the FFT code(resolution$\sim$250 kpc.)to the more accurate N-body code(softening radius$\sim$10 kpc,$\sim$6000 bodies)takes place at z=5.25. Their evolution is followed untill z=-0.2(h=1).

The particle distribution at z=0 exhibits a number of compact clumps. Their masses($\sim10^{12} M_\odot$)and their density(~0.01 Mpc^{-3})are consistent with the inferred masses and number densities of galactic halos. When one selects only collapsed portions of these objects(relative density$\sim$160),they appear to be relatively compact(sizes$\sim$100-200 kpc.)and have rotation parameters (λ)in the range 0.01-0.15. When these clumps are identified with galactic halos, one concludes the about 80% of the matter remains outside. This result -- if confirmed by further simulations -- would be of importance for biased galaxy formation but it should be treated with caution at this stage.

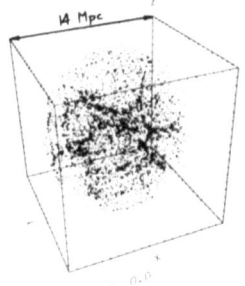
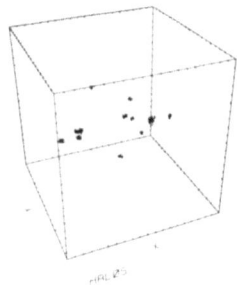

Fig: 1 The distribution of N-body   Fig: 2 The regions of figure 1 that
       particles at z=0.   have overdensity >160.

Blumenthal,G.R.,Faber,S.M.,Primack,J.R.,Rees,M.J.1984 Nature 311 517.
Davis,M.,Efstathiou,G.,Frenk,C.S.,White,S.D.M. 1985 Ap.J. 292, 371.
Peebles,P.J.E. 1982 Ap.J. 258 415.

J. Kormendy and G. R. Knapp (eds.), Dark Matter in the Universe, 285.
© 1987 by the IAU.

FORMATION OF LARGE SCALE STRUCTURE IN THE EXPLOSION SCENARIO

S. Saarinen[1], A. Dekel[1,2], B.J. Carr[3]
1) Yale Univ.   2) Weizmann Inst.   3) QMC, London

This is a progress report on a study of the formation of large scale structure in the explosive amplification scenario, using N-body simulations. The simulations start when galaxies of the last generation form. The galaxies are distributed at random in expanding shells around random seeds. They start with an expansion velocity 20% larger than the Hubble velocity, in accordance with the similarity solution that was valid before the gaseous shells fragmented into galaxies. The galaxies are treated thereafter as softened point particles that interact only gravitationally, embedded in an N-body background representing inter-shell gas and <u>dark matter</u> (in variable amounts).

Clustering on scales of 2-20 Mpc/h is investigated. Preliminary results indicate that a structure that may resemble the observed one is obtained by the models at an expansion factor, R, of order 10, where the interacting shells produce bound clusters in flattened superclusters separated by voids. Fig. 1. shows a projected distribution of galaxies in one model dominated by dark matter (at R=1 and R=11.3). Fig. 2 shows the correlation function of the galaxies versus the total mass. The functions steepen in time (as in pancake models), in a rate which depends on the amount of dark matter. A <u>bias</u> between the galaxies and the total mass is introduced by the initial distribution of galaxies in shells on top of an unperturbed dark matter, and it survives until R=10. This bias may reconcile the observations with a flat universe. The multiplicity function of clusters and the distribution of voids also evolve in time in a characteristic way, which may be consistent with observations only at a given time, thus providing constraints on the epoch of galaxy formation and on other parameters of the scenario.

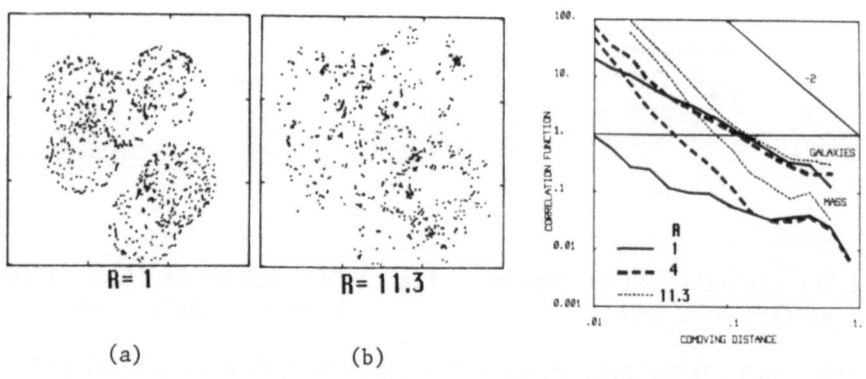

R= 1

R= 11.3

(a)   (b)

Figure 1.   Figure 2.

J. Kormendy and G. R. Knapp (eds.), Dark Matter in the Universe, 286.

PROFILES OF GALAXY CLUSTERS IN COSMOLOGICAL SCENARIOS

Michael J. West[1], Avishai Dekel[1,2], and Augustus Oemler, Jr.[1]
1) Yale University
2) The Weizmann Institute of Science

We have studied the properties of rich clusters of galaxies in various cosmological scenarios by comparing high resolution N-body simulations with observations of Abell clusters. The clusters have been simulated in two steps. First, protoclusters are identified in large-scale simulations which represent a wide range of cosmological scenarios (hierarchical clustering, pancake scenarios, and hybrids of the two, spanning a range of power spectra). Then the region around each protocluster is simulated with high resolution, the particles representing L[*] galaxies. The protoclusters have no spatial symmetry built into them initially. The final clusters are still dynamically young, and of moderate densities, which should be representative of Abell clusters of richness classes 1 and 2.

We find that the final cluster mass density profiles are quite similar in shape, independent of the initial conditions. The projected profiles are all well-fitted by the de Vaucouleurs $r^{1/4}$ law; their logarithmic slopes at the half-mass radius are about -1.8, and they steepen with increasing radius. The line-of-sight velocity dispersion profiles are also similar, and the velocities are quite isotropic. The existence of a universal profile suggests that violent relaxation is efficient at erasing traces of the initial conditions from the cluster profile during the first collapse, while secondary infall does not significantly affect it. Hence, the density profile is not a good indicator for the origin of the large-scale structure in the universe.

A comparison of the theoretical mass density profiles with the observed surface brightness profiles of a sample of 27 Abell clusters shows good agreement, suggesting that the radial cluster light distribution traces the mass distribution.

Other cluster properties (such as ellipticities, subclustering, binding energies, and others) are presently being examined, and preliminary results indicate that several of them may provide a more sensitive test for the formation of the large-scale structure. Detailed results of this work will be published elsewhere.

J. Kormendy and G. R. Knapp (eds.), Dark Matter in the Universe, 287.
© *1987 by the IAU.*

HALOS AND DISK STABILITY

Agris J. Kalnajs
Mount Stromlo and Siding Spring Observatories
Research School of Physical Sciences
Australian National University

ABSTRACT. The need for halos on stability grounds is not at all compelling. Compared to a bulge, a halo is not very efficient in stabilising a disk. Because the effect of a modest halo is small, it is difficult to infer its presence or absence from stability arguments.

1. INTRODUCTION

Historically the instabilities of disks have added much fuel to the argument for the existence of massive haloes around galaxies. We happen to live in that part of the Galaxy where the circular motion is much larger than the random motions of stars. Until fairly recently, we were ignorant about the velocity dispersion elsewhere in the Galaxy, and it was therefore quite reasonable to assume that the rest of the Galaxy, as well as other disk galaxies, would be mainly rotationally supported. Toomre's pioneering stability analysis (Toomre 1964) seemed to provide a reason for the small velocity dispersion: it was just enough to make the solar neighbourhood stable against axisymmetric instabilities. There was another, less worthy reason for wanting to think about cool disks: the dynamics could be worked out fairly simply if epicycles were to be good approximations to stellar orbits.

The first inkling that all was not well with this concept of a disk galaxy came with the advent of numerical simulations (Miller, Quirk, and Prendergast 1970; Hohl 1970). Disks with just enough velocity dispersion to make them stable against axisymmetric instabilities, evolved quite rapidly, chiefly through bar-making instabilities. These instabilities warmed up the disks to the extent that in the ensuing equilibrium the pressure from random motions became as important as rotation. The demonstration by Ostriker and Peebles (1973) that the random part of the kinetic energy should exceed the rotational part by about a factor of three in order to avoid the bar-making instabilities, seemed to rule out the idea that the Galaxy (where the local value of that energy ratio is two orders of magnitude less) could be a self-gravitating disk. One solution of the stability problem is to assume that the disk is surrounded

J. Kormendy and G. R. Knapp (eds.), Dark Matter in the Universe, 289–299.

by a rigid halo. In view of the virial mass discrepancies in clusters of galaxies, the idea of unseen matter surrounding galaxies is very appealing. It becomes even more appealing if the halos are needed to keep the galactic disks cool.

In this review I will attempt to convince you that the stability problems are associated with the inner parts of disk galaxies. They can be overcome by hot centers or small bulges. Halos will do it as well, if they provide a significant part to the equilibrium force field in the central regions. A halo with a scale length larger than the disk, and a comparable mass within a Holmberg radius, will not contribute significantly to the stability.

2. NUMERICAL EXPERIMENTS WITH DISKS

The problem faced by numerical simulations is the lack of knowledge about the orbital eccentricities of the stars within the disk. There is only a lower bound for the radial velocity dispersion, which is needed to insure axisymmetric stability (Toomre 1964). The earlier experiments all started out with just enough random velocities to satisfy this criterion.

The actual value of the radial dispersion measured in the units of the minimum Toomre denoted by Q. One can write Q as

$$1/Q = 2\pi G\mu(r)*(0.5345)/(\kappa re) \qquad (1)$$

if we express the radial velocity dispersion as (κre), where e is the mean eccentricity, κ the epicyclic frequency, and r the radius. In the denominator, μ is the surface density, and G the gravitational constant. The value of Q at the sun, which at first seemed to be tantalisingly close to 1.0, now appears to be in the range 1.2 - 2.0 (Toomre 1973). The corresponding eccentricity is around 0.16, which is a reasonably small number.

The Q = 1.0+ disks were quite lively in the non-axisymmetric sense, with the result that Q quickly rose to values in the range of 2 - 6, the highest values obtaining in the outer parts. Attempts to cool these hot disks were only moderately succesful (Hohl 1971). The value of Q could be reduced to about two, but then the bar reappeared and the subsequent stirring kept the value of Q in the range 2 - 3. Actually the measured velocity dispersions were annular averages, and as such they included any systematic motions due to the bar and spiral arms. Thus it is quite conceivable that the true dispersion may yet be a factor of 1.4 lower, implying final Q's in the range 1.4 - 2.1, which are not that discrepant with the solar value.

The cooling experiments probably deserve to be re-examined, in view of the fact that the bar is a negative energy creature, and hence can be provoked by indiscriminate cooling.

3. HALOS AND BULGES

If we start from the premise that disks are rotationally supported, we can already see that the central parts will have to contain stars in fairly eccentric orbits just to satisfy the axisymmetric stability criterion. Equation (1) can be rearranged slightly,

$$1/Q = [2\pi G\mu(r)*r/v^2]*[0.535*\Omega^2/\kappa^2]/e \qquad (2)$$

by bringing in the circular velocity v, the azimuthal frequency Ω, and recalling that $v = \Omega*r$. The ratio of the squares of the frequencies is a slowly changing function, which goes from 0.25 at the center, to 0.5 at the maximum of the rotation curve, and approaches 1.0 at large radii. This variation is much slower than that of the left-most bracket, and from now on we will pay most of our attention to the latter.

The denominator is the square of the rotation curve, produced by the combination of surface density and radius appearing in the numerator. Figure 1 shows the two quantities for an exponential disk. The ordinate is ln(r), because the relation between these two functions is independent of the radial scale. The relation is also linear, and hence they are related by a convolution on the ln(r) scale.

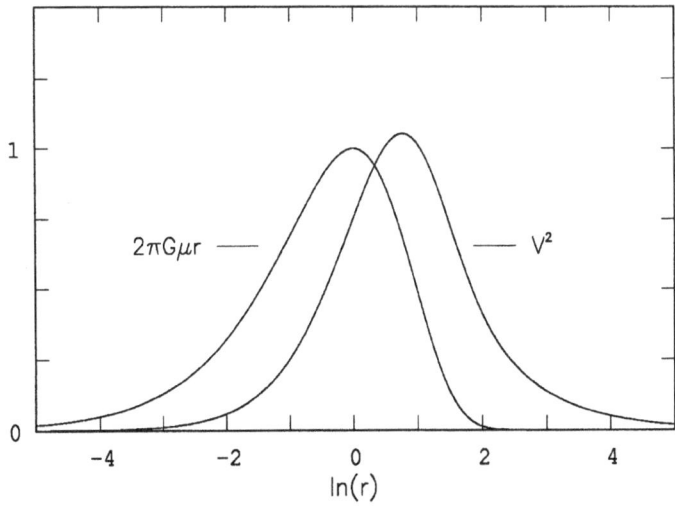

Figure 1. The radius-weighted surface density and the square of the circular velocity for an exponential disk of unit scale length.

The similarity in shapes and the relative displacement of the two curves is typical, provided the surface densities are as wide and smooth as the exponential.

The reciprocal Q has another interpretation: it is the maximum local response of the disk at zero temporal frequency. At a fixed radius and eccentricity the response is proportional to the surface density, and inversely proportional to the square of κ, which is a measure of the restoring force against radial displacement. Thus the surface density in the numerator promotes instability, while the frequency dependent terms in the denominator are stabilising. The two curves in Figure 1 play similar roles in the stability of the disk.

Axisymmetric stability requires a minimum eccentricity of 0.16 at the peak of the rotation curve in Figure 1. We can see from the ratio of the two curves that this value is more than adequate for stability at larger radii, but as we move towards the center the eccentricity has to increase to quite large values. While the quantitative aspect of this analysis becomes suspect for large eccentricities, the conclusion that the center must be hot remains.

If we want to keep the center of the disk cool, we must raise the rotational velocities there. This can be accomplished by introducing external masses in the form of a rigid bulge or halo. For example, by adding another rotation curve of the same shape, but shifted two units to the left, we obtain the curves shown in Figure 2a.

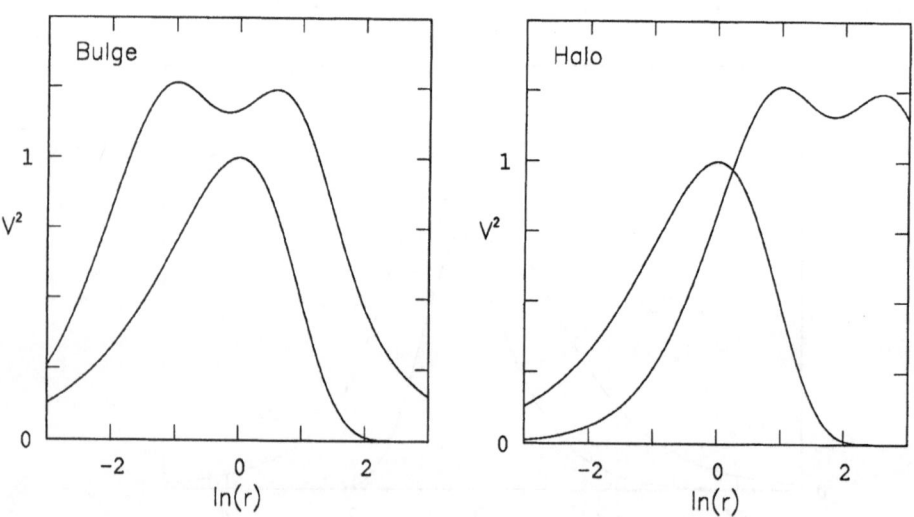

Figures 2a and 2b. The effect of adding a bulge (left) or a halo component (right) to the rotation curve shown in Figure 1.

The additional component could be produced by an exponential disk whose mass and scale length is reduced by exp(2) = 7.39, which we would call a bulge. This is to be contrasted with what we obtain if we add a similar velocity component, but now shifted to the right by two units. Figure 2b shows that such a halo-like component helps in the outer parts where help is not needed, but does nothing to the center. It is similar to the halos inferred from flat rotation curves. The mass of this component is 7.39 times that of the disk, or 55 times that of the bulge. It would have to be much heavier still to have any effect on the central part of the disk.

Figure 3 shows the contribution of the bulge to the rotation curve in the more conventional way. The halo version has the same shape, but a different radial scaling.

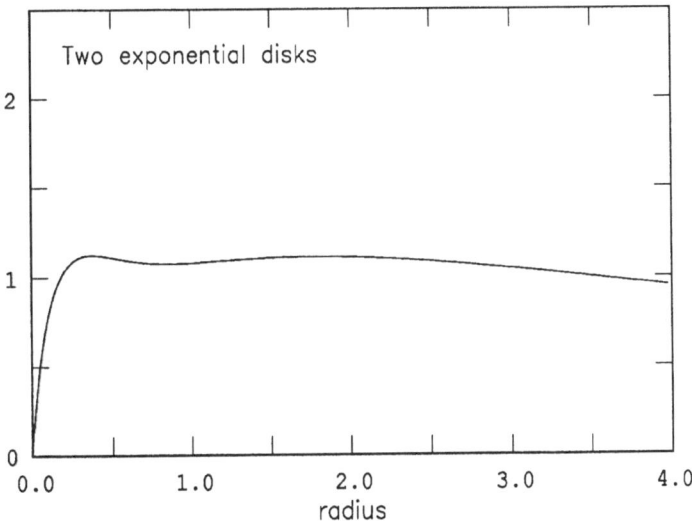

Figure 3. The rotation curve produced by a sum of two exponential disks with scale lengths and masses in the ratio of exp(-2) to one.

Real bulges and halos are not flat. However the same modifications to the rotation curves can be produced by spherical distributions, with very similar projected surface densities, provided the amplitudes are increased by about 50% . The mass ratio of halo to bulge remains the same.

The importance of Figure 1 lies in the relative displacement of the two curves: it shows that the outer regions benefit from the stabilising effect of the equilibrium mass distribution, to the detriment of the center.

The above discussion of stability is based on the axisymmetric criterion Q. There does not appear to be any simple stability criterion for the non-axisymmetric case, where new complications such as shear and resonances arise. Nevertheless the simple notion of keeping the disk stiff by means of external masses, is still correct. The stabilising effect of freezing a fraction of the disk, with the frozen part acting the role of the halo, has been illustrated by Toomre (1981). If the eccentricity in equation (2) is replaced by the reciprocal of the azimuthal wave number, the right side becomes proportional 1/X. X is an important parameter for the swing amplifier: large values of X will turn it off (Toomre 1981), which is in line with the above reasoning.

Of the many numerical experiments examining the stabilising effect of external masses, the recent study of the stability of the BSS model of our Galaxy (Bahcall, Schmidt, and Soneira 1982) by Sellwood (1985) is perhaps the most relevant. Much of the credit for the stability of this model goes to the combination of velocity dispersion and a central bulge component. The effect of the halo is minimal: removing it makes the disk lop-sided. The possibility exists that the lop-sidedness may have resulted from keeping the central component fixed during the simulation. The BSS model undoubtedly has some slack, which could be utilised to make it even more robust.

4. DISKS WITHOUT BULGES

The above discussion suggests that pure disk systems need massive halos if they are to remain cool. Lacking these, they must be quite hot, at least in the central parts. Surprisingly little work has been done on disks with hot centers, the notable exception being that by Athanassoula and Sellwood (1986). One reason for this is the lack of nice and simple equilibrium models, for there is little observational or theoretic basis for choosing the velocity structure.

The lack of specifications for a real disk galaxy is a big handicap in a talk of this nature. My own choice would be a stable differentially rotating disk which became cooler with increasing radius, and resembled the solar vicinity close to the peak of the rotation curve. It would be nice if I could produce such an example, for then I could quantify such phrases as "hot center", and provide the incentive for an observer to go out and prove me wrong.

The best I can offer at the moment is a model I will call JW , an abbreviation for "Jacobi waterbag". My purpose is to give you an idea of what the notion of a hot center may imply. JW is the coolest member of a family obtained by adapting Lynden-Bell's violent relaxation argument to disks (Lynden-Bell 1969). The phase space distribution is a function of Jacobi's integral, and has a constant value. It is truncated so that the disk is finite, and the mean and circular velocities become equal at the edge. The claim of stability rests on a 500-body numerical simulation.

Figure 4a summarises the kinematical data, and 4b shows the run of Q. The stellar disk rotates uniformly in the mean. The central velocity dispersion is about 66% of the maximum circular velocity. The velocity dispersion of the planetary nebulae and OH masers at the center of our Galaxy is 150 km/s, which is a similar fraction of the circular velocity

of 220 km/s at the sun. The ratio of random to rotational kinetic energy
is 3.6 - 20% higher than that implied by the Ostriker-Peebles criterion.
The vanishing of Q reflects the vanishing of the epicyclic frequency,
caused by a rapid decrease of the surface density near the edge. This is
a blemish, for it means that the outermost circular orbits are unstable.

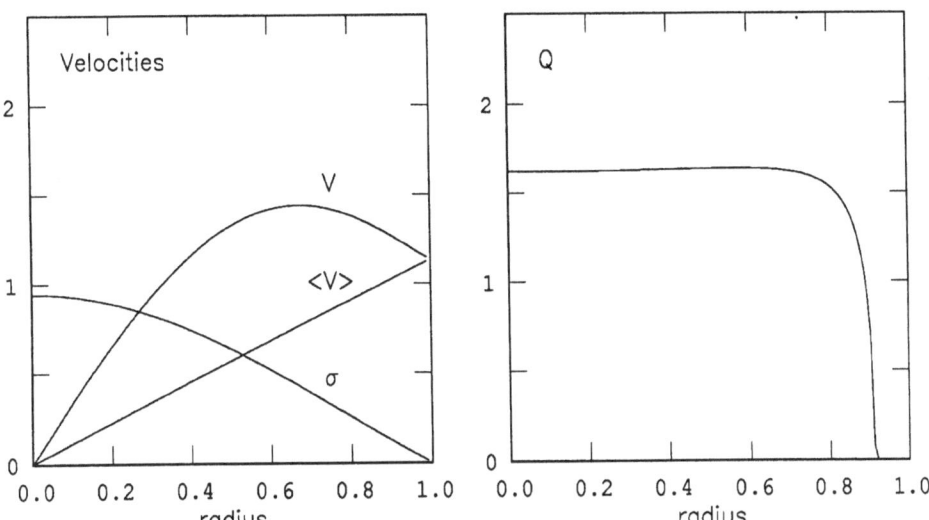

Figures 4a and 4b. Left: the rotation curve (V), mean velocity (<V>),
and velocity dispersion (σ) of the JW model. Right: the axisymmetric
stability parameter Q.

The JW model could conceivably be the first step in building a more
realistic disk galaxy. Starting from somewhere close to the peak of the
rotation curve, one would try adding a tapered ring, which should remain
cool enough to match the solar neighbourhood.

The JW model is one example of a stable disk. Undoubtedly there are
many more, and nicer examples.

5. CONCLUSIONS

The need for halos on stability grounds is not very compelling. The
fact that we have not come up with a decent halo-less Galactic model is
not an argument for the existence of halos. Small halos cannot be ruled
out, and might even be useful. At the other extreme, very massive halos
can be ruled out since they would make the disk dynamically dead. That
would eliminate such things as spiral structure, tidally induced spiral
structure, bars, lop-sidedness, and dynamical evolution.

REFERENCES

Athanassoula, E., and Sellwood, J. : 1985 (preprint)
Bahcall, J.N., Schmidt, M., and Soneira, R.M. : 1982, Astrophys. J.
      258, L23.
Hohl, F. : 1970, NASA Technical Report R-343.
Hohl, F. : 1971, Astrophys. J. 168, 343.
Lynden-Bell, D. : 1967, Monthly Notices Roy. Astron. Soc. 136, 101.
Miller, R.H., Prendergast, K.H., and Quirk, W.J. : 1970, Astrophys. J.
      161, 903.
Ostriker, J.P., and Peebles, P.J.E. : 1973, Astrophys. J. 186, 467.
Sellwood, J. : 1985, Monthly Notices Roy. Astron. Soc. 217, 127.
Toomre, A. : 1964, Astrophys. J. 139, 1217.
Toomre, A. : 1973, In Highlights of Astronomy, ed. G. Contopoulos,
      p. 457, (Dordrecht : Reidel)
Toomre, A. : 1981, In The Structure and Evolution of Normal Galaxies, ed.
      S.M. Fall and D. Lynden-Bell, p. 111, (Cambridge : Camb. U. Press)

DISCUSSION

DEKEL: How would you explain long-lived warps without massive halos?

KALNAJS: I probably would not. But the sort of halo needed to disturb the outer fringes of a galaxy will not contribute significantly to the force field in the central parts, and therefore is only of marginal interest to disk stability.

J. BAHCALL: Could you comment on Sellwood's numerical simulations of the Galaxy and how they affected your discussions?

KALNAJS: Sorry. I failed to mention the work by Sellwood, who simulated the BSS model for the Galaxy and got reasonable results. The model certainly doesn't have any of the obvious instabilities. At the moment, the rotation curve near the center is a bit high, but a little bit of tweaking will probably result in a fairly respectable dynamical model for our Galaxy. It is certainly a promising start.

OSTRIKER: You gave our T/W criterion much more credit than we ever intended. I have been surprised by how well this rule of thumb has worked for a variety of different stellar systems.
    Most models of the Galaxy have moderate halos, i.e., within several scale lengths a third to a half of the mass is in a hot component. This amount is quite adequate for removing gross instabilities, as Sellwood and others have found. You can throw away that component if you don't want the disks to be stable, and in any case you probably don't want to stabilize galaxies against all m = 2 modes. We suggested that galaxies are in fact moderately unstable and do produce bars. M31 has a bar, our Galaxy has a bar, and it may be that most ordinary spirals have a significant m = 2 component because they are not entirely stable.

KALNAJS: I agree with everything you say. At the time you did your work, chaps like me were thinking in terms of epicycles and rotationally supported disks, and your T/W criterion had a shattering impact on us, from which we have just about recovered.

TREMAINE: As a result of a great deal of work in recent years, we now have for a lot of disk galaxies good rotation curves, photometry, and a few measurements of velocity dispersions. It wasn't completely clear from your talk whether or not we should worry about the stability of those disks.

KALNAJS: They certainly should be stable on a timescale of a rotation or two. Halos can help, but are probably not necessary. You can have them, but not too much.

TREMAINE: The rotation curves say that you have to have them.

KALNAJS: Yes, but the mass required to explain the large rotation velocities in the outer regions of galaxies does not help with the

stability of disks. It is the inner parts you have to stabilize some
way if you want to have a cool disk. Otherwise the disk has to be very
hot in the center.

FABER: I'd like to ask about hot disks. The fact that edge-on spirals
have constant linear scale height as a function of distance from the
center must imply that at least the z dispersion σ_z is increasing toward
the center. Can you give us a rough estimate of σ_z and the scale height
in the inner part of the disk of our Galaxy? If someone measures a
velocity dispersion of 100 km s^{-1} for stars in Baade's window, does that
necessarily mean that they are measuring the spheroid or could they be
measuring a hot disk in the middle of the Galaxy?

KALNAJS: There is very little I can tell you about z motions; in this
business one ignores them. But if you do have a large velocity
dispersion in the plane, the disk will not remain thin. It will buckle
to such a height that the z velocity dispersion becomes 1/3 of that in
the plane. So that implies a minimum thickness for the disk. Presumably
you could have any additional amount of σ_z for other reasons.

SANDERS: Sellwood and I did a simulation of a model galaxy which
exactly resembled NGC 3198, discussed earlier by Sancisi. The model had
a maximum disk, the right amount of halo to give a flat rotation curve
and an initial Q of 1.5. It was violently unstable; it made a strong
bar within a few dynamical times. This is a practically bulgeless
galaxy, so if you want to stabilize the disk, it must have a hot
center.

GUNN: I would like to describe a set of simulations that Jens Villumsen
has just finished. This is a galaxy which at the end of the
calculations is cooked up to resemble our own, except that it doesn't
have a bulge. The simulations are designed to look at the effect of the
infall of cold matter. Apart from the fact that they are
three-dimensional, they are very much like the models that Carlberg and
Sellwood have made. The models evolve with essentially a constant value
of Q over the whole disk, and are never violently unstable. At the end,
the models have essentially a constant value of Q and a constant scale
height over the whole disk. The disk is thus quite hot at the center,
with the velocity dispersion rising like 1/r. I think that is more or
less in accord with the observations that Ken Freeman showed us. And
certainly this is what one expects σ_z to do, since the scale heights of
galactic disks are constant with radius.

VAN DER KRUIT: I would like to draw attention to a measurement by
Freeman and myself of the stellar velocity dispersion in NGC 7184. The
dispersions are determined from the measured asymmetric drift between
about one and two scale lengths from the center (the system has only a
minor bulge), and are increasing toward the center. The extrapolated
central value is over 100 km s^{-1}, hot enough to be significant in
contributing towards stability (van der Kruit and Freeman, Ap.J., in
press).

KORMENDY: I have measured stellar velocity dispersions in the disks of
two S0 galaxies, NGC 1553 and NGC 936 (Ap.J., 286, 116 and 132, 1984).
Both disks are very hot in their central parts. In fact, in NGC 1553,
the line-of-sight velocity dispersion rises from < 100 km s^{-1} at r $>$ 4
kpc to 179 ± 4 km s^{-1} at r = 1 kpc in the disk. This is the sort of
behavior inferred from the constant scale heights, but the magnitude of
the dispersion at small radii is remarkably large for a disk. The
density increases inward, too, and so Q $\simeq$ 2.5 varies little with radius.
Unlike NGC 1553, NGC 936 is barred. It has an even larger value of Q,
rising from ~4 at r = 8 kpc to ~7 at r = 2.5 kpc. Not surprisingly,
neither galaxy has any spiral structure. Since the inner parts of both
disks locally satisfy the Ostriker-Peebles criterion, it is tempting to
think that the dispersion is large enough to contribute to global
stability. However, the fact that NGC 936 is barred shows that even if
this argument is correct, it is not safe to use the present value of the
velocity dispersion to decide whether a disk was stable in the past.
The disk could have been cold, made a bar, and then used it to heat
itself up after the fact.

SANCISI: There is least one case in which we can rule out the
possibility that the halo is dominant in the inner parts of a galaxy.
This is NGC 2403, a normal spiral which I showed in my talk. If we try
to make a model with an insignificant disk, we find that M/L $\simeq$ 0.3 for
the disk. The maximum disk gives M/L = 1. Now unless you are willing
to believe that M/L $<$ 1, you are forced to conclude that inside 2 or 3
scale lengths the dark matter cannot be dominant.

RUBIN: There is not very much difference in what we are all saying. A
variety of approaches suggest M_{dark}/M_{lum} ~ 1 within the optical parts of
galaxies. No one claims that there are enormous amounts of dark matter
at these small radii. I have suggested that dark matter contributes,
but not that it is dominant there.

KALNAJS: I am not questioning that dark halos exist. I am only saying
that for stability arguments they are not necessary.

MASSIVE HALOS AND THE STABILITY OF HOT STELLAR DISCS

E. Athanassoula   and J.A. Sellwood
Observatoire de Marseille   Kapteyn Laboratory, Groningen

Bulge and halo material is often invoked to explain the absence of bars
in the majority of disc galaxies, although the amount required is not
known, in general, and is sensitive to the shape of the rotation curve.
Here we show that the necessary fraction of spherical material is also
affected by the degree of random motion amongst the disc stars. Quite
moderate velocity dispersion has a strong stabilising influence and our
hottest disc is stable without any halo (see also Kalnajs, this volume).

All our N-body models have the circular velocity rotation curve of
a Kuz'min-Toomre disc, but in some only a fraction, q, of the disc mass
was responsive. The remaining mass can be thought of as a bulge/halo
component having the density distribution of a Plummer sphere.

Our results are summarised in the figure below. The size of each
circle in (a) is proportional to the linear growth-rate of the bar mode
measured from each simulation. As a measure of the random motion, we
use the directly observable, dimensionless ratio $\sigma_u/\langle v\rangle$, i.e. the ratio
of the radial dispersion of velocities to the mean tangential streaming
speed, both measured at the "turn-over radius" of the rotation curve.
The points seem to lie on a simple surface in this space. The quality
of fit to a plane is shown in (b). (Error bars are internal estimates.)

The implied line of marginal stability is indicated in (a). If
this line is typical of mass distributions giving <u>gently rising</u> rotation
curves, it may be used to estimate the minimum fraction of mass in a
bulge/halo component needed to account for the absence of a bar in a
galaxy, once the velocity dispersion is known. (Steeply rising rotation
curves give different stability properties - see Sellwood, this volume.)

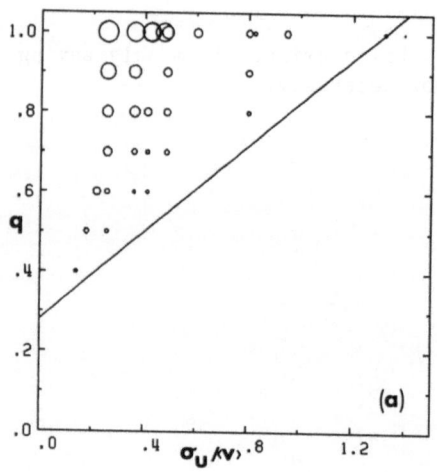
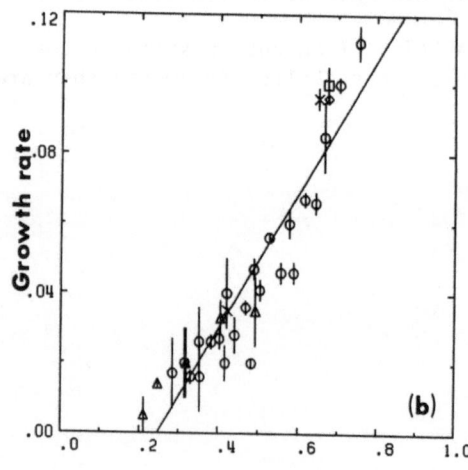

J. Kormendy and G. R. Knapp (eds.), Dark Matter in the Universe, 300.

THE GLOBAL STABILITY OF OUR GALAXY

J.A. Sellwood
Kapteyn Laboratory, Groningen

Many mass models of our Galaxy (see e.g. Schmidt 1985 for a review) assume the disc to be nearly axisymmetric. For consistency, therefore, no such model should possess gross non-axisymmetric instabilities. However, most attribute less than half the central attraction at the Sun to bulge and halo material – a situation where bar forming instabilities are frequently found (see e.g. Sellwood 1983).

Tests of bar stability have, hitherto, been largely confined to models lacking strong central concentration, whereas Toomre (1981) argues that a steeply rising rotation curve should inhibit bar forming modes. Our Galaxy may be just such a case, since the rotation curve appears to rise to an inner peak of some 220 km/sec at a radius of less than 2kpc. Thus more realistic mass distributions might require less spherical material than previous experience seems to indicate.

Accordingly, I have investigated the stability of the Bahcall, Schmidt and Soniera (1982) mass model of the Galaxy, which attributes the high orbital velocities near the centre to a small but dense component. The particles in my N-body simulations represent the disc only – the spheroidal components are included through an unresponsive additional central attraction. A special multiple time-step code was devised to cope with the large dynamic range in the forces acting on the particles.

The most realistic model exhibited faint, largely two-armed spiral structure, but seemed to be otherwise stable. In this case the disc was reasonably warm ($Q \geq 1.5$), as is consistent with solar neighbourhood data. Other models which were cooler or in which the dense central component was omitted quickly formed bars.

As the central attraction due to the dark halo, in this mass model, begins to dominate only outside the solar circle, it should have little influence on the stability of the disc centre. When this component was removed, the disc did not form a bar, but became extremely lop-sided rather quickly. Apparently, the dark halo inhibits one-armed modes in the disc, but has no effect on bar stability.

These results are reported fully in Sellwood (1985).

References

Bahcall, J.N., Schmidt, M. and Soniera, R. (1982) Ap. J. Lett. **258**, L23.
Schmidt, M. (1985) The Milky Way Galaxy, IAU Symposium **106**,
    eds. van Woerden, H., Allen, R.J. and Burton, W.B., Reidel.
Sellwood, J.A. (1983) Internal Kinematics and Dynamics of Galaxies,
    IAU Symposium **100**, ed. Athanassoula, E., Reidel.
Sellwood, J.A. (1985) MNRAS, **216**, in press.
Toomre, A. (1981) Structure and Evolution of Normal Galaxies,
    eds. Fall, S.M. and Lynden-Bell, D., Cambridge University Press.

J. Kormendy and G. R. Knapp (eds.), Dark Matter in the Universe, 301.
© *1987 by the IAU.*

OBSERVABLE CONSEQUENCES OF TRIAXIAL HALOS

James Binney
Department of Theoretical Physics,
Keble Road, Oxford, OX1 3NP,
England.

ABSTRACT. Heavy halos are probably not axisymmetric. During disk formation, the visible parts of disk galaxies are likely to have aligned their principal axes with those of the surrounding halos, but it is less clear that early-type galaxies are aligned with their halos. Triaxial halos may possibly sustain, or even excite, the warps often seen in the outer parts of disks. Polar rings and ripples around elliptical galaxies provide promising probes of the triaxiality of halo potentials.

1. Are Halos Triaxial and are they Aligned with their Galaxies?

Numerical simulations have shown that the collapse and virialization of a cloud of stellar objects from a wide variety of spherical (Barnes, Goodman & Hut 1985, Merritt & Aguilar 1985), and non-spherical (Binney 1976, Aarseth & Binney 1978, Miller & Smith 1981, Wilkinson & James 1982) initial configurations leads to the formaton of a triaxial system. Furthermore, encounters of stellar systems, especially nearly head-on encounters, often yield triaxial remnants (White 1979, Gerhard 1981, 1983, Negroponte & White 1982). Hence if, as seems overwhelmingly probable, massive halos formed by either the collisionless collapse of single systems or the merging of two or more collisionless systems, many halos should be markedly triaxial, and it is interesting to seek observational implications of this triaxiality.

The observational consequences of a halo's triaxiality depend very much on whether the halo's principal axes are aligned with those of the embedded galaxy. A *prima facie* argument against such alignment may be developed from the picture of galaxy and halo formation described by Gunn (1982). In this picture the galaxy's spheroidal component collapses first, and the disk and halo later, by way of infall. Hence the angular momentum and principal axes of the spheroid would have been determined earlier, and by sampling the ambient tidal field at smaller radii, than the principal axes of the halo and the angular momentum of the protodisk material. Consequently, we would not expect the *original* angular momentum vector of the

303

J. Kormendy and G. R. Knapp (eds.), Dark Matter in the Universe, 303–313.

spheroid to be parallel to that of the halo/disk. Indeed we might even anticipate a tendency to anti-alignment, since the spheroid will have acquired its angular momentum by exchange with the halo/disk. Hence elliptical galaxies, which are in this picture simply unreconstructed spheroids, are not expected to be aligned with their halos.

This argument does not apply to disk galaxies, since the spheroid of a disk galaxy must have been profoundly modified by the formation of the disk (Barnes & White 1984, Binney & May 1985). Estimates of the efficiency of angular-momentum acquisition from tidal interactions (Peebles 1969, Efstathiou & Jones 1979) suggest (Binney & Silk 1978, Fall & Efstathiou 1980) that the disks are made of gas left over from halo formation. Hence one expects the protodisk material to have had similar angular momentum per unit mass to that of the halo. Furthermore, any mismatch in the directions of their angular momenta should have been ironed out as the protodisk material spiralled through the halo towards the central spheroid. In Oxford we have recently analyzed the response of an axisymmetric spheroid to the steady accretion of material which is endowed with angular momentum about an axis that is skew to the spheroid's symmetry axis, and have concluded (Binney & May 1985) that the spheroid and any embedded disk would be dragged round into alignment with the angular momentum of the accreting material exactly as if it were a rigid body. Hence we expect disk galaxies to be aligned with their halos.

If, as White & Rees (1978) have argued, halos formed first, and both spheroids and the disks of galaxies formed from material that fell into preexisting halos, every galaxy would from the beginning have taken its cue from the dominant halo field, and galaxy and halo would always be aligned.

2. Halos and Warps

Toomre (1983) has reviewed the possible connection between halos and warped disks, and Sparke (1984a) has recently given a thorough discussion of the dynamics of self-gravitating disks in axisymmetric potentials. The simplest way in which a halo can contribute to the longevity of a warp, is by making the gravitational potential of the whole system more nearly spherical; the wind-up time of a warp is approximately proportonal to the inverse of the ellipticity of the equipotential surfaces. Tubbs & Sanders (1979) have argued that the best observed warps could last several billion years if the galaxies in which they occur have plausible massive halos.

Petrou (1980) has proposed a refinement of this idea by suggesting that the ellipticities of massive halos may frequently be increasing functions of radius. If this were so, the ellipticities of the overall galactic potentials might increase outwards fast enough to make the precession frequencies of disk stars roughly radius-independent notwithstanding the steep outward decrease in the overall orbital frequencies. If halos are constructed like massive elliptical galaxies, an outward increase in their ellipticities is certainly to be expected (Di Tullio 1979).

Another simple way in which a halo could sustain a warp, is by being mis-aligned with the galaxy (Dekel & Shlosman 1983, Toomre 1983, Marthur 1984). Deep inside the galaxy, gas will settle to the galaxy's equatorial plane, while far out, gas will follow a principal plane of the halo. The transition between these two regimes would be sharp and would be interpreted as the onset of a warp. If halos were expected to be misaligned with their embedded galaxies, this would be an attractive proposal. However, as explained above, one wonders whether such misalignment is truly possible, since the halo, through the agency of the protodisk material, should long ago have reoriented the galaxy parallel to the halo axes.

The warp theories described so far concern ways in which even an axisym-metric halo could increase the useful life of a warp that may not be caused by the halo itself. If the halo is triaxial, there are two ways in which the halo could generate a warp *ab initio*:

1) Binney (1978, 1981) pointed out that orbits in a fundamental plane of a tri-axial halo are unstable to vertical perturbations whenever the frequency at which orbiting stars pass density maxima is equal to twice the natural frequency of z-oscillations for these stars. In a non-rotating potential, this frequency condition is satisfied at all radii for orbits in the plane perpendicular to the intermediate axis of the potential (Heiligman & Schwarzschild 1979). In a rotating potential the condition is satisfied (i) on all retrograde orbits in an annulus that is located well inside the corotation radius of the potential, and (ii) on prograde orbits in an annulus that lies beyond corotation. When gas that is spiralling in from in-finity along a fundamental plane of a triaxial potential reaches the outer bound of such an unstable zone, it will tend to lift from the plane. Heisler Merritt & Schwarzschild (1982) and Mulder (1984) have followed the sequences of closed or-bits onto which gas in an otherwise empty galaxy would then move. The sequence of closed retrograde orbits tends ultimately to align with the plane perpendicular to the longest axis of the potential, while orbits of the prograde sequence become self-intersecting near corotation; presumably gas would drop back onto the plane once the self-intersecting orbits were reached. Observationally this might mark the onset of a warp.

2) Bertin & Mark (1980) have shown that corrugations in a thin cold disk that rotates through a slowly rotating axisymmetric halo, can grow unstably. A crude physical picture of this instability is as follows. We imagine the halo to be made up of a large number of streams of stars with the same velocity v. Each piece of the disk tries to focus these streams, much as in the standard picture of dynamical friction, a massive body focusses streams of background stars. When the disk is flat and axisymmetric, each element of the disk focusses upward- and downward-moving streams to the same degree, so no net vertical forces are generated. But if some piece of the disk is moving up, it focusses the upward-moving halo stars more successfully than the downward-moving ones, and hence produces a net wake above

it. This wake draws the offending section of the disk up still more, and overstable oscillations can set in.

Toomre (1983) has questioned the ability of these instabilities to generate warps in galactic disks, by noting that in a self-gravitating disk, corrugation waves have non-zero group velocities. Hence such waves carry energy from any region of overstable oscillations, and this will severely attenuate, or even eliminate, the growth of oscillations, unless something reflects the outgoing corrugation waves back into the unstable region. Recently Sparke (1984b) has investigated the effect of including realistic self-gravity on the Mathieu instability discussed by Binney. She finds that for realistic parameters, self-gravity eliminates the instability otherwise generated by a prograde halo, but the instability generated by a retrograde halo is only weakened by self-gravity.

Sparke interprets the survival into the self-gravitating regime of the instability associated with a retrograde halo in terms of a WKBJ analysis of tightly wound corrugation waves. This shows that the triaxiality of the halo causes leading waves, which propagate in one direction, to be converted into trailing waves, which propagate in the opposite direction, and *vice-versa*, thus enabling the wave energy to be fed again and again through the region of overstability. This is an interesting result with implications that extend beyond the Mathieu instability. Indeed it might even enable fuzzy-edged disks in triaxial halos, unlike similar disks in axisymmetric systems (Hunter & Toomre 1969), to admit discrete bending modes, and thus support long-lasting warps even in the absence of an exciting mechanism.

3. Polar Rings

The last few years have seen several observational studies of "spindle" galaxies (Schechter & Gunn 1978, Schweizer, Whitmore & Rubin 1983, Schechter et al.1984, Schechter, Ulrich & Boksenberg 1984). These systems consist of a central galaxy encircled by an bright annulus whose apparent major axis nearly coincides with the apparent minor axis of the central galaxy. The ratio of the outer and inner radii of the annulus can be as great as 3. The annuli often seem to be reasonably flat. In every case studied, spectroscopic and photometric observations have revealed the central galaxy to be a nearly edge-on S0 galaxy, and the bright annulus to consist of gas and young stars. Since the annuli, or *rings*, of these galaxies extend well beyond the visible part of the underlying galaxies, they provide excellent probes of halos.

There are two ways in which rings may place constraints on halos:

1) Dissipation tends to push gas towards closed loop orbits. In an axisymmetric potential, two types of closed orbit are possible; (*i*) orbits in the potential's equatorial plane, and (*ii*) orbits that pass through the potential's symmetry axis. Orbits within the equatorial plane are stable in the sense that if a particle on

such an orbit is deflected through a small angle, it subsequently librates about its original closed trajectory. Polar orbits, by contrast, are neutrally unstable; a small-angle deflection of a particle on one of these orbits causes the particle to diverge from its original trajectory at an approximately constant rate. In a slowly rotating triaxial potential such as that generated by a triaxial halo, three types of closed orbit are possible; (i) stable loop orbits that circulate about the potential's shortest axis; (ii) stable orbits that at low energies circulate about the potential's long axis, and (iii) unstable orbits that circulate around the potential's middle axis.

Gas that is injected into a halo on a non-closed orbit might be expected eventually to move to a stable closed orbit. Thus it is tempting to interpret spindle galaxies as galaxies with two orthogonal families of stable loop orbits (Tohline, Simonson & Caldwell 1982, van Albada, Kotanyi & Schwarzschild 1982). It would the follow that many S0 galaxies have triaxial potentials. Furthermore, the ring of a typical spindle galaxy rotates too slowly to follow the orientation of a bar in the visible galaxy, so any large-scale triaxiality of the potential would be generated by a triaxial halo.

This is an interesting line of argument, but two cautions are in order. (i) As Schweizer et al. (1983) have emphasized, the dynamical times of polar rings are very long, and the time required for nearly polar gas in an axisymmetric potential to settle to the equatorial plane can be very long. Hence it is not clear whether the observed rings really delineate families of stable loop orbits, or simply planes from which orbital decay is exceptionally slow. (ii) It is not clear that the differential precession, or libration periods of stellar orbits provide an adequate measure of the time required for a stream of gas to evolve from that orbit towards the relevant closed orbit. Convincing numerical simulations of this settling process, including realistic accounts of the action of viscosity and self-gravity, are not yet available, although studies have been published of the effects of self-gravity and viscosity taken in isolation.

Steiman-Cameron (1984) and Habe & Ikeuchi (1985) have presented numerical calculations of the settling of gaseous annuli in axisymmetric and triaxial potentials. Steiman-Cameron models the gaseous annuli with a series of thin rings. He finds that the time required for gas to settle to a preferred plane depends sensitively on the difference in the free precession or libration periods of stellar orbits accross the annulus. By contrast, large changes in the coefficient of viscosity lead to only small changes in the settling time. For relatively low coefficients of viscosity, Steiman-Cameron's annuli, unlike observed rings, become strongly twisted. The models of Habe & Ikeuchi employ extended particles. Hence they cannot simulate the low-viscosity regime. The annuli all settle to preferred planes in the appropriate precession or libration time.

Sparke (1984b) has recently discussed the effects of self-gravity on the evolution of a frictionless polar ring in an axisymmetric potential. She finds that self-gravity can stabilize a ring of realistic mass providing it warps towards the pole on the outside. This is exactly the reverse sense of warp to that predicted by assuming that the ring is made up of closed loop orbits in a tumbling triaxial potential. Sparke's work seems to be an important advance, which probably gives a truer account of the evolution of a real polar ring than do the calculations of Steiman-Cameron and Habe & Ikeuchi.

2) From the shapes and velocities of rings we may determine the flattening of the halo potential (Schweizer et al. 1983, Katz & Richstone 1984). The flatter the combined galaxy/halo potential is, the more the ring will be elongated parallel to the galaxy's shortest axis, and the slower the stream will move over the pole. For favourable ring orientations, these effects could be used to determine one axis ratio of the halo potential:

(i) If in some galaxy one finds no velocity gradient perpendicular to the line in which a polar ring cuts the equatorial plane of the embedded galaxy, the latter line must lie in the plane of the sky. The ring's intrinsic shape could then be determined from the apparent axis ratios of the ring and the disk. Application of the epicycle approximaton to polar rings, reveals that the ring ellipticity determined in this way would be, for a flat circular-speed curve, equal to the ellipticity of the potential (although the ring is extended parallel to the ring's *short* axis).

(ii) If the ring is seen nearly edge on, the axis ratio q_Φ of the equipotential surfaces could be determined by comparing the peak radial velocity v_{ring} in the ring with an estimate v_{disk} of the circular speed at a similar radius in the underlying galactic disk: For this case the epicycle gives for a flat circular-speed curve, $q_\Phi \simeq (2v_{ring} + v_{disk})/3v_{disk}$.

4. Ripples

In 1980 Malin & Carter drew attention to the presence of sharp steps in the outer brightness profiles of some giant elliptical galaxies. They called these features shells, but I shall follow Schweizer (1980, 1983) in referring to them as *ripples* since they are almost certainly not generated by shells of stars.

The survey of Malin & Carter (1983) has shown that on the order of 17% of isolated elliptical galaxies have ripples in their envelopes. Photometric observations (Carter, Allen & Malin 1982) have revealed that ripples are generated by abrupt steps in stellar density. A single ripple never seems to envelope the whole galaxy. They tend to cluster along the apparent major axis, and to be interleaved in radius on opposite sides of the nucleus. Ripples have been detected as far out as $3R_e$.

Quinn (1984) has investigated Schweizer's (1980) suggestion that ripples represent the debris of disk galaxies that have been eaten by a giant. This proposal hinges on the strong concentraton in phase space of the stars of any disk galaxy; in real space disk stars are concentrated to a plane, and at any point within this plane their velocity vectors cluster around the circular velocity. Hence the phase-

space density of disk stars is appreciable only near a two-dimensional subspace of
the full six-dimensional phase space. By Liouville's theorem, this concentration
of the disk stars is not affected by the cannibalization of the disk by the giant.
Consequently the projected image of the predator galaxy should subsequently be
marked by regions of enhanced luminosity wherever we view this phase-space sheet
edge-on.

The easiest way to picture the formation of high-density ripples around a
giant galaxy is to imagine the fate of a low-mass disk that approaches a giant on
a high-angular momentum orbit. The disk will be stretched and wrapped like a
bandage around the predator. High surface densities will occur where the line
of sight is tangent to this bandage. However, Quinn found that his test-particle
models generated the most convincing ripples when a disk was dropped into the
predator on a deeply plunging orbit. Beautiful ripples were then formed at the
radial turning points of the resulting stellar orbits. Stars linger at these turning
points, with the result that those orbits on which stars are currently at apocentre
stand out prominently. This model of ripples accounts very naturally for the
interleaving of the ripples on either side of the nucleus and for their tendency to
occur along the apparent major axes of galaxies. If Quinn is correct in maintaining
that ripples are generated by stars on very radial orbits and not by bandaging
giants with disks, we should think of ripples as being generated by phase-space
sheets rather than by real-space sheets of stars.

Ripples may soon prove very powerful probes of halo structure. The radial
density profile could follow from the radii r_n of successive ripples, these being
the radii at which n radial periods have elapsed since the victim fell in (Quinn
1984). The angular structure would follow from such things as whether pericen-
tre distances are marked by reversed ripples near the centre, and whether there
are ripples with angular structure characteristic of long-axis tubes in addition to
ripples generated by boxes and short-axis tubes.

However, it will not be easy to test these predictions because the existing
calculations of ripple formation are not very convincing, and it may not be easy to
improve on them in the near future. The difficulty is that the entrapment of the
victim by the tidal wave which the victim raises in the predator determines the final
energy distribution of the former disk stars. Hence credible calculations cannot be
performed with test-particle disks such as those employed by Quinn. On the other
hand, full n-body calculations are not only much more expensive, but are, as Quinn
(1984) discovered, also unlikely to produce nice ripples: Currently affordable n-
body calculations are always contaminated by a degree of collisional relaxation
(Norman, May & van Albada 1985). This artificially fattens an otherwise thin
phase-space stream of stars, and hence blurs any ripples to which that stream
gives rise. The solution to this problem would seem to require some cunning
combination of the n-body and test-particle techniques.

However, I think triaxiality of the halo can help us to a qualitative understanding of why so many giant ellipticals have swallowed their victims from highly

eccentric orbits: The fraction of phase space that is occupied by highly elongated orbits is simply higher in a triaxial potential than in an axisymmetric one. Consequently, if a halo is triaxial, the victim galaxy is likely to approach the predator on a highly eccentric orbit.

References

Aarseth, S.J. & Binney, J.J., 1978. *Mon. Not. Roy. astr. Soc.*, **185**, 227.

Barnes, J., Goodman, J. & Hut, P., 1985. *Astrophys. J.*, **000**, 000.

Barnes, J. & White, S.D.M., 1984. *Mon. Not. Roy. astr. Soc.*, **211**, 753.

Bertin, G. & Mark, J.W.-K., 1980. *Astron. Astrophys.*, **88**, 289.

Binney, J.J., 1976. *Mon. Not. Roy. astr. Soc.*, **177**, 19.

Binney, J.J., 1978. *Mon. Not. Roy. astr. Soc.*, **183**, 779.

Binney, J.J., 1981. *Mon. Not. Roy. astr. Soc.*, **196**, 455.

Binney, J.J. & May, A., 1985. *Mon. Not. Roy. astr. Soc.*, **000**, 000.

Binney, J.J. & Silk, J, 1978. *Comm. Astrophys. Sp. Sci.*, **7**, 139.

Carter, D., Allen, D.A. & Malin, D.F., 1982. *Nature*, **295**, 126.

Di Tullio, G.A., 1979. *Astron. Astrophys.Suppl. Ser.*, **37**, 591.

Efstathiou, G. & Jones, B.J.T., 1979. *Mon. Not. Roy. astr. Soc.*, **186**, 133.

Fall, S.M. & Efstathiou, G., 1980. *Mon. Not. Roy. astr. Soc.*, **193**, 189.

Gerhard, O.E., 1981. *Mon. Not. Roy. astr. Soc.*, **197**, 179.

Gerhard, O.E., 1983. *Mon. Not. Roy. astr. Soc.*, **203**, 19P.

Gunn, J.E., 1982. In: *Astrophysical Cosmology,,* eds H.A. Brück, G.V. Coyne & M.S. Longair, pp 23 (Vatican: Pontif. Acad.).

Habe, A. & Ikeuchi, S., 1985. *Astrophys. J.*, **289**, 540.

Heiligman, G. & Schwarzschild, M., 1979. *Astrophys. J.*, **233**, 872.

Heisler, J., Meritt, D. & Schwarzschild, M., 1982. *Astrophys. J.*, **258**, 490.

Hunter, C. & Toomre, A., 1969. *Astrophys. J.*, **155**, 747.

Katz, N. & Richstone, D.O., 1984. *Astron. J.*, **89**, 975.

Malin, D.F. & Carter, D., 1980. *Nature*, **285**, 643.

Malin, D.F. & Carter, D., 1983. *Astrophys. J.*, **274**, 534.

Marthur, S.D., 1984. *Mon. Not. Roy. astr. Soc.*, **211**, 901.

Merritt, D. & Aguilar, L., 1985. *Mon. Not. Roy. astr. Soc.*, **000**, 000.

Miller, R.H. & Smith, B.F., 1981. *Astrophys. J.*, **244**, 33.

Mulder, W., 1983. *Astron. Astrophys.*, **121**, 91.

Negroponte, J. & White, S.D.M., 1983. *Mon. Not. Roy. astr. Soc.*, **205**, 1009.

Norman, C.A., May, A., & van Albada, T.S., 1985. *Astrophys. J.*, **000**, 000.

Peebles, P.J.E., 1969. *Astrophys. J.*, **155**, 393.

Petrou, M., 1980. *Mon. Not. Roy. astr. Soc.*, **191**, 767.

Quinn, P.J., 1984. *Astrophys. J.*, **279**, 596.

Richstone, D.O. & Potter, M., 1982. *Nature*, **298**, 728.

Schechter, P.L. & Gunn, J.E., 1978. *Astron. J.*, **83**, 1360.

Schechter, P.L., Sancisi, R., van Woerden, H. & Lynds, C.R., 1984. *Mon. Not. Roy. astr. Soc.*, **208**, 111.

Schechter, P.L., Ulrich, M.-H. & Boksenberg, A., 1984. *Astrophys. J.*, **277**, 531.

Schweizer, F., 1980. *Astrophys. J.*, **237**, 303.

Schweizer, F., 1983. In: *IAU Symposium 100, Internal Kinematics and Dynamics of Galaxies*, (Dordrecht: Reidel), pp. 319.

Schweizer, F., Whitmore, B.C. & Rubin, V., 1983. *Astron. J.*, **88**, 909.

Sparke, L.S., 1984a. *Astrophys. J.*, **280**, 117.

Sparke, L.S., 1984b. *Mon. Not. Roy. astr. Soc.*, **211**, 911.

Sparke, L.S., 1985. *Mon. Not. Roy. astr. Soc.*, **000**, 000.

Steiman-Cameron, T.Y., 1984. *Ph.D. Thesis*, University of Indiana.

Tohline, J.E. & Durisen, R.H., 1982. *Astrophys. J.*, **257**, 94.

Tohline, J.E., Simonson, G.F. & Caldwell, N., 1982. *Astrophys. J.*, **252**, 92.

Toomre, A., 1983. In: *IAU Symposium 100, Internal Kinematics and Dynamics of Galaxies*, (Dordrecht: Reidel), pp. 177.

Tubbs, A.D. & Sanders, R.H., 1979. *Astrophys. J.*, **230**, 736.

van Albada, T.S., Kotanyi, C.G. & Schwarzschild, M., 1982. *Mon. Not. Roy. astr. Soc.*, **198**, 303.

White, S.D.M., 1979. *Mon. Not. Roy. astr. Soc.*, **189**, 831.

White, S.D.M. & Rees, M.J., 1978. *Mon. Not. Roy. astr. Soc.*, **183**, 341.

Wilkinson, A. & James, R.A., 1982. *Mon. Not. Roy. astr. Soc.*, **199**, 171.

DISCUSSION

PEEBLES: You described how a disk misaligned with an aspheric halo will precess. But I reckon that the precession rate is a function of radius, so a stellar disk would end up being thick. This is not acceptable, isn't that right?

BINNEY: If gas comes in on a skew axis, the stars that form out of it do indeed precess around and form a thick disk. What is interesting, though, is the response of the pre-existing galaxy. It turns out that an adiabatic invariance enables this stuff to turn right over in the new potential and stay thin.

GUNN: My impression is that the warps one sees begin at about the place where disks become non-self-gravitating. Could you comment on that?

BINNEY: I think you would say that warping is just a transient phenomenon - the junk falling in temporarily doesn't have the proper orientation because we are looking far out in the halo. I think I would agree. But then warps don't last long unless you have steady infall.

CASERTANO: It is true that most warps seem to start where there is little star light, but the local neutral hydrogen density is enough for self-gravity to be important. Besides, there is at least one case (NGC 4565, see poster paper by Casertano, Sancisi and van Albada) where the stars also take part in the distortion.

VAN WOERDEN: On the matter of polar rings: W. van Driel and I have mapped HI in 10 gas-rich S0 galaxies and 6 S0/a's. Among these, 6 and 1, respectively, have their HI in rings outside the optical body. At least three of these rings are polar. Hence the polar-ring phenomenon is quite frequent.

BINNEY: Perhaps these are S0's precisely because they have ceased to accrete material in their equatorial planes. Accretion over the pole is unlikely to be compatible with continued accretion in the equatorial plane, and spiral structure is then likely to die out.

WHITMORE: I would like to add a few items to the description of polar ring galaxies. About half of these systems actually have a thin ring rather than a broad annulus. In this case differential precession may be negligible. However, long-slit spectra show emission over a much wider region than the thin optical ring, suggesting that a disk is present. Perhaps some type of resonance phenomena results in star formation only at a certain radius, producing the thin ring. I hope theorists will address this question.
   I would also like to report that one of the candidate polar ring galaxies reported by Schweizer, Whitmore, and Rubin (A.J., 88, 909, 1983) turns out to be an elliptical rather than an S0. This system is described in a poster paper.

SCHWARZSCHILD: Earlier, Sancisi showed us one case where the HI was marvellously aligned but lopsided, meaning that phase mixing within the plane was not complete but that somehow orientation to the plane was well achieved. Can you do that with a highly flattened halo?

BINNEY: Yes, I think you can. If you rely on something like dynamical friction, then it is always the <u>smallest</u> component of the motion that's destroyed first. I don't know quantitatively how flat the halo would have to be.

RICHSTONE: James, can you expand on one of your earlier remarks? Why the preference for radial orbits in ripples and shells?

BINNEY: I think that triaxiality provides a very natural explanation, because it says that a very large portion of phase space consists of box orbits. Box orbits are all by definition extremely radial. In the transition from an axisymmetric to a triaxial potential, the most eccentric orbits go over into box orbits. So what used to take up a negligible proportion of phase space starts to take up a lot of phase space. I suspect that if you drop something into a triaxial halo even with a reasonable amount of angular momentum, it will still wind up in a box orbit - effectively, a zero-angular-momentum orbit.

LYNDEN-BELL: Assuming the Aarseth and Binney principle that anything that can happen does happen, would you not expect the barycenter of the halo to be initially offset from that of the visible galaxy? The initial sloshing of a galaxy within its halo may have interesting unsymmetrical consequences.

BINNEY: Gerhard (M.N.R.A.S., 203, 19P, 1983) and May, Norman and van Albada (Ap.J., in press) have found that n-body models sometimes continue oscillating for many dynamical times after violent relaxation would be expected to be complete. Hence it is indeed likely that halos, which must have very long dynamical times, are still shivering like cheap jellies. Perhaps a large-scale example of such a shiver is provided by the offset observed between the center of the Perseus x-ray source and the cooler x-ray source centered on NGC 1275 (Branduardi-Raymont et al., Ap.J., 248, 55, 1981).

OSTRIKER: Do you believe that either 1) the shapes of x-ray isophotes, or 2) the fact that the surface density needed for gravitational lenses does not coincide with the surface light density give evidence for triaxiality of dark matter accumulations?

BINNEY: Certainly I have long been an enthusiast for using x-rays to determine the shapes of clusters of galaxies; Sarazin yesterday mentioned the work Strimpel and I did a few years ago to develop this approach. It would certainly be interesting to apply that method to ellipticals. I have not thought about gravitational lenses, but it is tempting to blame some of the problems Turner discussed yesterday on the triaxiality of halos.

SELF-GRAVITATING POLAR RING MODELS

Linda S. Sparke
Institute of Astronomy, Cambridge, U.K.

In a number of SO galaxies, rings of gas, dust and stars are observed to lie roughly perpendicular to the galactic disc (e.g. Whitmore, these proceedings). The material sometimes forms a fairly broad annulus, and the ring appears nearly flat. Simple estimates suggest that differential precession will destroy a polar ring in much less than a Hubble time, implying that the observed structures have been formed only recently. It is then surprising that some of them are so regular, unless something acts to prevent their disruption.

A simple model for a self-gravitating polar ring is a collection of concentric but mutually tilted massive circular wires, in orbit about a flattened potential representing the body of the SO galaxy. Equilibrium states can be found in which the wires precess steadily as a unit, making up a coherent polar ring. There are stable equilibria near the pole provided that the ratio of the ring mass to the galactic quadrupole moment exceeds a minimum value (which increases as the ring becomes broader). Stable polar rings are nearly planar, but curve up towards the pole at their outer edge. This is the opposite sense of warping from that expected in a polar ring which is stabilised in orbit about the short axis of a rotating triaxial figure; these rings bend towards the equator (van Albada et al. 1982; MNRAS 198, 303). The required masses are comparable to those observed in neutral hydrogen (e.g. Schechter et al. 1984; MNRAS 208, 111).

Unless the polar ring is extremely heavy, stable states exist only near the pole or near the equatorial plane; rings at intermediate angles are unstable and will break up. Time-dependent calculations show that unstable rings split into a number of stable and independently precessing sub-rings. Thus a multiple polar ring system (such as that in ESO 474-G20: Schweizer et al. 1983; A.J. 88, 909) may result from capturing a single gas cloud.

A full account of this work will appear in Monthly Notices of the Royal Astronomical Society.

J. Kormendy and G. R. Knapp (eds.), Dark Matter in the Universe, 314.

DISTRIBUTION OF DARK MATTER IN POLAR RING GALAXIES

Bradley C. Whitmore and Douglas B. McElroy
Space Telescope Science Institute

François Schweizer and Vera C. Rubin
Dept. of Terrestrial Magnetism, Carnegie Inst. of Washington

The discovery of S0 galaxies with polar rings makes it possible to directly measure the gravitational potential of a galaxy in three dimensions. Schweizer, Whitmore and Rubin (1983) find a spherical potential in the case of A0136-0801. We have observed three more polar ring galaxies using the 4 meter telescope at CTIO. The following table summarizes the results for these three systems as well as A0136-0801, and figure 1 shows an example of the data.

Galaxy	V_{ring}/V_{disk}	σ_o	V_{disk}/σ_o
A0136-0801	0.94 ± 0.17	67 ± 7	2.3 ± 0.3
NGC 4650A	0.90 ± 0.11	77 ± 8	1.6 ± 0.2
ESO415-G26	1.07 ± 0.13	127 ± 6	1.4 ± 0.2
AM2020-5050	0.32 ± 0.07	159 ± 12	0.4 ± 0.1

For the top three galaxies, a comparison of the rotational velocity of the S0 disk to the perpendicular polar ring at the same radius (V_{ring}/V_{disk}) provides a measurement of the shape of the gravitational potential. The average value of V_{ring}/V_{disk} for these galaxies is 0.97 ± 0.08, corresponding to a nearly spherical shape for the halo.

The low value of V_{disk}/σ_o (where σ_o is the central stellar velocity dispersion) for AM2020-5050 shows that the inner component of this system is probably an elliptical galaxy rather than an S0 disk.

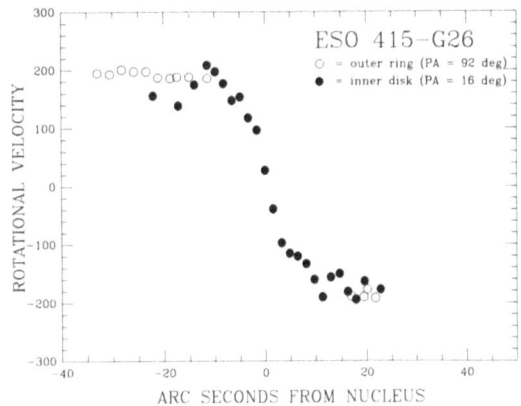

Figure 1 - Rotation curve for both components of ESO415-G26

REFERENCE

Schweizer, Whitmore, and Rubin 1983, A.J., **88**, 909.

J. Kormendy and G. R. Knapp (eds.), Dark Matter in the Universe, 315.

HALOS AROUND ELLIPTICALS AND THE ENVIRONMENT DEPENDENCE OF HUBBLE TYPE

W.H.Zurek, P.J.Quinn and J.K.Salmon
California Institute of Technology &
Los Alamos National Laboratory

I. Shells and Halos :

Shells are often observed around elliptical galaxies(Malin & Carter 1983). Stars in the shells can be regarded as test particles. Therefore for a simple shell morphology(Hernquist & Quinn 1985) the distances between shells allow one to trace the shape of the potential of the host galaxy(Quinn 1984). One can also deduce the distribution of gravitating material. Following this method one concludes that the distribution of luminous matter alone cannot account for the observed shell structure. Depending on the age of the outermost shell one can conclude that either the massive halo terminates at a radius $\simeq 5\ r_e \simeq 15$ kpc., or that the core of the halo $\gamma \simeq 2.5$ kpc, where the halo density is given by $\rho \simeq \{r^2 + \gamma^2\}^{-1}$. In either case, the central density of the halo is no less than an order of magnitude in excess of that expected for a spiral galaxy of comparable luminosity (Figure 1.).

The sample of elliptical galaxies for which shells provide sufficient information about their halos is at present limited. However, unless NGC 3923 is very unusual one is led to conjecture that a typical elliptical has a halo which is far more compact when compared with a spiral with a similar mass.If we accept this correlation between halo parameters and Hubble type as given, we can;(1)Enquire about the origin of this relation and;(2)Attempt to use it to explain other observational facts.

II. Elliptical Halos : Why are they compact ?

In the remainder of this note we accept the idea that galaxies have formed from primordial fluctuations through collapse. This leads one to expect that the baryonic mass of a galaxy(presumably related to its luminosity) is roughly proportional to the amount of dark material in its

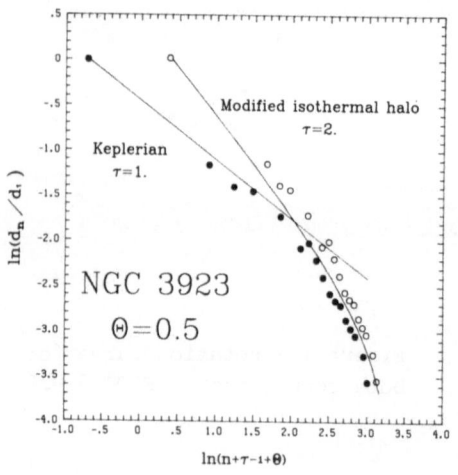

Fig. 1. The distribution of shells around NGC 3923. The outermost shell is labeled 1 and d_1 is its radius. The shell system age is measured in units of the period of the n=1 shell (τ) and Θ =0.5 means the merger occured at zero energy. The shell distribution is NOT consistent with the light distribution of N3923 but requires a halo component that either terminates near $5r_e$ ($\simeq 15$kpc) or has a small core radius depending on the age of the shell system.

J. Kormendy and G. R. Knapp (eds.), Dark Matter in the Universe, 316–317.
© 1987 by the IAU.

halo, since the baryonic and non-baryonic components of the initial
fluctuations have presumably collapsed together. Therefore, it is not
surprising that the baryonic material inside the more compact halos will
tend to form more compact, luminous ellipticals. What needs to be explained
is the difference in the value of the spin parameter (λ). It might be
tempting to speculate that more compact, dense halos have systematically
smaller values of λ. Such an effect is predicted by linear calculations
(Hoffman 1985). Our simulations show that it may exist(Fig.2.) but it
appears to be too small compared to the random scatter of the values of
λ and ρ to be decisive. It is more likely that the baryonic material has
initially similar λ both in the future spirals and ellipticals but compact
halos damp out the λ of the dissipative, baryonic material more readily.

III. The Environment Dependence of Hubble Type

     Given that for galaxies of comparable mass, elliptical halos are
more compact than spiral halos, one can understand the increase in the
relative fraction of ellipticals in those regions of the universe which
have an overall higher galaxy density. To this end let us consider two
galactic size perturbations. Suppose that their relative overdensities
with respect to their neighborhoods is the same, δ. Assume that one of the
two perturbations exists in a region which is also overdense on a larger
scale by a relative factor of Δ.(Such a region may eventually collapse to
form an Abell cluster.) Suppose that the other perturbation is inside a
region underdense by the same factor of Δ. In spite of the fact that the
two perturbations have locally similar sizes, their future fate will differ:
the relative overdensity of one of them is $\delta^+=\delta+\Delta$, while the other has an
overdensity $\delta^-=\delta-\Delta$. If we now assume that the spherical collapse approx-
imation(Peebles 1980)applies, then in an $\Omega=1$ universe the densities of
the objects they will form after collapse will be substantially different:

$$\frac{\rho^+}{\rho^-} \propto \left[\frac{\delta+\Delta}{\delta-\Delta}\right]^3$$

This strongly suggests that the relative abundance of compact halos will
be significantly higher in those regions which are also overdense on a
large scale. A more detailed, quantative discussion of this effect will
be given elsewhere (Zurek & Quinn 1985 in preparation).

Fig.2. The spin parameter λ as a function of
       the average halo density after collapse
       ρ. While there is an indication that λ
       decreases with increasing ρ, such
       systematic changes in λ are smaller
       than its random scatter. These results
       are from Quinn, Salmon and Zurek 1985
       (in preparation).

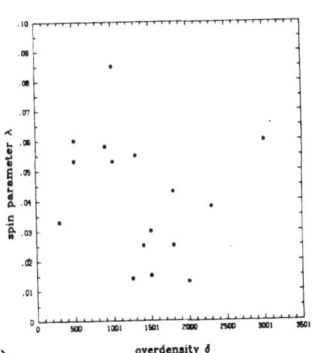

References
Hernquist,L. and Quinn,P.J. 1985 (in preparation).
Hoffman,Y. 1985 Ap.J. in press.
Malin,D.F. and Carter,D. 1983 Nature 285 643.
Peebles,P.J.E. 1980 The Large Scale Structure of the Universe,Princeton
Quinn,P.J. 1984 Ap.J. 279 596.

THE MODIFIED NEWTONIAN DYNAMICS AS AN ALTERNATIVE TO HIDDEN MATTER

Mordehai Milgrom *
Canadian Institute for Theoretical Astrophysics
University of Toronto
Toronto Ont. M5S 1A1 Canada

and

Institute for Advanced Study Princeton N J 08540 USA

Jacob Bekenstein
Department of Physics, Ben Gurion University
Beer-Sheva 84105, Israel

ABSTRACT. The mass discrepancy, which has led to the notion of dark matter may, in fact, be due to a breakdown of the Newtonian laws which are used to determine the masses of galactic systems. We describe a nonrelativistic theory which departs from Newton's in the limit of small accelerations. When one uses the modified dynamics to deduce gravitational masses, the need to invoke large quantitites of dark matter disappears. We outline the theory and give criteria for deciding which systems are expected to exhibit marked departures from Newtonian behaviour. The main body of the talk is a succinct description of the major predictions of the theory regarding dynamics within galaxies.

1. INTRODUCTION

We were asked to discuss an approach which is somewhat outside the mainstream of dark matter studies in that it advocates that dark matter does not actually exist (at least not in quantities as large as are required to bridge the galactic mass discrepancy). Instead we take the view that masses of galaxies and systems

---

* Permanent adress: Department of Physics, Weizmann Institute, Rehovot 76100 Israel

J. Kormendy and G. R. Knapp (eds.), Dark Matter in the Universe, 319–333.

of galaxies are grossly overestimated when they are deduced from Newton's laws (second law and law of gravity).

A modified set of dynamical laws (MOND) is used to describe the motions in galactic systems and, in particular, to obtain the masses of such systems. We find that the mass discrepancy disappears and a number of the observed traits of galaxies are unavoidable consequences of MOND.

The nature of MOND and its implications have been described in detail in a series of papers (Milgrom 1983a,b,c, Bekenstein and Milgrom 1984 and Milgrom 1984, 1985a,b, referred to hereafter as papers I-VII respectively). We thought it fit to concentrate, in the present talk, on the main predictions which MOND makes concerning dynamics within galaxies. We shall list these predictions and discuss each of them briefly. A more detailed discussion can be found in the references.

We would also like to bring to your attention the fact that there have been suggestions to explain away the mass discrepancy by adopting, unlike MOND, modified forms of the distance dependence of the gravitational force (e.g., Finzi 1963, Tohline 1983, Sanders 1984).

2. THE NONRELATIVISTIC FORMULATION

2.1 The Basic Postulates

Most of the major predictions of MOND follow from the following assumptions (Papers I, II).

(i) Newtonian dynamics breakdown when the accelerations involved are small.

(ii) The acceleration a of a test body at a distance r from a mass M is given by $a^2/a_0 \approx MGr^{-2}$ in the limit $MGr^{-2} \ll a_0$ (or $a \ll a_0$) .

Here a_0 is an acceleration constant which plays both the role of a transition acceleration from the Newtonian to the Non-Newtonian regime and the role of a proportionality constant in the modified equation of motion. The value of a_0 was determined (Paper II) to be about $2 \times 10^{-8}(H_0/50\,\mathrm{km\,s^{-1}\,Mpc^{-1}})^2\,\mathrm{cm\,s^{-2}}$. Interestingly, this value is very close to that of cH_0 .

2.2 The Theory We Now Use

Of the various interpretations of the basic assumptions and theories which may incorporate them, we have found the following the most appealing thus far (Paper IV). It is assumed that MOND signifies a breakdown of the Newtonian law of gravity (leaving the 2nd law intact). The gravitational acceleration field $\vec{g}$ is still taken to be derivable from a potential $\vec{g} = -\vec{\nabla}\varphi$. However, φ is now related to the density distribution ρ which induces it by the following field equation (derivable from a Lagrangian):

$$\vec{\nabla} \cdot [\mu(g/a_0)\vec{g}] = -4\pi G\rho, \tag{1}$$

instead of the Poisson equation. Here $\mu(x) \approx 1$ for $x \gg 1$ so that Poisson's equation is restored in the limit $g \equiv |\vec{g}| \gg a_0$, but $\mu(x) \approx x$ for $x \ll 1$ so that the desired low acceleration behaviour is obtained. Equation (1) is supplemented by the boundary condition at infinity: $\vec{g} \to 0$ for an isolated system and $\vec{g} \to \vec{g}_\infty$ for a system in a constant external acceleration field $\vec{g}_\infty$. Other than these requirements (and the monotonicity of μ which we always require) μ has remained undetermined.

We do not yet have a satisfactory relativistic extension of MOND.

2.3 A Simplified (approximate) Formulation

The theory given by eq.(1) is nonlinear and practically impossible to solve exactly for all but the simplest configurations. By eliminating ρ between eq.(1) and the Poisson equation for the Newtonian acceleration field $\vec{g}_N$, $(\vec{\nabla} \cdot \vec{g}_N = -4\pi G \rho)$ we get $\vec{\nabla} \cdot [\mu(g/a_0)\vec{g} - \vec{g}_N] = 0$. Equating the field in parentheses to zero (when in fact it is in general a non-zero curl field)

$$\mu(g/a_0)\vec{g} = \vec{g}_N, \tag{2}$$

seems to give a very good approximation for the field $\vec{g}$ when test particle motion is considered (see e.g. Paper VI). For example we find (paper VI) that a galaxy's rotation curve derived from eq.(2) differs by at most five percent from that which is derived from the exact eq.(1) for the many galaxy models which we have tried. Equation (2) is the formulation originally used for MOND in Papers I-III, and it is exact when the system has a plane cylindrical or spherical symmetry. The solution of eq.(2) is straightforward for an arbitrary mass distribution.

2.4 Some General Properties of the Field Equation

We have derived in paper IV the following results for systems which are governed by eq.(1):

(i) A system of gravitating masses with local accelerations given by the solution of eq.(1) conserves energy, momentum, and angular momentum.

(ii) A small, low mass, object in the field of a large massive body is accelerated like a test particle (irrespective of whether the accelerations within the object are large or small). Thus, for example, stars, binary stars, globular clusters etc. may, to a very good approximation, be regarded as test bodies when their motion in the field of a galaxy is considered.

(iii) The motions (relative to the c.o.m.) within a system s, which itself is in a field of a mother system S, are affected by the external field (Papers I, IV,VI) of S when the latter is not negligible compared with the internal acceleration. Thus let m and r be the mass and average radius of s and M and R those of S. Comparing the (Newtonian and thus not the actual) accelerations $g_{in}^N \equiv mG/r^2$ and $g_{ex}^N \equiv MG/R^2$ with each other and each of them with a_0 will help us decide

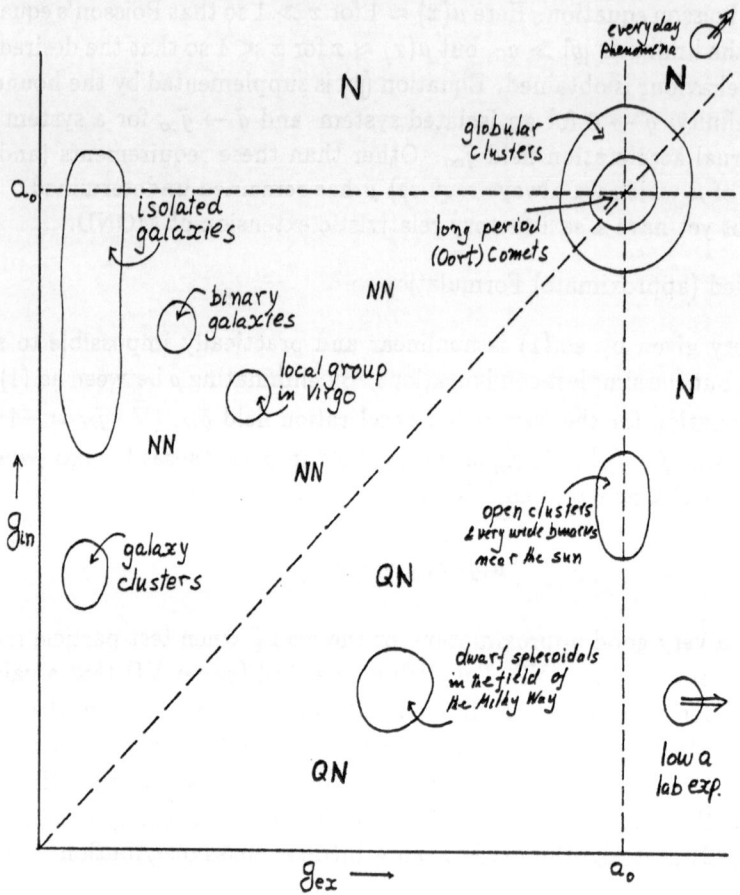

Fig. 1. A classification of various galactic systems according to their intrinsic and c.o.m. accelerations.

whether to expect strong departures from Newtonian laws in the internal dynamics of s . We should, strictly speaking, use the actual accelerations g_{in} and g_{ex} but these cannot be written a priori in simple terms of the sizes and masses. However, all strong inequalities between g_{in}, g_{ex} and a_0 are the same as those between g_{in}^N, g_{ex}^N and a_0. Figure 1 shows schematically where various systems of interest fall in the g_{in}, g_{ex} plane.

The general rules which apply when the inequalities between g_{in}, g_{ex} and a_0 are strong are as follows:

(i) When either $g_{in} \gg a_0$ or $g_{ex} \gg a_0$, the dynamics within s are Newtonian (region marked N in Figure 1).

(ii) When $g_{ex} \ll g_{in} \ll a_0$, the system is approximately isolated (in the MOND sense) and the small acceleration limit of MOND applies (region marked NN in Figure 1).

(iii) When $g_{in} \ll g_{ex} \ll a_0$, the dynamics are quasi-Newtonian but with a value of the effective gravitational constant being $G_{eff} = G/\mu(g_{ex}/a_0) \gg G$ (see detailed discussion in Paper VI). This case corresponds to the region marked QN in Figure 1.

3. PREDICTIONS WHICH ARE INCOMPATIBLE WITH DARK MATTER

One may ask whether all the predictions of MOND can be mimicked with hidden mass, maintaining Newtonian dynamics. This, as we shall now demonstrate, is not the case.

3.1 Negative "Dark Matter"

If a density $\rho(\vec{r})$ gives rise to an acceleration field $\vec{g}(\vec{r})$ according to MOND, the only way we could make the measured accelerations in this field consistent with Newtonian dynamics is by assuming that the actual density distribution is $\rho^* = -(4\pi G)^{-1}\vec{\nabla} \cdot \vec{g}$. However, it can be shown (Paper VII) that for various configurations $\rho^* < \rho$ or even $\rho^* < 0$. Insisting on describing such systems with Newtonian dynamics will imply that the system contains negative mass densities which is unacceptable. For example in any binary galaxy system there is a region (shown in Figure 2 for galaxies of equal mass) where one will find negative density if he insists on using Newtonian dynamics to explain the measured acceleration field.

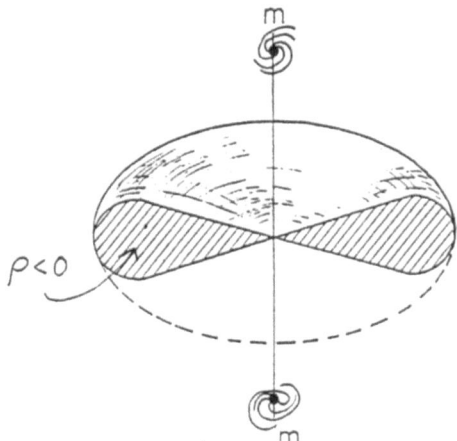

Fig. 2. The region of negative Newtonian density in a binary galaxy system.

3.2 Breakdown of the Strong Equivalence Principle (SEP)

MOND does not satisfy the strong equivalence principle (SEP) even in the nonrelativistic regime. Any observed manifestation of this fact will point to the breakdown of Newton laws. Examples are discussed in Papers IV and VI. For instance, a self-gravitating many-particle system (such as a gas cloud) with isotropic pressure (or velocity dispersion) in an external field $\vec{g}_{ez}$ and with $g_{in} < g_{ez} < a_0$ will not be spherical (as in the Newtonian case) even when $\vec{g}_{ez}$ is exactly constant. It will be an ellipsoid of revolution with its long (symmetry) axis along $\vec{g}_{ez}$, in conflict with the SEP.

3.3 Light Bending

As we do not yet have a relativistic generalization of MOND, we cannot predict the nature or strength of light bending in a gravitational field. There is however no reason we can think of why the same fictitious mass distribution will be required to explain the trajectories of light rays and those of massive particles in the field of an object such as a galaxy. In fact, in a toy relativistic model we studied, an attempt to explain light bending and say the rotation curve of a massive body (in the regime $g \ll a_0$), assuming the conventional dynamics, will fail.

4. OTHER PREDICTIONS CONCERNING GALAXIES

Many of the consequences of MOND which we list below involve galaxy properties that have already been observed. However, it has not been known how strong and general these observed characteristics are and, for that matter, what their exact nature is. These observations do not conflict with the dark matter hypothesis and, in fact, have been taken to reflect various properties of dark halos (for example, asymptotically flat rotation curves are interpreted as resulting from a r^{-2} behaviour of the dark halo's density law).On the other hand, those regularities in galaxy appearance are also not predictions of the dark matter hypothesis.In MOND they are exact general and unavoidable predictions.

4.1 Disc Galaxies

4.1.1. <u>Rotation curves.</u> The rotation curve of a galaxy deduced from the "observed" mass distribution using MOND should agree with the observed rotation curve. There are sub-predictions of this general one which do not require the full knowledge of the galaxy's rotation curve or mass distribution (see paper II).

(i) The velocity of a test body in a circular orbit around an isolated galaxy should become independent of the radius of the orbit at large radii.

(ii) The asymptotic circular velocity V_∞ depends only on the total mass M of the galaxy via $V_\infty^4 = a_0 GM$.

(iii) In high rotational velocity galaxies (such that $V_\infty^2/h > a_0$, where h is the galaxy's scale length), the local M/L value (as deduced from Newton's laws) should be constant at small radii and then start to increase around the radius where $V^2/r = a_0$.

(iv) Very low surface density (LSD) galaxies are particularly good test cases because of the following reasons:

a. When the average surface density is very small $\langle \Sigma \rangle \ll \Sigma_0 \equiv a_0 G^{-1}$, the accelerations are much smaller than a_0 and hence we predict large departures from Newtonian behaviour.

b. LSD'S tend to be bulgeless so there are fewer parameters involved (Carignan and Freeman 1985).

c. Since we are dealing with a system where $g \ll a_0$ everywhere we do not require the exact form of $\mu(x)$ and it is a good approximation to use $\mu(x) = x$.

d. All the uncertainties involved in comparing calculated and measured rotation curves (galaxy's distance, inclination, extinction, M/L, a_0 etc.) lump into one multiplicative factor and the test of MOND which such galaxies offer is much more clear-cut. [In Newtonian dynamics the dimensionless rotation curve $v(r) = V(r)/V_\infty$, depends only on the mass distribution in the galaxy but not on the total amount of mass say. In MOND $v(r)$ also depends on an additional parameter, say the average surface density. For instance, two pure exponential disc galaxies may have very different - looking rotation curves (see e.g. the model curves in paper II). The point we are making here is that when $\langle \Sigma \rangle \ll \Sigma_0$ there exist some similarity laws which eqs.(1)(2) obey and which make $v(r)$ depend on the mass distribution only. All this is discussed in detail in papers II,VI].

On the other hand, LSD disks tend to contain relatively large quantities of hydrogen which may contribute subtantially or even dominantly to the radial acceleration. Such cases may thus involve additional parameters and they provide less of a clear-cut test.

As a mere demonstration we give in Figure 3 the measured and calculated rotation curves for three galaxies. The data for NGC 3198 are taken from van Albada et al.(1985). This galaxy has one of the cleanest and the furthest reaching rotation curves to date (in optical radii). It is a relatively low-acceleration galaxy (maximum acceleration is $\approx .5a_0$). The data for NGC 247 and NGC 300 are taken from Carignan and Freeman(1985). These galaxies may be considered very low acceleration ones. In calculating the MOND curves we used the approximate eq.(2), assumed pure exponential discs, and used the values of the scale length s given in the above references. The choice of $\mu(x)$ does not affect the fit much. We have only one free parameter for each galaxy. This may be taken as the value of M/L assuming that all the others are as given by the observers (they all lump into one factor anyway when $g \ll a_0$). The value of M/L used is given beside each curve.

4.1.2. Surface densities. The constant a_0 defines a quantity with the dimension of mass surface density $\Sigma_0 \equiv a_0 G^{-1}$ which we predict to play an important role in galaxy dynamics. When a galaxy has an average surface density $\langle \Sigma \rangle \gg \Sigma_0$, its dynamics will be Newtonian out to large radii (compared say with the half-mass

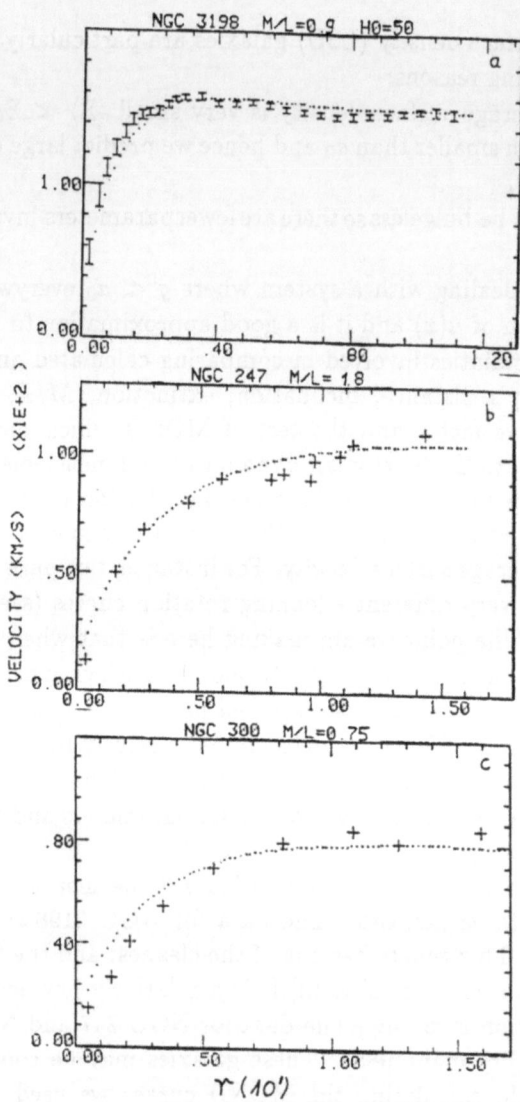

Fig. 3 MOND rotation curves compared with observations for three galaxies.

radius). In this case there will be a range of radii with a Keplerian decline of the rotational velocity before it reaches the asymptotic value V_∞. We found (Paper II) that for the velocity curve to remain approximately flat down to small radii we must have $\langle\Sigma\rangle \approx \Sigma_0$. When $\langle\Sigma\rangle \gg \Sigma_0$, the rotation curve should exhibit an appreciable hump. When $\langle\Sigma\rangle \ll \Sigma_0$, the velocity rises slowly, peaks at a few scale length s, and then decreases by a few percent to its asymptotic value. Thus *if galaxies are not observed to have considerbly humped rotation curves they should all have* $\langle \Sigma \rangle \lesssim \Sigma_o$.

4.1.3. The Oort discrepancy. Further predictions can be made which concern the dynamics of motions perpendicular to the plane of a thin planar galactic disc. Recent results of such analysis near the sun by Bahcall (1984) strengthen results from earlier reports starting with Oort(1960) showing that, near the sun, the dynamically determined mass is larger than that which is accounted for by known components. A detailed analysis of the expected fictitious "dark matter" distribution in very thin discs is given in Paper VII. Here we give only the predictions which are based on the further approximations that a. The accelerations perpendicular to the plane are small compared with the radial acceleration. b. The density in the disc is large compared with the density of the galaxy averaged within the galactic radius r ($\rho \gg M/4\pi r^3$). Both approximations are good near the sun up to a height of a few hundred parsecs above the galactic plane. Under these conditions we find that the dynamics are Newtonian but with an effective gravitational constant $G_{eff} = G/\mu(V_\odot^2/r_\odot a_0)$, where $V_\odot$ and $r_\odot$ are the galactic orbital velocity and radius at the sun's position. Thus we predict that when the above approximations are valid:

(i) The distribution of disc "dark matter" will be found to be the same as that of the visible mass.

(ii) The Oort discrepancy factor, which according to MOND is $1/\mu(V_\odot^2/r_\odot a_0)$, is the same as that for the total galactic mass discrepancy within the orbit of the sun.

(iii) The same factor appears (albeit multiplied by an additional parameter of order unity) in the dynamics of open cluster or very wide binaries in the solar neighbourhood (see Paper IV on the asymptotic field of a mass in a constant external field and also Paper VI on an N-body system in an external field).

4.2 Elliptical Galaxies

We work on the premise that galaxies contain no appreciable quantities of dark matter. We should thus be able to understand elliptical galaxies' light distribution and velocity dispersions selfconsistently. Unfortunately our analysis is beset by the same uncertainties which stand in the way of conventional analyses, i.e., those involved in deducing space mass distributions from surface brightnesses and velocity distributions from line-of-sight velocity dispersions. We can follow one of the two avenues which others before us have taken. We can make some assumptions about the stellar distribution function leaving certain parameters which specify it free. One then asks how such model systems look if they obey MOND instead of the Poisson equation, and to what extent they resemble the astronomical systems which they purport to represent.

It has been popular with model makers to describe ellipticals as some variety of isothermal spheres. We have studied in paper V self-gravitating, many-particle, spherical systems with radius independent radial and tangential velocity dispersions, assuming MOND. The following are the major traits of such models which are independent of the values of the parameters which determine their exact structure (velocity dispersion,ratio of radial to tangential dispersion, etc.).

(i) All such spheres have a finite mass (unlike their Newtonian cousins) and their density distribution tends to a power law asymptotically $\rho(r) \to r^{-\alpha}$, with $\alpha > 3$.

(ii) The surface density constant Σ_0 introduced earlier is an upper limit on the average surface density which such isothermal spheres can have.

(iii) The total mass M of a sphere is approximately proportional to the fourth power of the space velocity dispersion σ_s:

$$(a_0 G)^{-1} \leq M/\sigma_s^4 \leq 2(a_0 G)^{-1}. \tag{3}$$

Alternatively we may simply try to map the test-particle acceleration field of ellipticals and see if it agrees with that calculated from the observed light distribution, with reasonable M/L values, using MOND. In this conection we can make the following predictions:

(i) In ellipticals with test-particle gas discs the rotation curve of the disc will be that which MOND dictates for the observed light distribution.

(ii) Dwarf ellipticals or spheroidals with $\langle \Sigma \rangle \ll \Sigma_0$ will be found to contain large quantities of dark matter when treated Newtonically (see more details in paper II).

(iii)The observed temperature and density distributions in x-ray-emitting envelopes of ellipticals (such as described by Forman et al. 1985) will be those given by MOND (straightforwardly deduced from eq.(2), which is exact for spherical systems, or numerically from eq.(1) in the more general case).

The M/L values one obtains in this way, for individual galaxies from the existing data are rather uncertain. But it is interesting to see if there is a correlation of the Newtonian M/L values with the accelerations. We plot in Figure 4 Forman et al.'s values of M/L against the (Newtonian) accelerations g_N at R_{max}(value of radius where the M/L values are estimated, taken from Forman et al.) We use g_N rather than g because the first depends only on observed quantities and not e.g. on the form of $\mu(x)$ or the value of a_0. We take $g_N = L(M/L)_t G/R_{max}^2$. Here $(M/L)_t$ is an assumed stellar value of M/L.

The values of M/L given by Forman et al. are based on a single temperature of $T = 1$ KeV for all the galaxies. For some, an actual uncertainty range of T is given (in some cases not including 1 KeV).We also plot the range of M/L values which corresponds to the temperature uncertainties.

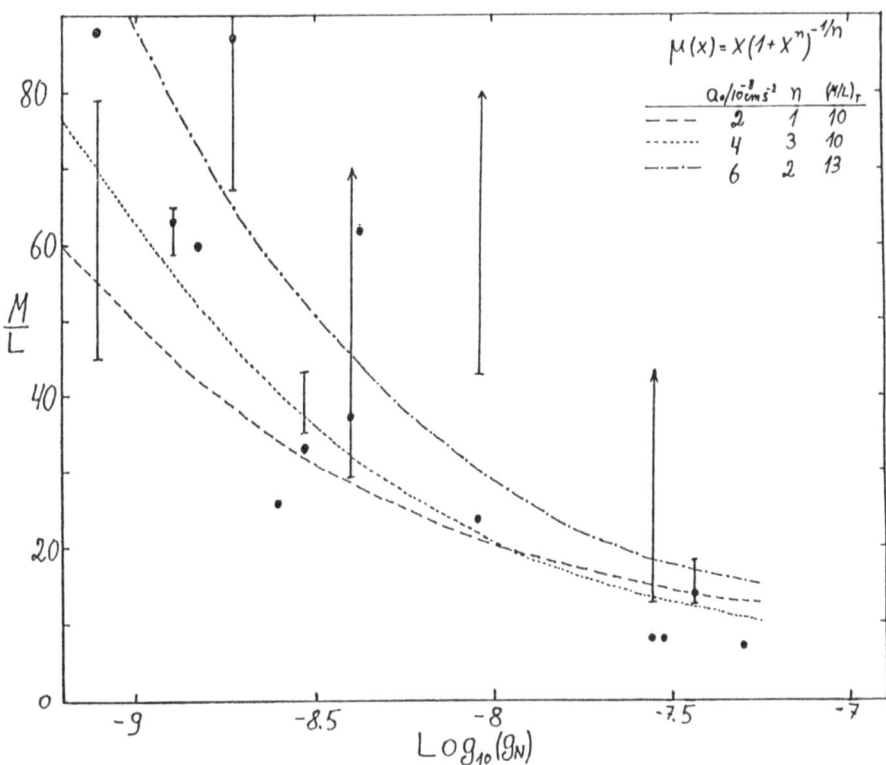

Fig. 4 Newtonian M/L values (as given by Forman et. al.1985) plotted against the Newtonian accelerations for early-type galaxies with x-ray envelopes ($\bullet$). Vertical bars indicate the range of M/L corresponding to the envelope's temperature uncertainty range. The lines show the predictions of MOND for point-mass galaxies with stellar M/L value $(M/L)_t$ for different forms of $\mu(x)$ and values of a_0.

REFERENCES

Bahcall, J.N. 1984. *Astrophys. J.* **276**, 169.

Bekenstein, J., and Milgrom, M. 1984. *Astrophys. J.* **286**, 7. Paper IV

Carignan, C., and Freeman, K.C. 1985. *Astrophys. J.*, In press.

Finzi, A. 1963. *Mon. Not. Roy. Astron. Soc.* **27**, 21.

Forman, W., Jones. C., and Tucker, W. 1985. preprint.

Milgrom, M. 1983a. *Astrophys. J.* **270**, 365. Paper I

Milgrom, M. 1983b. *Astrophys. J.* **270**, 371. Paper II

Milgrom, M. 1983c. *Astrophys. J.* **270**, 384. Paper III

Milgrom, M. 1984. *Astrophys. J.* **287**, 571. Paper V

Milgrom, M. 1985a. preprint, Paper VI

Milgrom, M. 1985b. preprint, Paper VII

Oort, J.H. 1960. *Bull. astr. Inst. Netherlands.* **15**, 45.

Sanders, R.H. 1984. *Astron. and Astrophys.* **136**, L21.

Tohline, J.E. 1983. In IAU Symp. No. 100 *Internal Kinematics and Dynamics of Galaxies.* Ed. Athanassoula (Dordrecht: Reidel).

van Albada, T.S., Bahcall, J.N., Begeman, K. and Sancisi, R. 1985. *Astrophys. J.* in press.

DISCUSSION

SCHECHTER: Would you tell us something about wide binaries?

MILGROM: For wide binaries, the acceleration is $< a_0$; in fact, the acceleration around a 1 $M_\odot$ star becomes equal to a_0 at a fraction of a parsec, $\sim$ 5000 AU radius. On the other hand, these binaries are in the solar neighborhood, so even if you make the separations very large, the size of the mass discrepancy is limited by the acceleration in the Galaxy. So you expect to find a mass discrepancy which is of the order of the Oort limit.

J. BAHCALL: Perhaps Scott Tremaine could say what is required observationally to test the modified dynamics using wide binaries.

TREMAINE: The Milgrom-Bekenstein theory predicts that there should appear to be "dark matter" in wide binary stars in the solar neighborhood. To measure the masses of wide binaries, you would need velocity measurements of $\sim 10^2$ binaries with a resolution of 0.1 km s^{-1} or better. The mass measurement is subject to most of the problems which plague mass measurements in binary galaxies; in addition, you have to worry about velocity perturbations due to dark companion stars.

OSTRIKER: What happens when we apply your formulation to the Local Group? It should apply, because the motions are non-relativistic and the Local Group is quite isolated. When I looked at this, I had trouble getting the orbits of Andromeda and the Galaxy to come out right in the time available.

MILGROM: So I'll answer you for the fifth time, Jerry, just for the record. What Jerry is saying is that, if we assume that M31 and the Galaxy are on radial orbits toward each other, and if we try to calculate, assuming modified dynamics, how long ago the Galaxy and Andromeda were close together, we find a few $\times$ 10^9 years, considerably less than the Hubble time. Jerry worried that this might be destructive to the two galaxies. My standard answer is that if the systems have a tangential velocity of only 50 - 60 km s^{-1}, they would never have gotten closer together than $\sim$ 150 kpc.

OSTRIKER: I just wanted to give you the opportunity to say it to this large audience (laughter).

VAN DER KRUIT: As a specific case of what Vera Rubin illustrated earlier, I would like to mention the two edge-on disk galaxies NGC 7814 and NGC 891. These have comparable distances and angular sizes. Between 1' and 6' radius, both systems have flat rotation curves of 220 km s^{-1} amplitude. Despite this, the light distribution of NGC 7814 consists almost entirely of a centrally concentrated $r^{1/4}$-law spheroid and NGC 891 almost entirely of a much more distended exponential disk. I do not understand how any gravitational law can give identical rotation curves if these light distributions trace mass distributions.

MILGROM: Vera mentioned two apparently odd phenomena: galaxies with similar light distributions but different rotation curves and galaxies like yours with different light distributions but similar rotation curves. The latter is actually not odd at all. You can have a spherical mass and a thin disk with arbitrary spheroid-to-disk mass ratio and with exactly the same rotation curve. This is true in Newtonian mechanics as well as in MOND. The first phenomenon, however, is truly puzzling. If you stick to Newtonian dynamics, you have to say that the two galaxies must have very different halos resulting in different rotation curves. MOND predicts exactly such different rotation curves for galaxies with similar light distributions, as I have explained in connection with the surface density constant Σ_0. You can see this effect clearly in Figure 2 of my paper on galaxies (Paper II of the references).

FELTEN: I think it should be mentioned that a theory like this causes big problems for cosmology.

MILGROM: It is not true at all that a theory like ours has problems with cosmology. What is true is that we cannot derive cosmology using Newtonian arguments, so we have to await a relativistic generalization of MOND. I do not forsee any potential problems.

FABER: Without a theory of relativity, it is not fair to ask you to do cosmology, but to what distances do you think you can apply your formulation? In particular, can you apply it to the Local Supercluster?

MILGROM: The formulation can be applied when the mass contained in the region of overdensity you're considering is large compared with the expected average mass within the same volume. The Local Supercluster is such a case.

FABER: Why is that a good criterion?

MILGROM: Because I can then assume that the acceleration is primarily determined by the mass that is actually there in the region of overdensity. Of course, another criterion is that velocities should be non-relativistic, which is certainly the case in the Local Supercluster.

FABER: So the Virgo Cluster meets your requirements.

MILGROM: Yes. I haven't kept up with the new developments, but when I wrote my first paper I analyzed the then-existing information on Virgocentric infall, and found M/L $\sim$ 1.

DAVIS: I don't think that the Virgo Cluster has a big mass overdensity.

MILGROM: Then the corresponding uncertainty in M/L is a factor of order 2. There are other factor-of-two uncertainties also, such as any non-Virgocentric acceleration components due to matter outside Virgo.

WHITE: The large hydrostatic atmospheres in galaxy clusters offer an interesting test of your theory. The x-ray data provide a good lower limit on the observed M(r), namely the observed mass in gas. The accelerations in clusters place them just inside your weak-acceleration regime, and your theory makes a prediction for the velocity dispersion of the galaxies and for the temperature of the gas. Have you made any models for the structure of these systems to check that you can get agreement with observations for the same value of a_0 that you need in galaxies?

MILGROM: No, we haven't looked at this question.

ISHIZAWA: An action principle has been applied to the gravitational model by Bekenstein and Milgrom (1984, Ap. J., 286, 7). Starting from the action of classical general relativity with a negative cosmological constant Λ in the weak limit of gravitation, we have obtained a system of the Poisson equation and natural boundary conditions. They are completely equivalent to an incompressible, irrotational flow surrounded by a constant-pressure gas. As a natural consequence, our model leads to a two-phase Universe, the gravitational channels in which the gravitational lines of force are confined (filamentary matter fields) and the gravitational vacuum (voids).

VISHNIAC: I wonder if you could comment on the constraints imposed on your theory by laboratory experiments designed to measure the gravitational constant.

MILGROM: A system which is embedded in a strong external acceleration field such as that of the Earth ($g \sim 10^{11} a_0$) is very nearly Newtonian. The deviation can be expressed by a slightly larger effective gravitational constant, $G_{eff} \approx G/\mu(10^{11})$. We have no idea how fast $\mu(x)$ approaches unity when its argument becomes large, but there are limits which we can put from the perihelion shift of Mercury and other solar-system measurements (Paper I). These imply that detecting the effect in laboratory experiments is far beyond present capabilities.

GRAVITATIONAL CHANNELS AND THE COSMOLOGICAL CONSTANT Λ

T. Ishizawa
Department of Astronomy, University of Kyoto

A new model of filamentary matter fields and voids is proposed. This is a gravitational version of the MIT bag model of hadrons (see a review of DeTar and Donoghue 1983). Bekenstein and Milgrom(1984) have first proposed a gravitational bag model. Their bag is closed but our bag, called a channel, is open.

We start from the action in the weak limt of gravity

$$I = \int_{-\infty}^{\infty} dt \int_{V} dx \left\{ \sum_{n} m_n \left[\tfrac{1}{2} \dot{x}_n^2 - \phi(x_n) \right] \delta(x - x_n) - \frac{1}{8\pi G} (\nabla \phi)^2 + B \right\}, \qquad (1)$$

where ϕ is the gravitational potential and $B = -c^4 \Lambda/(8\pi G)$. Varying $x_n(t)$, $\phi(x)$ and V under the conditions that $\delta x_n(t) \to 0$ as $|t| \to \infty$ and that $\delta\phi(x) \to 0$ as $|x^i| \to \infty$, we obtain the equations of motion for masses, Poisson equation for ϕ

$$\nabla^2 \phi = 4\pi G \sum_{n} m_n \delta(x - x_n), \qquad (2)$$

and the natural boundary conditions at the free surface (S_1 in Figure 1)

$$\frac{\partial \phi}{\partial n} = 0 \quad \text{and} \quad \frac{\partial \phi}{\partial \tau} = U, \qquad (2)$$

where $U = (8\pi GB)^{1/2} = c^2 (-\Lambda)^{1/2}$.

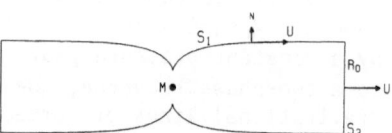

The condition $\frac{\partial \phi}{\partial n} = 0$ means that the lines of force are confined within the channels, i.e. filamentary structures. The potential ϕ is equivalent to the velocity potential in the incompressible irrotational flow

Figure 1. An axisymmetric channel of two opposite equal fluxes from a point mass.

which is surrounded by a gas of constant pressure and has the sources of total flux $4\pi Gm_n$ at $x = x_n$. The radius r_0 of the distant cross-section (S_2 in Figure 1) is obtained from Gauss's Theorem $2\pi r_0^2 U = 4\pi Gm$, so that $r_0 = (Gm^2/2\pi B)^{1/4}$.

The exact solution for the two-dimensional channel from a point mass has been obtained using the standard method of conformal mapping (Milne-Thompson 1938). The force F on a test particle at the center line changes from $F \propto 1/r$ (two-dimensionally Newtonian) to F=const. at a distance of the half-width of the channel. This would be the case with the three-dimensional channel from a disk galaxy.

Assuming that the law of force is $-Gm_1 m_2 (1 + r_{12}^2/A^2)^{1/2}/r_{12}^2$ between two masses in the disks of nearby spirals (M31, M33, NGC2403, NGC6946), we have found, from fitting the computed and observed rotation curves, that the scale length A satisfies a close relation $A = 2.7\text{kpc}*(m/10^{10}M\odot)^{1/2}$ where m is the total mass of a spiral. The acceleration $Gm/A^2 = 1.9 \times 10^{-8}$ cms^{-2} is coincident with Milgrom's(1983) limiting acceleration if $H_0 = 50$ kms^{-1}Mpc^{-1}. Equating $r_0 = A$, we have $\Lambda = -1.8 \times 10^{-57}$cm^{-2}.

References
Bekenstein, J. and Milgrom, M.:1984, Astrophys. J. 286, 7.
Milgrom, M.: 1983, Astrophys. J. 270, 365, 371, and 384.

J. Kormendy and G. R. Knapp (eds.), Dark Matter in the Universe, 334.
© 1987 by the IAU.

GALAXY FORMATION

Joseph Silk
Astronomy Department
University of California
Berkeley, California 94720 U.S.A.

ABSTRACT. Theories of galaxy formation via hierarchical clustering of cold dark matter and by fragmentation of gaseous pancakes or shells are reviewed and compared. Dissipative processes are crucial to all theories of galaxy formation, and are discussed in terms of a simple model involving multiple cloud interactions. Stellar energy input is found to play an important role in protogalaxies, both supporting the gas against collapse, thereby prolonging the duration of the active star formation phase, and driving winds from the less massive galaxies. The significance of such processes is explored both for chemical and dynamical evolution and for biasing galaxy formation towards density peaks in the primordial density fluctuation spectrum.

1. INTRODUCTION

What determines the masses and the binding energies of galaxies? These issues are the principal focus of galaxy formation theory. Once these are resolved, more detailed models can be constructed which may (or may not!) approximate the structural, morphological, dynamical, and chemical properties of galaxies.

There are two extreme viewpoints concerning the origin of galactic binding energy. It may derive from the cosmological initial conditions. Primordial curvature fluctuations, laid down at the inflationary epoch some 10^{-35} s after the big bang, develop adiabatically, and generate growing density fluctuations within the particle horizon on galactic and subgalactic scales. Small scales go nonlinear first, at redshift $z \sim 30$, and cluster hierarchically, merging together to form larger and larger systems. Gravitational clustering of weakly interacting dark matter determines the halo scale, and galaxies develop by baryonic infall into dark potential wells (Peebles 1984; Blumenthal *et al.* 1984). This scheme reduces to a resurrection of galaxy formation from primordial isothermal fluctuations (Peebles and Dicke 1968). The binding energy of a galaxy is effectively imprinted in the initial conditions.

In an alternative scenario, all primordial fluctuations on galactic and subgalactic scales are suppressed (Silk 1968; Bond *et al.*). Only fluctuations survive

335

J. Kormendy and G. R. Knapp (eds.), Dark Matter in the Universe, 335–359.

on galaxy cluster scales, collapsing anisotropically to form thin sheets or pancakes which subsequently fragment (Zel'dovich 1970). Dissipative processes are largely responsible for the binding energies of the galaxies that form (Larson 1975; Carlberg 1984). Galaxy halos are presumed to develop by secondary infall, if weakly interacting dark matter dominates the mass density of the universe (Gunn 1977). In a variation on this scheme, rare seed fluctuations of subgalactic mass explode, sweeping up a shell of matter that fragments on scales of mass greatly exceeding that of the initial seeds (Ostriker and Cowie 1981; Ikeuchi 1981; Carr and Ikeuchi 1985). Both pancake and explosive amplification schemes form galaxies by much the same process of dissipative fragmentation.

Although the physical processes that dominate hierarchical clustering and pancake fragmentation, namely gravitational clustering and N–body dynamics, as opposed to dissipative hydrodynamics and gaseous shocks, seem very different, one ends up forming the luminous cores of galaxies in much the same manner. Dissipation dominates in luminous cores, as is evident from the simple observation that galaxies have a much higher surface brightness than galaxy clusters. Put more quantitatively, the self–similar clustering hierarchy that describes the large–scale galaxy distribution is broken on the scale of galaxies. There is a unique scale where dissipation sets in, determined by the comparison of free–fall collapse and cooling time–scales Rees and Ostriker 1977; Silk 1977). The former is proportional to $n^{-1/2}$ and the latter is approximately proportional to $T^{-3/2}/(nZ)$, where n is mean gas density, Z is metallicity, and T is gas temperature, over the range 10^5–10^7 K and $Z \gtrsim 0.01\ Z_\odot$. Equating these two time–scales yields a critical mass of order $100\ \alpha^5 \alpha_g^{-2}$, for a primordial abundance gas, or more generally $\sim 10^{68}(Z/0.01\ Z_\odot)$ baryons, above which no cooling occurs. Here $\alpha = 1/137$ and $\alpha_g = Gm_p^2/\hbar c = 6 \times 10^{-39}$. By way of comparison, a similar argument utilizing pressure support yields the characteristic mass of a star as $\sim \alpha_g^{-3/2} \sim 10^{57}$ baryons.

These considerations might lead one to think that one could actually predict the existence of stars and of galaxies from fundamental physical principles. While this argument may be valid for stars, it is almost certainly specious when applied to galaxies, for the following reason. Solving the galaxy formation problem requires specification of initial conditions, which are then evolved forward in time according to the equations of cosmology, N–body dynamics and hydrodynamics. We have no a priori knowledge of initial conditions, although it is perhaps not inconceivable that quantum gravity may eventually provide them. For the present, however, one has to work within the framework of preconceptions or prejudices about the initial conditions. One can then test the resulting theoretical models of galaxies against the real universe, to see whether one's starting point has any possible validity.

To commence, I contrast the hierarchical clustering and fragmentation theories (§II). I then develop a simple model of dissipative galaxy formation, common to either theory, involving multiple cloud interactions (§III). The role of stripping by winds in chemical and dynamical evolution is described in §IV, and general implications of dissipative models are discussed in §V.

II. CLUSTERING OR FRAGMENTATION?

Galaxy formation scenarios are not sufficiently well defined that one can speak of any two, or even more, principal theories. There are innumerable variations, but as previously mentioned, one can at least discern the two main themes of hierarchical clustering and gaseous fragmentation, around one or other of which most theories are developed. To be specific, I shall choose the cold dark matter and pancake fragmentation theories, and discuss how they cope when confronted with observational and theoretical realities.

One of the more contentious issues concerns the epoch of galaxy formation. A possibly fatal blow against the pancake theory in a massive neutrino–dominated universe arises because the galaxy correlations on large scales can only be fit if pancaking occurred recently, at redshift $z \lesssim 1$, in which case galaxies are uncomfortably young (White et $al.$ 1983). Very few, if any, galaxies can have formed at $z \gtrsim 3$, despite the fact that most luminous quasars are found at high redshift. Moreover, galaxies observed at $z \sim 1$, inevitably very luminous ellipticals, appear to contain old stellar populations that must have formed much earlier, certainly at $z > 3$. The one ray of hope for pancake theories is that one can readily imagine variations on these theories that form rare objects early, with the bulk of galaxies forming late. The formation of most galaxies as recently as $z \sim 1$ cannot be excluded. Such variations presently seem rather ad hoc, involving speculations about isothermal seeds, strings, or an admixture of hot and cold dark matter. Even without such drastic measures, there are always a few, albeit very few, pancakes that collapse directly to form galaxies even in a massive neutrino–dominated universe. Hence it is premature to dismiss massive neutrinos, at least until the laboratory mass measurements are completed. Moreover, other more exotic possibilities exist for warm dark matter which would allow resurrection of a pancake theory. I shall subsequently refer to such theories as galaxy formation by generic pancake fragmentation. Included in this category for the purposes of the present discussion are the explosive amplification schemes for forming galaxies by fragmentation of swept–up shells.

There is also increasing evidence for the existence of relatively young galaxies. The prototype of such objects is the, hitherto unique, case of I Zw18, a gas–rich HII region–like galaxy with no evidence of an underlying old population and a metallicity of 1/30 that of the sun. Its current star formation burst is almost certainly its first (Kunth and Sargent 1985). Studies of deep counts show evidence for a relatively blue component in the distribution of distant galaxies (Kron et $al.$ 1985), and distant clusters contain many galaxies that appear to have the characteristics of ellipticals which have undergone recent star formation (Dressler 1986).

Then, of course, there are the cluster cooling flows. The central elliptical or cD galaxies at the centers of these flows appear to be undergoing considerable star formation (a few $M_\odot$ yr^{-1}) if a conventional solar neighborhood IMF is adopted, but the amount of gas apparently disappearing from the flow, presumably to form stars, is an order of magnitude larger (Fabian *et al.* 1984). These stars must have a lower supernova rate per unit star formation rate, and possibly also may be deficient in massive stars relative to the local IMF, in order to be consistent with observational constraints.

A generic pancake theory fares nicely with regard to forming galaxies in the very recent past: it provides an ample gas supply to explain even such possibly primitive objects as the intergalactic cloud in Leo and I Zw18. Hierarchical clustering theories, on the other hand, tend to use up the available gas supply during collapse on subgalactic scales: it is not at all obvious how much gas can avoid premature fragmentation into stars.

This difficulty arises again and again when other properties of galaxies are confronted with theoretical expectations. For example, the Hubble–de Vaucouleurs profiles of galaxy spheroids can be beautifully explained by gravitational collapse from cold but irregular initial conditions of some thousands of mass points (van Albada 1982). The extreme densities of elliptical galaxy cores (Fall 1979a) require some additional gaseous dissipation, however, as is also required to account for metallicity gradients in the old stellar population. That gas–rich systems are necessary throughout the build–up of structure from small to large galaxies, if mergers are important, is further indicated by the correlation between metallicity and luminosity (Mould 1984), which demonstrates that continuing star formation and gas recycling must have occurred.

Morphological correlations between galaxies and their environment (Dressler 1980) are most simply explained by appeal to tidal interactions, mergers or infall of gas–rich protogalactic clouds, although formation of galaxies biased towards rare, and hence clustered, fluctuations in cold dark matter offers some hope of a similar explanation. Globular clusters also are best explained if they formed at the same time as their parent galaxy. This is difficult to arrange in a cold dark matter theory, unless some mechanism such as heat input by low mass stars is postulated to prevent the protoglobular clusters from premature collapse (Silk and Norman 1981). However conditions in a generic pancake fragmentation theory seem favorable for the onset of thermal instability during gaseous protogalactic collapse, and this leads rather naturally to the observed mass–scales and distribution of globulars (Fall and Rees 1985). Gaseous collapse in a preexisting halo of cold dark matter would presumably also lead to a favorable environment for thermal instability, provided there was a suitable reservoir of gas.

Dark halos are a property of galaxies that find a relatively natural explanation in the cold dark matter theory, with dark matter defining the potential wells within which gas dissipates and fragments into stars (White and Rees 1978).

In the pancake model, halo formation is a secondary process, and it is by no means clear whether hot dark matter can avoid overdilution of phase space density to form the halos that are found around a number of galaxies, both large and small (Tremaine and Gunn 1979; Bond et al. 1983; Melott 1983). The rotation of spirals is a major success of hierarchical clustering models with dark halos. Tidal torquing between neighboring, asymmetric fluctuations of comparable mass induces an initial specific angular momentum in a newly formed protogalaxy that, after collapse through a dark halo, leads to rotationally supported disks with flat rotation curves and maximum rotational velocities that are in good accord with observations (Fall and Efstathiou 1980; White 1986). One possible flaw in this scenario concerns luminous ellipticals, which, despite having higher surface brightness than spirals, are flattened but not rotationally supported (Davies et al. 1983). Evidently, dissipation via collapse through dark halos cannot account for ellipticals. Possible formation mechanisms include loss of halos prior to dissipative infall (Fall 1979b), an option now considered unlikely in view of x–ray observations which indicate the presence of dark halos, or allowance for very different initial conditions, such as a much larger initial density as would happen if ellipticals formed early or preferentially in denser regions. Secondary infall of dark halos, as would happen in a pancake theory, seems more relevant for ellipticals, since it avoids excessive spin–up in dissipative collapse. Alternatively, mergers of gas–rich protogalaxies could plausibly lead to a satisfactory model. Adequate two–component models have yet to be constructed. To simultaneously obtain the $\sim r^{-3}$ profile of a flattened, anisotropy–supported luminous core coexisting with the $\sim r^{-2}$ halo profile will require models incorporating both dissipative and violent relaxation, over a greater dynamic range than hitherto available.

There are some properties that either class of theories seems capable of explaining. These include the galaxy luminosity and multiplicity functions, whose general features can be reasonably well understood in terms of processes involving dissipation (fixing L_*) and merging (yielding a characteristic power–law tail at low luminosity and exponential decline at high luminosity (Schaeffer and Silk 1985). The correlations between luminosity and either velocity dispersion (Faber and Jackson 1976) or maximum rotational velocity for various Hubble types (Tully and Fisher 1977; Rubin and Burstein 1985) can be interpreted in a rather simple framework involving dissipation and star formation (see below), and possibly also could arise in a purely dynamical model involving multiple mergers (Farouki et al. 1983).

The galaxy correlation function, however, has been one area where N–body simulations have come down rather strongly on the side of hierarchical clustering of cold dark matter. Both the correlation function slope and amplitude and the peculiar velocity field between galaxies can be explained if galaxy formation is biased towards the rarer, more tightly clumped fluctuations (Davis et al., 1985). By contrast, a neutrino pancake model yields excessive large–scale velocities, unless galaxy formation again occurs at an unacceptable recent epoch (Kaiser 1983). Biasing also allows $\Omega = 1$ to be reconciled with the observed velocity field, since the dark matter distribution defines the typical rms fluctuations, and is relatively uniform over supercluster scales, thereby not affecting local determinations of Ω. A simple physical mechanism for biasing (to be described below) utilizes both lack of dissipation and supernova–driven winds to inhibit formation of galaxies from low–σ fluctuations in a cold dark matter background.

While generic pancake fragmentation models are not as satisfactory for explaining galaxy correlations, nor do they have much to contribute towards allowing $\Omega = 1$, they do have an important advantage over cold dark matter in two other aspects. Primary halo formation can be somewhat of an embarrassment for binary galaxies and small groups of galaxies on account of dynamical friction (Barnes 1985). Excessive merging is avoided if galaxies orbit in a common dark halo, and both the lack of correlations found in binary samples (White *et al.* 1983) and the range in dwarf galaxy properties (Aaronson 1986) also support this idea. Communal halos are most easily produced in a warm or hot dark matter model for large–scale structure (e.g. Cowsik 1985).

Another observation best explained by pancake theories is that the large–scale structure of the universe appears rather inhomogeneous on scales of 10–100 Mpc. The microwave background anisotropy measurements, uniform to an upper limit of $\delta T/T \lesssim 3 \times 10^{-5}$ on scales from 5′ up to 30°, sample density fluctuations on larger scales, and cannot yet distinguish between rival dark matter–dominated theories (Vittorio and Silk 1985). However the fact remains that there may exist large voids (Kirshner *et al.* 1981), chains of clusters and superclusters of galaxies (Einasto 1986), enhanced cluster–cluster clustering (Bahcall and Soneira 1983), and very large–scale motions such as that of the local supercluster relative to the cosmic microwave background radiation (Aaronson *et al.* 1985). Confirmation of the reality of any of these phenomena would provide a severe blow to the biased cold dark matter theory.

I have summarized the pros and cons for the two rival theoretical approaches to galaxy formation in Table 1. A simple algorithm for assigning numerical significance to their qualitative goodness of fit to astronomical reality suggests that there is at present little to choose between them. Both theories have weak points, both have strong features, and the ultimate theory may well be a hybrid combination of both approaches. Strong motivation to choose one over the other may have to come from considerations outside the domain of astronomy, for example, by laboratory identification of a suitable dark matter candidate.

Table 1

Comparison of Clustering and Fragmentation Scenarios

Property	Clustering of Cold Dark Matter	Fragmentation of Gaseous Pancakes
Formation epoch	XX	?
Profiles of spheroids	X	XX
Metallicity correlations	X	XX
Environment correlations	X	XX
Globular clusters	X	XX
Halos	XX	?
Rotation of Sp's	XX	X
Anisotropy of E's	X	XX
Luminosity function	XX	XX
(L, σ, v_r) correlations	X	X
$\xi(r)$, $v(r)$	XX	X
$\Omega = 1$ and dwarf galaxies	XX	?
Binaries and groups	?	XX
$\delta T/T$	XX	X
Large scale structure	X	XX
SCORE	21	20

III. LUMINOUS CORES AND DISSIPATION

Despite the divergence of viewpoints about the cosmological initial conditions, reflected in the hitherto unsuccessful searches for protogalaxies and for cosmic background radiation anisotropy, there is reasonable unanimity that considerable dissipation occurred during the formation of the mostly baryon–dominated luminous regions of galaxies. The very fact that a galaxy stands out in surface brightness above the background clustering testifies to the different, non-gravitational, physics that must have played a role in its formation. This difference is readily quantified with the galaxy correlation function: the surface brightness of an elliptical within its half–light radius lies well above the extrapolation of $\xi(r)$ from larger scales.

The characteristic properties of a galaxy must have been frozen at, or soon after, the end of its gas–rich protogalactic evolutionary phase. During the dissipative gas–rich phase, the luminous mass, binding energy, half–luminous mass radius, density and metallicity would all have rapidly evolved. However, once stars formed, and the system was predominantly stellar, any ensuing evolution would have occurred on a very long time–scale. Indeed, apart from the occasional merger, one can now be confident that many Hubble times must elapse before any significant dynamical or chemical evolution could occur.

This observation provides a potential test of theories of galaxy formation.

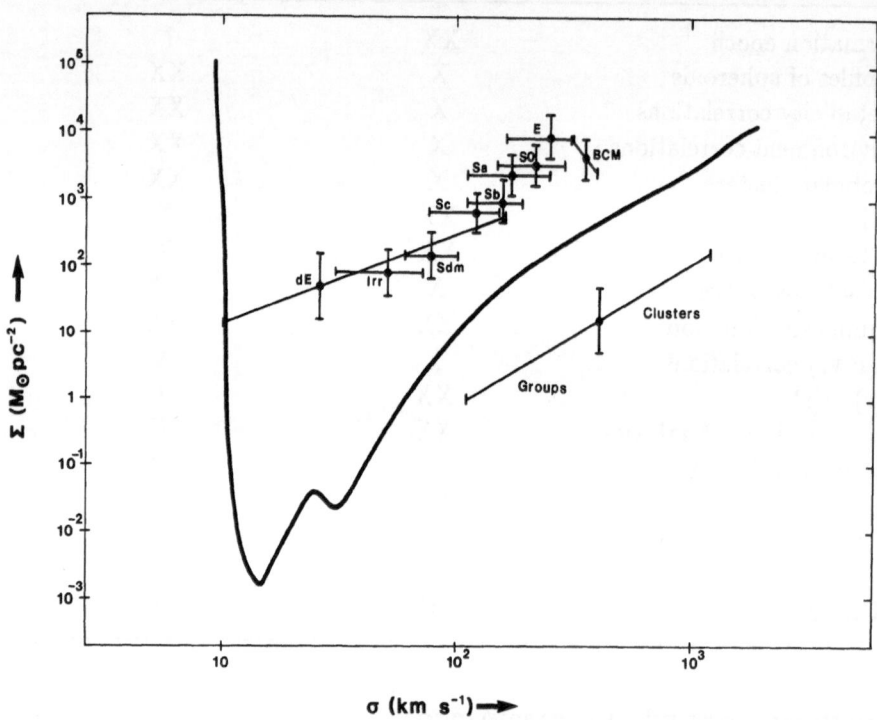

Figure 1. Surface density Σ ($M_\odot$ pc^{-2}) versus velocity dispersion σ (km s^{-1}) for different morphological types of galaxies and for galaxy groups and clusters. Plotted data is based on statistical correlations (see text for references) between luminosity L and σ (for ellipticals and brightest cluster members), and L and maximum rotational velocity (for spirals, S0's, irregulars), L and half–light radius (dwarf ellipticals), and number counts within the central megaparsec radius cores for groups and clusters (from the correlation found by Bahcall 1981). The velocity dispersion is the line–of–sight central velocity dispersion of a presumed isotropic velocity dispersion for the other systems. All correlations are converted to a (Σ,σ) relation by assuming that the systems predominantly consist of stars out to the half–light radii, with appropriate mass–to–light ratios taken from Faber and Gallagher (1979). The cooling curve shown is calculated for a collapsing cloud of primordial abundance. From Silk (1985).

It is well known that luminosity (L), one–dimensional velocity dispersion (σ), maximum rotational velocity (V_m), and surface brightness (Σ) all vary with Hubble type. A two–dimensional classification scheme of, say, Σ and σ serves to spread out Hubble types and, as will be seen shortly, has theoretical significance. I shall work with the old stellar populations, whether in disk or spheroid, and work with parameters averaged over the half–light radius. Virial equilibrium yields

$$\sigma^4 \sim G^2 \left(\frac{M}{L}\right)^2 \Sigma L,$$

so that specification of either $L(\sigma)$, $L(V_m)$, or $\Sigma(R)$, all well–observed statistical correlations, together with an adopted M/L appropriate to the Hubble type, suffices to yield $\Sigma(\sigma)$ for the various Hubble types. The results (Silk 1983a) are displayed in Figure 1. If, as I have argued, the old, spherically–averaged over R_e, stellar population bears some memory of the galaxy formation process, then galaxy formation theory must be capable of accounting for the location of the various galaxy types in the (Σ,σ) plane. I also indicate the position occupied by groups and clusters of galaxies, again using luminous mass: this reemphasizes the point that galaxies are well separated in binding energy and in density from large–scale structures.

A simple theoretical explanation of the general features of Fig. 1 is readily available. As noticed by Faber (1982) and by Gunn (1982), the condition that a uniform, spherically symmetric collapsing cloud can cool within a collapse time translates into a critical density at any given temperature, or equivalently, virial velocity dispersion which separates the regime occupied by galaxies from that occupied by clusters. Rees and Ostriker (1977) and Silk (1977) had originally applied this condition to deduce a critical mass of $\sim 10^{11}$-10^{12} $M_\odot$, which cannot be exceeded by any dissipative cloud.

This discussion was generalized by Silk (1984) to the following situation. Consider an ensemble of clouds interacting dynamically in the potential well of the protogalaxy. A necessary, although not sufficient, condition for dissipation of bulk orbital energy and star formation to occur is that the shocks induced in cloud–cloud collisions be radiative. The post–shock temperature is $kT = (3/16)m_p V_s^2$, where the shock velocity is due to virial motions in the galaxy potential well, namely $V_s = 6^{1/2}\sigma$. The pre–shock cooling time is $t_c = \rho^{-1}f(T)$, where $f(T)$ is the ratio of thermal energy density to cooling rate per gm and ρ is the mean cloud density. For the shock to be radiative, the cloud column density in a one-dimensional shock must exceed $\Sigma \equiv \rho v_s t_c$, where Σ_{cool} is a function only of σ. The formation of a dense layer in a radiative shock does not guarantee instability; in fact, three–dimensional compression of the cooled layer is necessary to ensure gravitational instability and fragmentation. But cooling is surely necessary to initiate this process.

Σ_{cool} is displayed in figure 1 for a primordial mixture of abundances. H_2 cooling has been neglected, and the steep rise of Σ_{cool} below $\sigma \sim 10$ km s^{-1} is due to suppression of Lyman alpha cooling. At higher σ, hydrogen and helium cooling by excitations and recombination is important, and eventually bremsstrahlung dominates. Evidently, Σ_{cool} cleanly separates galaxies, where dissipation occurred during formation, from groups and clusters, whose binding energy was evidently acquired with little or no dissipation. The theoretical justification for this conclusion is that once stars formed, dissipation effectively ceased.

How plausible is it that colliding clouds interact dissipatively and trigger star formation? While it is not straightforward to demonstrate this on theoretical grounds, for we are far from a complete theory of star formation, there are strong indirect arguments that support this outcome. Massive stars form contagiously in the spiral arms in giant molecular cloud complexes. The formation of these complexes from smaller clouds appears to be stimulated by coagulation in the spiral density wave (Kwan and Valdes 1983; Tomisaka 1984), and prolific star formation occurs at cloud masses above 10^5-10^6 $M_\odot$. It is likely that formation of OB stars is self–reinforcing, and smaller clouds, present throughout the disk, of mass 10^3-10^4 $M_\odot$ form too few OB stars for efficient formation of massive stars to occur (Silk 1986). Tidal interactions between nearby galaxies are similarly capable of inducing vigorous bursts of massive star formation (de Jong 1985). Again, enhanced collisions between molecular clouds seem the most likely mechanism (Icke 1985). Gas responds inelastically to any tidal perturbation with a greatly amplified response. The motivation for choosing cloud masses is also rather indirect. HI–H_2 cloud complexes in our own galaxy and in M31 have masses of 10^6 or even 10^7 $M_\odot$. Clumpiness seen in "young" galaxies (Lequeux and Viallefond 1980) and in at least one intergalactic cloud (Terzian et $al.$ 1986) suggests that primordial building blocks may have masses of $\sim 10^7$ $M_\odot$. Then there are the cosmological arguments: in a cold dark matter–dominated universe, while the dark matter clumps on all scales, baryonic cooling via H_2 formation and infall only occurs for clouds of mass in excess of $\sim 10^6$ $M_\odot$ (Bond and Szalay 1983). Even pancake fragmentation results in formation of baryonic fragments with mass $\lesssim 10^8$ $M_\odot$ (Sunyaev and Zel'dovich 1972).

One would expect an aggregation of many thousands of clouds to relax dynamically, much as van Albada (1982) found in his N–body simulations. Only in the core, where collisions occur within a dynamical time, would dissipative effects dominate and enhance the density. Clouds should survive in the outer regions, and coalesce or disrupt in the core, where star formation will be enhanced. This argument suggests that the mean surface density of a galaxy, if clouds are presumed to form stars mostly in the collision–dominated core, will be of order the mean surface density of the clouds, since the ratio of cloud–cloud collision to crossing times is of order the ratio of cloud to galaxy (averaged over cloud orbit) surface densities (Silk and Norman 1980). This takes us one step back in time from present luminous cores to protogalactic clouds, but a crucial question is, what determined the longevity of, and gas density in, these clouds? For if they were to have collapsed and fragmented into stars on a free–fall time scale ($\sim 10^6$ yr), practically all the gaseous dissipation would have terminated before a massive galaxy could have formed over the much longer time–scale of 10^8–10^9 yr. And moreover, knowledge of the cloud surface density is crucial to understanding the location of the old components of galaxies in the (Σ, σ) diagram. The key, it will now be argued, lies in understanding star formation in primordial clouds.

IV. PRIMORDIAL STAR FORMATION AND PROTOGALACTIC EVOLUTION

The best lever on the IMF of primordial stars, that is to say, the first generation of stars in a cloud of primordial abundance, is by inference from study of extreme population II stars. Heavy element abundance ratios in these stars are consistent with a precursor population of ordinary stars of intermediate to high mass (5–100 $M_\odot$). The fact that there is at least one star known (Bessell and Norris 1984) with $[Z] \approx -4.5$ suggests that the primordial IMF extended to below 1 $M_\odot$, for even a single generation of star formation should last long enough, by analogy with our own interstellar medium, for there to be some mixing of heavy elements at the 10^{-4} $Z_\odot$ level. Certainly from the theoretical viewpoint insofar as star formation is concerned, 10^{-4} $Z_\odot$ and zero metallicity should be indistinguishable: fragmentation and opacity effects only change appreciably at higher values of Z.

The number of surviving primordial or "population III" stars is very small. Hence if they did possess a normal, that is to say, solar neighborhood IMF, as is consistent with most other well–studied galaxies, very few must have formed. Indeed, the observations suggest that $dN/dZ \approx$ constant for $Z \lesssim 0.01$ $Z_\odot$ (Beers et al. 1985). This means that provided the star formation efficiency was initially very low, and recycling of gas ceased once Z rose to ~ 0.01 $Z_\odot$, one could readily account for the abundances of heavy elements in population II with a precursor population of stars spanning the 0.1 to 100 $M_\odot$ mass range. Alternative possibilities are that the yield of heavy elements was greatly enhanced in metal–poor stars (Jura 1985), that there was considerable and continuing infall of primordial material, as envisaged in some galactic disk models, or of course, that the population III IMF consisted exclusively of massive stars (Truran and Cameron 1971). There is no observational motivation to believe that population III might have primarily consisted of very massive objects or of Jupiter–mass bodies.

Theory, of course, is eminently qualified to give any answers whatsoever to the issue of the mass range of population III. Indeed, one can find suggestions in the literature that range from 10^{-3} $M_\odot$ up to 10^6 $M_\odot$. However the most naive theoretical approach concludes that the mass range of primordial stars should be similar to that of stars forming at present, with some possibility of enhancement of massive OB star formation. This arises from consideration of the role of cooling by molecular hydrogen during the fragmentation of clouds of primordial abundance. The H_2 initially forms via H^- or H^+ production from the residual post–recombination ionized fraction, followed by reaction with H to form H_2 at high density, when three–body formation of H_2 dominates. Consideration of the minimum Jeans mass attained during opacity–limited fragmentation of a spherically symmetric collapsing cloud suggests that, just as with conventional interstellar matter opacities, fragmentation proceeds to below 0.1 $M_\odot$ (Palla *et al.* 1983). If H_2 is destroyed in shocks, as might happen in cloud–cloud collisions in the potential well of a moderately massive protogalaxy, then with only Lyman alpha cooling, the fragmentation analysis favors predominantly massive (10–100 $M_\odot$) star formation (Silk 1985).

Hence it seems that star formation in a protogalaxy must result in formation of some massive stars, HII regions, and supernovae. This means that dynamical feedback of energy input from massive stars will inevitably accompany chemical enrichment by their ejecta. Now in a closed system, with no mass loss or infall, the metallicity asymptotically approaches the net yield, or about twice the solar metallicity. The existence of metal–poor galaxies, combined with the fact that many of these, the dE's, have very low surface brightness compared to luminous E's, strongly suggests that considerable mass loss has occurred. The existence of a correlation between Z and σ also supports the idea that chemical and dynamical evolution have been coupled together during the protogalactic phase.

A simple means of understanding this coupling arises if the stellar energy input from massive stars via winds and supernovae is sufficient to regulate the protogalactic gas density. In a quasispherical protogalaxy, the mean gas density would be similar to that in our present, relatively gas–poor, galactic disk, and the supernova rate per unit mass of gas which is forming stars may even be enhanced. If supernovae are capable of regulating the interstellar medium and star formation in our own galaxy, they should be equally capable of managing this task in a protogalaxy. Collapse induces more star formation, for example by enhancing the rate of cloud collisions, which in turn yield more supernovae that help support the gas, on the average, against further collapse. All of this can happen even when the mean cooling time t_c for the system as a whole exceeds the dynamical time–scale t_d, since stars can form within individual clouds, but one would expect the rate of dissipation and accompanying star formation to increase dramatically once $t_c <$ t_d. Figure 1 demonstrates that collapse inevitably brings a protogalaxy into this regime where one might expect bursts of star formation to be initiated.

A very simplified sketch of how self-regulation of the protogalactic gas density occurs may be constructed as follows. Let λ denote the energy input per gm of gas that forms stars associated with supernovae: for a standard IMF, $\lambda \approx 10^{51}$ ergs/100 $M_\odot \approx (700$ km s$^{-1})^2$. Then in order for the supernova energy input, neglecting radiative losses, to be balanced by cooling, one must have

$$\lambda \dot{M}_* = M_{gas} \tfrac{3}{2}\sigma^2 t_c^{-1} \tag{1}$$

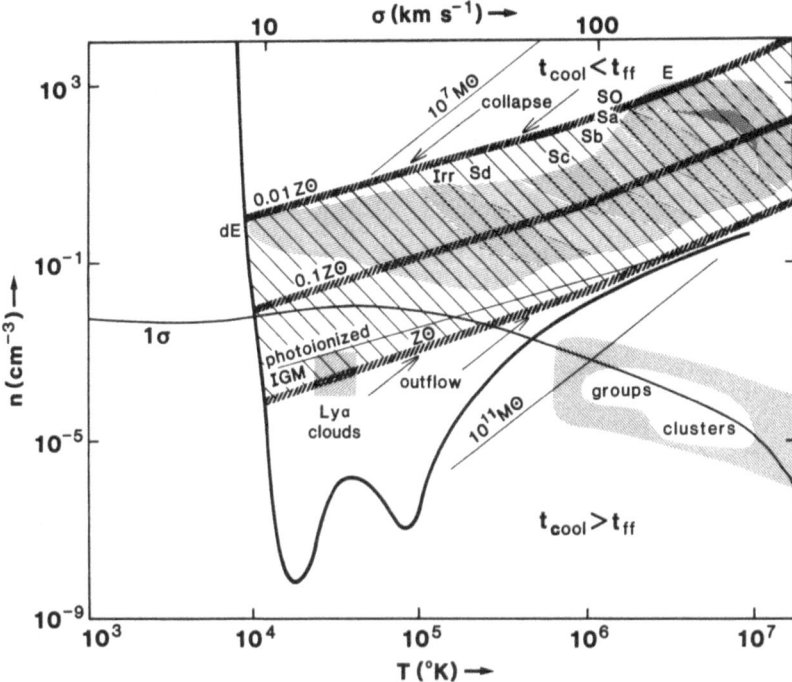

Figure 2. Constraints on protogalactic evolution. In this plot of baryon density versus virial temperature (or equivalently, velocity dispersion), the cooling curve for a collapsing protogalactic cloud or primordial abundance demarcates the region where cooling occurs rapidly or slowly with respect to a dynamical collapse time–scale. This curve reduces to the line marked "photoionized IGM" if the intergalactic medium is highly ionized, as is the case after quasars turn on at z = 2 or 3. The hatched area indicates the density range over which supernova–driven energy input can support a gaseous protogalaxy against collapse for several values of initial metallicity at onset of starburst. The various data points for galaxies and groups are taken from Figure 1; the region occupied by the Lyman alpha clouds seen in absorption against high redshift quasars is also shown. From Silk (1985).

where $\dot{M}_*$ is the average star formation rate, M_{gas} is the gas mass, and the mean cooling time–scale $t_c = \frac{3}{2}kTn^{-1}f^{-1}$, with

$$f \sim AZT^{-1/2} \quad \text{for} \quad Z \gtrsim 0.01Z_\odot, \quad 10^5 \lesssim T \lesssim 10^7 K.$$

Integrating (1) subject to the condition that $\dot{M}_* = -\dot{M}_{gas}$ yields, after one dynamical time has elapsed, the mean initial gas density

$$n \approx 0.01\sigma_{100}^{2.4}(Z_\odot/Z)^2 cm^{-3}, \tag{2}$$

where $\sigma_{100} \equiv \sigma/100$ km s^{-1}. At this stage, relatively little of the initial gas reservoir has been converted into stars. In fact, from (1), it is apparent that the gas supply is only exhausted if no mass loss occurs after a time of order

$$(\lambda/\sigma^2)t_c \sim 100\sigma_{100}^{-2}t_d. \tag{3}$$

This can mean that the protogalactic phase takes ~ 10 crossing times or $\sim 10^9$ yr in a typical massive elliptical galaxy.

Self–regulation becomes effective after roughly one dynamical time has elapsed. Condition (2) then yields the mean star density attained during the self–regulation phase, with some uncertainty depending on how much enrichment has occurred during the initial star burst. The higher the enrichment, the greater is the cooling rate and therefore the initial gas density that can be supported against collapse. Comparison of the predicted density range with observations of the density of the old stellar components of galaxies is shown in figure 2. The specific binding energy of galaxies can evidently be understood by the simple hypothesis of self–regulation of the gas density by supernova energy input. One can make this comparison a little more precise by assuming that the enrichment acquired after one dynamical time is incorporated into the ambient gas and determines its subsequent cooling efficiency. Now at t_d, a mass fraction

$$Z \equiv y\sigma^2/\lambda \equiv 0.01\lambda_{100}^2 \tag{4}$$

of heavy elements has been produced. Combination of (2) and (4) yields a dependence for the mass of a galaxy approximately as σ^4. Provided the normalization is adjusted to make some allowance for dark matter, and an IMF emphasizing massive stars is used, this accords well with the Faber–Jackson relation.

There is a further noteworthy implication of this model. If the protogalaxy potential well is sufficiently shallow, the supernova remnants can drive a wind and expel most of the gas. The condition for this to occur is that the expanding remnants overlap before decelerating below the protogalaxy velocity dispersion (Bregman 1978). In fact, a wind is driven if $\sigma \lesssim 60$ km s^{-1}, with remarkably little sensitivity to other parameters: for example, this critical velocity dispersion varies only as $n^{1/22}$ and as $\lambda^{3/11}$ (Dekel and Silk 1985).

The mass loss from low σ galaxies has two consequences. The metallicities will be determined by (4). Indeed, $Z \propto \sigma^2$ fits the metallicity over a wide range of elliptical galaxies, from dE's to luminous E's (Figure 3). A one–dimensional velocity dispersion of 60 km s^{-1} corresponds to $M_B \approx$ -18, and a plot of the nitrogen–to–sulfur abundance ratio reveals an intriguing change of slope at this magnitude: below $M_B \approx$ -18, it is flat; above this magnitude, it is linear. Wyse and Silk (1983) argue this may be indirect evidence for mass loss and failure to recycle stellar debris in low luminosity galaxies. A second consequence is that the drastic mass loss initiated at t_d will leave behind a remnant galaxy of greatly reduced binding energy. If the remnants are to be identified with dE's, then it is necessary to introduce sufficient dark matter to prevent overexpansion (Dekel and Silk 1985).

   In fact, there is a natural prescription for producing dE's with the observed range of surface brightness in a cold dark matter–dominated universe. Cold dark matter clusters as $M_H \propto R^{6/(5+n)}$ where n is the index of the initial power spectrum prior to galaxy formation and M_H is the halo mass on the newly bound scale R. Now on small scales, comparable in mass to dwarf galaxies, n $\approx$ -3, and so

$$M_H \propto R^3. \tag{5}$$

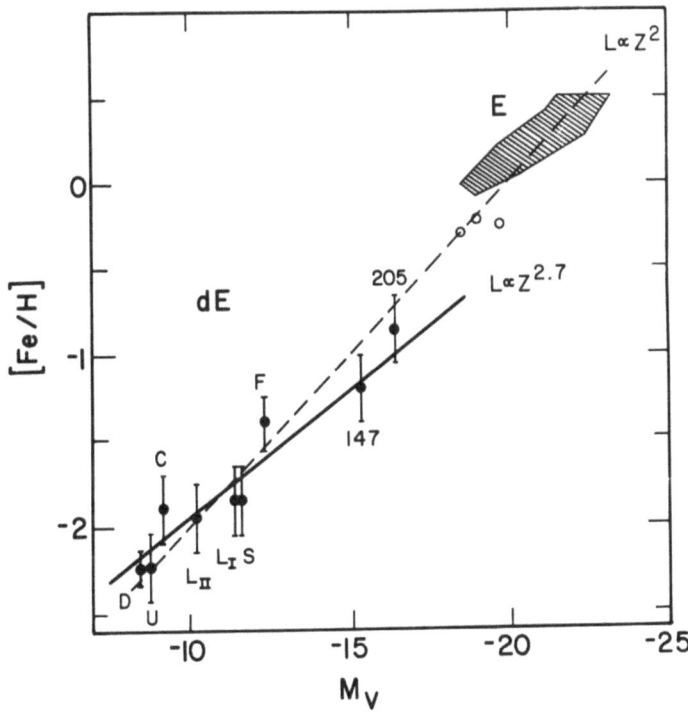

Figure 3. Metallicity versus luminosity for the local dwarf ellipticals and for "normal" ellipticals. Predictions are shown for the wind–driven mass loss model discussed in the text. From Dekel and Silk (1986).

Also, supernova heating is inefficient in dwarf galaxy potential wells, since remnants undergo strong radiative losses below an expansion velocity of about 100 km s^{-1}. Hence allowing for this inefficiency, the cumulative energy input from supernova, taken to be proportional to the stellar mass, is

$$L_* \propto M_{gas}\sigma \propto M_H\sigma, \tag{6}$$

for a universal gas–to–dark matter ratio. Finally, after mass loss, the stars are bound by the halo, and

$$\sigma^2 \sim M_H/R. \tag{7}$$

Combining (5), (6) and (7) yields

$$L_* \propto R^4 \propto \sigma^4 \quad \text{and} \quad M_H/L_* \propto L_*^{-1/4}. \tag{8}$$

In other words, the halos that naturally arise in the cold dark matter theory yield a run of surface brightness with luminosity similar to that observed for dE's (Figure 4), and predict a systematic increase of mass–to–light ratio for

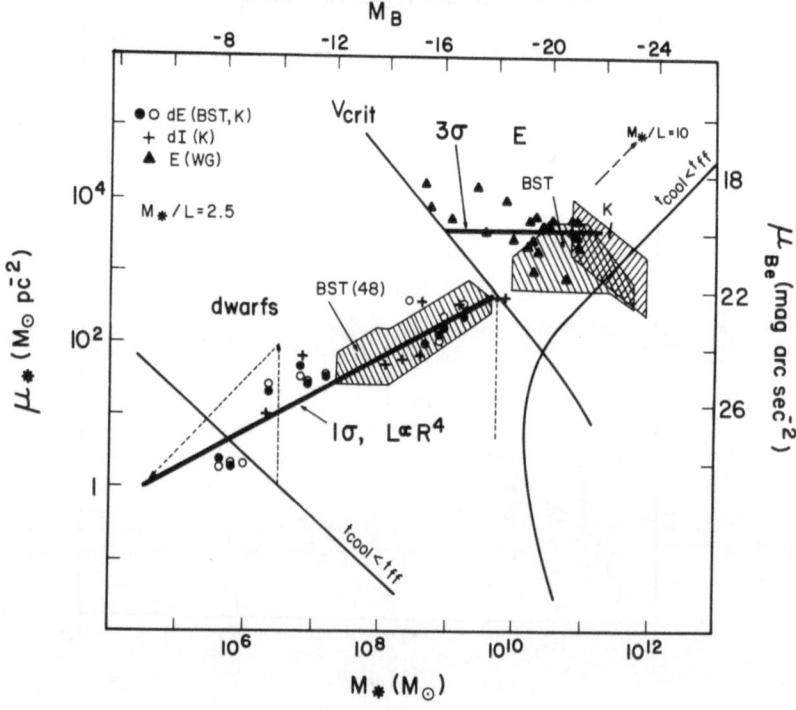

Figure 4. Surface–brightness within the effective radius plotted versus luminosity, for a compiled sample of dwarf ellipticals (in the Local Group and in Virgo), dwarf irregulars, and ellipticals. Predictions are shown for wind–stripping in a cold dark matter–dominated universe. From Dekel and Silk (1986).

the lowest luminosity dE's. The data fit these correlations when the cold dark matter theory is crudely normalized, although the low mass–to–luminosity ratio of the Fornax dwarf galaxy (Cohen 1983) remains somewhat puzzling.

V. TO BIAS OR NOT TO BIAS

Dwarf stripping solves an outstanding problem of the cold dark matter scenario for galaxy formation. If $\Omega = 1$, and this is the principal motivation for hypothesizing cold dark matter, one has to start with the universal mass–to–luminosity ratio

$$M/L_B \approx 1500\Omega h.$$

Then in a theory in which clustering occurs on all scales, one has to produce M/L_B values lower by a factor of 10 in galaxy clusters, groups and dark halos. This is accomplished, in principle, by arguing that the efficiency at which baryonic matter forms galaxies is enhanced by a factor of 10 or so in clusters and on smaller scales. If $\Omega_b/\Omega = 0.1$, then this enhancement explains the origin of the observed ratio of luminous to dark matter of 1/10 in galaxy clusters, groups and luminous galaxies (Faber 1982; Gunn 1982).

An additional bonus is that the clustered fluctuations are predominantly the rare peaks in an initially gaussian fluctuation spectrum. If only these form luminous galaxies, then the required amplitude of the average, less clustered, fluctuations can be lowered by about a factor of ν^2 (if one forms the observed luminous galaxies from $\sim \nu\sigma$ fluctuations) (Bardeen 1985; Kaiser 1985). Consequently, if $\nu \approx 3$, the dynamically measured value of Ω, on the scale of clusters or even superclusters, is lowered to about 0.1, while the true Ω, which samples matter that is uniform over these scales, is unity. Biasing not only solves these problems, but also yields acceptable correlation functions and peculiar velocities for the large–scale galaxy distribution (Davis *et al.* 1985). What remains unclear is *why* galaxy formation only occurs efficiently in the rare peaks of the primordial fluctuation spectrum.

Dwarf stripping by winds provides a physical biasing mechanism. Inspection of figure 2 reveals that 1σ fluctuations are caught between two limits. In order to be able to be in the dissipative regime ($t_c < t_d$), they must have velocity dispersion below ~ 100 km s^{-1}; however wind–stripping becomes effective in this range of velocity dispersion. Only the rare fluctuations find a niche in the (n,σ) plane, where $t_c < t_d$ and $\sigma > 100$ km s^{-1}, allowing the formation of luminous galaxies that have undergone efficient star formation by retention of the bulk of their initial gas supply. This result is displayed in Figure 4, which plots surface brightness (or mass density) versus luminosity (or mass). The unstripped ellipticals form a distinct sequence from the stripped systems, which have expanded as argued previously to form low surface brightness dE's. Wind–stripping can explain the bifurcation in the sequence of elliptical galaxies, in the context of a cold dark matter scenario for the initial fluctuations and halo formation. A consequence of this interpretation is that the dE's, which formed from the 1σ fluctuations, are the only true mass tracers in an $\Omega = 1$ universe.

6. CONCLUSIONS

The cold dark matter theory of galaxy formation by hierarchical clustering appears increasingly attractive in its explanations of many characteristics of galaxies and of galaxy clustering. Nevertheless, it also has serious flaws that generic pancake fragmentation theories are especially adept at explaining, despite their own weakness in other areas. This suggests to me that the ultimate theory of galaxy formation is going to contain elements of both approaches. Indeed, one lesson to be learned from the discussion presented here of the (n,σ) or (Σ,σ) diagram is that it is relatively easy to mask the initial conditions set by cosmology. Physical processes, especially involving star formation, are sufficiently ill understood that they may well play a dominant role in determining the principal characteristics of galaxies.

Cosmology is more intimately connected to the non–dissipative dark matter component, but here there is considerably more freedom for speculation because of inevitable uncertainty about its spatial distribution. Eventually, the dissipation diagram could become the Hertzsprung–Russell diagram of extragalactic astronomy, with evolution tracks linking cosmological initial conditions to fossilized properties of galaxies that were frozen at the end of the epoch of galaxy formation. For now, however, galaxy formation modellers would be well advised to concentrate more on accounting for galaxy characteristics by utilizing models involving dynamical relaxation and dissipation, and less on fidelity towards specific theories based on cold, warm, hot, or even no dark matter.

REFERENCES

Aaronson, M. 1986, these proceedings.
Aaronson, M, Mould, J. and Schechter, J. 1985, private communication.
Bahcall, N. 1981, *Ap.J.* **247**, 787.
Bahcall, N. A. and Soneira, R. M. 1983, *Ap.J.* **270**, 20.
Bardeen, J. M. 1985, in *Inner Space—Outer Space*, eds. E. Kolb and M. Turner (Chicago:University of Chicago Press) (in press).
Barnes, J. 1985, Ph.D. thesis, U.C. Berkeley (unpublished).
Beers, T. C., Preston, G. W. and Shectman, S. A. 1985, preprint.
Bessell, M. S. and Norris, J. 1984, *Ap.J.* **285**, 622.
Blumenthal, G., Faber, S., Primack, J. and Rees, M. 1984, *Nature* **301**, 584.
Bond, J. R., Efstathiou, G. and Silk, J. 1980, *Phys. Rev. Lett.* **45**.
Bond, J. R., Szalay, A. S. and White, S. D. M. 1983, *Nature* **301**, 584.
Bond, J. R. and Szalay, A. S. 1983, *Ap.J.* **274**, 443.

Bregman, J. 1978, *Ap.J.* **224**, 768.

Carlberg, R. 1984, *Ap.J.* **286**, 403.

Carr, B. J. and Ikeuchi, S. 1985, *M.N.R.A.S.* **213**, 497.

Cohen, J. G. 1983, *Ap.J. Letters* **270**, L41.

Cowsik, R. 1985, preprint.

Davies, R. L., Efstathiou, G., Fall, S. D. M., Illingworth, G. and Schechter, P. L. 1983, *Ap.J.* **266**, 41.

Davis, M., Efstathiou, G., Frenk, C. S. and White, S. D. M. 1985, *Ap.J.* **292**, 371.

de Jong, T. 1985, *Proc. Erice School on Spectral Evolution of Galaxies* (in press).

Dekel, A. and Silk, J. 1986 (in press).

Dressler, A. 1980, *Ap.J.* **236**, 351.

Dressler, A. 1986, in *Spectral Evolution of Galaxies*, ed. C. Chiosi and A. Renzini (D. Reidel:Dordrecht).

Einasto, J. 1986, these proceedings.

Faber, S. M. 1982, in *Astrophysical Cosmology*, ed. H. A. Brück, G. V. Coyne and M. S. Longair (Vatican:Pontificia Academia Scientrarum), p. 219.

Faber, S. and Gallagher, J. 1979, *Ann. Rev. Astr. Ap.* **17**, 135.

Faber, S. M. and Jackson, R. F. 1976, *Ap.J.* **204**, 668.

Fabian, A. C., Nulsen, P. E. J. and Canizares, C. R. 1984, *Nature* **311**, 733.

Fall, S. D. M. 1979a, *Rev. Mod. Phys.* **51**, 21.

Fall, S. D. M. 1979b, *Nature* **281**, 200.

Fall, S. D. M. and Efstathiou, G. 1980, *M.N.R.A.S.* **193**, 189.

Fall, S. D. M. and Rees, M. J. 1985, *Ap.J.* (in press).

Farouki, R. T., Shapiro, S. L. and Duncan, M. J. 1983, *Ap.J.* **265**, 597.

Gunn, J. E. 1977, *Ap.J.* **218**, 592.

Gunn, J. E. 1982, in *Astrophysical Cosmology*, ed. H. A. Brück, G. V. Coyne and M. S. Longair (Vatican:Pontificia Academia Scientiarum), p. 233.

Icke, V. 1985, *Astr. Ap.* **144**, 115.

Ikeuchi, S. 1981, *P.A.S.J.* **33**, 211.

Jura, M. 1985, preprint.

Kaiser, N. 1983, *Ap.J. Letters* **273**, L17.

Kaiser, N. 1985, in *Proc. Inner Space-Outer Space Symposium*, eds. E. Kolb and M. Turner (Chicago:University of Chicago Press) (in press).

Kron, R. G., Koo, D. C., and Windhorst, R. A. 1985, *Astr. Ap.* **146**, 38.

Kirshner, R. P., Oemler, A., Schechter, P. L. and Shectman, A. 1981, *Ap.J.* **248**, L57.

Kunth, D. and Sargent, W. L. W. 1985, preprint.

Kwan, J. and Valdes, F. 1983, *Ap.J.* **271**, 604.

Larson, R. 1975, *M.N.R.A.S.* **173**, 671.

Lequeux, J. and Viallefond, F. 1980, *Astr. Ap.* **91**, 269.

Melott, A. 1983, *M.N.R.A.S.* **202**, 595.

Mould, J. 1984, *P.A.S.P.*, **96**, 773.

Ostriker, J. P. and Cowie, L. 1981, *Ap.J. Letters* **243**, L127.

Palla, F., Stahler, S. and Salpeter, E. E. 1983, *Ap.J.* **271**, 632.

Peebles, P. J. E. and Dicke, R. 1968, *Ap.J.* **154**, 891.

Peebles, P. J. E. 1984, *Ap.J.* **277**, 470.

Rees, M. J. and Ostriker, J. P. 1977, *M.N.R.A.S.* **179**, 541.

Rubin, V. R. and Burstein, D. 1985, *Ap.J.* (in press).

Schaeffer, R. and Silk, J. 1985, *Ap.J.* **292**, 319.

Silk, J. 1968, *Ap.J.* **151**, 459.

Silk, J. 1977, *Ap.J.* **211**, 638.

Silk, J. 1983a, *Nature* **301**, 574.

Silk, J. 1983b, *M.N.R.A.S.* **205**, 705.

Silk, J. 1984, *The Big Bang and Georges Lemaître*, ed. A. Berger (D. Reidel:Dordrecht), p. 279.

Silk, J. 1985, *Ap.J.* (in press).

Silk, J. 1986, in *Luminous Stars and Associations in Galaxies*, I.A.U. Symposium **116**, ed. C. de Loore, A. Willis and P. Lazkarides (Dordrecht:D. Reidel)

Silk, J. and Norman, C. 1981, *Ap.J.* **247**, 59.

Sunyaev, R. A. and Zel'dovich, Ya. B. 1972, *Astr. Ap.* **20**, 189.

Terzian, Y., Schneider, S. E. and Salpeter, E. E. 1986, in *Structure and Evolution of Active Galactic Nuclei*, ed. G. Giuricin, F. Mardirossian, M. Mezzetti and M. Ramella (D. Reidel:Dordrecht)

Tohline, J. 1980, *Ap.J.* **239**, 417.

Tremaine, S. and Gunn, J. E. 1979, *Phys. Rev. Lett.* **42**, 407.

Truran, J. and Cameron, A. G. W. 1971, *Astrophys. Space Sci.* **14**, 179.

Tully, R. B. and Fisher, J. R. 1977, *Astr. Ap.*, **54**, 661.

van Albada, T. 1982, *M.N.R.A.S.* **201**, 939.

Vittorio, N. and Silk, J. 1985, *Ap.J. Letters* **293**, L1.

White, S. D. M. 1986, these proceedings.

White, S. D. M., Davis, M. and Frenk, C. 1983, *Ap.J. Letters* **274**, L1.

White, S. D. M., Huchra, J., Latham, D., and Davis, M. 1983, *M.N.R.A.S.* **203**, 701.

White, S. D. M. and Rees, M. J. 1978, *M.N.R.A.S.* **183**, 341.

Wyse, R. G. and Silk, J. 1985, *Ap.J. Letters* (in press).

Zel'dovich, Ya. B. 1970, *Astr. Ap.* **5**, 84.

DISCUSSION

RUBIN: It is nice to see your theory, which couples dynamical and chemical evolution. But you should be warned that the break in the slope of the [NII]/[SII] data may be an artifact of the observations. For high-luminosity galaxies, the values refer to the disk, excluding the nucleus. For the low-luminosity objects, the values generally come from a single aperture measurement centered on the nucleus. It is not clear how the plot would look with only disk data at all luminosities.

AUDOUZE: I would like to comment briefly about the concept of primary versus secondary elements. It might be more appropriate to speak about elements which are produced by high-mass stars and which look "primary" while those which are mainly produced in low-mass objects seem to be secondary.

SILK: I agree. The distinction I would make is that the wind-stripping model predicts that elements produced by massive stars will not be recycled through successive generations of star formation in galaxies which are fainter than $M_B \approx -18$, while elements produced by stars of all masses are recycled in the more luminous galaxies.

REES: You emphasised that the energy input from stars and supernovae could exert a negative feedback on the rate of star formation, making it impossible to convert all gas into stars on a protogalactic dynamical timescale. There are then in principle two possibilities: either the original gas is all eventually turned into stars, albeit slowly, or the first generation of stars generates enough energy to expel the bulk of the gas (rather then just slowing down further star formation). Your discussion of dwarf galaxies depended on the latter possibility being true. Is this just an assumption, or is there some physical reason why the negative feedback should "overshoot" as you envisage?

SILK: In a gaseous proto-galaxy whose one-dimensional velocity dispersion is less than about 60 km s^{-1}, I showed that a supernova-driven wind would be inevitable. This assumes that at least one supernova is produced for every several hundred solar masses of newly formed stars. If one could avoid making supernovae, then the evolution would proceed much more gently, until the gas supply is exhausted. Likewise, in massive proto-galaxies with deeper potential wells, where a wind is effectively quenched, the assumption of a standard initial mass function guarantees enough supernovae to maintain a negative feedback and prolong the duration of the star forming phase.

SPERGEL: I understand that your wind-stripping scenario predicts the existence of objects with a lot of mass but not much luminosity. In general, is it true that you predict a trend where lower-luminosity objects have higher M/L ratios?

SILK: That's right. In fact, it is necessary for this theory that dwarf ellipticals have high M/L ratios.

SPERGEL: So in some dwarf ellipticals, even though you have gotten rid of a lot of the gas, you might expect to have hot gas left behind. Would you predict that dwarf ellipticals might have x-ray halos?

SILK: I expect that the gas would be long gone. In any case, the potential well wouldn't be deep enough for the gas being blown out to have x-ray emitting temperatures.

REES: I'd like to ask at what redshift the first galaxies formed. You mentioned the Spinrad galaxy at z = 1.8 and quoted his argument that from its colors you can infer that it must have formed well before that redshift. I'd like to ask how strong that argument is, and how it depends on the initial mass function and on other assumptions.

SILK: That particular object does not give very strong constraints, because it has strong emission lines. But there is recent work by Hamilton and Kron in which they look at distant ellipticals not chosen from radio-source surveys. These turn out to have standard red populations. The galaxies are observed at redshifts of 0.8 - 0.9. Hamilton and Kron make the standard synthesis models and find rather large ages. The redshift of formation turns out to be at least 5. But let me emphasize that these are very rare, very luminous ellipticals and that the answer is model-dependent.

BERTSCHINGER: The coincidence in your diagram between the position of the cooling curves and the masses and radii of galaxies may be a red herring, particularly in the heirarchical clustering picture. For example, Simon White told us that Frenk et al., in their n-body simulations, were able to get reasonable maximum velocities of rotation curves in a situation where there was no cooling or dissipation. Furthermore, I have a poster paper where I consider just infall onto peaks in a cold dark matter spectrum, and I'm able to reproduce the Schechter mass function, again without any cooling. Cooling is not important in these cases because you are building up small structures first, so the cooling time is always shorter than the dynamical time.

SHAPIRO: I noticed that the cooling curve that you used to determine the range of virialized mass scales which would cool in less than the Hubble time does not take account of inverse Compton cooling by the background radiation. Since the rate of Compton cooling increases very rapidly with increasing redshift and is independent of gas temperature, your conclusions would be very different at high z if you included it. Have you assumed that all of the mass scales of interest condense out at redshifts z ≪ 10? If so, you would be self-consistent, although I would question whether your timescale arguments are as general as they might be.

SILK: In both cold dark matter scenarios and fragmentation scenarios, formation of luminous galaxies occurs at a redshift less than 10 when Compton cooling is unimportant. In the former picture, small mass scales can begin to condense out somewhat earlier, but Compton cooling

is irrelevant, for the following reason. The dark-matter potential
wells define the appropriate mass scales, and once sufficient
dissipation and density enhancement of the baryonic component occur,
local cooling processes dominate. There are, of course, more exotic
possibilities, such as the explosive amplification model of galaxy
formation, in which Compton cooling plays an important role in
determining the relevant fragmentation scales. My arguments could be
generalized to study fragmentation and dissipation at large redshift,
although I have not done so.

DRESSLER: In your scorecard comparing the hierarchical and
fragmentation models, you gave an edge to the latter for ease of
producing a correlation of galaxy morphology and environment. I'd
appeal that score on the grounds that once you have a mechanism that
preferentially produces spheroids in dense regions, such as biased
galaxy formation with cold dark matter, this may provide the primary
ingredient in galaxy differentation. Evidence for this is the fact that
Sa or S0 galaxies are, at least to first order, identical in and out of
clusters, which suggests that partitioning the baryons between spheroid
and disk may be more important than later environmental effects.

SCHECHTER: Joseph, I was jealous of your colorful viewgraphs and so
dazzled by them that I didn't notice that the ellipticals are 5σ away
from the cooling curves. Can you tell us how they got there?

SILK: There was a lot of dissipation as the galaxies formed.

SCHECHTER: So the number of σ away from the curves is a measure of how
much the space density has evolved.

SILK: Yes, although the amount of evolution depends on the theory - on
whether you think it took a 1σ or a 3σ fluctuation to make a galaxy.

PRIMACK: I'd like to comment on that. As we pointed out in our paper in
Nature, it is misleading to use these pictures in which you show 1σ and
3σ tracks if you are trying to figure out where a galaxy originally came
from. The problem is that there is some evolution in temperature, i.e.,
in (velocity dispersion)^2, as well as a great deal of evolution in
density. If you plotted mass versus temperature, then the ellipticals
would look like 3σ fluctuations when you normalize things so that
typical spirals are 1σ fluctuations.

SILK: I don't think that masses of ellipticals are known well enough to
do that.

E. TURNER: In the simple biased galaxy formation models, you have a
single free parameter: the number ν of sigma you associate with the
threshold for the formation of a visible galaxy. It must be adjusted to
give 1) the correct galaxy number density, 2) the right fraction of the
total (inflationary) $\Omega=1$ associated with galaxies, 3) the right galaxy
covariance function, 4) the correct depth of galaxy potential wells,

etc. Can this be done with a single value for the threshold? If not, is there any obvious second parameter you would call on?

KAISER: To get the required enhancement of clustering of galaxies relative to the matter (requirement 3), you need $\nu^2 \simeq 1.7\ (1+z_f)$, where z_f is the redshift of galaxy formation. Only for a moderately recent formation epoch, $z_f \simeq 3 - 5$, say, can the other requirements be satisfied. The most stringent requirement is that in rich clusters, where galaxy formation was by assumption ~ 5 times more efficient than average, we know that at least $\sim 10\%$ of the gas turned into galaxies. This means that the global efficiency is then at least 2%, implying $\nu \simeq 3$. For this choice of parameters, our estimates of number density and potential depth do not seem to be grossly unacceptable.

GUNN: I'd like to comment on the question of how much of the mass in the Universe is in galaxies, and of how this is related to the threshold ν for galaxy formation. Nick Kaiser and I have a small disagreement about this. I think that these questions are not necessarily closely connected. Suppose you assume that the mass is distributed around the high-ν peaks like the correlation function at that epoch. Then you can easily show that a large fraction $\sim 70\%$ of the time, if you have a 3ν fluctuation of $10^9\ M_\odot$, it will typically bind $10^{12}\ M_\odot$ to itself. Even though these 3ν fluctuations involve only 0.1% of the mass in the Universe, they each scoop up 1000 times their own mass. So you can gather up nearly all of the material around the 3ν peaks. I've made this argument just at one mass scale. I know that the real world is more complicated than this, but some such phenomenon must exist.

MELOTT: In response to the concerns voiced by Ed Turner and Jim Gunn, I'd like to report some results recently found by Jim Fry and me. In cold dark matter simulations, we calculated two-point and three-point correlations for biased sets. I should stress that this was done with a large smoothing window, corresponding to about $10^{12}\ M_\odot$, and that for the first time the correlations were calculated in a mass-weighted way, rather than including all connected regions with equal weight. We found that a simple 1σ bias was sufficient to boost the correlation amplitude to an acceptable level and to simultaneously make $Q \approx 1$, where Q is the reduced amplitude of the three-point function. An interesting extra benefit of this procedure is that voids are much more prominent than when using a larger bias threshold and a small smoothing window.

FABER: I'd like to ask the people who make n-body models whether large-scale structure makes a good case for the cold dark matter model. This is a point I was trying to raise in my introductory talk. Is the problem with the models that the correlation function doesn't stay positive at large enough radii to give the Kaiser effect, or is it that the predicted correlation function for galaxies stays high on large scales while the observed correlation function goes negative?

WHITE: In the cold dark matter picture the mass correlation function is supposed to go negative at a length scale of ~ 70 Mpc, well outside the range you can actually measure in the n-body calculations. The biasing, or Kaiser effect, will only amplify the underlying correlation. So an inconsistency only arises if it is really true that the cluster-cluster correlation function is positive at 100 Mpc; there isn't a problem with the fact that it is positive at 25 Mpc. I'd like to repeat that I don't think our present simulations can tell you much one way or the other about large-scale structure, because the regions of the Universe that we are simulating aren't large enough.

N. BAHCALL: I'd like to make a comment about large-scale structure to partially answer Sandy's question. There is structure seen not only in the cluster-cluster correlation function at large scales like 50 Mpc, but also in the supercluster correlation function. Here we see a positive correlation among superclusters at scales of $\sim 100 - 130$ Mpc, and the correlation is even stronger than the cluster-cluster correlation. This is of course somewhat uncertain because of the small size of the supercluster sample, but it does show that there is structure at these very large scales. Now there is a problem with the galaxy-galaxy correlation function going negative while the cluster-cluster function stays positive only if you believe the amplification model. Another point of view has been taken by Szalay and Schramm, who look at it as a scale-invariant clustering process. Looking at clusters of galaxies, the correlation strength appears to continuously increase with mean separation of the system for clusters of all richness classes. Now we have shown that the same correlation function exists for the superclusters, when normalized to the mean separation of the systems. In this case the possible slight difference of the galaxy-galaxy correlation function may be due to gravitational perturbations on top of some infrastructure which dominates the large-scale structure.

GALAXY FORMATION IN A UNIVERSE DOMINATED BY COLD DARK MATTER

Edmund Bertschinger
Department of Astronomy, University of Virginia

ABSTRACT

The mass spectrum of bound baryonic systems (galaxies and globular clusters) is computed as a function of redshift in an Einstein-de Sitter ($\Omega=1$) universe dominated by weakly interacting, cold dark matter. Baryons are assumed to fall into primordial density peaks in the cold particle distribution when the mass in the peaks exceeds the baryon Jeans mass. The distribution of peaks is computed using Gaussian statistics. As the universe expands the baryonic mass attached to a given peak increases because of infall (treated in a spherical approximation), and new peaks of lower amplitude become nonlinear. Globular clusters form first (by z~40 if the galaxies represent a biased mass distribution). The remaining gas may be reheated to ~10000 K if a few percent of globular cluster (or Pop. III) stars are very massive. Reheating increases the baryon Jeans mass and delays galaxy formation until z≲10. The present method reproduces the shape (but not the amplitude) of the Schechter galaxy mass function when merging of substructure is included in an approximate fashion.

J. Kormendy and G. R. Knapp (eds.), Dark Matter in the Universe, 360.

EMBEDDING OF GALACTIC SYSTEMS IN EXTENDED NEUTRINO CLOUDS

R. Cowsik and P. Ghosh
Tata Institute of Fundamental Research
Homi Bhabha Road,
BOMBAY 400 005
India.

ABSTRACT. In an expanding universe neutrinos of mass 10eV form conden-sates with typical mass 10^{16} $M_\odot$ and size ~ 0.5 Mpc. Visible matter like galaxies and stars are embedded in these and assuming a Maxwellian distribution and collisionless nature of these systems we solve the self consistent Poisson-Vlasov equations. These solutions correctly predict the profiles of luminosity of clusters of galaxies,dwarf,ellip-tic and spiral galaxies as well as their rotation curves; the embedding picture is supported by $M \sim R^3$ relation for astronomical systems as below.

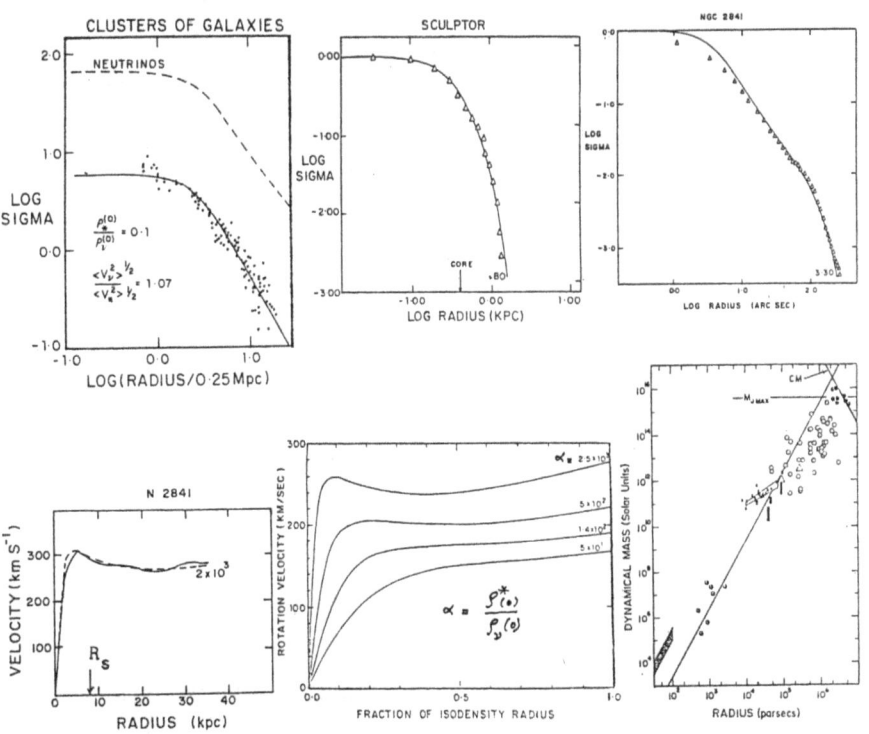

J. Kormendy and G. R. Knapp (eds.), Dark Matter in the Universe, 361.

DWARF GALAXIES, COLD DARK MATTER, AND BIASED GALAXY FORMATION

Avishai Dekel, Yale Univ. and Weizmann Inst.
Joseph Silk, U.C. Berkeley.

The formation of <u>dwarf, diffuse, metal-poor</u> galaxies, as a result of supernova driven winds, is reexamined in view of the accumulating data on dwarfs in the local group and in the Virgo cluster. The observed drop in both surface-brightness and metallicity with decreasing luminosity is not easily understood if the gaseous protogalaxies are self-gravitating (because they swell after gas-loss), but they are produced naturally inside <u>dominant halos</u>, with a mass-radius relation that indicates <u>'cold' dark matter</u>. The theory predicts for the faint dwarfs an M/L that increases with decreasing luminosity up to 10-100, and a corresponding slow decrease in velocity dispersion down to 5-10 km/s.

We find that the condition for global gas loss due to the first burst of star formation is that the virial velocity be below a critical value on the order of 100 km/s. In any hierarchical scenario for galaxy formation, this condition leads to <u>two distinct classes</u> of galaxies as observed: (a) the diffuse dwarfs (both dEs and dIs that have retained some gas) which mostly originate from typical (1 sigma) density perturbations, and (b) the normal, brighter galaxies (including compact dwarfs) which can come only from the highest density peaks (2-3 sigma). This provides a statistical <u>bias</u> for the formation of bright galaxies in denser regions, enhancing their clustering relative to the diffuse dwarfs. It may help reconcile the observed large scale universe with the flat model predicted by inflation. <u>The diffuse dwarfs are expected to trace the mass</u>; they should be present everywhere, including in the 'voids' which are deficient in bright galaxies. A substantial amount of lost gas is expected to be present in the 'voids'.

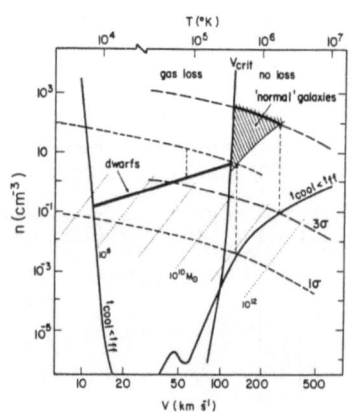 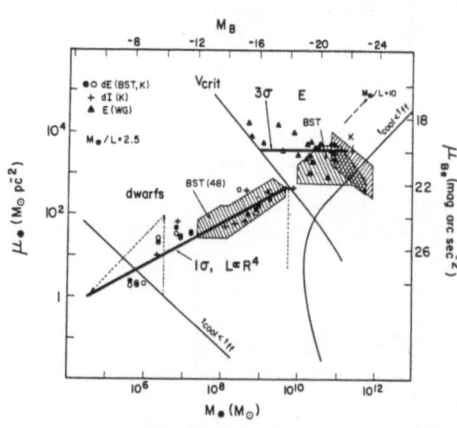

The critical curves for cooling and for gas-loss confine the locii for dwarfs and for 'normal' galaxies, that emerge from 1 and 3 sigma CDM perturbations. Excellent agreement between the theory and observations.

J. Kormendy and G. R. Knapp (eds.), Dark Matter in the Universe, 362.

HIERARCHICAL CLUSTERING: ANGULAR MOMENTUM DENSITY ANTI-CORRELATION

Yehuda Hoffman
Department of Physics, University of Pennsylvania
Philadelphia, Pennsylvania 19104

The growth of angular momentum in a general clustering scenario is analyzed and its dependence on the primordial density perturbation, δ_o, is calculated. High density peaks are found to evolve into low angular momentum and low spin parameter ($\eta = J\, E^{\frac{1}{2}}\, G^{-1}\, M^{-5/2}$) systems. The basic mechanism behind this anti-correlation is that anisotropic structure, which leads to the growth of angular momentum, is statistically uncorrelated with δ_o. The collapse time, on the other hand, depends on δ_o and it is this time scale which determines the angular momentum of bound objects. Three cases are studied: (i) Lagrangian sphere (quasi-linear evolution): $J(1\sigma)/J(3\sigma) = 3.3$ and $\eta(1\sigma)/\eta(3\sigma) = 1.9$ where δ_o is given in $\sigma(= <\delta^2>^{\frac{1}{2}})$ units.
(ii) Eulerian sphere (2nd order in δ): $J \propto (\delta_o/\sigma)^{-5/2}$ and $\eta \propto (\delta_o/\sigma)^{-2}$.
(iii) Arbitrary Lagrangian volume (1st order in δ): $J \propto (\delta_o/\sigma)^{-3/2}$ and $\eta \propto (\delta_o/\sigma)^{-1}$.

If, indeed, the morphological type of galaxies depends on their angular momentum and hence on their primordial density, then the dependence of the abundance of galaxies on morphology and on local density of galaxies is readily understood. The idea is that high σ fluctuations evolve into early type galaxies and low σ's into late types. In a clustering scenario in which the power index is less than 1, the hierarchical structure of clumps within clumps is unstable against tidal interaction. As the cross section for tidal disruption scales as the surface density (hence as δ_o^{-2}), one expects disruption in late-type galaxies more than in early types. In particular, in the violent environment of rich clusters only the high σ perturbations, i.e. early type galaxies, are likely to survive tidal interactions. This explains the variation of relative abundances of galaxies of various types with the mean local density. The enhanced clustering of the high σ peaks should further enhance this variation.

J. Kormendy and G. R. Knapp (eds.), Dark Matter in the Universe, 363.
© *1987 by the IAU.*

THE FORMATION OF GALACTIC HALOS IN UNIVERSES DOMINATED BY COLD DARK MATTER

Barbara S. Ryden and James E. Gunn
Princeton University Observatory

If the universe is dominated by cold dark matter (CDM), had an inflationary phase, and has $\Omega = 1$, then the spectrum of primordial density perturbations can be calculated. Using the spectrum calculated by J. Bardeen, we have found the structure of halos which form around peaks in the density. If the initial density is given by $\rho(\vec{x})=\rho_c[1+\delta(\vec{x})]$, where $\delta(\vec{x})$ is a Gaussian process, then the mean run of density around a peak is proportional to the correlation function of $\delta(\vec{x})$. If the mass about a peak comes to equilibrium in near-circular orbits, the exact equilibrium configuration can be calculated. The resulting rotation curve for a 1.5σ perturbation is shown as the curve HB in the figure below. As the baryon mass cools and condenses, it drags the CDM in with it, conserving the adiabatic invariants. Putting a 5×10^{10} $M_\odot$ exponential disk and a 2×10^{10} $M_\odot$ bulge at the center of the system, the rotation curve of the resulting compressed dark halo is given by the curve HA. D and B show the disk and bulge contributions, and T gives the net velocity. There are many arguments for wanting large galaxies to form from perturbations rarer than 1.5σ. Our spectrum was normalized by assuming that mass clusters as galaxies do on large scales; if $\Omega = 1$, however, galaxies cannot trace the mass and the amplitude must decrease. The flatness of the rotation curve from the baryon-dominated to the CDM-dominated region comes about, in this case, from the imposition of the phenomenological baryon distribution, but the following argument indicates that the result is general. The ratio of baryonic to dark matter is ~1/15; the rotation parameter Λ, the ratio of rotation velocity to halo dispersion, is also ~1/15. When baryons fall though the CDM to the point at which they are rotationally supported, their density increases by $(15)^3$. The CDM density increases by $(15)^2$. Hence, baryon and CDM densities are comparable at the mean baryon radius.

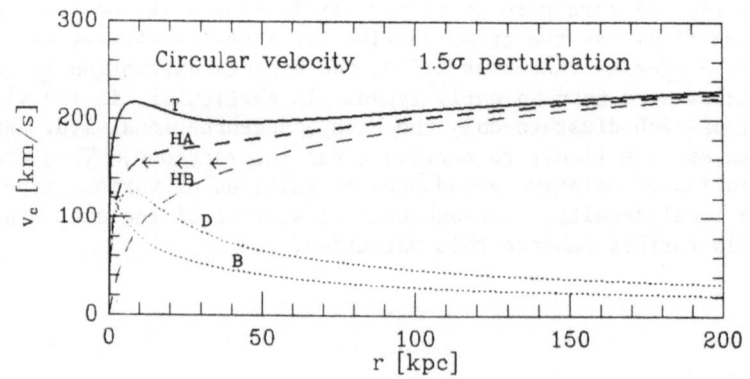

364

J. Kormendy and G. R. Knapp (eds.), Dark Matter in the Universe, 364.

HYDROGEN MOLECULES AND THE RADIATIVE COOLING OF PREGALACTIC SHOCKS

Paul R. Shapiro and Hyesung Kang
University of Texas at Austin

When a pregalactic gas of H and He is heated and reionized, as by a
shock wave occurring in the nonlinear collapse of density fluctuations
or in the case of explosions in the IGM, the gas cools radiatively and
recombines out of equilibrium. The temperature drops faster than the
ions can recombine. When the temperature falls below 10^4K, the
residual electron concentration is large enough, as a result, to form
H^- ions which form H_2 molecules, via $(H + e \rightarrow H^- + h\nu)$ and $(H + H^- \rightarrow$
$H_2 + e)$. Molecules also form via $(H^+ + H \rightarrow H_2^+ + h\nu)$ and $(H_2^+ + H \rightarrow$
$H_2 + H^+)$. As a consequence, H_2 can form with a sufficient concentration
$(\sim 10^{-3})$ to cool the gas further, by rot-vibrational line excitation and
the formation process itself, to $\sim 10^2$K. This has an important effect
on the Jeans mass and fragmentation. We show some illustrative results
below for the time-dependent cooling and non-equilibrium recombination
and molecule formation. The three cases are as follows: (1) isochoric
cooling at hydrogen number density 1 cm^{-3}; (2) isochoric cooling at
3×10^7 cm^{-3}; (3) isobaric cooling starting at initial density 1 cm^{-3}.
At high densities, molecular cooling is suppressed.

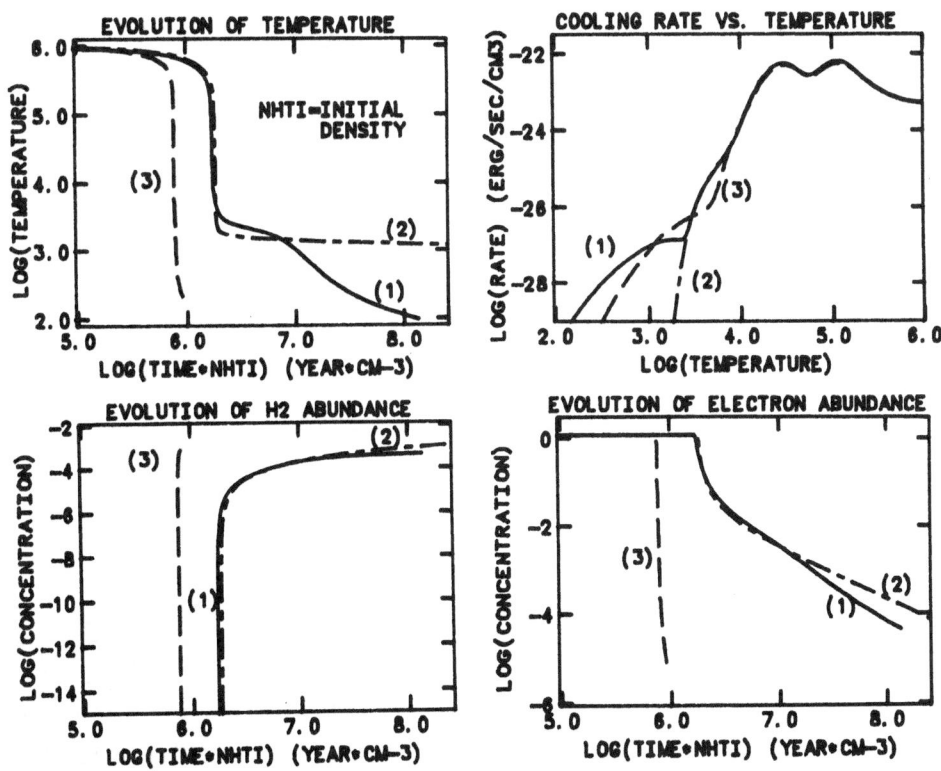

J. Kormendy and G. R. Knapp (eds.), Dark Matter in the Universe, 365.

COSMOLOGICAL HII REGIONS

Paul R. Shapiro[1], Ira Wasserman[2], and Mark L. Giroux[1]
[1]University of Texas at Austin
[2]Cornell University

We have generalized the classical description of ionization front propagation to the case of a point source in a uniform, cosmologically expanding gas. We present illustrative curves for the comoving radius and peculiar velocity for several turn-on redshifts, z_{ON}, for $\Omega_{tot} = 1$, $\Omega_b = 0.1$, $h = 1$. The quantity RS is the generalized Strömgren radius $[RS = RS_i (1 + z_{ON})/(1 + z)$, $RS_i = (3N_u/4\pi n_{H,i}^2 \alpha_2)^{1/3}$, $N_u =$ photoionizing number flux per source, $\alpha_2 =$ recombination rate to $n = 2$, $n_{H,i} = n_H^o (1 + z_{ON})^3]$. The quantity $T_{ON} = 2(1 + z_{ON})^{-3/2}/(3H_o)$. We also plot ζ, the value of $(2n_Q^o N_{ph,Q}/3H_o n_H^o)$ needed to ionize the IGM with overlapping QSO HII regions by redshift z_{OV} for QSO turn-on at various z_{ON}, where $N_{ph,Q} =$ ionizing photon luminosity per QSO, $n_Q^o =$ QSO number density (present co-moving value), $n_H =$ H density of IGM, and $n_H/n_H^o = n_Q/n_Q^o = (1 + z)^3$. From a recent preprint by Koo (1985), we estimate $\zeta \lesssim 1$ (for $\Omega_b = 0.1$, $h = 1$) for QSO's with $L \sim 10^{45}$ erg s^{-1}. In this case, the observed QSO's cannot be the sole source of the IGM ionization that is implied by the null detection of the Gunn-Peterson effect for QSO's with $z > 2$.

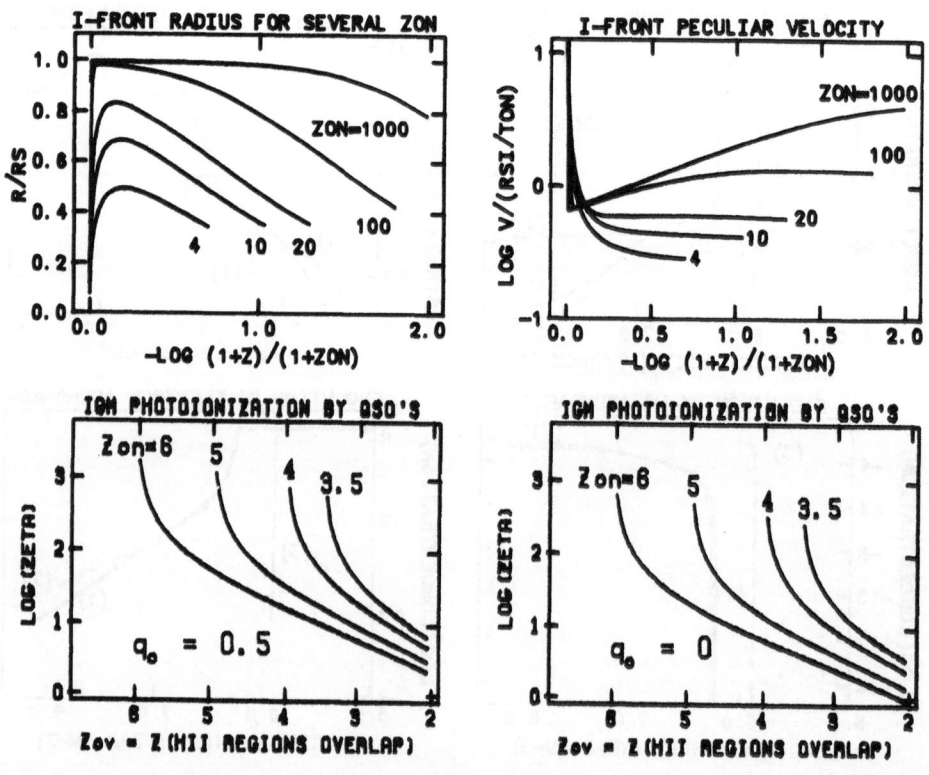

J. Kormendy and G. R. Knapp (eds.), Dark Matter in the Universe, 366.

A MODEL FOR THE SEQUENCE OF ELLIPTICAL GALAXIES

Rosemary F. G. Wyse,
University of California, Berkeley
Bernard J. T. Jones,
NORDITA, Copenhagen

We present a simple model for the formation of elliptical galaxies, based on a binary clustering hierarchy of dark matter, the chemical enrichment of the gas at each level being controlled by supernovae. The initial conditions for the non-linear phases of galaxy formation are set by the post-recombination power spectrum of density fluctuations. We investigate two models for this power spectrum – the first is a straightforward power law, $|\delta_k|^2 \propto k^n$, and the second is Peeble's analytic approximation to the emergent spectrum in a universe dominated by cold dark matter. The normalisation is chosen such that on some scale, say $M \sim 10^{12} M_\odot$, the objects that condense out have properties – radius and velocity dispersion – resembling 'typical' galaxies. There is some ambiguity in this due to the poorly determined mass-to-light ratio of a typical elliptical galaxy — we look at two normalisations, $\sigma_{1D} \sim 350 \text{kms}^{-1}$ and $\sigma_{1D} \sim 140 \text{kms}^{-1}$. The choice determines which of Compton cooling or hydrogen cooling is more important during the galaxy formation period. The non-linear behaviour of the perturbations is treated by the homogeneous sphere approximation.

Each power spectrum model is investigated by two techniques. A simple binary hierarchy which considers only *typical* 1σ fluctuations is compared with a more realistic approach where the amplitudes of the fluctuations on any scale are randomly sampled from a Gaussian of zero mean and variance set by the power spectrum. In the latter approach, the amplitudes of the fluctuations on all scales smaller than that destined to become a galaxy are self-consistently assigned from the largest scales down through the smallest, the technique being formally equivalent to a Hadamard decomposition of the density irregularities in that region. The amplitudes assigned to each level of the hierarchy enable us to calculate collapse times, and hence to know whether a spherical lump on a given level has time to complete its collapse before the next level turns around and collapses on top of it. We follow heating and cooling, and the chemical evolution, of lumps of gas due to the cumulative effect of supernovae in removing gas, or by the next level falling in on top. We allow for dissipation of energy as stars form by increasing the velocity dispersion of a given perturbation by an amount proportional to the mass of newly formed stars.

The simple model of a merging hierarchy presented here can account for the observed correlations between global properties of elliptical galaxies – metallicity, velocity dispersion, luminosity, surface brightness – perhaps favoring a flat spectrum of initial fluctuations, as predicted by models where the universe is dominated by cold, dark matter. Low luminosity (mass) ellipticals result from truncation of the hierarchy. The nature of the last merger (gaseous or stellar) and the stage along the hierarchy at which it took place are very important in determining the ultimate character of a galaxy. These are controlled in part by the statistical spread about the mean initial density fluctuation on a given scale, and on larger and smaller scales. If there is little gas around, either because the earlier stages have consumed it all or because it has been swept away in a cluster environment, the galaxy is an elliptical. If gas remains, but cannot cool rapidly to form stars, the galaxy will aquire a disk-structure.

J. Kormendy and G. R. Knapp (eds.), Dark Matter in the Universe, 367.

NON-LINEAR EVOLUTION OF INHOMOGENEITIES IN UNIVERSES DOMINATED BY DARK MATTER

S. F. Shandarin
Keldysh Institute of Applied Mathematics
Miusskaya Sq. 4, Moscow A-47, 125047

ABSTRACT. The astronomical data show that most of the mass in the Universe is dark. At present there are a few models of the dark matter: (i) the standard neutrino model, (ii) the model with unstable neutrinos, (iii) axion model. All models can be modified with Λ - term which plays role of the homogeneous component of the dark matter. In this paper the non-linear processes of gravitational instability are briefly discussed.

1. INTRODUCTION

The problem of dark matter has many different aspects, both observational and theoretical. In this report I consider one that concerns the non-linear evolution of density perturbations in their course to the formation of large scale objects: galaxies, clusters and superclusters of galaxies. It is assumed that most of the dark matter consists of some sort of weakly interacting particles: neutrinos, photinos, axions..., and the background universe model is based on the well known inflation scenario (for review see, for example [1,2]).

The theory of the non-linear processes must be an essential part of future full theory of the formation of structure in the Universe, because the observed objects like clusters and super-clusters are the inhomogeneities of density in the non linear regime of evolution: $\delta\rho/\bar{\rho} > 1$. In recent years the accuracy of theoretical predictions became so high that some models (for example the universe dominated by stable neutrinos with rest mass about 30 ev) are claimed to be rejected because they cannot match the observations within a factor only about 2 or 3. To make predictions of such quality one needs to possess the quantitative theory of non-linear processes. At present such theory is not fully developed. However the understanding of the non-linear stage of density perturbation growth has much improved recently. For this symposium the theoretical prediction that non-linear evolution of density perturbations goes different ways in the different models of dark matter is of particular interest, hence coupled with observations it can be a probe for these models.

J. Kormendy and G. R. Knapp (eds.), Dark Matter in the Universe, 369–377.

2. EVOLUTION OF PERTURBATIONS

The failure to measure angular variations in the temperature of relic
radiation (except the dipole component [3,4]) is interpreted as the
absence of any noticeable density perturbations at decoupling. On the
other hand at present time the Universe is highly inhomogeneous on
scales smaller than superclusters of galaxies $M < 10^{15} M_{\Theta}$. Thus there
must be considerable amplification of density fluctuations probably
present at decoupling. Remembering the fact that most of the mass in
the Universe consists of particles interacting essentially only due to
gravity one arrives to the conclusion that amplification must be due to
gravitational instability. There is a rather long stage of linear
growth of inhomogeneities when baryons, representing a small portion of
the total mass $\rho_b \sim (0.1 \rightarrow 0.01)\rho_t$ move in the gravitational field of
dark matter. Baryons can gravitationaly influence dark matter when
their perturbations at some place reach the strongly non-linear regime:
$\delta\rho_b/\rho_b \sim (\delta\rho_b/\rho_b)$ on the homogeneous background of dark matter or $\delta\rho_b/\rho_b$
$\sim (\delta\rho_\nu/\rho_\nu)(\rho_\nu/\rho_b)$ (here density of the dark matter is conventionally
marked by index "ν"). The latter probably takes place in the central
regions of galaxies but is hardly possible on scales of clusters and
superclusters of galaxies.

When density perturbations reach the non linear stage evolution can
go in two principally different ways, depending on the spectrum of
initial perturbations at the linear stage: they are called
"fragmentation"[5] and "hierarchical clustering"[6]. It is worth noting
that in Nature the structure formation can possess the features of both
processes, though it is useful to discuss them separately.

3. FRAGMENTATION SCENARIO

The standard inflation scenario of early universe predicts a Zeldovich
spectrum of primordial adiabatic fluctuations, which in long wavelength
limit (small k) has the form $\delta_k^2 \propto k^n$ (here δ_k^2 is spectrum of density
perturbations) independently of the type of collisionless particles
dominating in the dark matter. The short wavelength part of the
spectrum evolves differently depending on the kind of particles, that
enables Dick Bond to introduce the terms "hot", "warm" and "cold"
particles. These terms specify the largest scale of perturbations
suffering the free streaming dissipation.

The typical "hot" particles are electronic neutrinos with rest mass
about 30 eV. The dissipation scale in this case is about $M\nu \sim 10^{15} M_{\Theta}$
and approximately coincides with the masses of superclusters[7,8]. In the
"hot" particle model of dark matter the fragmentation scenario is
typical. Owing to the sharp cutoff in the spectrum on scales smaller
than M_ν, the objects formed first have just this size and the very
specific shapes of irregular pancakes. This was predicted by Zeldovich
in 1970[9]. Later it was found that other kinds of objects, both
elongated and compact ones form as well as pancakes[10]. They can be
classified as typical (generic) singularities in accordance with
catastrophe theory. Numerical simulations (2D[11,12] as well as 3D[13-16])
have shown that flattened, elongated and compact objects form a distinct

cellular or network structure depending on the density contrast level.

In the frame of this scenario it is natural to suppose that galaxies form later by fragmentation of the baryon component of the pancakes and/or filaments and compact clusters[5]. However at present there is no good quantitative theory for this process.

What we can be sure of is our understanding of causes of the cellular structure formation[17-19]. The cellular structure originates inevitably at the non-linear stage of gravitational instability if the spectrum of density perturbations at linear stage has a sharp cutoff on short wavelengths and the scale of the cutoff λ_c is much greater than the Jeans length λ_J ; $\lambda_c \gg \lambda_J$. As the smaller the temperature of the medium the less the Jeans scale becomes this inequality is easier to satisfy in cold medium at $T \to 0$. Thus one who likes paradoxes can say the cellular structure originates in the cold medium of "hot" particles.

Considering different scenarios one usually tries to find the stage in the evolution which is the best fit to the present Universe. In the fragmentation scenario the cellular structure is quite distinct at this stage. What will happen to it later? The answer to this question will help us to understand the process of hierarchical clustering better.

Evolution of density perturbations in a cold medium ($\lambda_c \gg \lambda_J$) can be expressed with the Zeldovich formalism[9]

$$r_i = a(t) \cdot (q_i - b(t) \, \nabla_q \phi(q)) \tag{1}$$

here r_i and q_i are Eulerian and Lagrangian coordinates of particles, $a(t)$ is a scale factor of Friedmann universe, $b(t)$ is the growing solution of the linear density perturbations (if $\Omega_0 = 1$, $\Lambda = 0$, $P = 0$, then $b(t) \propto a(t) \propto t^{2/3}$); $\phi(q)$ is a function conserving full information about the growing mode of perturbations. In the linear regime $\delta\rho/\rho = b(t)\Delta\phi$. At the non-linear stage

$$\rho/\bar{\rho} = (1 - b \cdot \alpha)^{-1} (1 - b \cdot \beta)^{-1} (1 - b \cdot \gamma)^{-1} \tag{2}$$

here α, β, γ are the eigenvalues of the tensor $\partial^2\phi/\partial q_i \partial q_k$. The important result of the Zeldovich formalism is that the structure forming at the beginning of the non-linear stage is determined by the spatial structure of the functions α, β, γ (mostly by the largest one)[5]. Unfortunately this formalism is not applicable at the non-linear stage when the cellular structure begins to disrupt.

4. HEIRARCHICAL CLUSTERING

The traditional approach[6] to the heirarchical clustering process is based on the calculation of the density perturbations at the scale $M \sim \rho \cdot k^{-3}$

$$\frac{\delta\rho}{\rho} (M) \propto (\int_0^k \delta_k^2 \cdot k^2 \cdot dk)^{1/2} \propto M^{-\frac{3+n}{6}} \tag{3}$$

if $\delta_k \propto k^n$ and $n > -3$. Combining this with the linear law of perturbation growth and assuming that after reaching the non-linear stage ($\delta\rho/\rho \sim 1$) the perturbation on the scale M virializes and ceases growing one easily finds the typical mass of the objects at time t ($\Omega_o = 1$, $b(t) = a(t)$)

$$M(t) \propto a^{\frac{6}{n+3}} \propto t^{\frac{4}{n+3}} \tag{4}$$

If $n > 4$ one must take into account the non-linear generation of long wave perturbations with spectrum $\delta_k \propto k^4$ that results in limit law $M \propto a^{6/7} \propto t^{4/7}$ even at $n > 4$.

However this approach is not fully non-linear, because it actually considers generation of long waves only "once" at some moment t_g and later they are assumed to grow in agreement with the linear law. But it cannot be true, because non-linear generation continues to work later at $t > t_g$ as well as at t_g.

Recently a rather simple but fully non-linear model has been developed (based on the Burgers equation, well known in the theory of turbulence) for the evolution of cellular structure at a stage when it disrupts[20]. Mass flows from the walls of the cells (originated as pancakes) to the ribs of the cells and from ribs to the compact concentrations in the apicies of the cells. Soon most of the mass concentrates in clusters whose mean mass grows continuously by the merging of clusters. This process is in fact hierarchical clustering.

Without going into details of the model I just enumerate the main features of the process. In contrast to the initial period of the non-linear stage when pancakes and the whole cellular structure is determined by the structure of the eigenvalues α, β and γ in Lagrangian space (2) the process of hierarchical clustering is determined by the structure of potential $\Phi(q)$ (1). The main result of this is a much stronger influence of the long wavelength part of the spectrum on the process of clustering – the so called long distance correlations. This becomes quite clear if one remembers that the spectrum Δ_k^2 of $\Phi(q)$ is $\Delta_k \propto \delta_k^2 \cdot k^{-4}$.

This non-linear theory of hierarchical clustering predicts a growth of the typical mass of clusters which coincides with the standard linear theory in the case of rather flat spectra

$$\delta_k^2 \propto k^n, \ M \propto a^{\frac{6}{n+3}} \quad \text{if } -1 < n < 1. \tag{5}$$

At $n > n_{cr} = 1$ there is a limit law $M \propto a^{3/2} \propto t$ independently of n. In terms of non-linear generation of long wavelength perturbations it takes place because of continuity of the process. It is interesting that n_{cr}

depends on the dimension of space: $n_{cr}(3D) = 1$, $n_{cr}(2D) = 2$ and $n_{cr}(1D) = 3$. If $n < -1$ the spectrum δ_k^2 must be bent down at some long wave λ_b, otherwise there is too strong divergence in spectrum of $\Phi(q)$ at $k \to 0$; because $\Delta_k^2 \propto k^{n-4}$. At $n < -1$ clustering is not actually pure hierarchical, because even at the time when masses $M \ll M_b \sim \rho \cdot \lambda_b^3$ decouple from Hubble expansion there is an ordering influence of scale λ_b. This influence is the stronger the steeper the spectrum is and at the limit of big negative n (n < 0 and $|n| \gg 1$) it becomes pure pancake picture with characteristic scale λ_b.

Spectrum with n $\approx$ -3 which naturally occurs in the axion model of the dark matter[22,23] is extremely difficult to analyse and at present there is no good theory for the clustering process in this case. Nevertheless it is clear that influence of the scale at the bend of the spectrum must be rather strong so the formation of the large scale network structure seems to be quite possible, however it is probably a much less distinct one than that in the standard neutrino model.

5. MODELS OF DARK MATTER AND SCENARIOS OF STRUCTURE FORMATION

Now let us briefly discuss present models of dark matter with respect to the structure formation. At present any baryonic model of the dark matter encounters probably unsolvable problems. Therefore I discuss only hypotheses assuming that most of dark matter is in a form of weakly interacting particles or a positive Λ - term.

As it has become clear from the above discussion the character of the large scale structure depends on the type of the spectrum of density perturbations at the linear stage after decoupling. In turn the spectrum of perturbations depends on the type of particles constituting the dark matter.

5.1 Stable electronic neutrinos

The model of the Universe dominated by stable electronic neutrinos with rest mass $\sim$ 30 eV (standard neutrino model[5,7]) seems to be the most economical of modern cosmological models in the sense of the number of ad hoc hypotheses needed. In this picture the formation of the large scale structure is mostly developed.

The spectrum of density perturbations is very simple: $\delta_k^2 \propto k$ on large scales M $> 10^{15}$ $M_\odot$ and very sharp cutoff at smaller scales[21] $\delta_k^2 \propto k^{-\mu}$ ($\mu \geqslant 12$) M $< 10^{15}$ $M_\odot$. For this reason the structure originating at non-linear stage is very distinct. It is probably more distinct than the real structure observed in the distribution of galaxies.

This model has several difficulties, widely discussed in the literature [2,5,24]. Remembering the uncertainties of astronomical data and poor understanding of the process of galaxy formation none of them seems to be fatal to the model[25]. However at present there is no positive solution to some of them. As the most severe of them I would like to mention: (i) the fast growth of the portion of the mass in the compact clusters[26] and (ii) too great M/L ratio in clusters. The problem of the correlation length becomes not so serious if the new results of Einasto et al.[27] are taken into account.

5.2 Unstable neutrinos

As an attempt to solve the problems of standard neutrino model
Doroshkevich and Khlopov[28] proposed the model with unstable neutrinos
($\nu_H \to \nu_L + f$). This model has succeeded in solving many problems of the
standard model including the ones mentioned above[29,30]. However it
operates with several additional parameters like: 1) ratio of mean
densities of unstable and stable components, 2) mean density of the
stable component, 3) ratio of masses of stable and unstable particles.
In addition it probably has difficulty with the age of the Universe and
appeals to the Λ -term. Compared with the standard model it is much
less economic and its basic "investment" is an appeal to "fantastic"
particles (term by A. Dolgov): unstable neutrino with rest mass about
100 eV, as well as familon.

Returning to the problem of the non-linear evolution one should say
that the process of structure formation in this model is very much
similar to that in the standard model. The only difference is that the
process of the evolution of cellular structure slows down due to the
decay of particles and the influence of a possible Λ-term.

5.3 Axion model

The other alternative to the standard model is the widely discussed
axion model[2,22,24]. The spectrum of density perturbations in this model
after decoupling in the long waves is the same $\delta_k^2 \propto k$; in the short
wavelength limit it is $\delta_k \propto k^{-3} \ln^2 k$[23] with a bend at $M \sim 10^{15} M_\Theta$. But
the bend is very smooth, extended more than an order of magnitude.

The process of structure formation starts from the origin of
objects about $10^6 M_\Theta$ (Jeans mass in baryons after decoupling) and
extends to $\sim 10^{15} M_\Theta$ at present time. This model has no principal
difficulties with the epoch of galaxy and quasar formation. It is more
economic in free parameters than the model of unstable neutrinos.
Nevertheless it also appeals to "fantastic" particles and for a better
fit of numerical parameters to Λ-term.

As follows from section 4 in this model a dim network structure can
naturally originate at the non-linear stage. 3D numerical simulations
perhaps give some positive evidence for this[31]. But to make adequate
simulations of this model is an extremely difficult task, because in
this case one needs to follow evolution of perturbations in very wide
range of scales which is beyond the possibilities of modern numerical
techniques. Nevertheless on the basis of available numerical data it
was reported that in this scenario there is a difficulty with
explanation of huge voids in distribution of galaxies in space[32]. This
must depend on the ratio of the power in k^{-3} part and the less steep
part of the spectrum at linear stage.

6. DISCUSSION

I have considered three of the presently most popular models of dark
matter from a dynamical point of view at the non-linear stage of the
density inhomogeneities. Comparison with the available astronomical

data shows that two of them: 1) model with unstable neutrinos and 2) axion model are in remarkably better positions than the standard neutrino model. However they have "missing" from the eyes of astronomers problems invoking "fantastic" particles like unstable neutrino, familon or axion as well as Λ-term of needed value. Will the "investments" in the form of additional hypotheses in this model bring enough "interest" in the form of better quantitative explanation of observational data to keep their positions of leaders on the "market" of models we shall see in a few years. New neutrino experiments and the Space Telescope will bring answers to many key questions.

Comparison of the structure originating at non-linear stage in different models with observational one seems to show that in neutrino models the structure is too distinct[40] and in axion model too dim[32] than real one. The former problem could be solved if one assumes some kind of explosions at the stage of galaxy formation. Physically the explosions could be similar to ones proposed by Ostriker and Cowie[33] but occuring only inside pancakes. This process can throw some galaxies out of the pancakes, making the structure dimmer, reducing the correlation length and slowing down evolution of the observable structure. In this case the distribution of galaxies would not be the same as that of the dark matter.

It is worth to mention the possibility of non-monotonic spectra of initial perturbations. Recently Kofman and Linde[34] found a mechanism to generate not only flat (in metrics) but also such kind of spectra during inflation. This gives new possibilities in the structure and galaxy formation but again by the price of introducing additional free parameters. The advantage of the spectra with two maxima is advocated by Dekel[35]. To make the structure comparable with the real one, one needs to have a rather steep short wavelength slope (probably $n < -3$) of the long wavelength maximum.

Finally I would like to make a short comment on works about percolation. I proposed[36] to use percolation parameter B_c as a quantity characterizing topological properties of the large scale distribution of galaxies. The application[40] of percolation technique to the data of CfA catalogue has shown its usefulness. Later Bhavsar, Barrow[37] and Dekel, West[38] criticized it as a cosmological test. However their criticism concerns mostly the method of estimation of B_c used firstly in[36,40]. That method was based on the calculation of the length of the largest cluster, which is simple and works well in the case of region of cubic shape analysed in[36,40], but for regions of more complicated shape considered in[37,38] it is not appropriate. Recently Klypin[39] has shown that the percolation parameter B_c can be easily estimated in regions of arbitrary shape using the approach of phase transitions. His analysis of different samples confirms our previous results[36,40]. Percolation analysis provides a new practically independent on correlation analysis parameter of the real distribution of galaxies characterizing its topology and any model of the large scale structure formation can be tested by it as well as by the correlation analysis.

I am grateful to A. Szalay for help.

REFERENCES

1. A.D. Linde Rep. Prog. Phys. 47, 925(1984).
2. J.R. Primack: Lectures presented at the International School of
    Physics "Enrico Fermi," Varenna, Italy, 1984.
3. I.A. Strukov, R.Z. Sagdeev, N.S. Kardashov, D. Skulachev, N.
    Eysmont: in Adv. in Space Res. Pergamon Press Proc. COSPAR, 1984.
4. J.M. Uson, D.T. Wilkinson: Astrophys.J. 283, 471 (1984).
5. S.F. Shandarin, A.G. Doroshkevich, Ya.B. Zeldovich: Sov. Phys. Usp.
    26, 46 (1983).
6. P.J.E. Peebles: The Large Scale Structure of the Universe,
    Princeton Univ. Press, Princeton 1980.
7. A.G. Doroshkevich, M.Yu. Khlopov, R.A. Sunyaev, A.S. Szalay, Ya.B.
    Zeldovich: in Proc. 10th Texas Sympos. on Relativistic Astrophys.,
    Ann. New York Acad. Sci 375, 32 (1981).
8. G.S. Bisnovatyi-Kogan, I.D. Novikov: Sov. Astron. 24, 516 (1980).
9. Ya.B. Zeldovich: Astron. Astrophys. 5, 84 (1970).
10. V.I. Arnold, S.F. Shandarin, Ya.B. Zeldovich: Geophys. Astrophys.
    Fluid Dynamics 20, 111 (1982).
11. A.G. Doroshkevich, E.V. Kotok, I.D. Novikov, A.S. Polyudov, S.F.
    Shandarin, Yu.S. Sigov: Mon. Not. R. Astr. Soc. 192, 321 (1980).
12. A. Melott: Mon. Not. R. Astr. Soc 202, 595 (1983).
13. S.F. Shandarin: The Origin and Evolution of Galaxies, eds. B.J.T.
    Jones, J.E. Jones, D. Reidel Publ. Comp. p.171, (1983).
14. A.A. Klypin, S.F. Shandarin: Mon. Not. R. Astr. Soc 204, 891
    (1983).
15. J. Centrella, A. Melott: Nature 305, 196 (1983).
16. C. Frenk, S.D.M. White, M. Davis: Astrophys.J. 271, 471 (1983).
17. Ya.B. Zeldovich: Sov. Astron. Lett. 8, 102 (1982).
18. S.F. Shandarin, Ya.B. Zeldovich: Comments on Astrophys. 10, 33,
    (1983).
19. S.F. Shandarin, Ya.B. Zeldovich: Phys. Rev. Lett. 52, 1488 (1984).
20. S.N. Gurbatov, A.I. Saichev, S.F. Shandarin: Preprint, Inst. Appl.
    Math. No. 152, (1984).
21. J.R. Bond, A.S. Szalay: Astrophys.J. 274, 433 (1983).
22. P.J.E. Peebles: Preprint, Princeton Univ. (1983).
23. V. Sahni, A.A. Starobinsky: Zh. Exp. Teor. Fiz. (1985).
24. G.R. Blumenthal, S.M. Faber, J.R. Primack, M.J. Rees: Nature 311,
    517 (1984).
25. A. Melott: Astrophys. J. 289, 2 (1985).
26. S.D.M. White: Preprint, NSF -ITP -84-103 (1984).
27. J. Einasto, M. Joeveer, E. Saar: in this volume (1986).
28. A.G. Doroshkevich, M. Yu. Khlopov: Mon. Not. R. Astr. Soc. 211, 277
    (1984).
29. M.S. Turner, G. Steigman, L.M. Krauss: Phys. Rev. Lett. 52, 2090
    (1984).
30. A.G. Doroshkevich, A.A. Klypin, E.V. Kotok: Preprint, Inst. Appl.
    Math. No. 120 (1984).
31. A. Melott, J. Einasto, E. Saar, I. Suisalu, A.A. Klypin, S.F.
    Shandarin: Phys. Rev. Lett. 51, 935 (1983).
32. M. Davis, G. Efstathiou, C. Frenk, S.D.M. White: Astrophys. J. 292,
    371 (1985).

33. J. Ostriker, L. Cowie: Astrophys. J. Lett. 243, L127 (1981).
34. L.A. Kofman, A.D. Linde: Nucl. Phys. B (1985).
35. A. Dekel: Astrophys. J. 264, 373 (1983).
36. S.F. Shandarin: Sov. Astron. Lett. 9, 104·(1983).
37. S.P. Bhavsar, J.D. Barrow: Mon. Not. R. Astr. Soc. 205, 61p (1983).
38. A. Dekel, M.J. West: Astrophys. J. 288, 411 (1985).
39. A.A. Klypin: Preprint, Inst. Appl. Math. No. 167 (1984).
40. J. Einasto, A.A. Klypin, E. Saar, S.F. Shandarin: Mon. Not. R. Astr. Soc. 206, 529 (1984).

ANISOTROPY OF THE MICROWAVE BACKGROUND AND COSMOLOGICAL DARK MATTER

V. N. Lukash
Institute for Space Research
Academy of Sciences of the USSR
Moscow 117810, USSR

ABSTRACT. Both large and small-scale $\Delta T/T$ anisotropies of the microwave background are reviewed in the context of the modern theories of the Universe structure origin. A number of primordial perturbation spectra and various types of cosmological missing mass are considered. Theoretical predictions are compared with the observational data. Importance of the $\Delta T/T$ measurements on large angular scales ($\theta > 5°$) is emphasized.

This year we observe the 20th anniversary of the microwave background radiation discovery which so splendidly confirmed a hot model of the Universe. Much has been done since then to study the degree of isotropy of the relic background. The experiment is already close to $\Delta T/T \sim 10^{-5}$. However, no angular variations of temperature beside the dipole component due to the motion of the Earth relative to the relic background [1], have been observed yet. Cosmological models with missing mass considered today predict the presence of fluctuations $\Delta T/T$ $\sim 5 \times 10^{-6}$ brought about by the embryos of the observed Universe structure. Thus it may well be that we are now on the verge of a new discovery which may confirm the validity of our concepts not only about the origin of galaxies, but also about the processes occurring in the Very Early Universe at the time primordial perturbations were only forming.

The modern theoretical and observational cosmology predicts that there were small primordial density and coupled gravitational perturbations in the early Universe. There also exists dark matter, whose nature is still unknown. The primordial perturbations start growing at the redshift $z_{eq} \sim 10^4$ and on due to the gravitational instability, become large and form the large-scale structure of the present Universe. Evolution of the small perturbations and the structure formation process depend upon the parameters of cosmological missing mass. However to detect these fundamental parameters from the observed structure is quite a difficult task due to nonlinear gas dynamical processes which accompanied the formation of galaxies.

Observation of the microwave background anisotropy provides an independent and much more accurate test of the processes which led to

J. Kormendy and G. R. Knapp (eds.), Dark Matter in the Universe, 379–392.

the formation of the Universe structure. Being small at the beginning, the primordial inhomogeneities perturbed the relic radiation which had been propagating freely since $z < 10^3$. These perturbations must manifest themselves at present as angular variations of the microwave background temperature.

Theoretical calculations of the $\Delta T/T$ fluctuations are carried out within the framework of linear theory. These calculations are simple and reliable. The $\Delta T/T$ amplitude is determined unambiguously by the primordial inhomogeneities and by the dynamics of their growth in the early Universe. Furthermore, the large-scale $\Delta T/T$ distribution is affected by the overall space curvature of the present Universe. Perturbation theory predicts for the open Universe ($\Omega_{tot} = \rho/\rho_{cr} < 1$) the existence of a characteristic correlation $\Delta T(\theta)/T$ structure with angular scale $\theta_c \sim \Omega_{tot}$. This structure is referred to as "spottiness". The discovery of the spotty structure (or its absence) in the $\Delta T/T$ large-scale anisotropy will enable a conclusion about the total matter density of the Universe, including the dark mass. To gain this sort of information in any other way is quite difficult.

Thus the measurements of the relic radiation anisotropy provide direct information about the primordial cosmological perturbations and about the parameters of the missing mass which govern the evolution of these perturbations.

On the other hand, the background anisotropy explorations are stimulated today by the theories of the Very Early Universe. Different theories predict different primordial perturbation spectra and different types of missing mass carriers. In particular, the standard inflationary theories predict both a flat primordial adiabatic perturbation spectrum and the overall matter density $\Omega_{tot}=1$. Thus, comparison of theoretical predictions for $\Delta T/T$ with the observational data provides a unique opportunity to test the theories of the Very Early Universe [2].

We shall outline below models to be considered and then proceed to the $\Delta T/T$ results. All figures are given for $H_o = 50$ km/s/Mpc and $T_\gamma = 2.7°$.

1. COSMOLOGICAL MODELS

The most attractive theoretically is a neutrino-dominated model of the Universe, since it has no free parameters and provides a natural cutoff in the primordial perturbation spectrum at wavelengths less than the supercluster scale. The laboratory experiment [3], which set the lower limit on the rest mass of the electron neutrino $m(\nu_e) > 10$ eV, has stimulated comprehensible investigations of the cosmological model with massive stable neutrinos [4]. Standard ν – models face well-known problems [5]: 1. Rapid nonlinear evolution ($z_s < 1$). 2. Small part of matter converted into galaxies ($M_{vir}/M_{gal} \sim 50$). 3. Large correlation scale of the dynamical structure (~ 20 Mpc). 4. Large peculiar velocities of galaxies. One can see that discrepancy between theory and observations is of the order of 2. We do not know yet how galaxies form and the relation between the distribution of galaxies and that of dynamical mass in clusters. Furthermore, according to [6] the

observational correlation scale of galaxies may increase from 10 Mpc to
20-30 Mpc. Thus, there are no convincing proof to turn down the ν -
model until these aspects of the structure are clarified (for
difficulties due to $\Delta T/T$ anisotropy see below). However we have
serious arguments to consider alternative models, which try to overcome
the above difficulties.

The missing mass is considered to consist of weakly interacting
particles. Here we list the alternative models according to how well
they are developed: (a) Models with unstable particles ($z_T \simeq 4-10$)
[7]. (b) Axion models and models with supermassive relic particles (m_R
> 1 KeV). (c) Models with Λ - term. (d) Open models $\Omega_{tot} < 1$.
Although the alternative models somewhat reduce the contradictions of
the ν - model, some subtle contradictions still remain. For example
models (b) poorly account for the homogeneous component of missing mass
($\Omega_h \simeq 0.8$), the age for models (a) is $\sim 10^{11}$ yr. The situation can be
improved by adding a small Λ - term, or by rejecting eq. $\Omega_{tot} = 1$ at
all and turning to open models.

The idea of models (a,c,d) is as follows: by introducing several
additional parameters to slow down the evolution at the nonlinear stage
and to decrease the gas temperature in superclusters. Models (a,c,d)
present a fine adjustment to the experimental data. Decay of massive
neutrinos into collisionless particles ($\tau_{dec} \simeq 10^{16} - 5 \cdot 10^{16}s$) or the
beginning of Λ - term or space curvature domination must be timed to
the nonlinear stage triggering. In addition such a small required
value of the Λ - term ($\Lambda = \rho_{vac} < \rho_{cr} \simeq 10^{-47}$ GeV^4) is a puzzle for the
elementary particle physics [8]. In this respect, the models (b) would
be more natural. The physical nature of supermassive particles is not
so much important. They may be primordial black holes, monopoles,
gravitinos and so on. What matters, in fact, is that they became
nonrelativistic long before the recombination epoch. In this case the
perturbation spectrum does not fall at shorter wavelengths, which
results in a more smeared structure.

2. PRIMORDIAL PERTURBATIONS

A field of primordial perturbations is described by a random function
$q = q(x)$ with Gaussian distribution of amplitudes. By definition:

$$\langle q_{\underline{k}} \, q_{\underline{k}'} \rangle = 2\pi^2 q_k^2 \delta(\underline{k} - \underline{k}'), \quad \langle q^2 \rangle = \int_{-\infty}^{\infty} q_k^2 \, d\ln k, \qquad (1)$$

where q_k is the Fourier transform of q ; $\langle ... \rangle$ means the average over
the state of the q - field; one-parameter family of q_k is called a
primordial spectrum; k is a modulus of the wave-vector. For
comparison, we give the relation of the gauge-invariant q - function to
metric perturbations $h_{\alpha\beta}$ in the synchronous reference system (t, x^α):

$$h_\alpha^\beta = q\delta_\alpha^\beta + 0 \ (t^2 q_{,\alpha}^{,\beta}) \qquad (2)$$

When relic nonrelativistic particles dominate the expansion, their density spectrum is as follows:

$$\delta_k = \frac{1}{20} (k\eta)^2 q_k \Psi (k) \tag{3}$$

where $\langle \delta_R^2 \rangle = \int \delta_k^2 \, d \ln k$, $\eta = \int dt/a$, $a = a(t)$ is the scale factor. The transfer function $\Psi (k)$ monotonically decreases with k growing ($\Psi(0) = 1$), its shape depends on the dark matter model considered.

We explored three simplest spectra of the primordial perturbations:

$$q_k \sim k^n \, , \; n = \begin{cases} -1 \\ 0 \\ 1 \end{cases} \tag{4}$$

The growing spectrum (n = -1, white noise) is taken from general considerations. The flat spectrum (n = 0) is predicted by standard models of the inflationary Universe [9]. The decaying spectrum (n = 1) originates in the parametric amplification theory [10], the scale of exponential cutoff in short wavelengths ($k^{-1} < 20$-600 Mpc) is an arbitrary model parameter. A combined spectrum with the changing slope $n = 0 \to 1$ is also allowed.

In the neutrino models the Ψ - function decays abruptly for large k [11], which allows for a simple spectrum normalization:

$$\langle \delta_R^2 \rangle = 1 \qquad \text{at} \qquad z = z_s \tag{5}$$

where at the redshift z_s the structure of the Universe emerges. The following results are given for $z_s = 3$.

In models (b) the Ψ - function damps rather slowly; $\Psi (k) \sim k^{-2} \ln k$ [12], which results in the logarithmic increase of the density perturbation growth at small scales. This demands a more subtle normalization of the primordial spectrum which is the weakest point of the linear theory (b). Further on we normalize the correlation function $\xi (r) = \langle \delta_R(0) \, \delta_R(r) \rangle^{1/2}$ by the correlation scale $r_c = 10$ Mpc ($\xi_c = 1$, $z = 0$) which corresponds to the formation at $z = 3$ of objects with masses $\sim 10^{12} M_\odot$ [13].

In models (a,c,d) subsequent evolution of the $\delta_k(\eta)$ function with massive particles no longer predominant at late stages, should be taken into account [14,15].

3. FORMATION OF $\Delta T/T$ ANISOTROPY

The possibility to compare theoretical $\Delta T/T$ calculations with observations is based on hypothesis (1), which is predicted by almost

all theories of the Very Early Universe. In fact, the reason is that
the primordial perturbations originate from quantum (or thermal)
fluctuations which obey, by definition, the Gaussian law of
distribution. The latter allows a theoretical estimate of the
confidence level for predicted quantities, originating from the fact
that we observe one of possible realizations of the Universe.

The theory gives $\Delta T(\underline{e})/T$ amplitude as a function of the unit
vector $\underline{e}$ along the line of sight. We can use it to calculate the
correlation function

$$\xi(\theta) = \langle \frac{\Delta T}{T}(\underline{e}) \frac{\Delta T}{T}(\underline{e}') \rangle^{1/2} \qquad (6)$$

and the root-mean square temperature fluctuation

$$\frac{\Delta T}{T}(\theta) = \langle (\frac{T(\underline{e}) - T(\underline{e}')}{T})^2 \rangle^{1/2} = (2(\xi(0) - \xi(\theta)))^{1/2} \quad (7)$$

which depend on the angle between the observation directions, $\cos\theta = \underline{e} \cdot \underline{e}'$. The ergodicity theorem allows identifying eqs. (6), (7) with
the observed distributions, where the average is taken over all
directions on the celestial sphere, with the fixed angle θ. For
confrontation, one should take into account the antenna beamwidth θ_a
which averages all fluctuations at $\theta < \theta_a$. (To make estimates, $\xi(\theta_a)$
function should be substituted in eq. (7) for $\xi(0)$).

The other way is to employ harmonic analysis (dipole component is
subtracted):

$$\frac{\Delta T}{T}(\underline{e}) = \sum_{l=2}^{\infty} \sum_{m=-1}^{1} a_{em} \Psi_{em}(\underline{e}), \qquad (8)$$

where $\Psi_{em}(\underline{e})$ are spherical functions. For the predicted multipole
anisotropy and dispersion we have:

$$(\frac{\Delta T}{T})_1 = (\sum_{m=-1}^{1} a_{em}^2)^{1/2}, \; \xi(\theta_a) \simeq (\sum_{l=2}^{[\theta_a]^{-1}} (\frac{\Delta T}{T})_1^2)^{1/2}. \qquad (9)$$

Hydrogen recombination dynamics in the Universe is well
investigated [16]. Cosmic plasma becomes transparent for relic photons

at $z_{rec} \simeq 10^3$ during the interval of redshifts $\Delta z/z \simeq 0.1$. The recombination scale and the transparent width correspond to the angles $\theta_{rec} \sim 5°$ and $\theta_\Delta \sim 10'$ on the celestial sphere ($\Omega_{tot} = 1$, $\Lambda = 0$).

For scales $\theta > 10'$ we can treat the recombination as an instantaneous process and explicitly calculate background anisotropy:

$$\frac{\Delta T}{T}(\underline{e}) = -1/2 \ e^\alpha e^\beta \int_{\eta_{rec}}^{\eta_o} \dot{h}_{\alpha\beta} \ d\eta + (\dot{v} + e^\alpha v_{,\alpha})_{\eta=\eta_{rec}} \quad (10)$$

where $h_{\alpha\beta}$, $\delta_b = 3\dot{v}$ and $u_b = (1; -v_{,\alpha}/a)$ are metric, baryon density and velocity perturbations in the synchronous, co-moving with relic particles, reference system; all the functions are taken on the light-cone; $(\cdot) = \partial/\partial\eta$; $(,\alpha) = \partial/\partial x^\alpha$. The first term in the right-hand-side of eq. (10) is due to the gravitational field perturbations [17]. The second and third terms represent distortions and motions of the sphere of the last scattering of photons, due to baryon density [18] and velocity [19] perturbations. On scales $\theta < 10'$ the recombination process via which the cosmic plasma transforms from the opaque to transparent state, should be taken into account.

The correlation scale of the ξ (θ) function appears to be $\theta_c \sim 20'$ [16]. At $\theta > 20'$, the dominant contribution in the relic-particle models comes from the integral (10), two next terms are essential for models (c,d).

In models with unstable particles the hydrogen recombination process may well be essentially slower due to a possible secondary ionization of plasma by the products of decay. This leads to weaker small-scale fluctuations $\Delta T/T$ ($\theta < 5°$) [20].

For large-scale calculations ($\theta > 5°$) only the integral in eq. (10) is important with the lower integration limit vanishing. Thus, $\Delta T/T$ predictions in this region do not depend at all on the recombination and secondary ionization dynamics, they are directly linked to the primordial metric perturbations.

4. RESULTS FOR $\Omega_{tot} = 1$

Fig. 1 presents the results of $\Delta T/T$ (θ) calculations for ν- and b-models with n = 0. The upper curves correspond to an ideal antenna, whereas the middle ones show an expected $\Delta T/T$ for the antenna beamwidth $\theta_a = 5°$ [2]. The dotted curves show the expected level of fluctuations in small and large angular scales when secondary ionization of cosmic plasma is taken into account [20]. (It is assumed here that massive neutrinos decay with $\sim 10^{-9} - 10^{-8}$ probability over the period of $10^{14} - 10^{16}$ s by producing γ - quanta with energy $10 - 10^2$ eV). Arrows ($\downarrow$) show the experimental restrictions [21-24]. Arrows ($\uparrow$) show the increasing large-scale anisotropy in model (b) due to possible growth of the correlation scale of galaxies. If $\xi_c \sim 30$ Mpc [6], then $\Delta T/T$ ($\theta > 5°$) fluctuations in ν- and b-models become of the same order.*
_____
* P.D.Naselskij, private communication.

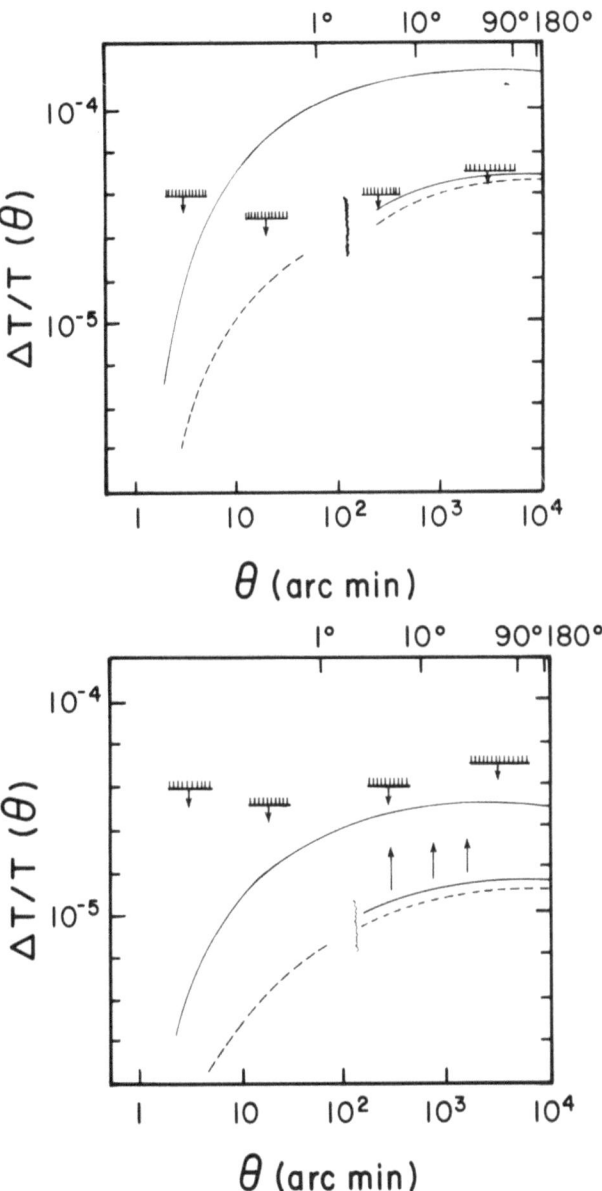

Fig. 1. The anisotropy of the microwave background $\Delta T(\theta)/T$ in stable neutrino model (upper panel) and in axion or very heavy particle model (lower panel). Observational restrictions: (4.5') [21]; (10') [22]; (6°) [23]; (10°-90°) [24].

Dark mass $\Delta T/T$	ν	a	b	c	Baryons; stable ν with $\Lambda \neq 0$
$1 = 2$ $((2\times10^{-5})$	2×10^{-5} [2,31]	2×10^{-5} [14]	4×10^{-6} [13,2,31]	10^{-5} [15]	$>5\times10^{-5}$
$\xi\,(6°)$ (5×10^{-5})	5×10^{-5} [2]	5×10^{-5} [14]	10^{-5} [2]	2×10^{-5} [15]	$>10^{-4}$
$\xi(20')$	10^{-4} [16]	2×10^{-5} [25]	2×10^{-5} [16,31,32]	8×10^{-5} [16,15]	$>10^{-4}$
$\xi(20')$ secondary ionization	$\sim10^{-6}$ [20]	–	$\sim2\times10^{-7}$ [20]	–	–
UW (3×10^{-5})	10^{-5}	3×10^{-6}	3×10^{-6}	2×10^{-5}	–
P (10^{-5})	5×10^{-5}	10^{-5}	10^{-5}	4×10^{-5}	–

Table I. $\Delta T/T$ anisotropy in different models of dark matter ($n=0$, $\Omega_{tot}=1$, ν – stable neutrinos, a – decaying neutrinos, b – axions or very heavy particles, c – models (b) with $\Omega_\Lambda = 0.8$). Observational data are given in parentheses (large-scale $\Delta T/T$ restrictions for quadrupole ($1=2$) and dispersion $\xi(6°)$ [24] and small-scale $\Delta T/T$ restrictions of [21] (UW) and [22] (P)).

1		$n = 0$	$n = -1$	$n = 1$
ν	1	10^{-2}	1.5×10^{-2}	10^{-2}
	2	2×10^{-5}	4×10^{-4}	$<10^{-5}$
b	1	2×10^{-3}	3×10^{-2}	2×10^{-3}
	2	4×10^{-6}	7×10^{-5}	$<10^{-6}$

Table II. Dipole ($1=1$) and quadrupole ($1=2$) ($\Delta T/T$) 1 amplitudes for different spectra of primordial perturbations.

In models with decaying neutrinos ($\tau_{dec} \simeq 1\text{-}5 \cdot 10^{16}$s, $z\tau \simeq 4\text{-}10$, $m_\nu \simeq 40\text{-}150$ eV) the level of small-scale fluctuations $\Delta T/T$ becomes lower by a factor of ~ 4 [25]. At large scales, $(\Delta T/T)_1$ anisotropy increases by a factor of 3 for $1 < z_\tau$ (and is of the same order for $1 > z_\tau$) compared to the models with stable neutrinos with the same m_ν, this may provide a test for the decaying neutrino model (a) [14]. The quadrupole component $\Delta T/T$ (l=2) recalculated for the experiment [24] (n=0), is as follows [14]:

$$(\frac{\Delta T}{T})_2 \simeq 8 \cdot 10^{-5} \; (\frac{24 \text{ eV}}{m_\nu}) \; . \tag{11}$$

The amplitude (11) becomes less than quadrupole anisotropy in the standard model ($2 \cdot 10^{-5}$) only if $m_\nu > 100$ eV. In models with Λ-term ($\Omega_\Lambda \simeq 0.8$) the amplitude of small-scale $\Delta T/T$ fluctuations increases by a factor of about 4, whereas large-scale fluctuations almost double [15,16]. The data about expected $\Delta T/T$ for n = 0 are summed in table I (the observational restrictions are given in parentheses). The last two lines show the value of $\Delta T/T$ for the experiment [21]

$$(\frac{\Delta T}{T})_{UW} = [2(\xi(1.5') - \xi(4.5')) - 1/2(\xi(1.5') - \xi(9'))]^{1/2}$$

and the expected value of $\Delta T/T$, which is compared to the experiment [22]

$$(\frac{\Delta T}{T})_P = [2(\xi(4.5') - \xi(9'))]^{1/2} \; .$$

Table II shows dipole and quadrupole components of $\Delta T/T$ for the ν- and b-models for various kinds of spectra [2].

5. SPOTTINESS

A radically new effect in the large-scale $\Delta T/T$ structure occurs in an open Universe. In this case the Universe 3-space curvature results in a new correlation scale $\Theta_c' \simeq \Omega_{tot}$ [26-28].
        The spotty structure effect may be qualitatively explained as follows. The field of random primordial perturbations $q(\underline{x})$ is homogeneous on the average, that is, the distribution does not depend on the $\underline{x}$ point. In the Lobachevski space (3-space of the homogeneous model) such fields may be presented as a random superposition of a full set of eigenfunctions, each with constant amplitudes over the space. These functions are plane-wave analogs in the Lobachevski space [27].

Each of such waves forms a "spot" in the large-scale $\Delta T/T$ distribution with an angular dimension $\theta_0 = 2 \tan^{-1} (\exp(-h\eta_0)) \simeq \Omega_{tot}$; here $h\eta_0$ is the ratio of today's horizon (η_0) to the curvature radius (h^{-1}). With $h > 0$ ($\Omega_{tot} < 1$) the random distribution of spots (with the fixed q_k spectrum) yields a statistical distribution $\delta T/T$ (θ) with the correlation scale $\theta_c' \simeq \theta_0$. If $h = 0$ ($\Omega_{tot} = 1$) the spot degenerates into a quadrupole structure ($\theta_0 = \pi/2$), and the spotty-structure effect is not present, since any superposition of quadrupoles is again a quadrupole.

The correlation scale θ_c' also depends on the primordial spectrum, hence if $\Delta T/T$ anisotropy over the range $\theta > 5°$ is found both the spectrum q_k and the total density of matter Ω_{tot} can be determined.

6. CONCLUSIONS AND DISCUSSIONS

The basic conclusions for the flat model of the Universe ($\Omega_{tot} = 1$) and the flat spectrum of primordial perturbations ($n = 0$) is as follows.

1. The standard model of the inflationary Universe with massive neutrinos (both stable and decaying) as missing mass contradicts the large scale $\Delta T/T$ limits [2]. The mismatching factor here is ~2 thus the final conclusion is only possible if the accuracy of measurements is improved.

2. The minimum predicted anisotropy level $\Delta T/T$ limits ($\theta > 6°$) may be detected if the RELIC experiment [24] sensitivity is improved to ~$6 \cdot 10^{-6}$.

3. The $\Lambda \neq 0$ models with baryon or stable neutrino missing mass disagree with current observations.

4. All models except those in item 3 do not contradict the limit [21], however, stable neutrino models and all models with $\Lambda \neq 0$ contradict the data [22].

5. The secondary ionization reduces $\Delta T/T$ ($\theta > 1°$) to the level $< 8 \cdot 10^{-6}$ [20]. A test for secondary ionization is to measure polarization which may reach here several tens percent with respect to $\Delta T/T$ anisotropy [29].

The flat spectrum ($n = 0$) is a crucial test in the $\Delta T/T$ problem. White noise ($n = -1$) practically contradicts the present – day observations while decaying spectra ($n \geqslant 1$) reduce the expected level of large-scale $\Delta T/T$ fluctuations. Note that spectra falling toward longer waves are predicted by the theory of parametric amplification of perturbations [10] and by inflationary theories of isothermal perturbations [28].

The optimal range of search for $\Delta T/T$ is $20' < \theta < 90°$. In case $\Delta T/T$ is detected at the angular scales $\Delta > 5°$ the spectrum of primordial perturbations q_k and the total matter density in the Universe Ω_{tot} could be determined [2].

References:

1. Smoot, G.F., Lubin, P.M. Ap.J. 234, L83, 1979.
    Fixsen,D., Cheng, E., Wilkinson, D. Phys. Rev. Lett. 50, 620, 1983.
2. Lukash, V.N., Naselskij, P.D., Novikov, I.D. Proc. Quantum Gravity-3, Moscow, 1984.
3. Kozik, V.S., Lubimov, V.A., Novikov, E.G., Nozik, V.Z., Tretyakov, E.F., Nuclear Phys. 32, 301, 1980.
4. Bisnovatyi-Kogan, G.S., Novikov, I.D. Astron. Zh. 57, 899, 1980.
    Doroshkevich, A.G., Khlopov, M.Yu., Zeldovich, Ya.B., Sunyaev, R.A. Astron. Zh. Lett. 6, 457, 1980.
    Doroshkevich, A.G., Khlopov, M.Yu., Sunyaev, R.A., Szalay, A.S., Zeldovich, Ya.B. Proc. X-Texas. Symp. Relat.Astrop. New York Acad. Sci., 32, 1980.
5. Klypin, A.A., Shandarin, S.F. MNRAS 204, 891, 1983.
    Frenk, C.S., White, S.D.M., Davis, M. Ap.J. 271, 417, 1093.
    Shapiro, P.R., Struck-Marcell, C., Melott, A. Ap.J. 275, 413, 1983.
    Kaiser, N. Ap.J. 273, L17, 1983.
    Hut, P., White, S.D.M. Nature 313, 637, 1984.
    Doroshkevich, A.G., Klypin, A.A., Kotok, E.V. Astron. Zh., 1985 (in press).
6. Einasto, J., Klypin, A.A., Saar, E. MNRAS, 1985 (in press).
7. Doroshkevich, A.G., Khlopov, M.Yu. MNRAS 211, 277, 1984.
    Turner, M.S., Steigman, G., Krauss, L.M. Phys. Rev. Lett. 52, 2090, 1984.
8. Dolgov, A.D. JETP Lett. 41, 280, 1985.
9. Gibbons, G.W., Hawking, S.W., Siklos, S.T.C. (eds.), The Very Early Universe, Cambridge Univ. Press, 1983.
10. Lukash, V.N. JETP 79, 1601, 1980.
    Kompaneets, D.A., Lukash, V.N., Novikov, I.D. Astron. Zh. 59, 424, 1982.
    Lukash, V.N., Novikov, I.D. In: Contr. paper 10-Int. Conf. GRG, ed. Bertotti, B. et al., 844, 1983.
11. Bond, J.R., Szalay, A.S. Ap.J. 274, 443, 1983.
12. Starobinski, A.A., Sakhni, V. JETP, 1985 (in press).
13. Peebles, P.J.E. Ap.J. Lett. 263, L1, 1982.
    Starobinski, A.A. Astron. Zh. Lett. 9, 579, 1983.
14. Kofman, L.A., Pogosyan, D.Yu., Starobinski, A.A. Astron. Zh. Lett., 1985 (in press).
15. Kofman, L.A., Starobinski, A.A. Astron. Zh. Lett., 1985 (in press).
16. Zabotin, N.A., Naselskij, P.D. Astron. Zh. 59, 447, 1982, 60, 467, 1983, Astron. Zh. Lett. 9, 643, 1983.
17. Sachs, R.K., Wolfe, A.M. Ap.J. 147, 73, 1967.
18. Silk, J. Ap.J. 151, 459, 1968.
19. Zeldovich, Ya.B., Sunyaev, R.A. Ap. Sp. Sci. 6, 358, 1970.
    Peebles, P.J.E., Yu, I.T. Ap.J. 162, 815, 1970.
20. Dorosheva, E.I., Naselskij, P.D. Ap. Sp. Sci., 1985 (in press).
21. Uson, J.M., Wilkinson, D.T. Ap.J. Lett. 277, L1, 1984.
22. Parijskij, Yu.N., Petrov, Z.N., Chernov, A.N. Astron. Zh. Lett. 3, 483, 1977.

Parijskij, Yu.N. et al. Communications Special Astron. Observatory, 1985.
23. Fabbri, R., Guidi, J., Melchiorri, F., Natali, V. Phys. Rev. Lett. 44, 1563, 1980.
Melchiorri, F., Melchiorri, B., Ceccarelli, C., Pietranera, L. Ap.J. 250, L1, 1981.
24. Strukov, I.A., Sagdeev, R.Z., Kardashev, N.S., Skulachev, D., Eysmont, N. In: Advances in Space Research, Pergamon Press, Proc. COSPAR, 1984.
25. Doroshkevich, A.G., Khlopov, M.Yu. Astron. Zh. Lett, 1985 (in press).
Doroshkevich, A.G., Astron. Zh. Lett., 1985 (in press).
26. Novikov, I.D. Astron. Zh. 45, 538, 1968.
27. Lukash, V.N. In: Contr. papers 8-Int. Conf. GRG, Waterloo Univ., 237, 1977. Proc. Symp. 104 IAU, ed. Abell, G., Chincarini, G., 149, 1982.
28. Bisnovatyi-Kogan, G.S., Lukash, V.N., Novikov, I.D. Proc. 5-Regional Meeting in Astr. (IAU/EPS), Liege, 1980.
29. Basko, M.M., Polnarev, A.G. MNRAS 191, 207, 1980.
30. Linde, A.D. JETP Lett. 40, 496, 1984.
Kofman, L.A., Phys. Lett. B, 1985 (in press).
31. Bond, J.R., Efstathiou, G. Ap.J. 285, L45, 1984.
32. Vittorio, N., Silk, J. Ap.J. 285, L39, 1984.

DISCUSSION

SILK: Could you give us some more details about the RELIC microwave background experiment, particularly about the confidence level of the measurements?

LUKASH: The first results of the RELIC experiment are as follows. The best scans cover about a fifth or a quarter of the sky. For these, contamination from the Moon and the Galaxy is less than 0.5 μK. The satellite was at about 10^6 km from the Earth and the scans were each about seven days long. During this length of time, the rotation of the satellite caused each point on the sky to be measured 10^6 times. This gives the very high sensitivity of the results. Each scan consists of observations of 120 points, where each point measures an area of the sky of ~6°×6°, the beamwidth of the antenna. On any scale ~6° or greater, the deviations from isotropy are found to be less than $\Delta T/T = 5\times10^{-5}$. If we assume that the primordial fluctuations were flat, this gives an upper limit to the amplitude of the quadrupole of 2×10^{-5}. This figure is a purely observational result not dependent on any theoretical assumptions.

LUBIN: What is the confidence level of the limit on fluctuations on the 6° scale? What is the value of the directly measured quadrupole?

LUKASH: There isn't a direct measure of the quadrupole term. The limit I gave is for temperature fluctuations on scales from 6° to 90°, but of course applies to any multiple harmonic, including the quadrupole term. If you assume Gaussian fluctuations, the upper limit on the quadrupole term is 5×10^{-5}. If you assume a flat spectrum, the limit drops to 2×10^{-5}. The confidence level of these values is 95%.

BURKE: When I look at the all-sky RELIC data, I agree that you have the accuracy you have quoted on 6° scales. But on the 90° scale I can see large-scale irregularities beyond the dipole term which look as though they are associated with the Galaxy. That is, there are fluctuations in the brightness of the background which are not the dipole term, but which have a scale larger than 6°, and which look as though they are along the galactic plane. I think that it is very hard to get an accuracy of 5×10^{-5} on the 90° scale when you can already see irregularities on a scale of ~60°.

LUKASH: Yes, of course, you are right. But as I said in my talk, data from only about a quarter of the sky were used to set the quoted limit. It was not found from the all-sky data. The influence of the Galaxy is quite subtracted from this result.

WILKINSON: The reason you can't get the large-scale measurement out of the RELIC data is that the sidelobes of the antenna are too large. For example, the Earth and Moon are seen even when the antenna is pointed far away from them. So there will be a lot of large-scale contamination in the RELIC data. I think that this is why you have to extrapolate from the 6° scale given by the antenna beamwidth to larger scales.

LUKASH: I agree, but the limits on the quadrupole and octupole anisotropies that I gave are quite adequate.

USON: I shall not discuss the validity of the results by Parijskii that you quote as he is not here to argue about them. But I would like to point out that when they were presented three years ago at IAU Symposium 104, there was quite a bit of controversy about them. The general consensus was that before they could be accepted, a thorough description had to be published. I have not seen this; has it been published yet?

LUKASH: Yes, the latest publication I know is that by Korolkov, D. V., and Parijskii, Yu. N. (1985), Communications of Special Astrophysical Observatory, 41, 42, 43. I'm still not sure what theoretical quantity should be compared with their results.

PEEBLES: You've given a very clear description of the value of $\Delta T/T$ expected from primeval adiabatic perturbations. Have you also considered the possibility of primeval isocurvature perturbations in which density is constant?

LUKASH: Yes. An analysis of the spectrum of isocurvature perturbations on all scales has been made by Starobinski and colleagues. If you take a flat spectrum, the predicted amplitude for isocurvature perturbations will be at least six times greater than that predicted for adiabatic perturbations. So in any case you should choose another spectrum for isocurvature perturbations. Alex Szalay has already mentioned the work of Kofman and Linde, who proposed an inflationary model in which they could get isocurvature perturbations which have a flat spectrum up to some scale and then an abrupt decay. This model is good for describing large-scale fluctuations.

FELTEN: You mentioned results of Kofman and Starobinski for a model with nonzero Λ. Are these limited to a specific model? For example, were they derived for a model with flat space and $\Omega_0 \approx 0.2$?

LUKASH: Yes. They consider flat space with arbitrary Ω due to Λ, but the figures I mentioned are for $\Omega_0 \simeq 0.2$ and $\Omega_\Lambda \simeq 0.8$.

DISTURBANCE OF THE CBR BY ISOTHERMAL PERTURBATIONS

Hideo Kodama, Yasushi Suto and Katsuhiko Sato
Department of Physics, University of Tokyo, Japan

Following the standard scenario of galaxy formation, density fluctuations with amplitude $\delta \sim 10^{-3}$ should have been present at the recombination time t_R in order that galaxies and clusters of galaxies can be formed. Recent observations of the anisotropy of the cosmic background radiation(CBR), however, indicate that δ is less than 10^{-4} at t_R for adiabatic perturbations in the baryon-dominated universe.

It is widely believed that this difficulty can be avoided if the initial density fluctuations are isothermal type. Here we show that this conventional prejudice is not correct. We only consider the pre-recombination stage in the baryon-dominated universe.

First we explain the essential feature of evolution of isothermal perturbations. In the linear perturbation theory baryon and radiation density contrasts, $\delta_b = \delta\rho_b/\rho_b$ and $\delta_r = \delta\rho_r/\rho_r$, are expressed in terms of $\delta = \delta\rho/\rho (\rho = \rho_b + \rho_r)$ and the perturbation to entropy, $S \equiv 3\delta_r/4 - \delta_b$ as

$$\delta_b = \frac{\rho}{h}\delta - \frac{4\rho_r}{3h}S, \qquad \delta_r = \frac{4\rho}{3h}\delta + \frac{4\rho_b}{3h}S, \qquad (1)$$

where $h = \rho_b + 4\rho_r/3$. Due to the strong coupling between photons and baryons S stays constant. On the other hand δ can be shown to remain much smaller than S on superhorizon scales. Hence δ_r/δ_b increases with ρ_b/ρ_r and eventually it may become greater than unity in the baryon-dominated stage. Thus on these scales the isotropy of the CBR constrains δ_b more strongly for isothermal perturbations than for adiabatic ones.

In order to obtain a precise constraint, we have numerically calculated the values of δ_r and δ_b at t_R and compared them with the observation of the CBR. In Fig.1 the resultant upper limits on δ_b at t_R are shown. From this figure we can conclude that it is difficult for the structures on scales, at least, larger than clusters of galaxies to form. This result holds even in a dark matter dominated case on supercluster scales. Further this figure indicates that even the formation of galaxies is difficult if δ_b on mass scale M at t_R obeys the power law $M^{-(n+3)/6}$ with $n \le 1$.

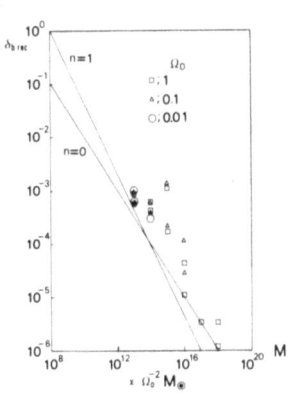

Fig.1

References
H.Kodama and M.Sasaki, Prog. Theor. Phys. Supple. No.78 (1984), 1.
Y.Suto, K.Sato and H.Kodama, Astrophys. J. Lett. 292 (1985), L1.
Y.Suto, K.Sato and H.Kodama, Prog. Theor. Phys. 73 (1985), 1151.

J. Kormendy and G. R. Knapp (eds.), Dark Matter in the Universe, 393.

POSSIBLE CONSTITUENTS OF HALOS

Martin J. Rees
Institute of Astronomy
Madingley Road
Cambridge CB3 0HA
United Kingdom

ABSTRACT. There still seem to be three serious contenders for the dark matter in galactic halos and groups of galaxies: (i) very low mass stars, (ii) black hole remnants of very massive stars or (iii) some species of particle (e.g. axions, photinos, etc.) surviving from the big bang. There are genuine prospects of detecting individual objects in all three of these categories, and thereby narrowing down the present range of options. If the Universe has the critical density (Ω = 1), rather than the lower value (Ω = 0.1 - 0.2) inferred from dynamical evidence, then the galaxies must be more clustered than the overall distribution even on scales 10 - 20 Mpc. "Biased" galaxy formation could account for this.

1. INTRODUCTION

At this conference, we have heard evidence for dark matter on various scales, which may implicate objects of different kinds. The local mass discrepancy within our Galactic disc probably involves low mass stars or white dwarfs, and I shall have little to say about it in this talk. On the scale of galactic halos and clusters, the evidence now points insistently towards the view that Ω (defined as the ratio of the actual mean density to the cosmological critical density ρ_{crit} = $(8/3\pi\ Gt^2)^{-1}$ is in the range 0.1 - 0.2, but that only 10 percent of this (Ω = 0.01 - 0.02) is definitely baryonic. However, there is no dynamical evidence for Ω = 1: there are no bound systems with M/L = 1000 h_{50} solar units, which would be the universal value if Ω = 1 (h_{50} denotes Hubble's constant H_0 in units of 50 km s^{-1}Mpc^{-1}).

The factor $\sim$ 10 discrepancy between the amount of "luminous" mass and the amount inferred from the dynamics of groups and clusters is the prime evidence for dark matter. This will be my main topic; in a concluding section I shall, however, comment on the theoretically-important issue of whether the Universe could have the critical density (Ω = 1).

The extensive menu of possible candidates could be shortened in several ways. For instance, progress in particle physics may give us

395

J. Kormendy and G. R. Knapp (eds.), Dark Matter in the Universe, 395–409.

firmer views on what particles should survive from the big bang, and
their expected contribution to Ω.   The predictions of cosmogonic
models,   particularly   regarding   clustering   scales,   halo density
profiles, etc., can tell us whether or not the dark matter has
undergone dissipative processes.   There is a clear distinction between
so-called 'hot' and 'cold' non-baryonic matter. The former, typified by
10 ev neutrinos, would have had sufficiently high thermal velocities in
the early universe for phase mixing to have smeared out fluctuations on
scales up to that of a galaxy cluster. In contrast, cold matter, such
as axions or GeV super-symmetric particles, would be sufficiently
slow-moving   that   primordial   fluctuations   would   survive   on   all
interesting scales, leading to a hierarchical picture for the buildup
of gravitationally-bound cosmic structures.
        The most clear-cut way to settle the nature of the hidden
mass would of course be to detect the objects that make it up. It is
on this aspect that I will concentrate in the present paper. Baryonic
systems - stars or their remnants - are already severely constrained by
the fact that the dark mass is so inconspicuous.   More remarkably,
there are genuine prospects that elementary particles of the kind that
could dominate the halo hidden mass may be individually detectable by
terrestrial experiments.

2. BARYONIC DARK MATTER: FAINT STARS, MASSIVE STELLAR REMNANTS, ETC.

The constraints from primordial nucleosynthesis are reviewed by Audouze
(these proceedings). The baryonic contribution to Ω in a "standard"
model is constrained to lie in the range

$$0.04 \lesssim \Omega_b \, h_{50}^2 \lesssim 0.15 \tag{1}$$

This restriction comes primarily from the measured D and ^{3}He abundance;
values outside these limits cannot be excluded, but require
modification of the standard homogeneous hot big bang, or some
alternative (non-cosmological) mechanism for producing light elements
such as deuterium.   The inequalities (1) imply that for a low Hubble
constant, some dark mass must be baryonic, and everything that is
dynamically inferred could be. Contrariwise, a standard hot big bang
with a high Hubble constant requires non-baryonic matter even to
account for galactic halos and the virial equilibrium of clusters.
        If H_0 = 50 km s^{-1} Mpc^{-1}, all the dynamically-inferred dark matter
could be baryonic.   The astronomical constraints on this option
have been recently discussed in detail by Carr, Bond and Arnett (1984)
and are summarized by Carr (these proceedings). The import of these
studies is that stars or their remnants cannot contribute $\Omega \gtrsim 0.1$
unless they are either predominantly "Jupiters" (stars of below
0.1 $M_\odot$) or else black holes which are the remnants of very massive
objects ("VMOs") with masses between a few hundred solar masses and
$10^6 M_\odot$.

Low mass stars

The main present constraint on low mass stars in galactic halos comes from limits to the observed optical and infrared emission. A very faint optical halo in M87 has been traced out to 300 kpc (Arp and Bertola 1969). In the edge-on spiral NGC 4565, there are limits on the near infrared emission, corresponding to 76 solar units in the I-band (Hegyi and Gerber, 1977), and 38 solar units in the K-band (Boughn, Saulson and Seldner 1981). The constraints thereby imposed on the slope of the initial mass function (IMF) have been discussed by Peebles (1985), and by Hegyi and Olive (1985). If the entire halo mass were contributed by stars with an IMF of the form

$$\frac{dn}{dm} \propto m^{-(1+x)} \tag{2}$$

down to some minimum mass M_{min}, then if M_{min} exceeds 0.007, $x \gtrsim 1.9$. (Salpeter's classic (1955) study derived $x = 1.35$ for our galactic disc.) In fact the infrared colours are not specially helpful in pinning down x, since for any $x < 2$, they are dominated by red giants. Note that the power-law example (2) is only illustrative; the same amounts of mass could equally be concealed by a population with a log gaussian distribution peaking below 0.1 $M_\odot$.

One thing is clear: one cannot invoke a smooth extension of the IMF that is observed below 1 $M_\odot$, which actually seems to flatten off below $\sim$ 0.3 $M_\odot$. The objects constituting the dark matter must in some sense by a "special creation". However, this is perhaps not a cogent objection to the idea. After all, the IMF derived by Miller and Scalo (1979) and Scalo (1985) for the solar neighbourhood does not look like a single power-law, but rather resembles two superposed log gaussian distributions. In an interesting recent discussion, Larson (1985) suggests that these represent the products of two distinct modes of star formation, and that the relative importance of these two modes may have varied over galactic history. Conceivably the star formation process relevant to the halo involved a third mode with a different characteristic mass. If the halo objects formed at an early pregalactic epoch, and subsequently clustered non-dissipatively into galaxies and clusters, then there would be even less reason to suspect that their IMF should resemble that of stars forming here and now. Indeed, we should remain open-minded even about the IMF of stars forming within galaxies. Cooling flows in clusters of galaxies (Fabian, Canizares and Nulsen 1984, Fabian, Arnaud and Thomas 1986), where the gas pressure is $\sim$ 100 times higher than in our Galactic Disc, have been interpreted as implying star formation with a very steep IMF. When stars in the halo formed, conditions maybe resembled those in cooling flows more than they resembled those in our Galaxy now. Perhaps the Salpeter-Miller-Scalo function pertains only in environments of atypically low pressure.

Improved infrared limits to the brightness of halos (e.g in NGC 4565) can in principle constrain the properties of these hypothetical "Jupiters", as of course can IRAS-type searches for high-proper-motion objects in our own Galaxy. However, because the luminosity is such a steep function of mass below 0.1 $M_\odot$, even a substantial improvement of such tests only tightens the IMF constraints slightly.

VMO remnants

Heavy elements are expelled from massive stars in their terminal phases unless they are so massive that they end their lives by collapsing to black holes after the pair-production instability (Truran and Cameron 1971; Woosley and Weaver 1982; Carr, Bond and Arnett 1984 and references cited therein). Collapse rather than explosion is thought to occur for core masses above $\sim$ 200 $M_\odot$. If the hidden mass were in VMOs, then the requirement that heavy elements be not overproduced therefore requires a very flat IMF. If this were actually a power law, the value of x (in equation (2)) would actually have to be negative. Moreover, there must be a cutoff above $\sim 10^6 M_\odot$, at least for objects within individual halos, because dynamical friction would have increased the velocity dispersion of disc stars to an excessive degree (Carr 1978; Lacey 1984). Within the context of VMO theories, we have little evidence on whether the preferred mass is closer to 10^3 or to $10^6 M_\odot$. The upper mass limit could be pushed downward if we had a firmer understanding of what luminosity would result from accretion onto black holes passing through the Galactic Disc (Ipser and Price, 1977, 1982; McDowell 1985; Lacey and Ostriker 1985).

VMO remnants, black holes in the mass range 10^3 to $10^6 M_\odot$, could reveal their presence by accretion of surrounding gas. The accretion rate for supersonic motion at speed V through gas of density n is proportional to $nV^{-3}M^2$. The luminosity depends on the accretion rate, and also on the efficiency ε. The latter is the least sure thing. For spherical accretion, where the efficiency is low because the radiative cooling time is long compared to the free-fall time, ε should scale with $\dot{M}$, making the luminosity proportional to $\dot{M}^2$. For disc-like accretion, ε may be as much as 0.1, independent of $\dot{M}$. The spectrum of the emergent radiation is also uncertain. The case of spherical inflow has been considered by Ipser and Price (1977), who argue that the radiation emerges mainly in cyclotron harmonics. These would typically peak in the infrared. Disc-type accretion could yield a high luminosity, predominantly thermal radiation in the ultraviolet.

The most conspicuous holes would be those which were passing through dense gas clouds, and which had V much less than the mean velocity. Although the number of these scales with V^3, the resultant higher $\dot{M}$ ($\propto V^{-3}$) makes them more readily detectable, despite the fact that the nearest one would be more distant from us (distance $\propto V^{-3/2}$).

The detectability of massive holes in our galaxy has been discusssed by McDowell (1985). He shows, following the assumptions of Ipser and Price, that if the typical mass is $10^5 M_\odot$ or more, the nearest objects passing through a dense interstellar cloud would be at 1 kpc distance, and would contribute 400 Jy flux at 100 microns, well above the IRAS detection limit; the same object would have an optical magnitude V = 10 (plus some correction for absorption). Lacey and

Ostriker (1985), assuming disc-mode accretion, predict even higher luminosities, but suggest that this would be UV radiation giving rise to an HII region.   The Ipser/Price estimates of luminosity are indeed rather conservative even on the basis of their assumed spherical infall, since a possible non-thermal tail of electrons is neglected. Unfortunately, the distinctive signature of an accreting black hole is hard to estimate, and so one cannot at the moment place firm limits on the number or mass of putative halo objects of this kind. Nevertheless, it already seems unlikely that the bulk of the mass could be in objects that are individually as heavy as $10^6 M_\odot$.

Gravitational "minilensing"

At the moment, it is conceivable that halos are made up of compact objects whose masses range from $10^6 M_\odot$, down to Jupiters, about 10^8 times smaller. One way of discriminating between these options is by searching for manifestations of gravitational lensing. The probability of seeing lensing due to an object in our own halo is only of order 10^{-6}. However, it is, ironically, much easier to detect objects in the halos of galaxies half way out to the Hubble radius. As was first clearly realized by Refsdal (1970), the probability that a compact source at a redshift $z > 1$ is significantly lensed by objects along its line of sight is of order Ω_{lens}, independent of the individual lens masses involved. However, the angular separation θ of the lens images is a diagnostic of the masses. For a path length of order the Hubble radius

$$\theta \cong 10^{-6} \left(\frac{M_{lens}}{M_\odot}\right)^{\frac{1}{2}} \text{ arc sec.} \qquad (3)$$

For $M_{lens} > 10^5 M_\odot$, very long baseline radio interferometers provide adequate resolution. For $M_{lens} < 0.1 M_\odot$ ("Jupiters") the angular scale is $< 10^{-6}$ arc sec. This cannot be directly resolved by any technique, until optical interferometers are deployed in space.   There is nevertheless a genuine prospect of detecting lensing of this kind because of the variability that would ensue if the lens were to move transversely (Gott 1981, Young 1981). It takes only a few years for an object at the Hubble radius moving at -10^3 km per sec to traverse an angle 10^{-6} arc seconds. Another possibility, emphasised by Canizares (1982) is that "minilensing" might be detectable because it would affect the optical continuum of quasars but not the spectral lines, since the latter come from a more extended region. If there were a firm observational limit to the scatter in the equivalent widths of the lines from quasar to quasar (i.e. in the line/continuum ratio) this would constrain the value of Ω contributed by small compact objects.

To detect very small compact objects via lensing requires bright background sources whose intrinsic angular size is well below the value of $\theta(\propto M_\ell^{\frac{1}{2}})$ given by (3). The optical continuum of quasars probably comes from a region small enough to be lensed by Jupiters ($\sim 10^{-2} M_\odot$); its typical size, is, however, uncertain, and could be anywhere in the range $10^{14} - 10^{16}$ cm. Although no conventional astrophysical process could predominantly produce macroscopic discrete masses $\ll 10^2 M_\odot$, such objects could be the outcome of, for instance, phase transitions at early epochs. Is there any class of source, detectable out to large z, that could be even more compact than quasars, and thereby able to lens such masses? One such candidate would be supernovae, whose effective radius at peak light is a few times 10^{14} cm. A significant contribution to Ω in $\sim 10^{-6} M_\odot$ objects would prevent supernovae from behaving as standard candles; the light curve of an individual supernova would also be distorted because the magification (along a typical line of sight) would change as its surface area expands.

3. EXOTIC PARTICLES

Provided we know the mass and annihilation cross-section of an elementary particle, we can in principle calculate how many of them survive from the big bang, and the resultant contribution to Ω. Progress in experimental particle physics may therefore reveal a particle which must contribute significantly to underline{Ω}, unless we abandon the hot big bang theory entirely. No such definite candidate is known at present; the masses of known particles such as neutrinos are not well enough determined experimentally; and there are many possible species whose existence is still conjectural. The idea of non-baryonic dark matter is nonetheless attractive, especially because the clustering properties of such matter could mimic the inferred mass distributions in galaxies and clusters in a gratifying way (Blumenthal et al. 1984 and references cited therein; Frenk et al. 1985).

Many ways have been recently proposed for detecting, or at least constraining, candidate particles. If the "inos" were unstable, and photons were among the decay products, there may be observational traces even for a decay timescale as long as 10^{24} seconds. This is because limits to the hard radiation background amount to only 10^{-8} of the critical density. Antiprotons observed in the cosmic radiation may even be decay products of "inos" (Silk and Srednicki 1984).

There has recently been a spate of interesting suggestions about how "inos" might reveal their presence relatively close at hand. Weakly interacting massive (GeV) particles would have cross sections σ of order 10^{-36} cms^{-2} for interactions with nucleons. The "optical depth" of the Sun is of order $(\sigma/10^{-36} \text{cm}^{-2})$. Such a particle, scattering elastically off a nucleon in the Sun would lose energy via the recoil, and could thereby become trapped (Steigman et al. 1978, Press and Spergel 1985). Over the lifetime of the Sun, an accumulated isothermal core of "inos" could build up a mass of $10^{-12} M_\odot$ if annihilations did not occur. However, annihilations would restrict this buildup, unless one adopts a rather artificial model in which the cross section for annihilation is far below that for scattering (Krauss et al. 1985a). However, even though annihilations may prevent a dense enough core building up to effect the standard solar neutrino problem, high energy neutrinos from these annihilations may reveal their presence in the underwater detectors developed to search for proton decay. Already, scalar or Dirac neutrinos with mass exceeding

6 GeV can be excluded. Analogous limits come from considering annihilations in the Earth rather than the Sun, as discussed by Silk, Olive, and Srednicki (1985) and Krauss, Srednicki, and Wilczek (1985).

Goodman and Witten (1985) and Drukier, Freese and Spergel (1985) have discussed <u>direct</u> <u>detection</u> of "inos" in the laboratory, using a so-called "super CD" - an array of superconducting grains maintained just below the transition temperature. The heat deposited by a single "ino" could raise the temperature of one of these grains above the critical value, thereby allowing magnetic flux to penetrate in a manner that could be detected. If the "inos" were, for instance, scalar neutrinos of mass > 5 GeV, the halo density would yield up to 10^4 counts per day per kilogram of detector. (For photinos, the expected rates are $\sim 10^3$ lower). The thermal noise in the system could perhaps be lowered sufficiently to detect particles with masses down to 2 Gev. The count rate is proportional to the 7th power of the velocity. By adjusting the threshold, one could thereby, if such effects were detected, determine the velocity distribution of the halo particles, and see if the halo were rotating. (Alternative schemes are discussed by Moody (these proceedings) and by Krauss <u>et</u> <u>al</u>. (1985b)).

Witten (1984) conjectured that grains or nuggets of "strange matter", containing up, down, and strange quarks, may survive stably from the quark hadron transition at $t = 10^{-4}$ seconds. Such objects, in some sense intermediate between elementary particles and lumps of astrophysical size, would count as non-baryonic matter in the context of nucleosynthesis. Recent work (Applegate and Hogan 1985, Alcock and Farhi 1985) suggests that neutrino heating would destroy nuggets unless they had a mass of planetary order, and it is unclear that any larger than this would even form, since this would involve coordination over a scale larger than the particle horizon at the relevant epoch. I find the "demise" of Witten's nuggets disappointing for two reasons. First, they might, as De Rujula and Glashow (1984) have suggested, have been detectable: interesting constraints could be set from the results of monopole searches, proton decay experiments, from the number of meteor showers, and from limits on the frequency of small-scale seismic events. A second appealing feature of the nugget concept is that it leads naturally to a universe where the respective contributions of ordinary and dark matter to Ω do not differ by more than an order of magnitude. If the dark matter were in, for instance, axions, some "fine tuning" must be invoked to prevent these contributions from differing by many powers of 10.

So there are at least three serious candidates for the dark matter in galactic halos and clusters: low mass stars; black hole remnants of very massive objects; or non-baryonic matter, in the form of supersymetric particles or axions. I would myself lay even odds between these three options at the moment. However, it is gratifying that we can expect the odds to change quite rapidly, owing either to (i) improved observational and experimental searches for candidate objects, (ii) progress in particle physics, or (iii) clearer evidence on how the dark matter is distributed (is it really present, for instance, in dwarf galaxies?).

4. A FLAT UNIVERSE?

At an IAU Symposium held in Poland in 1973, the mean density of the universe was a topic of discussion. In the concluding session the Chairman, Profesor Wheeler, conducted a poll among the audience to seek the favoured value of Ω. A gratifying feature of this poll was that a majority of participants accepted that Ω was unknown. Many, however, shared Professor Wheeler's aesthetic preference for a closed universe (or even an ensemble of closed universes) with Ω well in excess of unity. A similar poll taken today would doubtless reveal a "reasoned prejudice" in favour of $\Omega = 1$, this being the value favoured by inflationary cosmology. Maybe it is worth spelling out the basis for this attitude.

For all the present observable universe to have evolved from a region that was in causal contact at the earliest times, inflation by a factor of at least $\sim 10^{30}$ is required. In most versions of inflation, the exponential growth, once started, readily continues for many expansion timescales: it is likely to overshoot, stretching any small part of an initial chaotic hypersurface so that it becomes essentially flat over our present horizon scale. This would yield $\Omega = 1$, with a precision of order 1 part in 10^5 (the expected fluctuation amplitude). For inflation to yield the dynamically preferred value $\Omega = 0.1$, the inflation factor would have to be "just" $\sim 10^{30}$, making the present Robertson-Walker curvature radius of order the Hubble radius. This would demand some coincidence. Moreover, there is an additional requirement: our present universe would have to arise from a segment of the initial hypersurface with the special property that its curvature was uniform to one part in 10^5 - otherwise the curvature fluctuations that would produce quadrupole effects in the microwave background would not be 10^5 times smaller than the overall Robertson-Walker curvature. Our universe could thus not have inflated from a typical element of an initial chaotic hypersurface: the required region would have to be special, rather as the surface of a sphere would be special if the perturbations amounted to 10^{-5} of the mean curvature. (The alternative formulation of inflation due to Gott (1982) actually fulfils this latter requirement quite naturally, though it still requires fine turning of the amount of inflation.)

If the universe were indeed flat, what could make it so? Recall that most of the dynamical evidence suggests that Ω is only 0.1 - 0.2. Moreover, our infall towards Virgo (relative to the Hubble flow) is only ~ 250 km s^{-1}. This tells us, essentially, the amount of excess mass within a sphere centred on the Virgo cluster and whose surface lies near the local Group. The galaxies within this sphere are ~ 3 times more close-packed than in a typical volume of space, and the relatively low infall velocity is then inconsistent with $\Omega = 1$, unless for some reason the galaxy distribution is more clumped than the mass in general.

There are two ways of reconciling the observations with a flat universe, both of which require that the dominant hidden mass must be more smoothly distributed on larger scales than are the galaxies, or at least the conspicuous galaxies included in surveys. (Note that, if the conventional hot big bang model is correct, the nucleosynthesis constraint (3) favours a non-baryonic form for the dynamically dominant constituent of an $\Omega = 1$ universe):

(i) The Universe may be dynamically dominated by ultrahot weakly interacting particles which do not cluster. One difficulty here is that if such particles had always been present they would have inhibited gravitational clustering altogether, as well as yielding an unaceptably fast expansion timescale at the era of nucleosynthesis. This problem is eased if the hot particles represent decay products of massive particles with a lifetime $\sim 10^9$ years. A non-zero cosmological constant (Λ-term) is an alternative hypothesis whose consequences are similar (but to postulate a value of Λ such that it is dynamically competitive with matter at the present epoch introduces the kind of unappealing fine tuning that inflationary cosmology seeks to avoid).

(ii) Some kind of biasing in the galactic distribution might render galaxies more clumped than the overall mass distribution even on scales as large as 20 Mpc. Were this so, voids would not be as empty as they look, and the local "Virgo Supercluster" would not be a threefold enhancement in the total density. It is unlikely, especially in the otherwise attractive dark matter cosmology discussed in section 3, that material (even baryonic material alone) could be pushed over distances exceeding 20 Mpc. So could the efficiency with which baryons transform into luminous galaxies be patchy? If so, the ratio of baryons to cold dark matter could be constant on all scales larger than one or two megaparsecs (up to which we expect some segregation due to cooling flows, etc.).

The latter possibility involve a simple consistency requirement. If clusters such as Coma embody a fair sample of the contents of the Universe - i.e. if their ratio of baryonic to total mass equals Ω_b/Ω_{total} - then the fraction of baryons in clusters cannot exceed the value of Ω naively estimated from Coma-like systems, i.e. 0.1 or 0.2. So if

$$(M/L)_{Universe} \cong 1000\ h_{50}\ \Omega_{total} \qquad (4)$$

then $\Omega_{total} = 1$ and $\Omega_b = 0.1$ are compatible with $(M/L)_{cluster} = 100$ provided that M/L for galaxies is less than 10; these illustrative "round numbers" do indeed seem marginally consistent. This idea suggests that up to 90 per cent of baryons may remain as diffuse gas in voids, or else in faint or low surface brightness galaxies; the mean (M/L) for baryonic matter would then be ~ 100. In this connection, one wonders whether there might, in some voids, be dark halos with no luminous galaxies in them. Such objects might account for double quasars with no sign of a galaxy to act as a gravitational lens.

Physical mechanisms for bringing about this biasing have been discussed by Rees (1985), Silk (1985) and others. A feature common to several such mechanisms is that the first galaxies to form would exert negative feedback on the formation of later ones. This has the advantage that the resulting galaxies would then automatically display enhanced clustering, for reasons decribed in detail by Kaiser (1984). No mechanism has yet been worked out in convincing detail. However, the idea of biasing is not just an ad hoc contrivance, introduced to shore up the philosophically attractive $\Omega = 1$ model against apparently conflicting evidence. It would be astonishing if no such mechanism were important — if no large scale environmental effects influenced galaxy formation, and if light did indeed trace mass on all scales > 1 Mpc. Any convincing determination of Ω must await much further data on galactic morphology and evolution, on the content of voids, and on the nature of the hidden mass. In the meantime, the virial evidence does not seem a severe embarassment for advocates of $\Omega = 1$.

Acknowledgements. I am grateful to many colleagues for helpful discussions and to G. Steigman for historical perspectives.

REFERENCES

Alcock, C. and Farhi, H. 1985. Phys. Rev. D. (in press).

Applegate, J. and Hogan, C.J. 1985. Phys. Rev. **D31**, 3037.

Arp, H. and Bertola, F. 1969. Astrophys. Lett., **4**, 23.

Blumenthal, G., Faber,S.M., Primack, J.R., Rees,M.J. 1984. Nature, **311**, 517.

Boughn, S.P., Saulson, P.R., Seldner, M. 1981. Astrophys. J., **250**, L15.

Canizares, C.R. 1982. Astrophys. J., **263**, 508.

Carr, B.J. 1978. Comm. Astrophys., **7**, 161.

Carr, B.J., Bond, J.R. and Arnett, W.D. 1984. Astrophys.J., **277**, 445.

Davis, M., Efstathiou, G., Frenk, C.S. and White, S.D.M. 1985. Astrohys. J., (in press).

De Rujula, A. and Glashow, S.L. 1984. Nature, **312**, 734.

Drukier, A.K., Freese, K. and Spergel, D.N. 1985. preprint.

Fabian, A.C., Arnaud, K. and Thomas, P. 1986. these proceedings.

Fabian, A.C., Nulsen, P.E.J., Canizares, C.R. 1984. Nature, **310**, 733.

Frenk, C., Davis, M., Efstathiou, G. and White, S.D.M. 1985. Astrophys. J. (submitted).

Goodman, M.W. and Witten, E. 1985. Phys. Rev. **D31**, 3059.

Gott, J.R. 1981. Astrophys. J. **243**, 140.

Gott, J.R. 1982

Hegyi, D.J., Gerber, G.L. 1977. Astrophys. J., **218**,L7.

Hegyi, D.J. and Olive, K.A. 1985. Astrophys. J. (in press).

Ipser, J.R. and Price, R.H. 1977. Astrophys. J., **216**, 578.

Ipser, J.R. and Price, R.H. 1982. Astrophys. J., **255**, 651.

Kaiser, N. 1984. Astrophys. J. (Lett), **284**, L9.

Krauss, L., Freese, K., Spergel, D. and Press, W.H., preprint.

Krauss, L., Cabrera, B., and Wilczek, F. 1985b Phys. Rev. Lett. (in press)
Lacey, C.G. 1984. in "Formation and Evolution of Galaxies and Large
     Structures in the Universe", eds. J. Audouze and J. Tran Thanh Van
     (Reidel, Dordrecht).
Lacey, C.G., Ostriker, J.P. 1985. Astro. Phys. J. (in press).
Larson, R.B. 1985. MNRAS (in press).
McDowell, J. 1985. MNRAS (in press).
Madsen, J. and Epstein, R.I., 1984. Astrophys. J., **282**, 11.
Miller, G.E. and Scalo, J.M. 1979. Astrophys. J. Suppl., **41**, 513.
Peebles P.J.E. 1985. in "Theoretical Aspects of Astrophysics and
     Cosmology" ed. J. Sanz (World Scientific Publishers, Singapore).
Press, W.H. and Spergel, D. 1985. Astrophys. J. (in press).
Rees, M.J. 1985. MNRAS, **213**, 75P.
Refsdal, S. 1970. Astrophys. J., **159**, 357.
Salpeter, E.E. 1955. Astrophys. J., **121**, 161.
Scalo, J.M. 1985. Fundam. Cosmic Phys. (in press).
Silk, J.I. 1985. Astrophys. J. (in press).
Silk, J.I., Olive, K. and Srednicki, M. 1985. preprint.
Silk, J.I. and Srednicki, M. 1984. Phys. Rev. Lett., **53**, 624.
Steigman, G., Sarazin, C.L., Quintana, H. and Faulkner, J. 1978.
     Astron. J., **83**, 1050.
Tremaine, S. and Gunn, J.E. 1979. Phys. Rev. Lett., **42**, 407.
Truran, J.W. and Cameron, A.G.W. 1971. Astrophys. Sp.Sci., **14**, 179.
Uson, J. and Wilkinson, D.T. 1984. Nature, **312**, 427.
Witten, E. 1984. Phys. Rev., **D30**, 272.
Woosley, S.E. and Weaver, T.A. 1982 in "Supernovae: a Survey of
     Current Research", eds. M.J. Rees and R.J. Stoneham (Reidel,
     Dordrecht).
Yang, J., Turner, M.S., Steigman, G., Schramm, D.N. and Olive, K.A.
     1984. Astrophys. J., **281**, 493.
Young, P.J. 1981. Astrophys. J., **244**, 756.

DISCUSSION

SHAPIRO: In your table in which you considered the feasibility of
different candidates for dark matter, you excluded hot dark matter as a
possible explanation for galactic halos. I am not aware of any argument
that excludes hot dark matter except for the the still-uncertain
inferences from the observed stellar velocities in dwarf elliptical
galaxies. Were you using these results? If not, I do not think that
either the phase-space density arguments or the numerical simulations
yet exclude massive neutrinos, for example, as constituents of ordinary
galactic halos.

REES: The table I showed (which comes from Bernard Carr's poster paper
at this conference) should really be depicted in shades of gray rather
than in black and white! The dwarf galaxy data, as we've heard from
Kormendy and Aaronson, are still tentative. Even if there is dark matter
in dwarfs (which we know couldn't be low-mass neutrinos), this still
doesn't necessarily exclude neutrinos as the main contributors on larger
scales. My personal view is that the main problem for neutrino-
dominated cosmogony (with adiabatic fluctuations) is to understand how
bound systems can form early enough to account for high-z quasars.

CARR: My constraints diagram excludes hot ino's from comprising the
closure or cluster dark matter on the basis of the numerical simulations
reported by White at this conference. However, the associated regions
are only shaded lightly in view of the uncertainty in this conclusion.
Hot ino's are excluded from comprising galactic halos on the basis of
the Tremaine-Gunn argument. In fact, numerical simulations indicate
that this conclusion need not apply in some circumstances, so that
region should also be shaded lightly.

MELOTT: Since a number of independent numerical studies of the collapse
of pancakes in hot particle models have shown that at least 10% of the
particles wind up with a low velocity and high phase-space density, I
would maintain that such particles as 30 eV neutrinos could comprise the
material of halos around normal galaxies, even in the context of
adiabatic perturbations. The formation of individual galaxies can be
driven by thermal instabilities inside the pancakes where the neutrino
condensate exists.

REES: The simulations that you and your collaborators have done
certainly show that the phase-space dilution is less catastrophic than
naive arguments suggest, especially when the collapse is essentially
one-dimensional rather than quasi-spherical. Until we understand the
gas-dynamical aspects of galaxy formation, I agree that we must be
cautious in our claims that neutrino-dominated models run into problems
when confronted with the clustering data. Your comment also highlights
the important question of whether all galaxies have similar dark halos,
or whether some might have formed from squeezed clouds of baryons whose
location isn't necessarily correlated with potential wells dominated by
non-baryonic matter.

J. JONES: One of the problems with putting the dark matter in abnormal-mass stars is that there are quite strong constraints on the proportion of such objects that can be accommodated in the disk. What mechanism could be responsible for causing the majority of the mass to go into such stars in the halo but not in the disk?

REES: Several authors have conjectured how the IMF might change from place to place, with different conclusions. I'm not sure we know enough about star formation even to decide whether a very different IMF for Population III is likely or unlikely.

SILK: You have presented with more-or-less equal emphasis three different possibilities for the nature of dark matter. Two, namely Jupiters and supermassive black holes, involve extreme and ad hoc extrapolations beyond any directly measured aspects of star formation. The third, exotic particles, may involve similar extrapolations by the particle physicist. Would you care to indicate your ranking of these options in terms of plausibility? (laughter)

REES: I gave them equal emphasis because I am genuinely agnostic. In particular, I am very unconvinced by theoretical arguments that claim to prove that the first stars "must" or "cannot" have such-and-such a mass. To be specific, I would assess the three options as having 25% probability each, leaving the last 25% for things that we haven't thought of yet. But what is most encouraging is the prospect of observational and/or experimental discrimination between the options over the next few years. We won't stay "in the dark" forever.

FABER: Suppose it were to be shown that dark matter really exists in a galaxy like Ursa Minor. What additional constraints might you have on massive objects from dynamical friction in such a system?

REES: Naively, there is a dynamical friction problem for any mass above $\sim 200\ M_\odot$, as you yourself have pointed out. Now, I think there is an escape clause, which Lacey and Ostriker have pointed out. Unless we really know the density profile, can we really rule out the presence of, say, one $10^6\ M_\odot$ black hole in the middle of such a system, or a few in orbit around it? It is certainly the case that if you have massive black holes, then dynamical friction is important; it will for these systems tend to make the black hole go outward and the ordinary stars go inward, contrary to the way things usually happen. But that might still leave you with one or two in the center and a few around the outside. Before you can shoot down the massive black hole model using this line of argument, you need to know not merely the overall velocity dispersion but something about its distribution with radius.

PACZYNSKI: It might be easier to detect Jupiters, if there are any, in the halo of our Galaxy rather than in halos at cosmological distances. If you put a Jupiter at a cosmological distance, you have to wait an unreasonably long time before any lensing variation is observable. Besides, the events are not frequent. But if you calculate the optical depth to gravitational lensing in galactic halo objects, it is about one part in 10^6. This is small but not hopeless. So if you put in the background a large number of point sources, like in the Magellanic Clouds, one out of 10^6 would be a minilens at any given time, if the objects in our halo had the right mass. There is a lower mass limit of $\sim 10^{-8}$ $M_\odot$, at which the splitting is comparable to the size of a dwarf star in the Magellanic Clouds, and the event would last for a fraction of an hour. The upper end of the suitable mass range is $\sim 10^3$ $M_\odot$, where the variation time is ~ 10 years and we run into the limit set by our own lifetimes.

REES: I agree. The situation for cosmologically distant objects may not be quite as bad as you think, because the velocity you should use is not the transverse velocity of a typical star in the halo but the velocity of the galaxy as a whole. That could be large.

E. TURNER: Alcock and Anderson have pointed out that the dispersion in time-delay H_0 determinations for different gravitational lens systems is related to the fluctuations in the total mass along various lines of sight out to cosmological distances. In principle, this could be compared to the observed galaxy clustering amplitude and thus directly test the hypothesis that the total mass is more uniformly distributed than the luminous matter.

REES: Yes, this is a good test. Gravitational lensing is also a possible probe for "failed galaxies" in voids, i.e., halos of dark matter without luminous cores. It also offers a way of testing the "pre-Newtonian" theories of Milgrom and others, unless these theories predict exactly the same relationship between the bending angle for light rays and the gravitational acceleration of ordinaty matter as does standard physics.

PEEBLES: Martin, I might remind you that astrophysically biased galaxy formation could go either way - it could depress Ω as well as raise it. When I look at the data, it suggests, if anything, that Ω has been pushed down. The Coma Cluster has surely been accreting material. If Ω were unity, that material would have to have a high mass-to-light ratio. But when you look at the data you find, if anything, that the mass-to-light ratio decreases with increasing radius.

REES: I'm reluctant to dissent from any of that. Let me just say that we do not have any quantitative picture for biasing, we just have lots of rather poor and vague ideas. The best idea is in fact your own, which is that we do another unattractive thing and abandon Gaussian random phases. Then you could imagine that the Universe is inherently more perturbed in some places than in others. This could prevent galaxy formation. But there is no lack of ideas, and I think that all one can say is that on general grounds we cannot rule out a model where in 90% of the Universe we have uncondensed baryons while in the other 10% we have baryons turned into galaxies.

NOLTHENIUS: Simon White and Marc Davis pointed out that both their $\Omega = 0.2$ unbiased and $\Omega = 1$ biased simulations provide reasonable fits to the two- and three-point correlation functions and overall sky appearance. However, my latest results from looking at their simulations favor the biased scenario: the detailed properties of the biased-simulation groups - M/L ratios, percentage of galaxies in groups, many other measures, and their trends with selection cutoffs - are all in agreement with the CfA data, while the $\Omega = 0.2$ unbiased simulations are not. The unbiased catalogs give too few galaxies in groups, M/L's that are too high, and trends with cutoff that differ from the observations.

B.J.Carr
School of Mathematical Sciences, Queen Mary College, London

There is evidence for four types of dark matter: (1) the local d.m. in the galactic disc; (2) the d.m. associated with galactic halos; (3) the d.m. in clusters; and (4) a background closure density of d.m. required if the Universe undergoes an inflationary phase. There are three types of explanation: (1) remnants of a first generation of Population III stars, including black holes (SMOs, VMOs or MOs), neutron stars, white dwarfs, or LMOs (M-dwarfs and Jupiters); (2) elementary particle relicts of the Big Bang (inos), usefully classified - according to their mass - as hot, warm, or cold, since this determines the scale on which they can cluster; and (3) primordial black holes, formed from density perturbations or phase transitions in the early Universe. Various constraints on the d.m. candidates are indicated by the shaded regions in the Figure below. The conventional model of cosmological nucleosynthesis precludes Population III remnants providing the closure and perhaps cluster d.m., while stellar nucleosynthesis constraints preclude neutron stars from explaining anything and allow white dwarfs to provide only the local d.m. Source counts exclude M-dwarfs from providing the local or halo d.m., while gravitational lensing effects exclude SMOs larger than $10^8 M_\odot$ from explaining anything and LMOs or VMOs from having the closure density. Dynamical considerations imply $M < 2 M_\odot$ for the local d.m., $M < 10^6 M_\odot$ for the halo d.m., and $M < 10^9 M_\odot$ for the cluster d.m.; they also imply that the local d.m. cannot be inos and that the halo d.m. cannot be a hot ino. The table suggests the following conclusions: (1) no single candidate can explain all four d.m. problems; (2) the best candidate for the closure d.m. is an ino; (3) the best candidates for the local d.m. are white dwarfs or Jupiters; (4) the halo (and possibly cluster) d.m. could plausibly be black holes or Jupiters.

		LOCAL	HALO	CLUSTER	CLOSURE
POPULATION III	SMO				
	VMO				
	MO				
	NS				
	WD				
	LMO				
	PBH				
INOS	cold				
	warm				
	hot				

$\uparrow M$

J. Kormendy and G. R. Knapp (eds.), Dark Matter in the Universe, 410.
© 1987 by the IAU.

LIMITS ON THE MISSING MASS IN DARK STELLAR REMNANTS

Janet E. Jones*,   Philip Mellor**   and   Jesper Storm*
*NORDITA, Blegdamsvej 17,   **Institut d'Astrophysique,
DK-2100 Copenhagen Ø,   98 bis, Bd Arago,
Denmark   75014 Paris, France

A set of comprehensive computer models for the chemical evolution of galaxies have been used to determine the limits on the amount of mass that could exist in the form of dark stellar remnants deriving from normal stellar evolutionary processes. In these models, the instantaneous recycling approximation is not assumed: stars are binned into 10 mass intervals, with different lifetimes, yields and remnant masses. The models were run using many different values for the IMF (including non-Salpeter and varying IMFs), star formation rates, yields, remnant masses, gas infall and outflow rates, primordial metalliciy and initial conditions. The Galaxy is described by a two-zone halo-disk system, where gas from the halo falls onto the disk. Elliptical galaxies are described by single-zone models.

The results of the computed models were compared to detailed observational data, using both globular cluster and halo field stars for the halo zone of our Galaxy, and solar neighbourhood stars for the disk. The errors for the parameters were determined by
(a) considering samples with different selection cutoffs, and
(b) calculating the effect of an artificially constructed selection bias.

The most striking feature of the models is the relative insensitivity to the many parameters of the amount of material left as dark stellar remnants. The total amount of dark matter depends sensitively on the observed metallicity of the system, and no amount of fiddling the other parameters can vary that mass by very much. The reason for this is very clear: in order to make a lot of stellar remnants, there must be a lot of cycles of stellar evolution, and this produces a lot of high metallicity gas. Even when the value of the remnant mass is made artificially high (which produces inconsistencies both with stellar evolution theory and with chemical evolution models in the solar neighbourhood) the relationship between metallicity and total dark matter remains strong: metal rich systems have more stellar remnants than metal poor ones because they have undergone more cycles. Thus metal-rich elliptical galaxies are found to have a large amount of dark matter: 10 to 1000 times that of our Galaxy. Metal-poor ellipticals have less: dwarf ellipticals cannot have massive haloes composed of normal dark stellar remnants, regardless of the amount of gas lost by stripping. Similarly, the halo of our Galaxy is found to have a low mass compared to the disk: $M(disk)/M(halo) = 10.6$. The addition to the models of a dominant population of low mass "Jupiters" does not help the situation, as they increase the hidden mass in both regions without changing the mass ratio significantly.

The total amount of mass in dark stellar remnants in our Galaxy was found to be: halo 27%, disk 36%, Galaxy 35% for the best fit model. Note that the value of 36% for the disk accords well with the estimated amount of dark matter in the solar neighbourhood.

J. Kormendy and G. R. Knapp (eds.), Dark Matter in the Universe, 411.

MASSIVE BLACK HOLES IN GALACTIC HALOS?

C. G. Lacey & J. P. Ostriker
Princeton University Observatory

We consider the idea that galaxy halos are composed of massive black holes, as a possible resolution of two problems: the composition of dark halos, and the heating of stellar disks. Scattering of disk stars by halo black holes with mass M_H, velocity dispersion σ_H and number density n_H causes the stellar velocity dispersion to increase with time t as $\sigma \approx (Dt)^{1/2}$ for t large, where $D \propto n_H M^2_H \ln \Lambda / \sigma_H$, and $\ln \Lambda$ is the Coulomb logarithm. This time-dependence is in good agreement with observations, as is the prediction for the axial ratios of the velocity ellipsoid $\sigma_u : \sigma_v : \sigma_w$. To account for the magnitude of the disk velocity dispersion in the solar neighbourhood, we require $M_H \approx 2 \times 10^6 M_\odot$. The stellar distribution function is predicted to be approximately isothermal at low epicyclic energies, in the Fokker-Planck regime in which the effect of the many distant, weak encounters dominates, but with a power-law tail at high energies produced by the relatively rare close encounters. This tail has the form $N(E) \propto E^{-2}$, where E is the horizontal or vertical epicyclic energy, and $N(E)$ is the number of stars per unit area of the disk, per unit E. The fraction of stars in this power-law tail depends only on the value of $\ln \Lambda$, and is about 1% for typical values. This provides a possible explanation for the high velocity A stars found in the solar neighbourhood. This disk heating mechanism can also account for the approximate constancy of the disk scaleheight with radius that is observed in other spiral galaxies, although this does not result as naturally as the other properties.

We have calculated the effects of dynamical friction on the black holes, and find that ~100–1000 should have spiralled into the Galactic center over the life of the Galaxy. However, gravitational three-body interactions are likely to be effective in ejecting black holes from the center by the slingshot effect, so that at any given time there will probably be no more than two black holes at the center. Thus we naturally account for the compact object postulated to be at the center of the Galaxy by some observers. We have calculated the effects of gravitational lensing by the black holes, and find that they are probably consistent with observational limits. Approximately 1% of sources at redshift $z \approx 1$ will be significantly lensed, with typical image splittings of several milliarcsec, which should be detectable by VLBI. We have also considered the radiation emitted by the black holes as they accrete gas from the interstellar medium, which should make some of them very luminous. The absence of clear evidence for accreting black holes may pose a problem for the scenario.

REFERENCE

Lacey, C.G., & Ostriker, J.P. 1985, Ap.J. 299.

J. Kormendy and G. R. Knapp (eds.), Dark Matter in the Universe, 412.

BIMODAL STAR FORMATION AND REMNANT-DOMINATED GALACTIC MODELS

Richard B. Larson
Yale Astronomy Department

Current data on the luminosity function of nearby stars allow the possibility that the stellar initial mass function (IMF) is double-peaked and that the star formation rate (SFR) has decreased substantially with time. It is then possible to account for all of the unseen mass in the solar vicinity as stellar remnants. A model for the solar neighborhood has been constructed in which the IMF is bimodal, the SFR is constant for the low-mass mode and strongly decreasing for the high-mass mode, and the mass in remnants is equal to the column density of unseen matter; this model is found to be consistent with all of the available constraints on the evolution and stellar content of the solar neighborhood. In particular, the observed chemical evolution is satisfactorily reproduced without infall. The total SFR in the model decreases roughly with the 1.4 power of the gas content, which is more plausible than the nearly constant SFR required by models with a monotonic IMF.

Similar models can account for the high formation rate of massive stars in the inner disks of our Galaxy and M83 without predicting more mass in low-mass stars than is allowed by the rotation curve. In these regions, the high-mass mode of star formation is more dominant than in the solar neighborhood and stellar remnants account for a large fraction of the mass. Bimodal remnant-dominated models also account better than conventional models for the colors, mass-to-light ratios, and gas contents of spiral galaxies. The colors of the bluest galaxies can be explained without requiring extreme youth or bursts, and the mass-to-light ratio as a function of color can be accounted for with a simple two-population model. Because of the decreased importance of low-mass stars and the increased importance of recycling from massive stars, the time-scale for gas consumption is larger than previously estimated, and is consistent with a simple exponential decay of the gas content. The increase of both metallicity and mass-to-light ratio with mass among giant elliptical galaxies can also be accounted for by a bimodal model if the characteristic mass and the relative amplitude of the high-mass mode both increase with increasing galactic mass.

All of the data are consistent with a picture in which the formation of massive stars is strongly favored in regions where the SFR is high. An extension of such a picture to the earliest stages of star formation in galaxies suggests that rapid formation of large numbers of massive stars took place, and that the dark matter in galactic halos may consist of the remnants of these early generations of massive stars.

J. Kormendy and G. R. Knapp (eds.), Dark Matter in the Universe, 413.

BACKGROUND LIGHT FROM POPULATION III STARS

Jonathan C. McDowell
Institute of Astronomy, University of Cambridge

It has been proposed (e.g. Carr,Bond and Arnett 1984) that the first generation of stars may have been Very Massive Objects (VMOs, of mass above 200 $M_\odot$) which existed at large redshifts and left a large fraction of the mass of the universe in black hole remnants which now provide the dynamical 'dark matter'. The radiation from these stars would be present today as extragalactic background light. For stars with density parameter Ω_* which convert a fraction ϵ of their rest-mass to radiation at a redshift of z, the energy density of background radiation in units of the critical density is $\Omega_R = \epsilon \Omega_* /(1+z)$. The VMOs would be far-ultraviolet sources with effective temperatures of 10^5 K. If the radiation is not absorbed, the constraints provided by measurements of background radiation imply (for H =50 km/s/Mpc) that the stars cannot close the universe unless they formed at a redshift of 40 or more. To provide the dark matter (of one-tenth closure density) the optical limits imply that they must have existed at redshifts above 25.

There are several opacity sources which could redistribute the background radiation energy to longer wavelengths. If the stars are surrounded by dense clouds of hydrogen ($n > 10^4$ cm^{-3}) the ionizing continuum could be reemitted as recombination radiation. In this case the Lyman-alpha flux could produce a near infra-red background whose spectrum would depend sensitively on the redshift range of VMO formation (Carr, McDowell and Sato 1983). If this absorption does not occur, such stars would ionize any intergalactic medium and the major remaining opacity source would be dust.

Intergalactic dust, if it existed at high redshift, could degrade the radiation into a distortion to the microwave background, or produce a separate far-infrared background. I have calculated the expected background with a code which models evolution of the background spectrum and the dust temperature in the expanding universe. The IRAS limits provide a good constraint if the far-IR background is well separated from the microwave background, true for heating at low redshift ($z \lesssim 5$); for higher redshift dust emission the limits are much weaker and cannot exclude $\Omega_* = 0.1$. If intergalactic dust formed well after the VMOs, for instance in a model where galaxy formation occurs at redshifts of about 3, dust absorption is only significant if the amount of dust is within a factor of a few of the maximum amount allowed by measurements of quasar reddening.

I acknowledge the support of an SERC studentship.

REFERENCES

Carr, B.J., Bond, J.R., and Arnett,W.D., 1984. Astrophys. J. 277,445
Carr, B.J., McDowell,J.C., and Sato,H., 1983. Nature 306,666.

OBSERVATIONAL SEARCHES FOR DARK HALOS

P.C. van der Kruit
Kapteyn Astronomical Institute
P.O. Box 800
9700 AV Groningen
The Netherlands

ABSTRACT. A review of observational searches reveals the following constraints for the constituents of dark halos: (1) Optical searches show that these halos are not for a large fraction of their mass made up of dwarfs of spectral type M5 or earlier. (2) K-band (2.2 μ) searches virtually rule out all H- burning Main Sequence stars. (3) IRAS upper limits are consistent with black dwarfs of any age or Jupiters. (4) The inferred metallicity and M/L variations in the spheroid of NGC 7814 are consistent with the hypothesis that the dark matter consists of low mass objects that formed along with the luminous population II.

1.   WHERE DO WE LOOK AND FOR HOW MUCH?

This review of observational searches for dark halos will be devoted entirely to spiral galaxies. Current best estimates show that within the cut-off radius R_{max} of the disk only about one-third of the total mass within this radius as indicated by the rotation curve is contained within the disk itself. There are three general methods to estimate the disk mass: (i) The disk surface density can be estimated at various radii by comparing the vertical velocity dispersions (generally measured in nearly face-on spirals) to the vertical scaleheights (measured in edge-on galaxies). The components usuable for this are the HI-gas, which has a constant velocity dispersion of 7-10 km s^{-1} but increasing thickness with radius, and the old disk populations, which have a constant <u>exponential</u> scaleheight of 0.35±0.1 kpc but its vertical velocity dispersion decreasing with radius (see e.g. van der Kruit and Shostak, 1984; van der Kruit, 1981; van der Kruit and Searle, 1982; van der Kruit and Freeman, 1985). (ii) The rotation curve may contain a "truncation" feature at the cut-off radius, whose amplitude is a measure of the disk to halo mass-ratio (Casertano, 1983). (iii) Rotation curves can be analysed in terms of an exponential disk determined from surface photometry and a specified halo density distribution (Bahcall, 1983; Carignan and Freeman, 1985; van Albada et al., 1985). This gives a somewhat wide range of mass-ratios for the extreme possibilities of "maximum" and "minimum" disk (see also Sancisi and van Albada in this volume).

415

J. Kormendy and G. R. Knapp (eds.), Dark Matter in the Universe, 415–424.

The first method appears to be the most restrictive and widely
applicable in practice and currently suggests within R_{max} that M(halo)/
M(disk) = 1.5 - 3. Also over a limited range in R it is consistent with
a constant <u>old disk</u> M/L with radius with a value of 6 ±2 $M_\odot/L_B$ for H =
75 km s^{-1} Mpc^{-1}. The three methods indicate that a halo of dark matter
must exist that is significantly less flattened than the disk. In what
follows I assume it to be spherical (see also Rubin in this volume) and
to have roughly a density distribution $\rho \propto R^{-2}$. On the sky this implies
a surface density distribution $\sigma \propto R^{-1}$.

2.   OPTICAL AND NEAR-INFRARED SEARCHES

The most important searches reported in the literature are listed
in Table 1. A number of remarks can be made:
- The first study by Freeman et al. (1975) on NGC 253 already showed
that any dark matter must have an $(M/L)_B$ in excess of several hundred.
This photographic study therefore already indicated that dark halos
cannot be made up for a considerable fraction of their mass of main
sequence stars earlier than spectral type about M5.

Table 1 - Searches for dark halo's

Authors	Year	Galaxies	Type	Bands
Freeman et al.	1975	N253	Sc	B
Frankston & Schild	1976	N4565	Sb	VI
Gallagher & Hudson	1976	I2233	Sd?	BVi
Hegyi & Gerber	1977	N4564	Sb	VI
Kormendy & Bruzual	1978	N4565	Sb	V
Spinrad et al.	1978	N4565, 4594, 253	Sb, Sb, Sc	Br
Hegyi & Gerber	1979	N4565	Sb	RI
Davis et al.	1980	N4565	Sb	B
Hohlfeld & Krumm	1981	N2768, 4762, 4203, 4565	So, So, So, Sb	JK
Boughn et al.	1981	N4565	Sb	K
Jensen & Thuan	1982	N4565	Sb	BVrI
Skrutskie et al.	1985	N2683, 4244, 5907	Sc	VK

- Later studies in the optical bands (B, V, r, etc.) using different and
often very clever and innovative techniques have not been able to significantly improve on this. This is mainly a result of the very sharp rise
at optical wavelengths in M/L for dwarfs later than M5. For example we
may look at Van Biesbroeck 10 which is believed to be very close to the
minimum mass necessary for H-burning (Greenstein et al., 1970):

M5   0.22 $M_\odot$ 400 $M_\odot/L_{\odot,B}$ 60 $M_\odot/L_{\odot,V}$ 9.6 $M_\odot/L_{\odot,K}$

VB10   0.09 1.7×10^5 3.8×10^4 35

- These numbers also show that one can profitably observe at near-IR wavelengths such as K-band (2.2 μ). There the limits are a few tens $M_\odot/L_\odot$, which essentially rules out all H-burning Main Sequence stars as a major constituent.
- Almost everybody's favorite appears to be NGC 4565. Kormendy (1982) has produced from all the data an accurate minor axis profile which extends to about 6.5 arcmin where the surface brightness is reported as ~31 B-mag arcsec^{-2} (see Fig. 1). Two remarks can be made: (i) Beyond the "kink" at about 50 arcsec the profile has $I \propto r^{-2.55}$ which is very much steeper than expected from a dark halo with constant M/L. (ii) Such a halo with $(M/L)_B$ = 75 (which fits the observed surface brightness at ~50 arcsec) still would have a surface brightness of about 27 B-mag arcsec^{-2} at 7'. Now this is the distance at which observers usually have assumed to observe pure sky!

This last point is worth investigating somewhat further. Most authors have been aware of this and have made remarks to this extent in their papers. E.g. Hegyi and Gerber (1977) "have briefly taken obser-

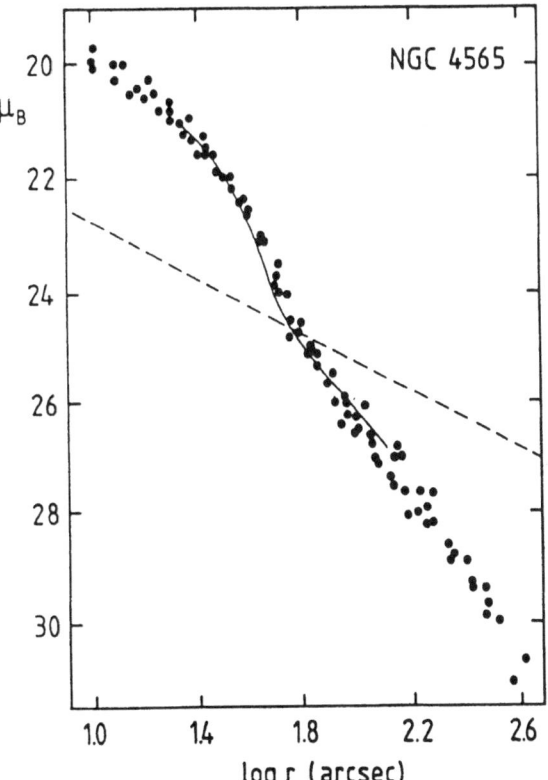

Fig. 1 - The minor axis profile of NGC 4565 as compiled by Kormendy (1982). The dashed line indicates a $\sigma \propto R^{-1}$ halo as expected from the rotation curve with M/L $\approx$ 75 (and assumed spherical). Beyond 50 arcsec the observations show $I \propto R^{-2.55}$.

vations" at somewhat larger distances and Davis et al. (1980) note an absence of a slope in their data between 5 and 8 arcmin. However, a detailed comparison shows that their limits are only marginally inconsistent with the variations expected from an $I \propto R^{-1}$ halo with $(M/L)_B \approx$ 75. The point here is of course that in such a halo the surface brightness will become fainter by only 1 mag when the distance from the centre is increased by a factor 2.5. The non-existence of such halos is certainly not as secure as a cursory glance at the literature would suggest. On the other hand it may be too severe to suspect Kormendy's profile as very wrong, since a serously wrong sky level would result in a strong decline at 5-6 arcmin, which is not observed.

3.   LIMITS IN NGC 5907

Obviously the analysis of surface brightness distributions is seriously complicated by the presence of a spheroid, and there are clear advantages in chosing an essentially bulge-less, edge-on system. Now except for the last entry in Table 1 only Galagher and Hudson (1976) have chosen such a system; however IC 2233 is a dwarf galaxy. The most favorable choice is NGC 5907 for which there exists both surface photometry and mass models. As the picture in the Hubble Atlas shows, it has at most a very tiny bulge. I will proceed to derive the formal upper limits and will also use data at much larger radii than in NGC 4565 to determine the sky level.

I use a distance of 11 Mpc ($H = 75$ kms^{-1} Mpc^{-1}), so that the galaxy has a disk cut-off at 19 kpc (6 arcmin). Estimates for the disk mass have been given as 8×10^{10} $M_\odot$ by van der Kruit and Searle (1982; corrected to an $(M/L)_{old\ disk} = 6$ $M_\odot/L_{\odot,B}$) and 9×10^{10} $M_\odot$ by Casertano (1983). This leads to halo masses within 19 kpc of 14×10^{10} $M_\odot$ and 13.5 $\times 10^{10}$ $M_\odot$ respectively. If spherical the dark halo has ρ ($M_\odot$ pc^{-3}) ≈ 5.9 $\times 10^{-1}$ R^{-2} (kpc) and surface density σ ($M_\odot$ pc^{-2}) $\approx 1.8 \times 10^3$ R^{-1} (kpc) if extending to infinity.

For the optical surface photometry I use the data from van der Kruit and Searle (1982) in the J and F bands. The plates therein were digitized over an area of 17×34 arcmin ($\alpha \times \delta$) and have been analysed in 4-arcsec square pixels. A new sky fit was performed to a linear polynomial (sloping plane) using only pixels more than 14 arcmin from the centre of NGC 5907. Over area's of the plate values for the extended surface brightness were determined from an analysis of histograms of the pixel values. The bright pixels with stars where ignored and a Gaussian centered on the median pixel value was fitted to the peak. The widths of these correspond to r.m.s. values of ~1.2% of sky surface brightness.

The resulting medians in rings on the sky are shown in Fig. 2. To obtain these for R < 6' only pixels with distance z > 2' from the plane of NGC 5907 were taken. It is clear that for z > 3' there is no halo in excess of 0.2% of sky, which corresponds to surface brightnesses of 28.8 mag arcsec^{-2} in J and 28.0 in F. Since the difference of 0.8 mag is

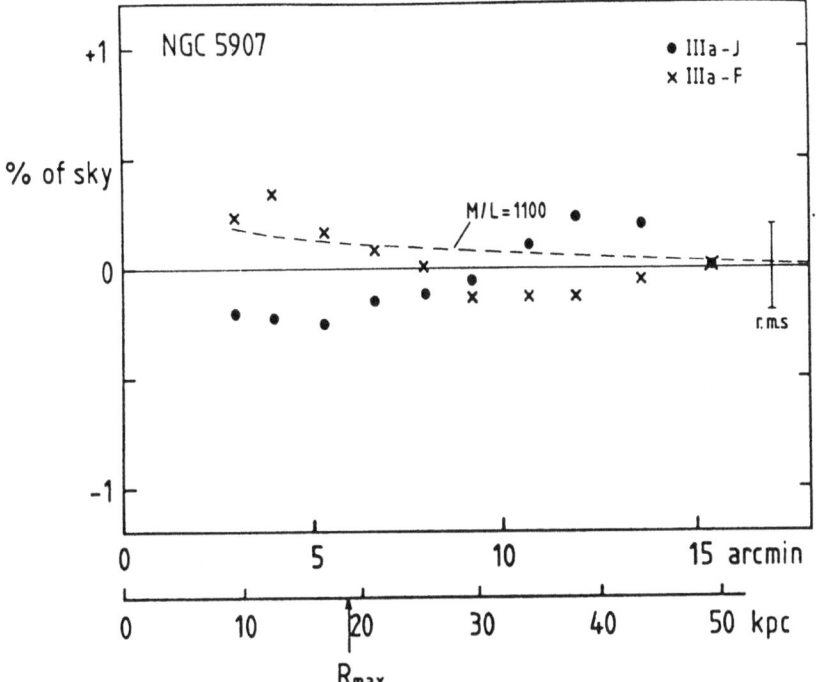

Fig. 2 - Median pixel values, expressed in percent of sky surface brightness in rings centered on NGC5907. The dashed line shows an I ∝ R^{-1} halo with M/L = 1100.

equal to the sun's colour index in these bands I find the same M/L-values in both colours.

We may integrate the data over the area R < 6', z > 2' to find that L < 1.2×10^8 $L_\odot$ or M/L > 760. Also we may take an upper limit at R = 3' as above to find M/L > 1100. For comparison:

M5-dwarf	$(M/L)_J = 280$ $M_\odot/L_{\odot,J}$	$(M/L)_F = 80$ $M_\odot/L_{\odot,F}$
VB10	1.2×10^5	1.8×10^4

Next we take the K-band upper limits from Skrutskie et al. (1985) along the minor axis. These lead to the following results:

z = 1.0 arcmin (3.2 kpc)	M/L > 36 $M_\odot/L_{\odot,K}$
z = 1.5 arcmin (4.8 kpc)	M/L > 20 $M_\odot/L_{\odot,K}$

Again for comparison

M5-dwarf	M/L = 10 $M_\odot/L_{\odot,K}$
VB10	M/L = 36 $M_\odot/L_{\odot,K}$

The values derived here are based on a definite mass model that incorporates a measured disk mass independent of the general shape of the rotation curve. Also the optical data utilise a sky-fit at more than 2 disk radii (the K-band data are obtained wilth a wobbling secondary with a throw of 8 arcmin; an $I \propto R^{-1}$ halo would have a surface brightness at the sky position 6-8 times fainter than at z = 1-1.5 arcmin). We probably have here the best defined upper limits. They show that dark halos cannot be made up of H-burning Main Sequence stars, at least not for a considerable fraction of their mass. It would be profitable to push the limits in the K-band further by a factor two or so.

4.   IRAS MEASUREMENTS

Deep maps of a number of edge-on galaxies (NGC 4565, 4244, 891, 5907, etc.) have been obtained with IRAS and put at my disposal (T. de Jong and R. Wainscoat, private communication). None of these show evidence for emission at large distances from the plane. I give below again detailed numbers for NGC 5907. For some of the data see Fig. 3.

For definiteness I compare the data with dark halos made up of the following objects:
- Main Sequence dwarfs of mass ~0.09 M$_\odot$ (such as VB10)
- Black dwarfs. Here I use Tarter's (1975) cooling curves as adapted by Staller and de Jong (1981):

$$L/L_\odot = 8.57 \ (M/M_\odot)^{1.91} \ (t/yr)^{-0.836}$$

$$R/R_\odot = 0.23 \ (M/M_\odot)^{-0.22} \ (T_e/K)^{-0.096}$$

- "Jupiters". I take the relevant values for the planet, keeping in mind that it radiates more energy than it receives from the sun.

Some relevant parameters are summarized in Table 2. The black dwarfs are assumed to be either primordial (t = 15 Gyr) or young (t = 1 Gyr). The radiation is assumed purely black body and λ_m is the wavelength of the maximum in the Planck curve.

Table 2 - Parameters of dark objects

Object	age(Gyr)	M(M$_\odot$)	L(L$_\odot$)	R(R$_\odot$)	T$_e$(K)	$\lambda_m(\mu)$
RD		9×10^{-2}	9.7×10^{-4}	0.023	2600	1.1
BD	15	7×10^{-2}	1.7×10^{-5}	0.28	700	4.1
BD	1	7×10^{-2}	1.6×10^{-4}	0.26	1270	2.3
BD	15	1×10^{-2}	4.0×10^{-7}	0.48	210	13.6
BD	1	1×10^{-3}	3.9×10^{-6}	0.45	380	7.5
BD	15	1×10^{-3}	4.9×10^{-9}	0.91	51	56
JUP		9.5×10^{-9}	1.0×10^{-9}	0.10	100	28

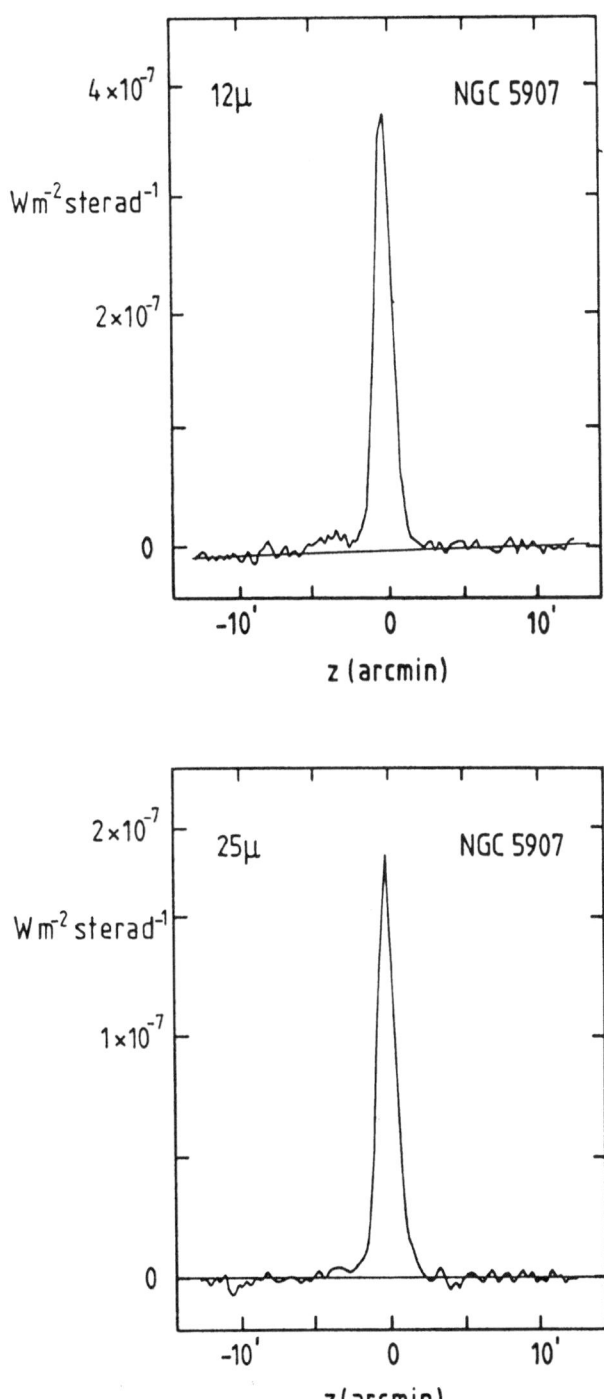

<u>Fig. 3</u> – Minor axis scans obtained by IRAS of NGC 5907 at 12 and 25 micron. No significant emission is detected at |z| ≳ 2 arcmin.

The predicted fluxes and the upper limits are compared in Table 3. The IRAS results are consistent with the halo being made up of any of the objects used in the comparison, i.e. all kinds of collapsed objects less massive than Main Sequence stars.

Table 3 - Comparison of IRAS data on NGC 5907 with predictions of halos made up of the various objects given in Table 2.

λ	12 μ	25 μ	60 μ	100 μ
z	2' (6.4 kpc)	2' (6.4 kpc)	4' (12.8 kpc)	5' (16.0 kpc)
$\log f(IRAS)$ Wm^{-2} sterad^{-1}	$\lesssim$ -8.3	$\lesssim$ -8.7	$\lesssim$ -8.7	$\lesssim$ -8.7
Object (t,M)				
RD	-10.9	-11.8	-13.1	-15.1
BD (15,0.07)	- 9.4	-10.3	-11.4	-13.4
BD (1,0.07)	- 9.1	-10.0	-11.2	-13.2
BD (15,0.01)	-10.0	-10.0	-10.8	-12.7
BD (1,0.01)	- 8.9	- 9.4	-10.5	-12.4
BD (15,10^{-3})	-12.9	-11.6	-11.8	-13.6
JUP	-16.1	-12.2	-11.0	-12.4

5.   EVIDENCE FROM COLOUR GRADIENTS IN SPHEROIDS

There are two ways in which colour gradients in spheroids can tell us anything about the constituents of dark halos. In the first place, a change of the make-up of the spheroids towards very low-mass stars would reveal itself by a gradual reddening with radius. Close scrutiny in this respect has again been performed in the spheroid of NGC 4565. No clear, significant change has been reported (Hegyi and Gerber, 1979; Boughn et al., 1982; Thuan and Jensen,1982).

A second way in which the dark halo may manifest itself is by a radial gradient towards lower metal abundance which would give rise to a increasingly bluer colour with radius. This would result when an in-creasing amount of low mass objects have been formed along with the luminous population II at larger galactocentric distance, while the initial mass function of the metal producing stars is invariant. This is so because these low mass objects act as sinks for the products of nucleosynthesis, leaving the stellar population with a lower mean abun-dance when the fragment of the protogalaxy has evolved to completion (see e.g. van der Kruit and Searle, 1982b). Simple models indicate that this mean metallicity should be proportional to the "yield" of heavy elements and therefore to the local M/L.

Such colour changes are often seen in the spheroids of spiral gala-xies (Wirth, 1981; Wirth & Shaw, 1983; van der Kruit and Searle, 1981; 1982b), and are generally believed to be due to a change in metallicity (see also Mould, 1984). In this respect the Sab galaxy NGC 7814 offers a unique possibility, because it not only displays a sizable colour gradient, but also has an insignificantly small disk (Van der Kruit and Searle, 1982b). Its flat HI rotation curve can uniquely be interpreted in terms of a spheroid mass distribution. The $R^{1/4}$ surface brightness distribution and the R^{-2} density distribution indicate an increase in M/L by a factor 10 between 3 and 21 kpc (major axis distance). At the same time the colour changes of $\Delta(U-B) = 0.3\pm0.3$ mag. and $\Delta(B-V) = 0.7\pm 0.3$ mag span the total range of colour index variation among Galactic globular clusters. This also implies a change in metallicity of an order of magnitude, in agreement with the predictions.

Other interpretations may certainly be invoked; nevertheless, the spheroid of NGC 7814 displays the signs expected if the dark matter consists of low mass objects formed along with the luminous constituents of population II.

As a final point I note the interesting fact that the two edge-on galaxies NGC 891 and 7814 have identical rotation curves which are flat between about 3 and 20 kpc at 220 km s^{-1}. Yet NGC 891 is strongly disk-dominated and NGC 7814 spheroid-dominated in their light distributions. Constant M/L ratio's of the luminous components would give rise to grossly different rotation curves (see Fig. 1 in Van der Kruit, 1983).

Acknowledgement

I thank T. de Jong for making the IRAS data available and R. Wainscoat for help with the reduction.

References

Albada, T.S. van, Bahcall, J.N., Begeman, K., Sancisi, R. 1985, Ap. J. (in press)
Bahcall, J.N. 1983, Ap. J. 267, 52
Boughn, S.P., Saulson, P.R., Seldner, M. 1981, Ap. J. 250, L15
Carignan, C., Freeman, K.C. 1985, Ap. J. (in press)
Casertano, S. 1983, M.N.R.A.S. 203, 735
Frankston, M., Schild, R. 1976, Astron. J. 81, 500
Freeman, K.C., Carrick, D.W., Craft, J.L. 1975, Ap. J. 198, L93
Gallagher, J.S., Hudson, H.S. 1976, Ap. J. 309, 389
Greenstein, J.L., Neugebauer, G., Becklin, E.E. 1970, Ap. J. 161, 519
Hegyi, D.J., Gerber, G.L. 1977, Ap. J. 218, L7
Hegyi, D.J., Gerber, G.L. 1979, Photometry, Kinematics and dynamics of Galaxies (Austin, Texas), p. 119
Hohlfeld, R.G., Krumm, N. 1981, Ap. J. 244, 476
Jensen, E.B., Thuan, T.X. 1983, Ap. J. Suppl. 50, 421
Kormendy, J. 1982, Saas-Fee conference, p.144
Kormendy, J., Bruzual, A.G. 1978, Ap. J. 223, L63
Kruit, P.C. van der 1981, Astron. Astrophys. 99, 298
Kruit, P.C. van der 1983, Proc. Astron. Soc. Austr. 5, 136

Kruit, P.C. van der, Freeman, K.C. 1985, Ap. J. (in press)
Kruit, P.C. van der, Searle, L. 1981, Astron. Astrophys. 95, 116
Kruit, P.C. van der, Searle, L. 1982a, Astron. Astrophys. 110, 61
Kruit, P.C. van der, Searle, L. 1982b, Astron. Astrophys. 110, 79
Kruit, P.C. van der, Shostak, G.S. 1984, Astron. Astrophys. 134, 258
Mould, J.R. 1984, Astron. J. 96, 773
Skrutskie, M.F. Shure, M.A., Beckwith, S. 1985, Ap. J. (in press)
Spinrad, H., Ostriker, J.P., Stone, R.P.S., Chiu, L.G., Bruzual, A.G.
    1978, Ap. J. 225, 56
Staller, R.F.A., Jong, T. de 1981, Astron. Astrophys. 98, 140
Tarter, J.G. 1975, Ph.D. Thesis, Berkeley
Wirth, A., 1981, Astron. J. 86, 981
Wirth, A., Shaw, R. 1983, Astron. J. 88, 171

DISCUSSION

ROBERTS: In NGC 5907 you have a value of M/L in the K band of > 36, and you gave a value of M/L = 36 in the K band for van Biesbroeck's star. Is this the basis for your eliminating all main sequence stars?

VAN DER KRUIT: Yes. You find similar limits in other galaxies, notably NGC 4565, in the literature. The limits also depend on the fraction of the mass in the disk, and on whether the halo is spherical. The conclusion is not very strong. But the observations rule out main sequence stars unless you make models that are very contrived.

TYSON: On faint photographic surface photometry: Photographic plates have served astronomy well, but it is dangerous to attempt surface photometry at 30 mag arcsec^{-2} due to systematic errors over large spatial wavelengths. Whereas the sky noise decreases like the number of pixels and plates averaged, systematic noise does not decrease like $1/\sqrt{N}$.

BURSTEIN: The reflection nebulae which create the "galactic cirrus" place a fundamental limit on the uniformity of the sky background, as pointed out 10 years ago by Sandage and more recently by de Vaucouleurs. A reddening of $E(B-V) \approx 0.02 - 0.04$ mag corresponds to reflection nebulae of $\sim 27 - 28$ mag arcsec^{-2}. Since much (but not all!) of the sky has reddenings of at least this value, and since IRAS measures have emphasized the intrinsic patchiness of the cirrus, these reflection nebulae will place an upper limit on the uniformity of the sky background at all optical wavelengths.

VAN DER KRUIT: Indeed these reflection nebulae are a fundamental limitation. However, NGC 5907 seems to sit in a relatively uniform area of the sky, so I was able to derive an upper limit that is rather faint. I think that one should worry seriously about galactic cirrus if ever a positive detection of a halo is reported.

FAINT PHOTOMETRY OF EDGE-ON SPIRAL GALAXIES:
A SEARCH FOR MASSIVE HALOS

Michael F. Skrutskie, Mark A. Shure, Steven Beckwith
Department of Astronomy, Cornell University

Upper limits have been set to the luminosity from the massive halos of three late-type edge-on spiral galaxies: NGC 2683 (Sb), NGC 4244 (Scd), and NGC 5907 (Sc). The limits resulted from simultaneous photometry in the visual (V) and 2.2μm (K) photometric bands which is sensitive to both luminosity and color changes along the minor axes of the three galaxies. The 3σ lower limits to the mass-to-light ratios for the halo of NGC 5907 are the largest ever recorded: $M/L_V > 2000$ and $M/L_K > 64$ in solar units. Since K band M/L for M-dwarf stars lying just above the hydrogen-burning limit is about 35, the results virtually eliminate the possibility that hydrogen-burning stars comprise more than a fraction of the halo masses. If the halos contain a more realistic spectrum of stellar masses, for example Population II, the visual band measurements imply that these stars account for less than one percent of the halo mass. Similar limits were obtained for NGC 4244 and NGC 2683. Variations of the V-K colors along and perpendicular to the disks show no sign of population changes toward redder objects at large galactocentric radii.

The nucleus of NGC 5907 contains an unresolved source less than 330pc in size with a 2.2μm luminosity of order 5×10^9 $L_\odot$, and may be an example of a star-burst galactic nucleus overlooked by visual observations.

J. Kormendy and G. R. Knapp (eds.), Dark Matter in the Universe, 425.

AN INFRARED SEARCH FOR SUBSTELLAR OBJECTS

F. J. Low
Steward Observatory
University of Arizona
Tucson, Arizona 85721

ABSTRACT. Preliminary results from a systematic search for nearby substellar objects in the IRAS data bases has revealed only a single candidate among the 12 μm sources in the region of the polar caps. This object appears to be a distant carbon star. All 5700 sources were positionally associated with stars or galaxies.

INTRODUCTION

Studies of low mass stars in the solar neighborhood show that their space density decreases with mass below about 0.2 $M_\odot$. Extrapolation of this trend to masses below the end of the hydrogen burning main sequence (0.08 $M_\odot$) fails to predict enough unobserved objects to account for the local mass deficit discussed by Bahcall (1985, reference these proceedings for the most current discussion). It is quite possible, however, that substellar objects (SSO), such as the newly discovered companion to VB 8 (McCarthy et al, 1985), exist in very large numbers as individual systems with masses in the range 0.05 to 0.001 $M_\odot$; furthermore, these objects could radiate enough power at infrared wavelengths to be detected (Stevenson 1978). In this report preliminary results are given on a project to use the data bases from the IRAS sky survey in a systematic search for the closest members of this as yet undiscovered class of ubiquitous brown dwarfs.

THE IRAS DATA BASES

In 1983 the Infrared Astronomy Satellite (IRAS) performed the first deep survey of the sky in the infrared and produced a raw data base of more than 30 gigabytes. The IRAS catalog contains nearly 250,000 point sources detected at 12, 25, 60 and 100 μm and was released in November 1984 (Beichman et al.). New software is under development at Jet Propulsion Laboratory which will be used to co-add the redundant survey scans to produce a new data base with sources 2 to 3 times fainter. Both of these large scale data products are derived from the

427

J. Kormendy and G. R. Knapp (eds.), Dark Matter in the Universe, 427–434.

"all sky survey" which attained sky coverage of 96 percent, but, because of source confusion, it is unlikely that the co-addition will be extended to galactic latitudes below 30 degrees. A very much smaller data base is under construction which will contain sources about 5 times fainter than those in the point source catalog and will cover only a few percent of the sky. This "serendipity survey" utilizes the IRAS pointed observations of specific fields and should be released by mid 1986.*   Table 1 summarizes the parameters of the three data bases as they pertain to the search for substellar objects.

When combined with Table I, Table II shows the applicability of the IRAS data products to a systematic search for substellar objects. Table II gives magnitudes at key wavelengths based on black body predictions for objects of the expected diameter at a distance of 1 pc. The V and R magnitudes include a possible 3 magnitude deficit below the blackbody values based on observations of the lowest mass M dwarfs (Probst 1983).   For objects in the temperature range of interest the 12 μm IRAS band is the most sensitive. The reason that the higher galactic latitudes are much preferred over lower latitudes is two fold, first, the number of sources that must be processed is much smaller and, hence, more tractable and , second, the number of sources with circumstellar dust shells which mimic the temperatures of interest is found to be very much lower as the galactic latitude increases.

In order to estimate the number of SSO that might be within the roughly two cubic parsec volume surveyed by IRAS, the volume density of unobserved material from Bahcall's analysis can be combined with the ad hoc assumption of a flat mass spectrum. Then the total mass of $0.2\,M_\odot$ might be contained in approximately 5 objects with masses in the range 0.010 to 0.050 $M_\odot$ and temperatures in the range 500 to 1500 K. This is an optimistic view in the sense that the population must be quite young. If, instead, a vast population of SSO formed very early in the life of the galaxy, then their temperatures today would be too cold for IRAS to detect. For this reason Table I includes estimates of the performance of two possible searches in the IR to much deeper limits.

THE SEARCH METHOD

Since the 12 μm band of IRAS is the most sensitive for objects warmer than 200 K, and since even the closest SSO is unlikely to be detected at both 12 and 25 μm, it is necessary to reduce the number of possible candidates by a large factor prior to detailed studies from the ground. Fortunately, it has proved possible to positionally associate most 12 μm point sources detected by IRAS with optically bright sources, predominantly stars. The procedure used was as follows: (1) the machine readable catalogs of stars were matched with the IRAS sources of interest (2) overlays were made for the northern and southern photographic surveys for the remaining sources to carry out associations on an individual basis.

TABLE I. Properties of Infrared Sky Surveys at 12 and 2.2 μm

	Usable Sky Coverage (Sq. deg.)	No. of Pt. Sources	Limiting Magnitude
IRAS 12 μm			
Pt. Src. Catalog	>2 E 4	> 2 E 4	4.5
Coadded Survey	2 E 4	> 4 E 4	5.5
Serendipity Survey	500	1500	6.3
FUTURE			
2.2 μm	>2 E 4	> 1 E 6	13 to 15
SIRTF 12 μm	2 E 4	2 E 6	9 to 10

TABLE II. Brown Dwarf Magnitudes vs Temp. and Wavelength
Distance = 1 pc

T (K)	Wavelength (μm)			
	0.5	0.85	2.2	12
200			38	11
300			26	9
500		37	17	7
800	38	24	11.5	6
1000	30	19	9.5	5.5
1500	20-23	13-15	7	5
2000	15-18	10-13	6	4.5
2500	12-14	8-9.5	5	4

The reliability of this method remains to be demonstrated through further tests. However, it is already clear that the number of candidates is small enough to permit individual follow up from the ground. Of more concern is the possibility that a miss identification with a randomly placed star in the field will disguise the detection of a much cooler object. For these reasons the results reported here must be used as preliminary.

Once a candidate has been identified by its optical/IR properties, it must be observed from the ground to obtain accurate positions for proper motion and parallax determinations. Photometry in the near IR and spectroscopy at appropriate wavelengths may also serve to classify the source. Clearly, all objects found by this search are potentially interesting and should be followed up.

PRELIMINARY RESULTS

Under the direction of T. Chester at the IRAS Processing and Analysis Center (IPAC), the 12 μm point sources in the polar caps (defined here as |b| > 50 deg.) were searched for substellar candidates. The results are summarized in Table III. It is important to note that with special study of a few individual sources it was possible to positionally associate all 5776 12 μm sources with their optical counterparts. Only one source emerged as a clear substellar candidate and was observed from the ground to determine its spectral energy distribution, its proper motion and its spectral characteristics. The most likely interpretation of this singular object is that, despite its unusual temperature of 1300 K and its very high galactic latitude, 86 deg., it is probably a Carbon Star.

CONCLUSIONS

Preliminary results from analysis of the sources in the polar caps indicate that substellar objects are not easily found in the IRAS

TABLE III.   IRAS 12 Micron Point Sources $|b| > 50^{\circ}$

Total Sample	5776
Unmatched	444
$T_c < 2500$ K	63
$2000 > T_c > 1000$	1
Sub-Stellar	0

point source catalog. However, it has also been shown that a viable method exists for extending the search to at least another quarter of the sky using the published data base. Two new data bases will become available for deeper searches and it seems clear that by exhaustive optical identification of the infrared sources a number of interesting objects will be found and even if no "ubiquitous brown dwarfs" are found, useful limits on their numbers will be determined.

ACKNOWLEDGMENTS

This work was supported by the National Aeronautics and Space Administration and represents the combined efforts of many of the author's colleagues on the IRAS Science Team.

* Carried out by: R. Cutri, F. Gillett, S. Kleinmann, F. Low, E. Young.

REFERENCES

Bahcall, J. N. 1984, Ap. J., **276**, 169.

McCarthy, D. W., Probst, R. G., and Low, F. J. 1985, Ap. J. Letters, **290**, L9.

Beichman, C. A. et al., 1984, NASA.

Probst, R. G. 1983, Ap. J., **274**, 237.

Stevenson, D. J. 1978, Proc. Astr. Soc. Australia, **3**, 227.

DISCUSSION

PACZYNSKI: A comment on brown dwarfs: AA Doradus is an eclipsing
binary system whose dark companion has a mass of only 0.04 $M_\odot$
(Kudritzki et al. 1982, Astron. Astrophys., 106, 254; Paczynski 1980,
Acta Astron., 30, 113). This is a spectroscopic binary; the dark
companion is responsible for eclipses and the hemisphere heated by the
companion is eclipsed. So the mass is very well established.

        I also have a question to do with your last statement that there
are so many bright infrared galaxies, which may be more numerous than
quasars. Is it clear from the IRAS data that what is seen in the
infrared is not nuclear activity? That it must be a starburst
phenomenon?

LOW: I believe it to be a starburst phenomenon in an extreme limit in
which so much gas has fallen into the nucleus that other processes, such
as the Seyfert or quasar phenomenon, must be very closely related.
Notice that the volume within ~ 100 Mpc of us contains several galaxies
emitting $\geqslant 10^{12}$ $L_\odot$, but, as far as I know, not a single quasar.

J. BAHCALL: Isn't the IRAS spatial resolution inadequate to resolve the
region from which the bright infrared emission comes? You don't know
whether it comes from the nucleus or is distributed in the disk.

LOW: In the case of Arp 220, the 25 μ emission region is extremely
compact. Data on that were presented at the Noordwijk symposium. It is
also known that the emission from Mk 231 is extremely concentrated to
the nucleus. So what I'm saying is that a large part of the mass
reservoir, which is somehow preferentially present in the nucleus, gets
converted extremely rapidly into stars to give the high luminosity that
we see.

LYNDEN-BELL: Do you know that it comes from stars rather than something
else?

LOW: I don't know that it comes from stars. But let me call to your
attention something that I consider to be a very good test of this.
That is the relationship between the total infrared brightness and the
radio continuum emission. This relationship has been around for fifteen
or twenty years, but the IRAS data have allowed it to be refined. We
have found that the relationship is very tight all the way up to very
luminous objects like Arp 220 or Mk 231 for a very large fraction of the
total sample. If this is a true indicator of the conversion of mass
into very luminous stars that produce supernovae, then I think that it
is an indication that star formation is still at work.

OSTRIKER: A comment and a question on whether the emission is extended
or nuclear. At least some subset of the infrared-bright galaxies must
overlap with the galaxies observed by Condon, because he picked out a
group of continuum-bright spiral galaxies, which I believe have shown
up as IRAS sources. These galaxies were also studied spectroscopically

and show emission extending over a good fraction of their disks. So it is known that at least that subsample has starburst activity going on over much of the disk. My question is: Have you cross-correlated your infrared-bright galaxies with the EINSTEIN results to see if there is any correlation with x-ray brightness?

LOW: I certainly have not, and I don't recall anything recent on that subject. I don't think that there is much of a correlation. Certainly we know for NGC 1068 that 95% of the infrared output comes not from the nucleus but from a very small, centrally condensed region of radius ~ 500 - 1000 pc. NGC 1068 is by far the closest and most easily studied infrared-bright galaxy, with an infrared luminosity about half that of the most luminous galaxies observed.

SOLOMON: Most of the very luminous infrared galaxies ($L_{IR} > 3\times10^{11}$ $L_\odot$) have strong CO emission, indicating an origin for the infrared luminosity in active star formation within molecular clouds and not in very small sources in the galactic nucleus. However, these starburst regions may be concentrated in the inner galactic disk or even in the inner 1 or 2 kpc of the galaxies.

LOW: Yes. Mk 231 is an exception. It is also right at the top of the luminosity range and may be part quasar and part starburst galaxy. That might tell us something. The weight of evidence now is that the starburst phenomenon, if that's the term you like to use, takes place up to this $> 10^{12}$ $L_\odot$ range. One of the interesting things the IRAS data tell us is that we can't stop there, we've got to study objects which put out hardly any optical photons at all. At Noordwijk, Mike Rowan-Robinson seemed to be convinced that he could build a model in which the luminosity from these systems is adequate to explain the 100 μ background of a few MJy steradian^{-1}. This could be a very important subject in the future. At the moment it is still premature to attach high significance to the 100 μ background measurement.

FABIAN: If Ω in baryons is several tenths and if this is uniformly distributed and makes the x-ray background, then the hot gas Compton scatters the microwave background, and we predict about 2 MJy steradian^{-1} at 100 μ. So you could rule out the x-ray background if you could explain the 100 μ background some other way.

BURSTEIN: Marcia Rieke and I have analyzed the IRAS fluxes for 2000 UGC Sc galaxies with diameters $\leqslant$ 3!5. To make a long story short, we find that the detection of an Sc galaxy is a function of the inclination of the galaxy to the line of sight, as well as of wavelength. A chain of argument leads to the conclusion that most of the 60 μ and 100 μ flux from Sc's comes from near their nuclei, and is extincted by the dust in the disk when the galaxy is edge-on.

LOW: Yes, I think that is consistent with what we know from resolving individual galaxies of various luminosities. But I really can't overemphasize the fact that for our Galaxy we need to have a better

handle on the outer parts in the infrared. Just because the nucleus
contains so much luminosity doesn't mean we can neglect the infrared
emission from processes going on in the outer parts that are so
important for the rotation curves. For the most part IRAS is capable of
telling us the answer. In some cases, we just have to open the
catalogue.

DRESSLER: Did I understand you to argue that the idea that the blue
luminosity is extinguished in infrared galaxies, resulting in a high
ratio of infrared to blue luminosity, is ruled out by a correlation of
this ratio with luminosity?

LOW: To my knowledge no-one has excluded the possibility of very dark,
cold galaxies, i.e., objects which emit primarily in the infrared and
have modest luminosities. These are hard to find. What leaps out at us
are these extreme cases, the very high-luminosity objects. There will
probably be a believable relationship between L_{IR}/L_B and L_{Tot} coming out
of the data shortly. But this doesn't mean that there might not be
large numbers of galaxies with low optical and infrared luminosities.

DRESSLER: This sounds like a classic Malmquist bias - those objects
which are very luminous in the infrared are very distant.

LOW: I agree.

DRESSLER: May I point out something else? I was intrigued by your
comment that most of the subluminous stars seem to be very young. I
want to mention a paper by Poveda which suggests the possibility that
these stars don't achieve fusion, and are all young and cooling down. I
haven't heard that mentioned very much, but it's a way to have the mass
function continue to climb below 0.2 $M_\odot$ even if the luminosity function
turns over.

LOW: It's ideas like that which keep us sifting through the data.

CONTRIBUTIONS TO THE LOCAL GRAVITATIONAL FIELD
FROM BEYOND THE LOCAL SUPERCLUSTER

A. Yahil
Astronomy Program
State University of New York
Stony Brook, NY 11794-2100, USA

ABSTRACT. IRAS 60μ sources are used to map the local ($\lesssim 200h^{-1}$ Mpc, $H_o=100h$ km s^{-1} Mpc^{-1}) gravitational field, and to determine its dipole component, on the assumption that the infrared radiation traces the matter. The dipole moment is found to point in the direction of the anisotropy of the microwave background radiation. Comparison of the two anisotropies, using linear perturbation theory, yields an estimate of the cosmological density parameter, $\Omega_o=0.85\pm0.16$, with nonlinear effects increasing Ω_o by ~15%. The quadrupolar tidal field within the Local Supercluster, due presumably to the same density inhomogeneities, is detected in a kinematical study of the velocity field.

1. INTRODUCTION

The velocity field in the Local Supercluster (LSC) has been extensively studied by many authors (e.g., the review by Yahil 1985). There is general agreement that the infall velocity of the Local Group (LG) toward Virgo is 250±50 km s^{-1}. This is different from the velocity of the LG relative to the microwave background radiation (MBR): $u_{MBR}=$ 600 km s^{-1} in a direction ~45° away from Virgo (Lubin et al. 1983; Fixsen et al. 1983). The difference between the two velocities is most easily understood as the bulk motion of the LSC, induced by density inhomogeneities on scales larger than the LSC.

The IRAS catalogue offers an opportunity for identifying a complete sample of galaxies, which are calibrated homogeneously over almost the entire sky, range in distance far beyond the limits of present redshift surveys, and are unaffected by extinction. As detailed in § 2, a dipole anisotropy is detected in the surface brightness of the IRAS 60μ sources, which is aligned with the anisotropy of the MBR (Yahil et al. 1985). The luminosity function of the IRAS galaxies is used to convert the angular dipole moment into the gravitational force with which these sources attract the LG. Comparison with u_{MBR} then yields an estimate of the cosmological density parameter Ω_o, on the assumption that the infrared radiation traces the total mass-energy.

The anisotropy provides only the dipole moment of the density

435

J. Kormendy and G. R. Knapp (eds.), Dark Matter in the Universe, 435–443.

distribution of the IRAS galaxies.   Higher moments are unlikely to be
determined for the IRAS galaxies before a complete redshift catalogue
becomes available. These moments, however, result in a shear velocity
field within the LSC.   The measurement of the quadrupolar component of
this shear field (Lilje et al. 1985) is reported in § 3.

2. IRAS DIPOLE ANISOTROPY

The IRAS point source catalog contains ~250,000 sources, of which only
~20,000 are galaxies.   It is very easy to discriminate spectrally
against the hotter stellar sources: for sources with high quality
detection in the 60μ band, the condition $S_{25} < 3S_{60}$ eliminates all but few
of the IRAS sources identified with stars.   The main problem is
contamination by the infrared "cirrus" emission from interstellar dust
in our own Galaxy, which is spectrally similar to the emission of
external galaxies. The solution adopted involves masking the part of
the sky in which cirrus is suspected.   The preferred mask, so-called
n=1, covers about half the sky (for details see Yahil et al. 1985).
     It is assumed that there exists a universal luminosity function
Φ(L) for the IRAS galaxies (Yahil et al. 1980), so the number of
galaxies observed in a luminosity range dL, and in a volume element d^3r,
can be written as

$$dN = D(\vec{r})d^3r \ \Phi(L)dL \qquad , \qquad (1)$$

where $D(\vec{r})$ is the local relative density function, D=1 corresponding to
the mean density of the universe.
     The luminosity function of the IRAS galaxies is well represented by
a two-power function

$$\Phi(L) = CL^{-2}(1 + L/\beta L_*)^{-\beta} \qquad (2)$$

(the limit $\beta \to \infty$ is a Schechter function).   Lawrence et al. (1985) give
C=(11.5±0.4)x10^6h L_0 Mpc^{-3}, for a fit of Φ(L) of the form of equation
(2), where $L = \nu L_\nu (60\mu)/L_0$.   As the area which they study has a source
density which is 18% higher than that for the whole sky (outside the
mask), their value should be corrected to C=(9.7±0.3)x10^6h L_0 Mpc^{-3}.
     For this particular luminosity function, the product of the flux
and the dipole moment of the surface brightness, hereafter loosely
referred to simply as the dipole moment, is given by

$$4\pi S \vec{\sigma}(S) = 12\pi S^2 \Delta S^{-1} \ \Sigma \ \hat{r}_i = \frac{3C}{4\pi} \int D(\vec{r})(\vec{r}/r^3)(1+r^2/\beta r_*^2)^{-\beta}d^3r \quad (3)$$

Here the surface brightness is differential in flux, the sum over the
unit vectors in eq. (3) being only over the sources in a flux bin ΔS.
The distance $r = \sqrt{(L_*/4\pi S)}$ is that at which a source with luminosity L_*
is seen with flux S. Except for the cutoff factor, $(1+r^2/\beta r_*^2)^{-\beta}$, to be
discussed below, and the parameter C, which characterizes the luminosity
function at the faint end, this dipole moment is seen to be identical to
the density moment

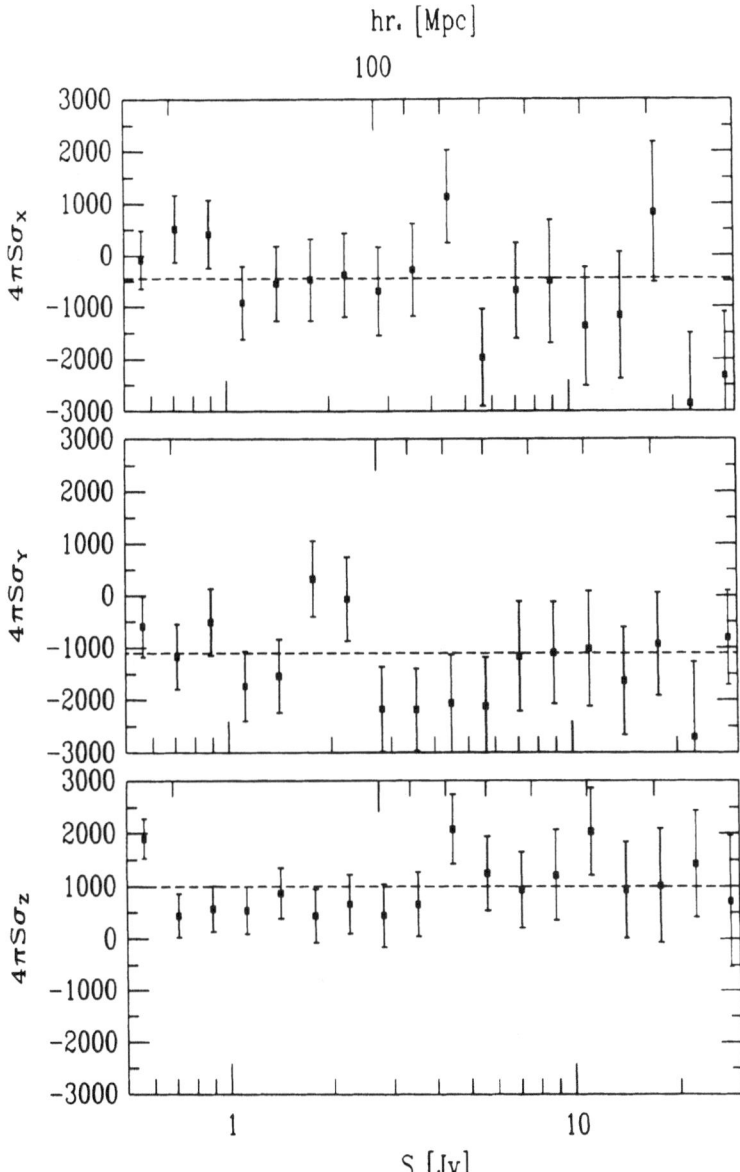

Fig. 1: Components of the dipole moment $4\pi S\vec{\sigma}(S)$. Data points at different flux bins are statistically independent; errors are *statistical* sampling ones. The dashed lines are the means of the data points, weighted by the inverse of the sum of the variances of the three components for each flux bin. The upper axis shows r_*, the distance at which a source with luminosity L_* is seen with flux S.

$$\vec{G} = (3/4\pi) \int D(\vec{r})(\vec{r}/r^3)d^3r \quad , \tag{4}$$

which is proportional to the peculiar acceleration.

Fig. 1 shows the three components of the dipole moment $4\pi S\vec{\sigma}(S)$, derived in a harmonic analysis including terms through quadrupole, using the mask n=1. It is assumed in the analysis that the spherical harmonic expansion derived in the unmasked part of the sky can be extrapolated without modification to the entire sky. (This is different from assuming that the masked area is isotropic, and does not contribute to the dipole and higher moments.) Points in different flux bins in Fig. 1 are statistically independent, and each one measures the dipole moment in its flux bin. Each point is therefore an independent estimate of the density moment in equation (4), with a cutoff at distance $r_*(S)$; the r_* scale (corresponding to $L_* = (5.0 \pm 0.9) \times 10^{10} h^{-2} L_\odot$, Lawrence et $al.$ 1985) is marked on the top axis.

While there is scatter in the data, it can be seen in Fig. 1 that the dipole moments are consistent with being independent of S, presumably because the density inhomogeneities giving rise to the dipole terms occur over distances that are smaller than the appropriate r_*, and the cutoff term has little effect. If this interpretation is correct, then the dipole moments measure $\vec{G}$ itself, without the cutoff factor. The average over flux, $\langle 4\pi S\vec{\sigma}(S)\rangle$, can therefore be taken, yielding a better estimate of $\vec{G}$. This average is shown as dashed lines in Fig. 1, and is given in Table 1, together with those of the masks n=0 and n=2. Only $statistical$ sampling errors are quoted. To these must be added probably comparable errors due to residual cirrus contamination, and extrapolation into the masked area of the sky, as well as systematic observational errors.

TABLE 1

Average Dipole Moment[a]

Mask	X-comp.	Y-comp.	Z-comp.	Mag.	l	b	θ_{MBR}	Ω_o [b]
n=0	-580±210	-910±200	700±150	1290±190	237±11	33±10	34±12	1.15±0.29
n=1	-450±190	-1100±180	990±140	1550±170	248±9	40±8	26±10	0.85±0.16
n=2	-300±180	-1240±170	960±140	1590±160	256±8	37±8	19±9	0.81±0.14

[a]Flux averaged dipole moment, $\langle 4\pi S\vec{\sigma}(S)\rangle$, in units of Jy str^{-1}. Each flux bin is weighted by the inverse of the sum of the variances of its three components. All errors are $statistical$ IRAS errors only.

[b]Determined from linear perturbation theory. Nonlinear effects increase these values by ~15%.

The direction of the velocity of the LG relative to the MBR (l=277, b=29) is close to the one determined for the IRAS dipole moment, as shown in Table 1. Given the statistical and systematic errors in the IRAS dipole moment, the difference in direction is quite acceptable. The direction of the velocity of the LG relative to the MBR is also uncertain by a few degrees, due to measurement errors, uncertainties in the solar velocity relative to the LG, and the neglect of the random ("thermal") peculiar velocity of the LG relative to nearby galaxies.

If the IRAS dipole moment is measuring the gravitational field responsible for the MBR anisotropy, then its magnitude can be used to determine Ω_o. In linear perturbation theory (Peebles 1980), the peculiar velocity is parallel with, and proportional to the peculiar acceleration, and hence to $\vec{G}$:

$$\vec{u} = \frac{2}{3}\Omega_o^{-0.4} H_o^{-1} \vec{g} = \frac{1}{3}\Omega_o^{0.6} H_o \vec{G} \tag{5}$$

Substituting, $u_{MBR}=600$ km s^{-1}, and using $C=(9.7\pm0.3)\times10^6 h\ L_o$ Mpc^{-3} to convert $\langle 4\pi S\sigma(S)\rangle$ to $\vec{G}$, yields $\Omega_o=0.85\pm0.16$ (statistical IRAS error only). This is inconsistent with the dynamical estimates obtained from the Virgocentric flow model and the cosmic virial theorem, $\Omega_o=0.1-0.2$ (e.g., the review by Yahil 1985).

The difference between the determinations of Ω_o from the Virgocentric infall and the cosmic virial theorem on the one hand, and the dipole anisotropy of the IRAS galaxies on the other hand, may be due to a number of causes. The *statistical* IRAS errors quoted for Ω_o may seriously underestimate the total error. The extensive mask used in this investigation may hide considerable structure, whose contribution to the gravitational field is not given by a simple extrapolation.

The IRAS galaxies seem to extend far enough to cover all the superclusters giving rise to the local gravitational field, as witnessed by the insensitivity of the dipole moment to the cutoff r_* (200h^{-1} Mpc for S=0.5 Jy). The distances of these local superclusters are more likely to be of the same order as deduced from the kinematical study of the shear velocity field in the Virgo supercluster (§ 3), i.e., $R\sim50h^{-1}$ Mpc. It is therefore unlikely that the inclusion of more distant galaxies by deeper surveys will change the IR dipole moment.

In fact, the above guess of the distance of the perturbations giving rise to the dipole moment can be used to estimate the size of the nonlinear corrections. Yahil (1985) shows that, for a spherically symmetric perturbation, $\vec{u}$ is smaller than the linear estimate, eq. (5), by a factor $(1+\langle D\rangle)^{-0.25}$, where $\langle D\rangle=|\vec{G}|/R$ is the mean density in a sphere centered on the perturbation, with us at the periphery. For the mask n=1, $\langle D\rangle\sim0.40$, so the nonlinear effects increase Ω_o by ~15%. The actual perturbations giving rise to the dipole moment undoubtedly deviate significantly from this simple spherically symmetric model, but the estimate of both the sign and the size of the nonlinear correction are probably reasonable.

The IRAS galaxies are different from optically selected galaxies in the absence of elliptical galaxies, and in the predominance of emission line galaxies. A preliminary study of the two-point correlation function of the IRAS galaxies (Rowan-Robinson and Needham 1985)

indicates that it is similar to that deduced from optically selected
surveys, i.e., $\xi(r) \alpha r^{-1.8}$, but the normalization constant is lower by a
factor ~2. This is not too different from the diminution by a factor
~1.5 found by Davis and Geller (1976) for spiral galaxies in general.
It is not clear, however, how this translates into density estimates of
specific structures. In fact, Yahil et al. (1980) find the same
Virgocentric density profile for early and late type galaxies.

There remains the possibility that the difference between the
Virgocentric and IRAS results is real, showing that either the optically
selected galaxies, or the IRAS galaxies, or both, do not trace the total
mass-energy, and the determinations of Ω_o are therefore biased.

3. TIDAL VELOCITY FIELD IN THE LOCAL SUPERCLUSTER

By the equivalence principle, the mean gravitational field in the LSC
can not be determined by measurements within it. An external reference
frame, such as the MBR, is required in order to measure the bulk free-
fall velocity which this mean field imparts to the LSC. Thus, the
dipole moment of the density structure outside the LSC, which has
presumably been measured from the distribution of the IRAS galaxies,
leads to no observable consequences within the LSC.

The higher moments of these same density inhomogeneities, however,
also result in a tidal field within the LSC (Binney and Silk 1979;
Palmer 1983). The leading quadrupolar tidal acceleration is given by

$$\vec{g}_t = \Sigma_t \cdot \vec{r} \tag{6}$$

where Σ_t is a symmetric traceless shear matrix.

Except near the central Virgo cluster, the growth of density
perturbations in the LSC can be approximated by the linear theory. The
systematic peculiar velocity field should therefore be parallel with,
and proportional to, the local gravitational acceleration, eq. (5).
Since the gravitational acceleration is the sum of the ones due to the
LSC and the external tidal field, it follows that the total peculiar
velocity can be well approximated as a sum of the two peculiar
velocities due to each field separately.

Lilje et al. (1985) have performed a Tully-Fisher fit to the
velocity field in the LSC, along the lines of Aaronson et al. (1982),
but adding the shear velocity field which follows from eq. (6). They
find that at the distance of the Virgo Cluster the eigenvalues of the
tidal field are ~200 km s^{-1}, but the component in the direction of Virgo
is only 46±70 km s^{-1}. The determination of Ω_o from the Virgocentric
infall is therefore little affected by the addition of the tidal field.
The residual random ("thermal") velocity of the LG relative to its
nearest neighbors is 72±37 km s^{-1}, which is not statistically
significant.

The validity of the tidal field fit has been checked in a variety
of ways, including separate fits to subsets of the data in the distance
ranges 300 km s^{-1}<v<1000 km s^{-1}, 1000 km s^{-1}<v<2000 km s^{-1}, and
2000 km s^{-1}<v<3000 km s^{-1}. They yielded identical results within the

errors. This is a sensitive test of the form of the tidal velocity
field, which is expected to grow linearly with distance.

It is interesting to note that the eigenvector corresponding to the
largest positive eigenvalue of Σ_t points toward the Hydra-Centaurus
supercluster (Chincarini and Rood 1979; Hopp and Materne 1985), the
nearest supercluster to the LSC. Although the tidal field is not due to
a single nearby supercluster (the eigenvalues are not properly related),
a rough estimate of the distance of the perturbers can be made by
comparing the r.m.s. of the eigenvalues of the tidal velocity field with
the bulk velocity of the LSC. This gives

$$R \sim (500/165)R_V \sim 50h^{-1} \text{ Mpc} \tag{7}$$

For details see Lilje *et al.* (1985).

4. ACKNOWLEDGMENTS

This research was supported in part by USDOE grant DE-AC02-80ER10719 at
the State University of New York, by SERC grants at the Institute of
Astronomy, Cambridge, and at Queen Mary College, London, and by a
Guggenheim Fellowship.

5. REFERENCES

Aaronson, M., Huchra, J., Mould, J., Schechter, P. L., and Tully, R. B.
    1982, *Ap. J.*, **258**, 64.
Binney, J., and Silk, J. 1979, *M. N. R. A. S.*, **188**, 273.
Chincarini, G., and Rood, H. J. 1979, *Ap. J.*, **230**, 648.
Davis, M., and Geller, M. J. 1976, *Ap. J.*, **208**, 13.
Fixsen, D. J., Cheng, E. S., and Wilkinson, D. T. 1983, *Phys. Rev.
    Lett.*, **50**, 620.
Hopp, U., and Materne, J. 1985, *Astr. Ap. Suppl.*, **61**, 93.
Lawrence, A., Walker, D., Rowan-Robinson, M., Penston, M., and Leech, K.
    1985, *M. N. R. A. S.*, submitted.
Lilje, P. B., Yahil, A., and Jones, B. J. T. 1985, *Ap. J.*, submitted
Lubin, P. M., Epstein, G. L., and Smoot, G. F. 1983, *Phys. Rev. Lett.*,
    50, 616.
Palmer, P. L. 1983, *M. N. R. A. S.*, **202**, 561.
Peebles, P. J. E. 1980, *The Large-Scale Structure of the Universe*,
    (Princeton: Princeton University Press), § 14.
Rowan-Robinson, M., Chester, T., Soifer, T., Walker, D., and Fairclough,
    J. 1985, *M. N. R. A. S.*, submitted.
Rowan-Robinson, M., and Needham, G. 1985, *M. N. R. A. S.*, submitted.
Yahil, A. 1985, in *The Virgo Cluster of Galaxies*, eds. O. G. Richter and
    B. Binggeli, (Garching: ESO), p. 359.
Yahil, A., Sandage, A., and Tammann, G. A. 1980, *Ap. J.*, **242**, 448.

DISCUSSION

N. BAHCALL: Aren't you seeing mostly late-type galaxies in your sample?
What is the effect of omitting essentially all the early-type galaxies?

YAHIL: The problem is somewhat worse than that. We don't use regions
with large cirrus indicators, i.e., regions with a large density of 100
μ sources. The result is that the centers of rich clusters look like
cirrus. For example, look at the region of the Virgo Cluster in the
northern-hemisphere map I showed. You notice one or two "cirrus" bins,
but they are not cirrus, they are the Virgo Cluster. So there is a bias
both against early-type galaxies and against the cores of rich clusters.
All I can say is that, since most of the mass in superclusters is not in
the central clusters, I hope we won't be too badly affected by these
biases when we use large areas of the sky to measure the dipole moment.

E. TURNER: Given the differences between the IRAS "colors" of stars and
galaxies, and given the huge range of L_{IR}/L_B quoted earlier by Frank
Low, it would appear that the IRAS fluxes are determined by a galaxy's
dust content and/or current star formation rate (and distance, of
course). It seems unlikely that either is particularly well correlated
with the total mass distribution or even the total baryonic mass. Is
this a fundamental limitation on the use of the otherwise excellent IRAS
data set for this method of determining Ω_0?

YAHIL: The only question is whether the IRAS galaxies are good tracers
of the total mass-energy on a very large scale. Nobody knows the answer.
A redshift survey of IRAS galaxies might be helpful.

DAVIS: A correlation analysis has been made of the IRAS galaxy list.
We found the angular correlation to be consistent with that of nearby
spiral galaxies, but scaled to a distance of approximately 100 $(h_{100})^{-1}$
Mpc. This is also consistent with the observed surface density of the
IRAS galaxies and their overlap with the CfA catalog.

FABER. Did you discover any new nearby clusters near the galactic plane?

YAHIL: Not yet. We look so deep that a nearby cluster is only a small
perturbation. It's like looking at the poster by Peebles and Groth and
trying to find the Virgo Cluster. It isn't easy. We are now searching
the sky for regions of high surface brightness to try to find objects
and to measure their redshifts. (I suspect that people will only be
satisfied with our results when we have measured redshifts for at least
a fair sub-sample of the objects.)

SCHECHTER: I don't think that it is enough to look at the shape of the
spatial correlation function. You also want to know the amplitude. And
isn't that your entire paradox: you have measured it and it is different
for galaxies and for the 60 μ sources?

YAHIL: Yes, you're absolutely right. All I can say at the moment is
that Mike Rowan-Robinson tells me that the angular correlation function

is intermediate between those found for the Zwicky and Shane-Wirtanen catalogues. The characteristic L_* is also intermediate. So it's in the right ball park. But whether the normalization comes out exactly right or whether we're missing a factor of two is going to be critical.

FELTEN: What is the sign of the probable effect on your results of masking a large region of the sky?

YAHIL: The effect could go either way. A larger IRAS anisotropy for the same MBR anisotropy implies a smaller Ω_0 (see equation 5).

GUNN: You have quoted us statistical errors, but could you comment on the possibility of systematic errors introduced by the width of the luminosity function of IRAS galaxies? The observed optical luminosity function always looks rather like a Gaussian around L_*, but the infrared luminosity function must be much broader than that. Since the correlation function takes a rather different moment of the luminosity function than the counts do, I would suspect the possibility of large systematic errors.

YAHIL: I don't know how big the effect would be.

LOW: Much as I hate to pour cold water on such an exciting IRAS result, I have to ask what you have done in your analysis about three major causes of systematic error in the production of the catalog. The first is the South Atlantic anomaly, which is unfortunately in the South Atlantic (laughter). The second is that the satellite scanned in one direction across the sky. As a result, it always went through the galactic plane in the same direction. We applied a significant hysteresis correction after the threshold detection, because this effect was not discovered until late in the analysis. The third effect is just the effect of radiation. The polar horns, as they're called, are brighter in one hemisphere than the other at a given time of year. IRAS detected sources which were bright above the local noise, and that noise was often produced by activity in the van Allen belts. These effects will certainly affect the accuracy and precision of your result, and may change it altogether.

YAHIL: Let me deal with your points one by one. We did nothing about the South Atlantic anomaly and just took the IRAS fluxes as they are given in the catalogue. We are aware of the errors introduced by crossing the plane, and threw out all the data within 5° of the plane. The masks actually throw out much more than that, sometimes up to b = 30°. But we were also careful to throw out the areas with hysteresis problems. Perhaps you think the effect is larger than we took into account - that should be checked. As for problems caused by the van Allen belts, we didn't do anything about those either. But I want to point out that our result is independent of the flux limit that we used; we get the same answer for the dipole moment even if we use a fairly high flux limit. I would expect effects of the type you mentioned to be very sensitive to the flux level. But that's my only defense.

A COSMOLOGIST'S TOUR THROUGH THE NEW PARTICLE ZOO (CANDY SHOP?)

Michael S. Turner
Departments of Astronomy and Astrophysics and Physics
The University of Chicago
Chicago, IL 60637

NASA/Fermilab Astrophysics Center
Fermi National Accelerator Laboratory
Batavia, IL 60510

ABSTRACT. Recent developments in elementary particle physics have led to a renaissance in cosmology, in general, and in the study of structure formation, in particular. Already, the study of the very early (t $\leq$ 10^{-2} sec) history of the Universe has provided valuable hints as to the 'initial data' for the structure formation problem -- the nature and origin of the primeval density inhomogeneities, the quantity and composition of matter in the Universe today, and numerous candidates for the constituents of the ubiquitious dark matter. I review the multitude of WIMP candidates for the dark matter provided by modern particle physics theories, putting them into context by briefly discussing the theories which predict them. I also review their various birth sites and birth processes in the early Universe. At present the most promising candidates seem to be a 30 or so eV neutrino, a few GeV photino, or the 'invisible axion' (weighing in at about 10^{-5} eV!), with a planck mass monopole, quark nuggets, and shadow matter as the leading 'dark' horse candidates. I also mention some very exotic possibilities -- unstable WIMPs, cosmic strings, and even the possibility of a relic cosmological term.

1. INTRODUCTION

The hot big bang model (also known as the standard model of cosmology) is almost universally accepted -- and for good reason. The model provides a reliable description of the evolution of the Universe from the epoch of primordial nucleosynthesis (t $\simeq$ 10^{-2}sec, T $\simeq$ 10 MeV) until the present (t $\simeq$ 15 Byr, T $\simeq$ 3 K). [For a review of the standard cosmology and primordial nucleosynthesis, see Audouze (1986), Boesgaard and Steigman (1986), and Steigman (1986).] Within the context of the standard cosmology there is a general picture of how the structure in the Universe which is so conspicuous today formed -- small primordial density inhomogeneities ($\delta\rho/\rho$ $\simeq$ 10^{-4}-10^{-5}) began to grow via the Jeans

445

J. Kormendy and G. R. Knapp (eds.), Dark Matter in the Universe, 445–488.

instability when the Universe became matter-dominated, eventually becoming the highly nonlinear structures we observe today, galaxies, clusters of galaxies, etc. [For a recent review of structure formation see Efstathiou and Silk (1983).]

The structure formation problem can be viewed as an initial data problem. The initial epoch being the onset of matter domination

$$R_{eq} \approx 3.5 \times 10^{-5}(T_{2.7}^{4}/\Omega h^2) , \tag{1a}$$

$$T_{eq} \approx 6.8 \text{ eV } (\Omega h^2/T_{2.7}^{3}) , \tag{1b}$$

$$t_{eq} \approx 3 \times 10^{10} \text{sec } (\Omega h^2/T_{2.7}^{3})^{-2} , \tag{1c}$$

where $R(t)$ is the cosmic scale factor (normalized so that R=1 today), $2.7T_{2.7}K$ is the present temperature of the microwave background, $\Omega = \rho_{TOT}/\rho_{crit}$, $\rho_{crit} = 1.88h^2 \times 10^{-29}g \text{ cm}^{-3} = 1.05 \times 10^4 h^2 \text{ eV cm}^{-3}$ is the critical density, and $H_0 = 100h \text{ km sec}^{-1} \text{ Mpc}^{-1}$ is the Hubble parameter today. The initial data consist of: (i) the spectrum and type (adiabatic or 'isothermal') of density perturbations present; (ii) the amount of matter in the Universe (quantified by Ω); (iii) the composition of the matter--fraction (Ω_{baryon}) that is baryonic, fraction (Ω_{WIMP}) that is exotic Weakly-Interacting, Massive Particles (or WIMPs), etc. In principle, once armed with a possible set of initial data for the problem one can numerically simulate the formation of structure, and compare the results with the observed Universe to test the viability of those initial data. [For a recent review of numerical simulations of structure formation see White (1986a,b).] Until recently progress towards filling in the details of structure formation suffered severely from lack of knowledge of the initial data for the problem. Simply put, there was just too much phase space to explore!

The renaissance in cosmology initiated by the infusion of new ideas in theoretical particle physics has also revitalized the study of the formation of structure in the Universe. Preliminary forays into the very early Universe (t < 10^{-2}sec) have provided a number of important hints as to the initial data for the structure formation problem. Baryogenesis, the theory of the origin of the baryon number of the Universe, all but precludes the possibility of baryonic isothermal density perturbations (Turner and Schramm 1979; also see Barrow and Turner 1981; Bond, Kolb, and Silk 1982; and Kolb and Turner 1983). In addition to solving the homogeneity, isotropy, flatness, and monopole problems, the inflationary Universe scenario (Guth 1981, Linde 1982, and Albrecht and Steinhardt 1982) leads to calculable primordial density perturbations. Quantum fluctuations during inflation result in adiabatic perturbations with the Zel'dovich spectrum (Bardeen, Steinhardt, and Turner 1983, Hawking 1982, Starobinskii 1982, and Guth and Pi 1982) and in an inflationary Universe with axions, isothermal axion perturbations with the Zel'dovich spectrum also arise (Steinhardt and Turner 1983, Linde 1985, and Seckel and Turner 1985). A class of Grand Unified Theories (or GUTs) lead to the production of topological entities which are line singularities and are referred to as cosmic strings. The production of cosmic strings in the very early Universe leads to isothermal perturbations in the matter of a definite spectrum and amplitude. Cosmic strings have recently been reviewed by Vilenkin (1985).

Since primordial nucleosynthesis constrains the fraction of critical density contributed by baryons to be

$$0.014h^{-2} \leq \Omega_{baryon} \leq 0.035h^{-2}$$

and the inflationary Universe scenario (as well as other theoretical prejudices) strongly suggest that $\Omega = 1$, the early Universe seems to be telling us that most of the matter in the Universe is non-baryonic (which is not inconsistent with the fact that most of the matter in the Universe is dark). Of course one of the currently fashionable (and I believe very attractive) possibilities is that the constituents of the dark matter are relic WIMPs left over from the very hot, early epoch of the Universe. The early Universe and modern particle theories working together have provided a very generous list of candidates for the dark matter, most of them hypothetical particles and other hypothetical entities (for a partial listing, see Fig. 1).

This will be the focus of my article. To place the candidates in their proper context I will begin with a very brief and superficial review of modern particle theory. Next I will discuss the production of relic WIMPs in the early Universe. Although in many respects the various WIMPs are interchangeable, there are some very important differences, differences which bear on the details of structure formation and the possible detection of the cosmic reservoir of WIMPs which may surround us; this will be the focus of the next section. As if a Universe dominated by WIMPs is not exotic enough, I will go to discuss some very exotic solutions to the Ω problem (the discrepancy between theory and observation with regard to the value of Ω). I will conclude with some prognostications and summarizing remarks!

Let me end the introduction with a set of conversion factors and useful formulae. Every problem has it's natural set of units; for the early Universe it is the so-called natural units of particle physics where $\hbar = c = k_B = 1$. In this system, the fundamental unit is the GeV = 10^3 MeV = 10^6 keV = 10^9 eV, and

$$1 \text{ GeV}^{-1} = 1.97 \times 10^{-14} \text{cm} ,$$

$$1 \text{ GeV}^{-1} = 6.58 \times 10^{-25} \text{sec} ,$$

$$1 \text{ GeV} = 1.16 \times 10^{13} \text{K} ,$$

$$1 \text{ GeV} = 1.77 \times 10^{-24} \text{g} ,$$

$$G_{Newton} = 1/m_{pl}^2 \quad (m_{pl} = \text{'planck mass'})$$

$$m_{pl} = 1.22 \times 10^{19} \text{GeV} ,$$

$$1 \text{ pc} = 1.5 \times 10^{32} \text{GeV}^{-1} ,$$

$$1 \text{ M}_\odot = 1.1 \times 10^{57} \text{GeV} ,$$

$$H_o = 2.2 \times 10^{-42}h \text{ GeV} .$$

DARK MATTER CANDIDATES

(<u>VERY</u> EXOTIC CANDIDATES NOT LISTED)

FOR REFERENCE:

$$\rho_{crit} \simeq 10^{-29}\, g\, cm^{-3} \simeq 10^4\, eV\, cm^{-3}$$
$$n_\gamma \simeq 400\, cm^{-3}$$
$$n_{baryon} \simeq 10^{-7}\, cm^{-3}$$

'BEWARE OF THE DARK SIDE'

NECESSARY FOR $\Omega_x \simeq 1$

WIMP	MASS	'ABUNDANCE'	BIRTH SITE	
'INVISIBLE' AXION	$10^{-5}\,eV$	$10^9\,cm^{-3}$	$10^{-30}\,sec$	$10^{12}\,GeV$
'LIGHT' NEUTRINO[†]	$30\,eV$	$109\,cm^{-3}$	$1\,sec$	$1\,MeV$
PHOTINO / GRAVITINO / SIMPSON NEUTRINO / MIRROR PARTICLE / AXINO	keV_s	$10\,cm^{-3}$	$10^{-5}\,sec$	$300\,MeV$
PHOTINO / SNEUTRINO / HIGGSINO / GLUINO / HEAVY NEUTRINO / SHADOW MATTER	GeV_s	$10^{-6}\,cm^{-3}$	$10^{-4}\,sec$	$50\,MeV$
SUPERHEAVY MAGNETIC MONOPOLE	$10^{16}\,GeV$ ($\simeq 10^{-8}\,g$)	$10^{-22}\,cm^{-3}$	$10^{-34}\,sec$	$10^{14}\,GeV$
PYRGONS / MAXIMONS / PERRY-POLES / NEWTORITES / SCHWARZSCHILDS	$\gtrsim 10^{19}\,GeV$ ($\gtrsim 10^{-5}\,g$)	$\lesssim 10^{-25}\,cm^{-3}$	$10^{-43}\,sec$	$10^{19}\,GeV$
QUARK[†] NUGGETS	$\simeq 10^{15}\,g$	$10^{-44}\,cm^{-3}$	$10^{-5}\,sec$	$300\,MeV$
PRIMORDIAL BLACK HOLES[†]	$\gtrsim 10^{15}\,g$	$\lesssim 10^{-44}\,cm^{-3}$	$\gtrsim 10^{-12}\,sec$	$\lesssim TeV$

(left margin labels: SUSY $\neq 0$ · GUTs · VERY SPECULATIVE THEORIES)

<u>NB.</u> [†]SIGNIFIES A SPECIES KNOWN TO ACTUALLY EXIST

Figure 1 - A partial listing of dark matter candidates (or WIMPs) provided by modern particle theories. The abundance listed is the average cosmic abundance required to provide $\Omega = 1$. Note that if the WIMPs also provide the halo density their local abundance should be about a factor of 10^4 higher.

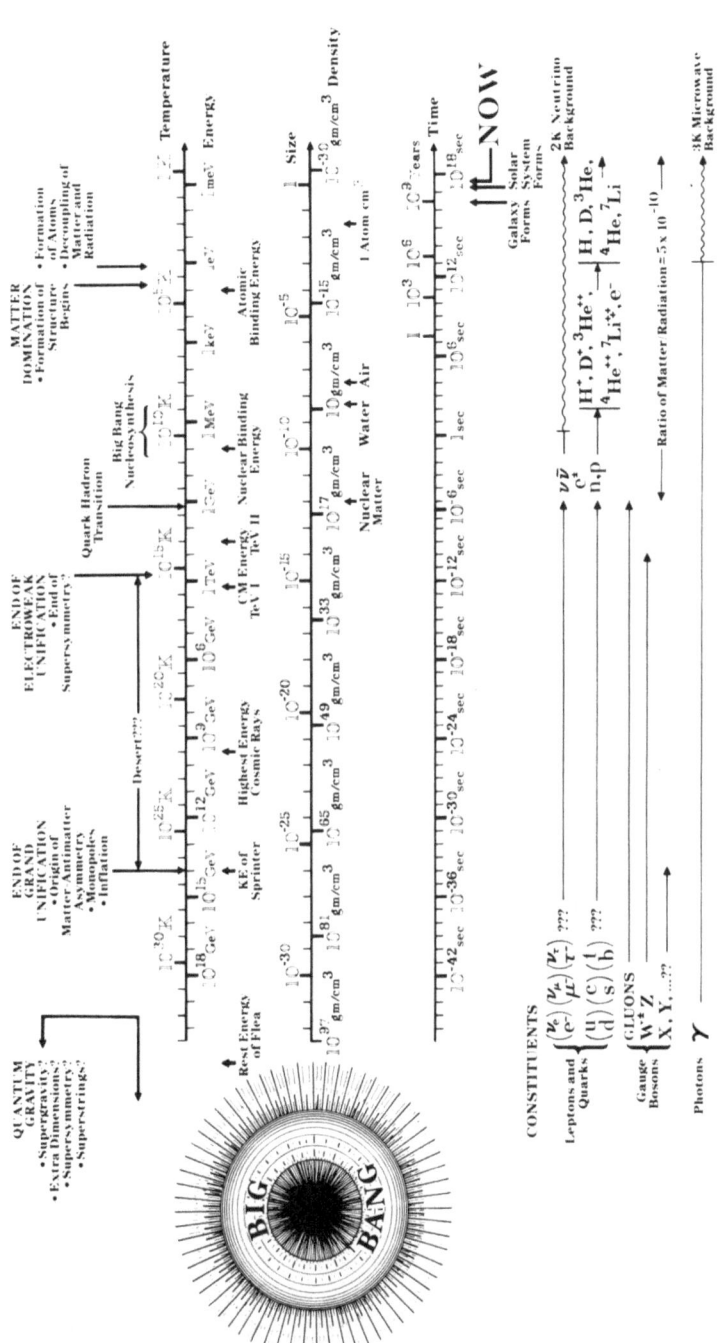

Figure 2 - The complete history of the Universe according to the standard hot big bang cosmology (also see the Figure in the Introduction to these proceedings).

During its earliest moments the Universe was radiation-dominated; i.e., for $t \lesssim 3 \times 10^{10}$ sec $(\Omega h^2/T_{2.7}^3)^{-2}$. During this period the evolution of the cosmic scale factor R(t) and the temperature T are given by

$$R(t) \propto t^{1/2}, \tag{2a}$$

$$H = 1/2t = 1.66g_*^{1/2}T^2/m_{pl}, \tag{2b}$$

$$T = 1.5g_*^{-1/4}\text{GeV } (t/10^{-6}\text{sec})^{-1/2}, \tag{2c}$$

$$s = (2\pi^2/45) g_* T^3, \tag{2d}$$

where g_* counts the total number of relativistic degrees of freedom (i.e., species with mass << T), and s is the entropy density of the Universe. The complete history of the Universe (according to the hot big bang model) is summarized in Fig. 2, as well as in the introduction to these proceedings.

2. 'FOUR TRANSPARENCY' COURSE IN MODERN PARTICLE THEORY

2.1 Their Standard Model

Particle physics has its standard model also. It is the SU(3)xSU(2)xU(1) gauge theory of the strong (or color), weak, and electromagnetic interactions. It is every bit as successful as the standard model of cosmology, providing an accurate and consistent description of elementary particle physics at energies of up to about 1000 GeV (corresponding to distances as small as about 10^{-17} cm!).

The fundamental constituents of matter are the quarks and leptons (see Fig. 3). Each quark flavor (6 are known: up, down, charm, strange, top, bottom) comes in three colors. [Color is a 3-dimensional charge to which the strong (or color) force couples.] The leptons are colorless. The color force is so strong that at low temperatures (T < few 100 MeV) the only finite energy configurations in the theory are 'colorless' -- quark-antiquark states known as mesons, triplets of quarks (one of each color) known as baryons, and the colorless leptons. The quarks and leptons seem to come in families -- a pair of quark flavors and a pair of leptons in each family or generation. So far three families have been discovered. At present there is no understanding of the number of families that exist, or how many should exist altogether. [Cosmology strongly suggests that there are less than or equal to 4 families (with light neutrinos) and the width of the recently discovered Z boson indicates that the number must be less than of order 10. See Schramm and Steigman (1985) for further discussion.]

Symmetry is a guiding principle in modern particle physics. The fundamental interactions of the quarks and leptons are described mathematically by an SU(3)xSU(2)xU(1) gauge theory, a theory based on the symmetry group SU(3)xSU(2)xU(1). In gauge theories particles exist in multiplets, members of which are related to each other by symmetry operations. The interactions are mediated by gauge bosons and the gauge bosons are the physical manifestations of the symmetry transformations.

The color force is described by the SU(3) part of the gauge group; the quarks come in color triplets (say, red, green, and blue); and there are 8 massless gauge bosons, called gluons, which mediate the color force and rotate one color quark into another (see Fig. 3). [Note the gluons themselves possess color and form an octet multiplet. The 'strong nuclear force' is now generally believed to be the residual color force felt between color neutral states, in analogy with the van der Waals force.]

In the standard model the electromagnetic and weak forces are unified in the framework of the so-called electroweak interaction, which is described by the SU(2)xU(1) part of the model. The particle multiplets are the quark and lepton 'flavor' pairs (or doublets), e.g., u-d and ν_e-e; the gauge bosons are the photon and W and Z bosons (which form a triplet of particles under SU(2)xU(1)). The unified electroweak theory is also known by the names of its inventors: The Weinberg-Salam-Glashow theory.

2.2 Hidden Symmetry (Spontaneous Symmetry Breaking, or SSB)

As you must well know the W and Z are not massless bosons, having masses of about 81 and 93 GeV respectively. How can that be in a unified gauge theory where their sibling the photon is massless? This brings us to one of the most fundamental ideas in modern particle physics, 'Hidden Symmetry' or SSB. The basic idea is that the theory possesses more symmetry than its solutions do. The theory does indeed have the full symmetry of SU(3)xSU(2)xU(1) in spite of the massive W and Z bosons. The full symmetry however is not possessed by the lowest energy solution or vacuum state of the theory.

In gauge theories the free energy (per unit volume) can be expressed in terms of one or more of the scalar fields which are also part of the theory (often called Higgs fields). The free energy $V_T(\phi)$ is often referred to as the effective potential or Higgs potential. At low temperatures the free energy is minimized by the Higgs field having a non-zero value (see Fig. 3). This vacuum expectation value (or vev) of the Higgs field(s) acts as an order parameter whose non-zero value signals SSB. In particular, the masses of the W and Z bosons are proportional to $\langle\phi\rangle$:

$$M_W, M_Z \approx g\langle\phi\rangle$$

where g is the gauge coupling constant and $\langle\phi\rangle \approx 300$ GeV. The vacuum state only possesses a U(1) symmetry, which corresponds to electromagnetism. At low temperatures the SU(2)xU(1) theory is said to be spontaneously broken to U(1). At high temperatures, finite temperature effects change the shape of the Higgs potential, so that its minimum occurs at $\langle\phi\rangle = 0$, and at high temperatures the full symmetry of the theory is restored (see Fig. 3).

The symmetry restoration temperature for the electroweak theory is about 300 GeV. While 300 GeV is a very high temperature by laboratory standards, such high temperatures (and up to 10^{19} GeV) should have existed during the earliest moments of the Universe. Thus spontaneously broken symmetries should have been restored in the early Universe, and

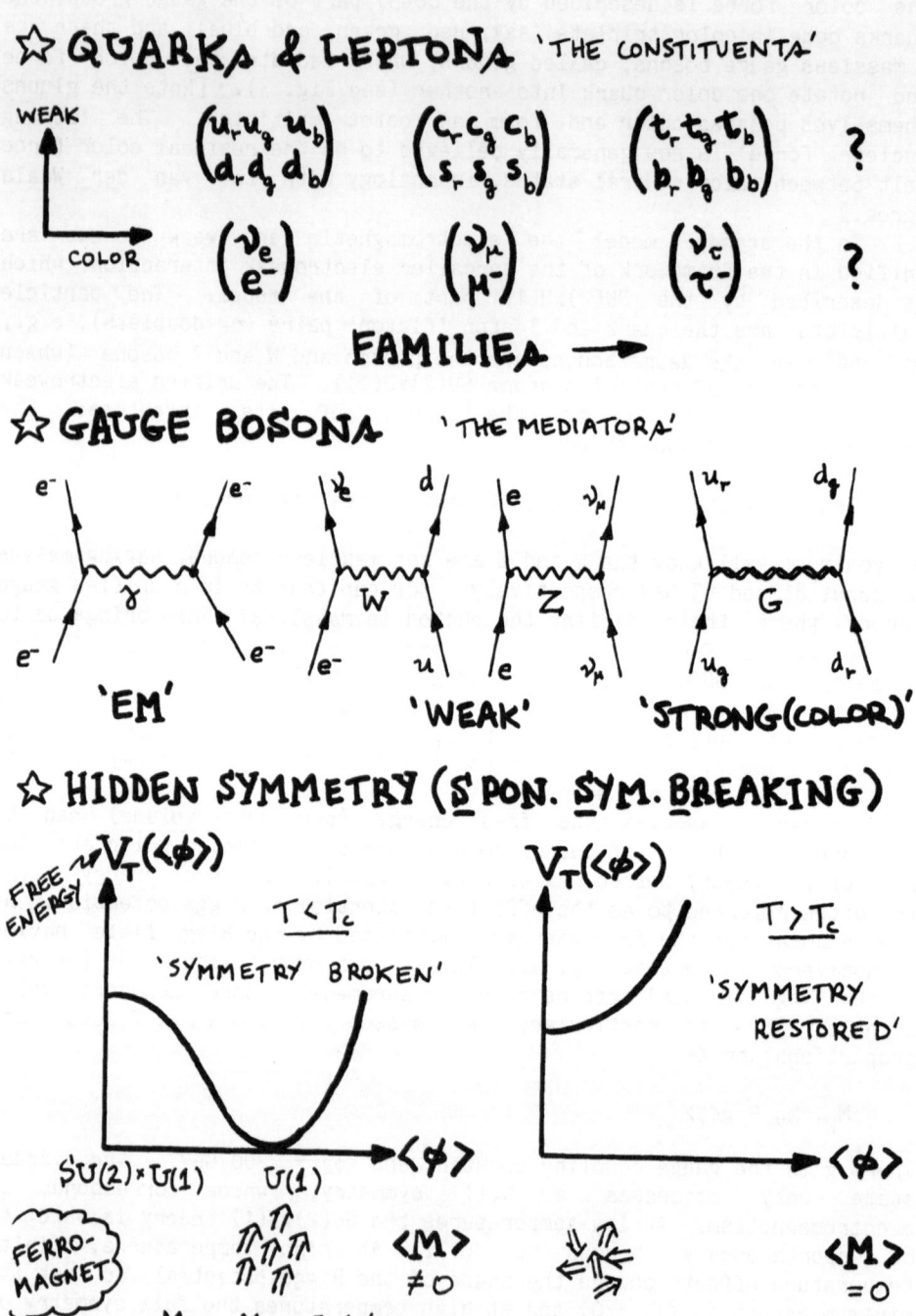

☆ QUARKS & LEPTONS 'THE CONSTITUENTS'

FAMILIES ⟶

☆ GAUGE BOSONS   'THE MEDIATORS'

'EM'   'WEAK'   'STRONG (COLOR)'

☆ HIDDEN SYMMETRY (SPON. SYM. BREAKING)

Figure 3 - Summary of key aspects of the standard model of particle physics, the SU(3)xSU(2)xU(1) gauge theory.

broken as the Universe cooled. SSB phase transitions then are a natural consequence of modern particle theory applied to the early Universe.

Analogous phenomena exist in more familiar settings. Consider a ferromagnet. The spin interactions of the individual atoms are described by Maxwell's equations, which of course possess rotational symmetry. However, at low temperatures, rotational invariance is no longer manifest as the lowest energy configuration of the system (spins aligned) does not possess rotational symmetry -- rotational symmetry has been spontaneously broken. At high temperatures (i.e., temperatures above the Curie temperature) the symmetry is restored as the spins are no longer aligned in the configuration with the minimum free energy.

2.3 Beyond the Standard Model--Why?

Their standard model then is a gauge theory which undergoes SSB

$$SU(3) \times SU(2) \times U(1) \xrightarrow[T_c \approx 300 \text{GeV}]{} SU(3) \times U(1) \, ,$$

reflected in the fact that only the photon and gluons are massless and the fact that the weak force is very short-ranged ($r \approx M_W^{-1}$, $M_Z^{-1} \approx 10^{-16}$ cm). Their standard theory is well supported by all the experimental data, which at present involves experiments done at energies $\lesssim 1000$ GeV.

It is not without its shortcomings however. Although it is clear that a key feature of the theory is SSB, the scalar (or Higgs sector) of the theory is totally unexplored. No Higgs particles have been discovered. Indeed, these fields were put into the theory for the express purpose of SSB and to give quarks and leptons masses. [Owing to their expected masses and the weakness of their interactions, Higgs particles will only be accessible to experiments at the next round of new accelerators -- the Tevatron p̄p collider at Fermilab, the Stanford Linear Collider (or SLC) at SLAC, or perhaps not until the Superconducting SuperCollider (or SSC) is built.] While no Higgs particles have yet been discovered, almost all high energy theorists are confident that something like the Higgs mechanism must exist.

There are many indications in the standard model that there must be some more fundamental theory beyond the standard model. The theory is not truly unified in the sense that it is based upon a group which is a (direct) product of groups. There are too many particle multiplets and the quarks and leptons exist in separate multiplets. Because the quarks and leptons exist in separate multiplets there is no reason that their electric charges be related in a simple way (and of course we know that they are, the charge of the proton and of the positron are equal to a high degree of precision). The standard model casts no light on the values of quark and lepton masses or on why quarks and leptons come in families. Gravity is not included in the theory. There are many other more technical problems which also point to the fact that there must be something beyond the standard model.

Just as cosmologists suspect that something interesting must have happened during the first 10^{-2} sec after the bang, particle physicists strongly suspect that there must be a more fundamental theory which incorporates and goes beyond their standard model. With the exception of quark nuggets (stable, macroscopic aggregates of quark matter with nuclear density) the standard model of particle physics supplies no candidates for dark matter beyond ordinary baryons in some non-luminous guise (e.g., jupiters, primordial black holes, massive black holes, etc.). As we shall soon see virtually all the extensions of the standard model provide us with a generous supply of dark matter candidates. Next I briefly review some of the currently fashionable theoretical ideas in particle physics, emphasizing the dark matter candidates which are predicted (also see Fig. 3). Beware! My review does not do justice to these theoretical ideas. I refer the interested reader to the following very readable literature: Quigg (1983), Georgi (1984), Langacker (1980), Ramond (1983), Ross (1984), Green (1985), and Greenberg (1985).

2.4 Minimal Extensions of the Standard Model (or Messing with the Higgs Sector)

Since so little is known about the Higgs sector it seems like a natural place to start tinkering around. In the standard model there is one complex doublet of Higgs fields. By adding a triplet of Higgs which also develops a vev (albeit a very small one, $\langle\phi_{triplet}\rangle \lesssim$ MeV) Gelmini and Roncadelli (1981) constructed a model (the so-called 'majoron model') in which neutrinos have a (majorana) mass and additional interactions which violate lepton number (which is spontaneously broken in this theory). Other similar models exist. The new interactions which violate lepton number have all kinds of interesting astrophysical effects which have recently been reviewed by Kolb (1984).

Peccei and Quinn (1977) proposed adding one additional Higgs doublet to the theory, so that the theory would have an additional symmetry (now known as PQ symmetry) which is also spontaneously broken. Why would they do such a thing? [As you are beginning to see, symmetry is a guiding principle in modern particle theory. Since we only see a few symmetries at the energy scale at which we operate ($\lesssim$ TeV), essentially all the new symmetries introduced into the theory must be spontaneously broken!]

Although essentially all particle physicists believe that SU(3) or QuantumChromoDynamics (QCD) is the correct theory of the strong interactions, it has one, very bad problem: non-perturbative effects in the theory violate CP (charge conjugation combined with parity) and T (time reversal) invariance (leaving CPT intact) and should lead to an electric dipole moment for the neutron which is a factor of 10^9 or so larger than the present experimental limit (unless the non-perturbative effect is 'fine-tuned' away). The PQ symmetry solves this problem by effectively making the coefficient of the offending term in the Lagrangian a dynamical variable, whose potential has a minimum at a value where CP and T are very nearly conserved. Wilczek (1978) and Weinberg (1978) pointed out that the existence of such a broken symmetry would lead to a new, light psuedoscalar boson, which they dubbed the axion. The mass of the axion, its lifetime, and its coupling to ordinary matter are all determined by the symmetry breaking scale of the PQ symmetry, f_{PQ}.

$$m_a \approx 10^{-5} eV \, (10^{12} GeV/f_{PQ}) \, ,$$

Figure 4 - Guide to the theoretical ideas beyond the standard model of particle physics and the candidate WIMPs they predict.

$$\tau(a \rightarrow 2\gamma) \simeq 10^{41} \text{yrs} \ (f_{PQ}/10^{12} \text{GeV})^5 \ ,$$

$$g_{aee} \simeq m_e/f_{PQ} \qquad (g_{aee} = \text{coupling of the axion to } e^-)$$

Originally, Peccei and Quinn proposed that f_{PQ} be the same as the weak symmetry breaking scale (economy of scales so to speak). However such an axion (mass of a few 100 keV) was quickly ruled out by laboratory searches and astrophysical arguments. [Because of its short lifetime ($\simeq$ few sec) such an axion would not be of any interest as a dark matter candidate.] The requirement that the cooling of various kinds of stars by axion emission not be too efficient leads to a lower bound to f_{PQ} of about 10^8 GeV -- very far from the weak scale, but as we shall see cosmologically very interesting for $f_{PQ} \simeq 10^{12}$ GeV. The so-called 'strong CP problem' is solved regardless of the value of f_{PQ}. In fact, at present despite the lack of any experimental evidence for its existence, the axion remains the most attractive solution for this nagging problem.

2.5 Grand Unification

The first step towards unification of all the forces is the unification of the strong, weak, and electromagnetic forces, or so-called grand unification. Grand Unified Theories (or GUTs) are usually based upon a gauge group which is not a product of separate factors, and have quarks and leptons in the same multiplets. The simplest GUT is based upon the group SU(5), although its viability is in doubt as its prediction for the proton lifetime is about a factor of ten shorter than the present experimental lower limit (Perkins 1984). A multitude of other groups have been proposed including SO(10), E6, SO(18), and E8, to mention just a few.

Generically, GUTs makes several predictions: interactions which violate B and L (afterall quarks and leptons are in the same multiplets), the existence of stable, superheavy magnetic monopoles, and an additional scale of SSB, typically of order 10^{14} GeV. The gauge (and Higgs) bosons which mediate proton decay obtain masses of this order so that the processes which violate B and L are very, very weak, leading to a rather longlived proton, $\log (\tau_p/\text{yr}) \simeq 0(30)$. Most GUTs also predict that neutrinos have small masses (much smaller than those of the other quarks and leptons and very often $\ll$ 1 eV). Some GUTs also predict the existence of cosmic strings. I will discuss monopoles and cosmic strings again later. GUTs can also incorporate PQ symmetry and therefore axions. In fact grand unification provides another natural scale for f_{PQ}, the grand unification scale, or about 10^{14} GeV.

2.6 Supersymmetry/Supergravity (SUSY/SUGR)

Supersymmetry is the symmetry which interchanges fermions and bosons. In a supersymmetric theory there is a bosonic counterpart for every fermion and vice versa. We certainly see no evidence for such a symmetry in the world around us, e.g., there is no massless fermionic partner for the photon, or scalar partner for the electron. What is the motivation for supersymmetry then? Mathematically supersymmetry is very elegant, and it is the last symmetry one has available to invoke! In a more serious vein, when it is made a gauge symmetry (this is called supergravity), supersymmetry leads to a generally covariant theory, that is, it automatically incorporates general relativity into the theory. Thus it offers the hope of unifying gravity with the other forces. Supersymmetry also offers the hope of cleaning up a technical problem which all GUTs have in common: the discrepancy of the weak and GUT symmetry breaking scales, some 12 or so orders of magnitude in a typical GUT. Although one is free to set these scales to very different energies, quantum corrections spoil this, and tend to raise the weak scale up to the GUT scale (or the highest scale in the theory). Supersymmetry can be used to stabilize the discrepancy once it is initially set, "set it and forget it", so to speak.

Since we see no evidence of SUSY in our world it too must be a broken symmetry. In order to stabilize the weak scale, SUSY breaking must effectively occur at the weak scale. This means that the SUSY parnters, or spartners, of all the known particles must have masses of order the weak scale, where "of order" means between a few GeV and a TeV. The scalar partners of the quarks are called squarks; the scalar partners of the leptons are called sleptons; the fermionic partners of the photon, gluon, W, Z, and graviton are the photino, gluino, Wino, Zino, and gravitino respectively. The fermionic partners of the Higgs particles are known as Higgsinos.

Because of an additional symmetry that most SUSY/SUGR models have (called R-parity) the lightest spartner is stable. Because the effective SUSY breaking scale is of order the weak scale, the interactions of spartners with ordinary particles are about as strong as the usual weak interactions. This makes the lightest spartner (or LSP) an ideal candidate WIMP. In different models different spartners turn out to be the LSP; the most popular LSPs are the photino, sneutrino, and Higgsino. Typically, the LSP has a mass of order a few GeV.

GUTs can be supersymmetrized and in fact almost all SUSY/SUGR models are SUSY/SUGR GUTs. The unification scale in these theories is higher, more like 10^{16} GeV and these theories are supposed to describe physics at energies up to 10^{19} GeV. Therefore, SUSY/SUGR models also predict all the additional particles that GUTs do -- magnetic monopoles, massive neutrinos, axions, and cosmic strings (in some cases).

2.7 Kaluza-Klein Theories

Another approach to unification is through geometry (in analogy to general relativity). Indeed this approach dates back to work done by Kaluza, and Klein in the 1920's (and also caught Einstein's fancy). The basic idea of Kaluza-Klein theories is that space-time has more than the 3+1 (3 space, 1 time) dimensions that we are familiar with, say, 3+N space and 1 time dimensions. Space-time in these theories is supposed to be a 4-dimensional manifold cross an N-dimensional compact manifold which we haven't yet noticed (typical dimensions of the order of the planck length, 10^{-33} cm). The symmetries (more precisely, the isometries) of the compact manifold give rise to the gauge symmetries we observe in our 4 dimensions. The natural energy scale of these theories is 10^{19} GeV. In order to accommodate the gauge symmetry of the

SU(3)xSU(2)xU(1) model N must be ≥ 7. These theories need not necessarily incorporate grand unification (although that possibility is not precluded). Such theories predict the existence of stable planck mass objects, sometimes called pyrgons. In addition, these theories suggest that during its earliest history (t < 10^{-43} sec), the Universe might have had all its 3+N space dimensions equally accessible. Of course one has to explain why the vacuum state of the theory has N space dimensions curled up (or alternatively, why 3 of the spatial dimensions are so large).

The concept of additional space-time dimensions has become increasingly popular in recent years (because all the physics of 4 dimensions has been done!), while the popularity of the Kaluza-Klein idea has waned. Although conceptually very attractive, there are many serious difficult problems, including incorporating chiral (particles whose right-handed and left-handed components have different interactions) fermions (the kind we know and love), keeping the compact dimensions compact, and constructing a quantum theory which is at least renormalizable.

2.8 Superstring Theories

Superstring theories incorporate every trick in the book -- gauge symmetry, supersymmetry, extra dimensions and one new one, strings. The basic idea is that the fundamental particles are not point-like, but rather are string-like, 1-dim entities. Such theories can only be consistently formulated in 10-dimensions with either the gauge group E8xE8' or SO(32).

Particle theorists are extremely excited about superstring theories as they unify all the forces of nature (including gravity) in a finite quantum theory and are almost unique (only five string theories are known to exist). [The usual gauge theories are not finite, but rather are only renormalizable, i.e., infinities can be consistently swept under the rug.] In principle, starting from the superstring (which describes physics at or above the Planck scale) one can calculate everything -- the masses of all the fermions, the GUT, etc.

When viewed at large distances the loops look like point-like particles (large distances here means large compared to the Planck length, 10^{-33} cm). The so-called point-like (or field theory) limit of a superstring theory is supposed to be a SUSY/SUGR GUT. All the WIMP candidates predicted by SUSY GUTs are also predicted by superstring theories.

If the symmetry group of the point-like theory is E8xE8', there is an interesting new possibility for dark matter. In this case there are two sets of particles, those whose interactions are described by E8 and those whose interactions are described by E8', which only interact with each other via gravity. Assuming that this is the case, at low energies one would have baryons, mesons, and leptons and their analogous (say, shadow) counterparts, alike in every respect, same masses, same interactions, etc, but only interacting with each other via gravity. Shadow matter is the perfect (but as it turns out also the perfectly implausible) candidate for the dark matter. [For further discussion of the shadow world see Kolb, Seckel, and Turner (1985).]

While particle physicists are very optimistic about the superstring (and rightfully so, a quantum theory of gravity which is finite, almost unique, and in principle predicts everything doesn't come along every day), solid results have been few and far between thus far.

Notice the progression of theoretical ideas here. GUTs are supposed to describe physics up to around 10^{14} GeV or so, SUSY/SUGR GUTs up to 10^{19} GeV, and superstring theories at energies above 10^{19} GeV.

2.8 Composite Models

Another, somewhat orthogonal, approach to going beyond the standard model involves the idea that many of the fundamental objects of the standard model, e.g., quarks, leptons, Higgs bosons, etc. are not fundamental, but rather are themselves made up of more fundamental entities, named variously as preons, rishons, etc. Indeed the number of quarks and leptons and their pattern of increasing masses suggests that they might actually be bound states of more fundamental objects. These models all face one fundamental difficulty: to have objects of a given size whose mass is much, much smaller than 1/size (i.e., have a Compton wavelength much bigger than their size; in all previous experience the reverse is true). Experiments indicate that the scale of compositeness (if there is one) must be greater than about a TeV. Various tricks (including chiral symmetry, the Nambu-Goldstone mechanism, and SUSY) have been used to keep the masses of the composite objects small compared to the scale of compositeness. A number of such theories have been proposed, including technicolor, the preon model, etc. Thus far none of them have proven to particularly elegant or compelling. Most of these theories predict some exotic, stable states which could be dark matter candidates.

2.9 'The Program'

The following is a brief summary of which theories predict which dark matter candidates.

Axion -- Simple extensions of the standard model, GUTs, SUSY/SUGR, Superstrings.

Massive neutrinos -- Simple extensions of the standard model, GUTs, SUSY/SUGR, Superstrings.

Spartners -- SUSY/SUGR, Superstrings

Monopoles and Cosmic Strings -- GUTs, SUSY/SUGR, Superstrings.

Quark Nuggets -- All of the above (potentially).

Pyrgons -- Models with extra dimensions.

3. THE PRODUCTION OF RELICS IN THE EARLY UNIVERSE

3.1 Hot and Cold Running WIMPs

Because the Universe was very hot during its earliest epoch, all kinds of interesting particles were present in great abundance. When the temperature of the Universe is >> the mass of a given species, that species (if in equilibrium) should be present in almost equal numbers as the photons,

$$n_x/n_\gamma = (g_{xeff}/2) ,$$

where g_{xeff} is 1(or 3/4) times the number of degrees of freedom (g_x) for a boson (or fermion). At temperatures << the mass of the species the equilibrium abundance relative to photons is exponentially small (assuming the species does not have a chemical potential)

$$n_x/n_\gamma \simeq (\pi/8)^{1/2}(g_x/2\zeta(3))(m_x/T)^{3/2}\exp(-m_x/T) .$$

The abundance of a given species can only track its equilibrium abundance so long as the interactions which allow it to adjust its number per comoving volume (decays and annihilations) are occurring rapidly on the expansion timescale. For stable particles this means so long as the annihilation rate ($\Gamma \simeq n_x(\sigma v)_{ann}$) is greater than the expansion rate H.

Once the annihilation rate drops below the expansion rate the number of particles per comoving volume remains constant. If this occurs while the species is still relativistic, its abundance relative to the photons freezes out at a value of order unity. Such relics are often referred to as hot relics. If this occurs when the species is nonrelativistic, its abundance relative to photons freezes out at a value much smaller than that of the photons. Such relics are often referred to as cold relics.

The freeze-out temperature (T_f) depends upon the annihilation cross section and is given by

$$x_f \equiv m_x/T_f \simeq \ln[(n+1)a\lambda] - (n+1/2)\ln[\ln[(n+1)a\lambda] , \tag{3a}$$

$$a = 0.15(g_x/g_*) , \tag{3b}$$

$$\lambda = 0.264 \ g_*^{1/2} \ m_{pl} \ m_x(\sigma v)_o , \tag{3c}$$

where the annihilation cross section has been parameterized by

$$(\sigma v)_{ann} = (\sigma v)_o(T/m_x)^n.$$

Since entropy per comoving volume remains constant (assuming the expansion is isentropic), the number of WIMPs per comoving volume is simply proportional to Y, where

$$Y = n_x/s$$

and s is the entropy density of the Universe ($\approx 0.44 g_* T^3$). The final abundance Y_{final} is determined by when the annihilations freeze out.

For hot relics, $x_f \ll 3$ ($T_f \gg m_x$) and

$$Y_{final} \approx 0.278 \ (g_{xeff}/g_*). \tag{4}$$

On the other hand, for cold relics $x_f \gg 3$ ($T_f \ll m_x$) and

$$Y_{final} \approx (n+1) x_f^{n+1}/\lambda \ . \tag{5}$$

Since the entropy density today (photons and 3 neutrino species) is about 7.04 times the number density of photons

$$(n_x/n_\gamma)_{today} \approx 7.04 \ Y_{final} \ .$$

The contribution of a given relic species to Ω is then

$$(\Omega_x h^2/T_{2.7}^3) \approx 2.67 \times 10^8 (m_x/GeV) \ Y_{final} \ . \tag{6}$$

[For further details and references see, Scherrer and Turner (1986).]

Light neutrinos ($\lesssim$ few MeV) with the usual weak interactions freeze out when they are still relativistic and so are hot relics, with

$$(n_{\nu\bar\nu}/n_\gamma)_{today} = 3/11 \ , \tag{7a}$$

$$(\Omega_\nu h^2/T_{2.7}^3) \approx m_\nu/96 eV. \tag{7b}$$

For heavy neutrinos ($\gtrsim$ few MeV), the annihilation cross section is of the order $G_F^2 m_\nu^2$, so that $x_f \approx 20$ ($T_f \approx m_\nu/20$), and so they are cold relics with

$$(\Omega h^2/T_{2.7}^3) \approx (m_\nu/2 GeV)^{-1.9} \tag{8}$$

(Lee and Weinberg 1977).

For the lightest spartner (LSP), the annihilation cross section depends upon the masses of the other spartners. Because all the spartner masses are typically of the order of the weak scale, the annihilation cross section is also of the order of $G_F^2 m_{LSP}^2$, implying that the LSP will also be a cold relic, with Ω_{LSP} given by a formula similar to that for a heavy neutrino. [For further discussion see Ellis etal. (1984).]

3.2 Topological Relics (Monopoles and Cosmic Strings)

In spontaneously broken gauge theories there are, in addition to the fundamental particles of the theory, topological entities, monopoles, strings, and domain walls. These objects correspond to classical configurations of the gauge and Higgs fields. Let me be a little more specific.

In general the Higgs field has many components and the minimization of the free energy may not uniquely specify all the components. Say for instance that the magnitude of the Higgs field is specified, but not the direction of the Higgs field in group space. In this case there will be a set of Higgs field values which minimize the free energy, but differ in the direction they point in group space (they comprise the vacuum manifold). Consider the possible ways that the Higgs field can be laid out in physical space. One way is for the Higgs field to be laid out uniformly (boring!). Another way is to be laid out in different directions in different places in physical space as in Fig. 5. In the configuration shown in Fig. 5a the Higgs field must necessarily vanish at a point. This configuration corresponds to a stable, magnetic monopole. The energy associated with the configuration (part in potential energy as ϕ deviates from the minimum at the center and part in magnetic field energy) is

$$m_M \simeq M/\alpha ,$$

$$\simeq 10^{16} \text{GeV} \ (M/10^{14} \text{GeV}) ,$$

where $\alpha = g^2/4\pi$, g is the gauge coupling constant (typically $\alpha \simeq 10^{-2}$) and M is the symmetry breaking scale. The size of the region where ϕ vanishes is of order M^{-1}, i.e., the monopole is not point-like, but has a finite size. In configuration 5b, there is necessarily a line along which ϕ vanishes; this object corresponds to a cosmic string. The width of the string is of order M^{-1} or 10^{-28} cm $(10^{14} \text{GeV}/M)$ and the mass per unit length is of order M^2 or 10^{18}g cm^{-1} $(M/10^{14} \text{GeV})^2$.

Whether or not a given gauge theory has monopole or string solutions depends upon the structure of the vacuum manifold. Whenever a semi-simple group (i.e., a group without an explicit U(1) factor) breaks down to a group with an explicit U(1) factor monopole solutions exist. Since we know that the 'low energy' (low here means of order a TeV or so) group is SU(3)xSU(2)xU(1), which has an explicit U(1) factor, any GUT based upon a semi-simple group will have monopole solutions. Thus monopoles are a very generic prediction of GUTs.

The condition to have string solutions is that the group which the theory breaks down to must have a discrete symmetry. Some, but certainly not all GUTs have this feature. [There are also topological configurations which correspond to two-dimensional sheets where ϕ vanishes; these are called domain walls. Walls are cosmologically disastrous. For further discussion of all of these topological objects, see Vilenkin (1985).]

The primary way that topological objects are produced is via the so-called Kibble (1976) mechanism during a SSB phase transition. Recall that at high temperatures symmetries are restored; as the Universe cools below the critical temperature for a given SSB transition ($\simeq M$) the Higgs field takes on a non-zero vev. The standard cosmology has particle horizons ($d_H \simeq ct$) and so causality prevents any physical process from operating on scales greater than order the horizon distance. Clearly the Higgs field cannot become correlated on scales larger than the particle horizon (and often the microphysics sets an even smaller correlation scale). If the theory permits such configurations, of the order of 1 monopole or string will be formed per horizon volume just due to the fact that the Higgs field cannot be correlated on larger scales (see Fig. 5).

SSB IN THE EARLY UNIVERSE

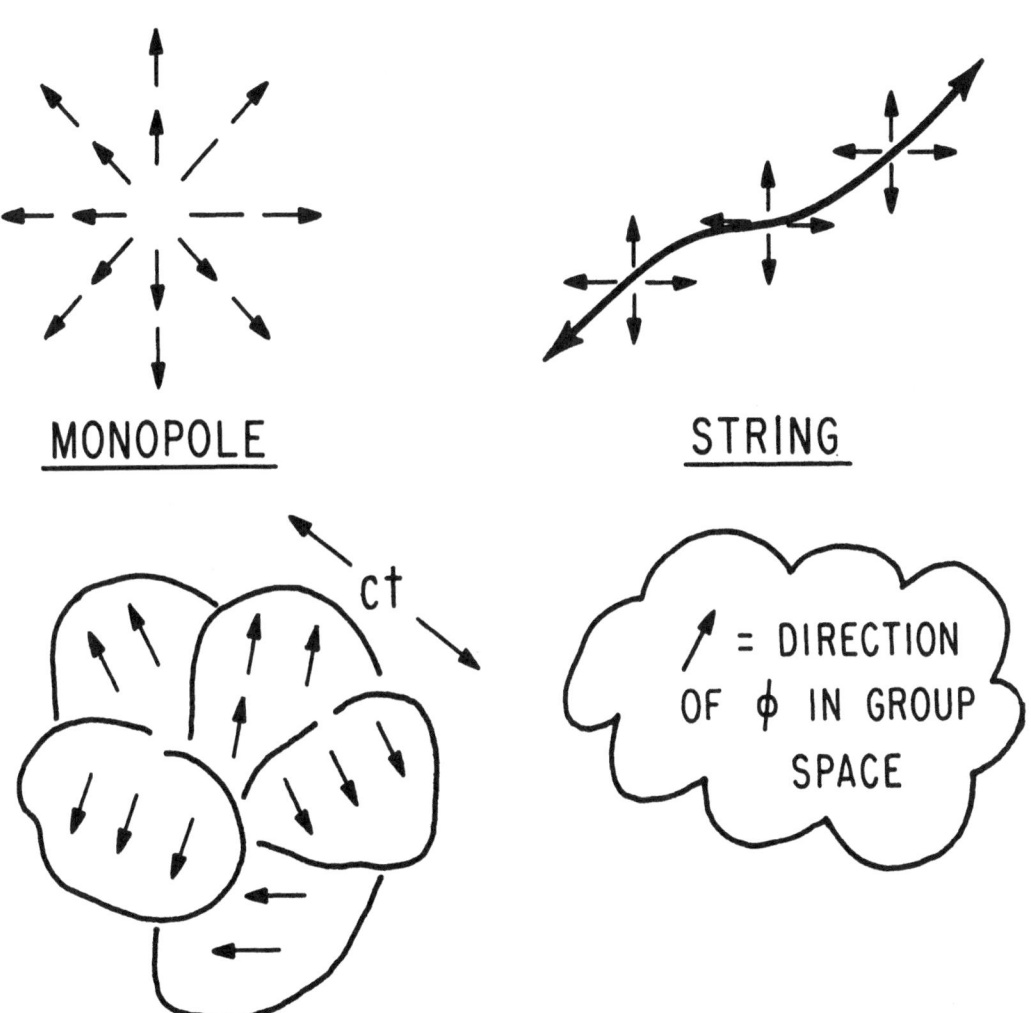

Figure 5 - Schematic representation of two of the topological objects predicted by GUTs, the monopole and cosmic strings, and their production by the Kibble mechanism in the very early Universe. These objects correspond to non-trivial configurations of the Higgs and gauge (not shown) fields. Because of the existence of particle horizons in the early Universe, of order 1 of these topological objects is produced per horizon volume during SSB.

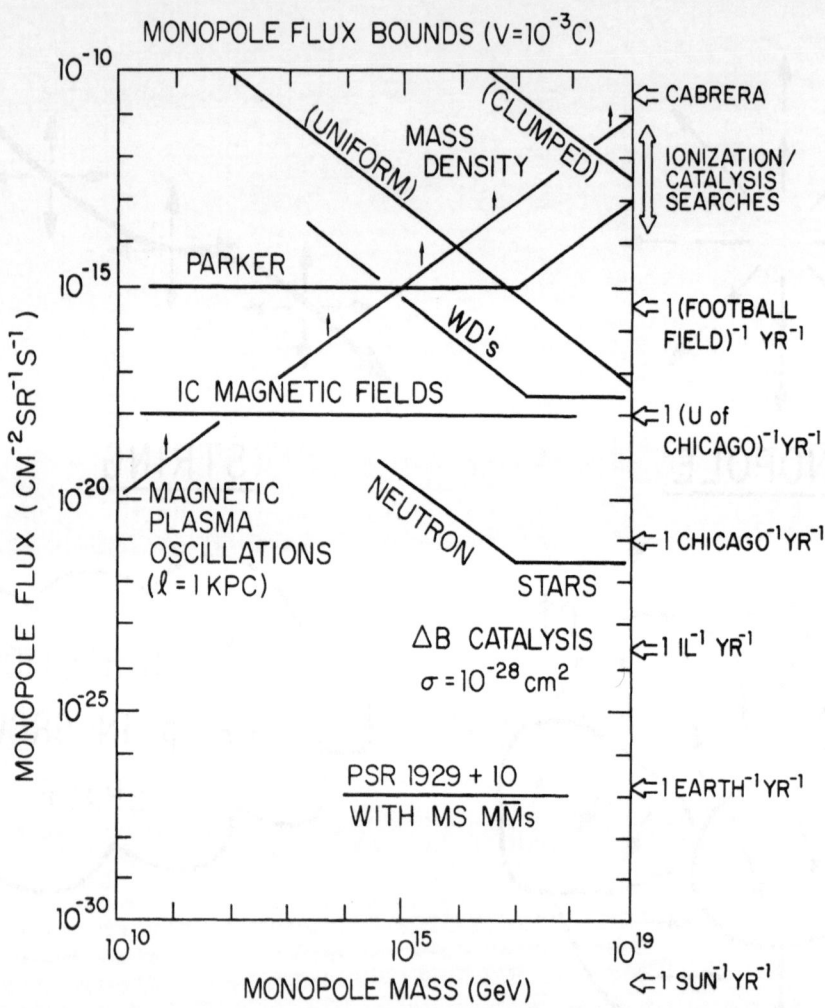

Figure 6 - A summary of the astrophysical and cosmological bounds to the local flux of superheavy magnetic monopoles (from Turner 1986b).

For the monopole and the usual GUT scale of 10^{14} GeV, this leads to a disastrous overproduction, so many monopoles that the Universe would reach a temperature of 3 K at the tender age of 30,000 years! If the overproduction can be avoided (e.g., by a complicated symmetry breaking pattern, or by inflation followed by thermal pair production or production at a much later phase transition) monopoles could possibly be a viable dark matter candidate. Of course the relic monopole abundance is also severely constrained by various astrophysical arguments, especially if they catalyze nucleon decay (see Fig. 6). The lack of any sensible theoretical guidance as to their primordial abundance and the very stringent astrophysical constraints on their relic abundance make monopoles a less than attractive dark matter candidate. [For a recent review of magnetic monopoles see Preskill (1984) or Turner (1986).]

While cosmic strings do not behave like relic particles (e.g., they eventually cut themselves, forming loops which can evaporate by the emission of gravitational waves as they oscillate), they can move matter around and induce isothermal perturbations with unusual properties. For $M \approx 10^{16} - 10^{17}$ GeV, cosmic strings may be able to trigger a viable scenario for structure formation and offer an intriguing alternative to the scenario of a WIMP-dominated Universe with adiabatic perturbations with the Zel'dovich spectrum (for further discussion see Vilenkin 1985).

3.3 Quark Nuggets

In a very interesting and thought-provoking paper Witten (1984) raised (and also all but dismissed) the possibility that the Universe could be baryon-dominated and flat ($\Omega = 1$). He supposed that for very large baryon number (>> 100) the stable configurations of matter were quark matter rather than nuclear matter (at present there is no evidence for this supposition).

He then investigated the formation of big globs of quark matter (hereafter referred to as quark nuggets) during the quark/hadron transition (t $\approx 10^{-5}$ sec, T $\approx$ 200 MeV). He concluded that were it not for the fact that cooling is a very efficient process in the primordial plasma, most of the quarks in the Universe might have formed into quark nuggets of size 0.1-100 cm and nuclear density, leaving only about 10% of the quarks in the form of free nucleons. Since quark nuggets would presumably not have participated in primordial nucleosynthesis, one could have Ω_{baryon} = 1, with 0.1 in free nucleons and 0.9 in nuggets. The formation of nuggets has been studied further, by Degrand and Kajantie (1984) and very recently by Applegate and Hogan (1985) and Alcock and Farhi (1985), who all also conclude that quark nuggets are not a very likely candidate for the dark matter (although they could possibly have some interesting effects on primordial nucleosynthesis).

3.4 Cosmic Harmonic Oscillations (AKA Axions)

I have already discussed the motivation for the axion, now I will discuss how cosmic axions come into being. For the allowed values of the PQ symmetry breaking scale ($f_{PQ} \geq 10^8$ GeV) axions interact so weakly that they should never have been in thermal equilibrium. They are however produced in another very novel and interesting way (see Fig. 7).

COSMIC AXIONS

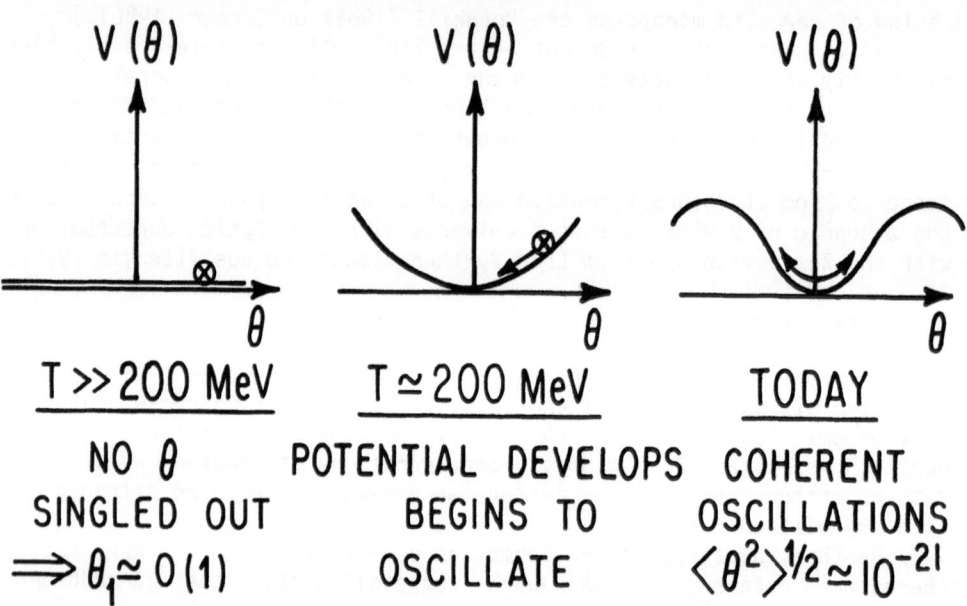

$$V(\theta) \qquad V(\theta) \qquad V(\theta)$$

$T \gg 200$ MeV	$T \simeq 200$ MeV	TODAY
NO θ SINGLED OUT $\Rightarrow \theta_1 \simeq O(1)$	POTENTIAL DEVELOPS BEGINS TO OSCILLATE	COHERENT OSCILLATIONS $\langle \theta^2 \rangle^{1/2} \simeq 10^{-21}$

Figure 7 - Thermal history of the axion potential, $V(\theta)$. At high temperatures ($f_{PQ} \gg T \gg$ few GeV) the potential is flat and no value of θ is preferred. At low temperatures ($T \cong$ few GeV) the potential develops a minimum due to instanton effects. Because of its initial misalignment, θ begins to oscillate. These oscillations have an amplitude of about 10^{-21} today and correspond to a condensate of very cold axions.

Denote the axion degree of freedom as θ and its potential as $V(\theta)$. At low temperatures ($\ll$ GeV) $V(\theta)$ is periodic and has its minimum at $\theta = 0$ (the CP and T conserving minimum). The axion's potential develops due to non-perturbative QCD effects (so-called instanton effects). These non-perturbative effects vanish at high temperatures (T $\gg$ GeV). That is at high temperatures, the axion potential is flat and has no minimum (and the axion is massless).

Now let's follow the birth of the cosmic axions. At temperatures much greater than a few GeV, but below the PQ symmetry breaking scale, the PQ symmetry is broken but there are no dynamics to determine θ since its potential is flat. Denote its initial value by θ_1. There is no reason that θ_1 should be 0; in general, one would expect it to be of order unity, i.e., misaligned with the soon-to-be-determined minimum of its potential. Due to this initial misalignment, once the potential does develop (when non-perturbative QCD effects become important, T $\simeq$ 0(GeV)), θ will begin to oscillate. These cosmic coherent, classical oscillations of θ have energy density associated with them. In fact they behave just like NR matter. From the particle point-of-view they correspond to a condensate of very cold (NR) axions.

The energy density in these oscillations can and has been calculated by Preskill, Wilczek, and Wise (1983), Abbott and Sikivie (1983), Dine and Fischler (1983), and Turner (1986):

$$(\Omega_a h^2/T_{2.7}^3) \simeq 1.0(f_{PQ}/10^{12}\text{GeV})^{1.18}(N/6)^{0.83}\theta_1^2 , \qquad (9)$$

here N is an integer which depends on precisely how the PQ symmetry is implemented (in the simplest models N = 6). Note that Ω_a increases with f_{PQ} and depends on the square of the initial misalignment angle, θ_1. If the Universe never inflated or inflated before PQ symmetry breaking, then it is the RMS value of θ_1, $\theta_{1RMS} = (\pi/N)/\sqrt{3}$, which should be used (when the oscillations commence, θ_1 is uniform on the scale of the horizon, but is uncorrelated on larger scales). In the inflationary case, θ_1 takes on different values inside different bubbles (or fluctuation regions), so that we do not know what value θ_1 takes on within our bubble (of course, averaged over all bubbles $\theta_{1RMS} = (\pi/N)/\sqrt{3}$). For $\theta_1 = \theta_{1RMS}$, $\Omega_a = 1$ requires a PQ breaking scale of about 10^{12}GeV, corresponding to an axion mass of about 10^{-5}eV.

3.5 Dark Matter and a New Dimensionless Number

As physicists we are all aware of the importance of dimensionless numbers. There are already a handful of important dimensionless numbers in cosmology, the net baryon number to entropy ratio ($n_B/s \simeq 10^{-10}$), the fractional primordial abundances of the light elements (D, ^{3}He, ^{4}He, ^{7}Li), the horizon crossing amplitudes of adiabatic density perturbations, to mention a few. For the first two we believe that we have a fundamental understanding of their origin, and with the inflationary scenario we may be on the way to understanding the third. Martin Rees has emphasized the existence of yet another dimensionless number if the Universe is not baryon-dominated. That number is the ratio of mass density in ordinary baryonic matter to that in exotic matter. If we adopt $\Omega = 1$ and $\Omega_{baryon} = 0.1$, then this ratio r is about 0.11.

Is this a significant dimensionless number whose eventual understanding may provide us with a new insight to the Universe and the laws of nature? Why is its value so close to order unity and not say 10^{-30} or 10^{30}? Is it yet another example of the anthropic principle at work (God forbid!)? If quark nuggets are the dark component, then a value of about 0.1 arises quite naturally: it is the fraction of quarks that remain free, which based upon simple physics Witten (1984) estimated to be of order 0.1. In this case there is good reason for r to be of order unity.

What about the other dark matter candidates? According to the very attractive scenario of baryogenesis the net baryon number to entropy ratio evolved during the very early history of the Universe due to non-equilibrium interactions which violate B, C, CP (for a recent review, see Kolb and Turner 1983). The baryon to entropy ratio which evolves can be written as

$$n_B/s \approx \varepsilon/g_* \ (10^{14}\text{GeV or so}) \tag{10}$$

where ε parameterizes the C, CP violation and must, based upon very general arguments be less than about $(\alpha/\pi) \approx 10^{-3}$.

Now let's consider the relic abundance of a WIMP whose annihilations proceed via an interaction with roughly the strength of the weak interactions (e.g., a heavy neutrino, photino, sneutrino, Higgsino, etc.). The ratio of its relic abundance to the entropy density is given by Eqn. (5):

$$(n_x/s) \approx x_f/(0.15g_*^{-1/2}m_{pl}m_{WIMP}(\sigma v)_o) \tag{11}$$

where $n = 0$, and $x_f \approx 0(20)$ depends logarithmically upon $(\sigma v)_o$, etc. This together with the baryon to entropy ratio allows us to calculate Martin Rees's ratio r

$$r = 0.15 \ \varepsilon \ g_*^{-3/2} \ m_{pl}m_{nuc}(\sigma v)_o/x_f \ . \tag{12}$$

For simplicity, if we take the WIMP mass to be of order m_{nuc}, $(\sigma v)_o \approx G_F^2 m_{WIMP}^2$, $x_f \approx 20$, $g_* \approx 100$, and $\varepsilon \approx 10^{-4}$, then the ratio r is just

$$r \approx 10^{-9}m_{nuc}^3m_{pl} \ G_F^2 \ . \tag{13}$$

That r be of order unity (and not say $10^{\pm 30}$ then requires a large discrepancy between the weak scale ($G_F \approx \langle\phi\rangle^{-2} \approx 10^{-5}\text{GeV}^{-2}$, where $\langle\phi\rangle \approx$ 300 GeV is the Higgs vev and sets the weak scale) and the scale of particle masses relative to the planck scale ($\approx 10^{19}$ GeV).

Now consider the case of relic axions. Using Eqn. (9) it is straightforward to compute Martin Rees' ratio r:

$$r \approx 0.1 \ \varepsilon \ g_*^{-1/2} \ (m_{pl}/f_{PQ})^{1.2}\theta_1^{-2} \ . \tag{14}$$

From this expression for r it is clear that in order to have r be of order unity, the PQ symmetry breaking scale shouldn't differ from the planck scale by too many orders of magnitude. Put another way, it means that f_{PQ} of the order of the GUT scale results in r of order unity, whereas $f_{PQ} = 300$ GeV results in $r = 10^{-17}$ (assuming that the axion were stable for such a value of f_{PQ}).

Is there any significance to these relations or to the Martin Rees' ratio r? I don't know. In the case of quark nuggets, however, it is clear that r is quite naturally of order unity. In the other two cases, r being of order unity can be traced to relationships between fundamental scales in particle physics.

4. IMPLICATIONS

Although in many regards the different WIMP candidates are interchangeable, there are several important differences -- how they process primeval adiabatic density perturbations, the scales upon which there is enough phase space for them to cluster, and the prospects for their detection. I will summarize those differences here.

4.1 Freestreaming WIMPs

It has long been realized that density perturbations in a self-gravitating fluid in which the mean free path of the fluid constituents is finite will undergo Landau damping. For WIMPs, this effect is particularly severe as they are always effectively collisionless. Until the Universe becomes matter-dominated and WIMP perturbations can start to grow via the Jeans instability, perturbations can be damped by freestreaming of the WIMPs out of the perturbations. Following Davis, Lecar, Pryor, and Witten (1981) one can define the characteristic freestreaming scale

$$\lambda_{FS} = \int_o^{t_{eq}} v(t')dt'/R(t') . \tag{15}$$

Physically, λ_{FS} is the comoving distance that a WIMP could have traveled since the bang. Most of the contribution to the integral arises during the epoch when the WIMPs are relativistic (once they become NR, $v \propto R(t)^{-1}$ and λ only grows logarithmically). Clearly, λ_{FS} defines the characteristic damping scale for primeval perturbations: WIMP perturbations on scales smaller than the scale λ_{FS} will be strongly damped by the streaming of WIMPs out of the overdense regions and into the underdense regions. Careful calculations of the damping effects of collisionless WIMPs have been performed by Bond and Szalay (1983), Bond, Szalay, and Turner (1982), Peebles (1982), and Blumenthal and Primack (1984).

Assuming that the Universe is WIMP-dominated and flat, and that the WIMPs are characterized by a temperature T_{WIMP} (which is not necessarily the same as the photon temperature T), it is straightforward to compute λ_{FS}:

$$\lambda_{FS} = 1 \text{ Mpc } (1\text{keV}/m_{WIMP})(T_{WIMP}/T) \times \tag{16}$$

$$[1 + \ln\{6(m_{WIMP}/\text{keV})^{1/2} (T/T_{WIMP})^{1/2}\}]$$

Note that for heavy WIMPs the damping scale is smaller; this is because they become NR earlier during the history of the Universe and hence cannot stream as far. The Table below shows the damping scale for a few of the dark matter candidates.

WIMP	MASS	T_{WIMP}/T	λ_{FS}(Mpc)
Neutrino	light	$(4/11)^{1/3}$	40 Mpc/(m/30eV)
Axion	10^{-5}eV	$< 10^{-14}$	$< 10^{-5}$Mpc
Axino/RH Neutrino/Light Gravitino	keV	1/4	1 Mpc
Heavy Neutrino/ LSP	GeV	1	10^{-5}Mpc

The scale 1 Mpc corresponds to a galactic scale. The relationship of λ_{FS} to the galactic scale neatly divides the WIMPs into three categories: (i) Cold, $\lambda_{FS} \ll 1$ Mpc -- the characteristic damping scale is much smaller than a galactic scale, and galactic-sized perturbations survive freestreaming; (ii) Warm, $\lambda_{FS} \simeq 1$ Mpc -- the characteristic damping scale corresponds to a galactic scale; (iii) Hot, $\lambda_{FS} \gg 1$ Mpc — only perturbations on scales much larger than a galactic scale survive freestreaming. Almost all of the WIMPs fall into the category of cold dark matter. Only the neutrino is a hot WIMP. At present there are a couple of warm dark matter candidates -- a 1 keV gravitino, 1 keV right-handed neutrino, or a 1 keV axino (supersymmetric partner of the axion).

By far the damping effect of the WIMPs on the primordial spectrum of adiabatic density perturbations is the most important implication of the different candidates on structure formation. The damping mass determines which structures form first: for cold and warm WIMPs it's galactic-sized objects or smaller; for hot WIMPs it's very large structures (superclusters).

4.2 The Tremaine-Gunn Constraint

In a very nice paper Tremaine and Gunn (1979) discussed a kinematical constraint on dark matter candidates. In brief, they pointed out that for a gravitationally-bound system characterized by mass M, velocity dispersion σ, and size r, there is only so much phase space available

$$\mathcal{U}_{ph.sp.} \simeq f\,\mathcal{U}_x\,\mathcal{U}_p/(2\pi)^3 \simeq f m_{WIMP}^3\,\sigma^3\,r^3 \qquad (17)$$

where f is the possible quantum occupancy of each state. For fermions, f is at most the number of spin degrees of freedom; in fact, this is true for any particle which decouples while still in thermal equilibrium. Based on the amount of phase space available it follows that there is a maximum to the mass in WIMPs such a system can have

$$M_{max} \simeq m_{WIMP} \bigcup{}_{ph.sp.} \simeq f m_{WIMP}^4 \sigma^3 r^3 , \qquad (18)$$

For all the WIMPs, except axions which are born with a very high quantum occupancy ($\geq 10^{52}$!), f is of order unity.   Eqn. (18) then implies a constraint on the minimum WIMP mass required such that the system could be WIMP-dominated.   Taking $f \simeq 2$, the Tremaine-Gunn bound is

$$m_{WIMP} \geq 100eV \; (100 \; kms^{-1}/\sigma)^{1/4} \; (1kpc/r),^{1/2}$$

$\qquad \geq O(5eV) \qquad$ Rich cluster ,

$\qquad \geq O(10eV) \qquad$ Small Group ,

$\qquad \geq O(30eV) \qquad$ Healthy-sized Galaxy ,

$\qquad \geq O(150eV) \qquad$ Dwarf Galaxy.

Only for the neutrino is the constraint interesting (recall because of their high quantum occupancy, these bounds do not apply to axions), and only for small systems, such as dwarf galaxies. If dwarf galaxies are WIMP-dominated and characterized by the parameters used to obtain the above bound, then neutrinos cannot be the dark matter (at least in dwarf galaxies), based upon kinematical grounds alone. [Madsen and Epstein (1985) have recently reexamined this constraint in light of better determinations of σ and r.]
    The Tremaine-Gunn constraint is basically a kinematical constraint. If the mass of a WIMP exceeds their bound for a given system, that does not guarantee that such systems are WIMP-dominated -- that is a question of dynamics, it merely implies that it is kinematically possible.

4.3 The Search for WIMPs

From the above discussion, it is apparent that many of the WIMPs (essentially all the cold ones) have identical implications for structure formation, and therefore cannot be distinguished on that basis alone. Until recently, it was generally thought that in spite of the great reservoir of WIMPs in which we are swimming (see Fig. 1 and remember for WIMPs which cluster with galaxies the local density is about a factor of 10^4 or so higher) it would be impossible to detect their presence because of the feebleness of their interactions. It now appears that this pessimism was somewhat premature. A number of very clever ideas have been proposed for detecting the presence of WIMPs in the halo of our galaxy. I will briefly summarize this very exciting work.
    Axions -- If axions are the halo dark matter, then their local number density is enormous, 3×10^{13} or so cm^{-3}. Sikivie (1983) proposed an idea which exploits the axion coupling to 2 photons to convert halo axions into photons.   Because of this coupling, in the presence of a strong, inhomogeneous magnetic field axions will convert to photons. Because the halo axions are very NR ($v/c \simeq 10^{-3}$) the width of the line should be very narrow ($\Delta\lambda/\lambda \simeq 10^{-6}$). In a large high-Q, microwave cavity these photons might be detectable. Several groups are designing and/or building experiments based upon Sikivie's idea.

WIMP Heat :- In a very interesting paper, Goodman and Witten (1985) discussed the possibility of using supercooled, ultra-low heat capacity bolometric detectors to detect the small amount of energy (of order keV) deposited by a variety of WIMPs (photinos, sneutrinos, heavy neutrinos) when they interact in matter. A 10 kg detector operating at a temperature of order a few milliKelvin would register a count or so per day (depending upon the couplings of the WIMP). Drukier, Freese, and Spergel (1985) have followed up this idea in more detail. Cabrera etal. (1985) have proposed a bolometric detector which may be suitable for this purpose.

WIMP Annihilations -- If the dark matter in our halo is photinos, sneutrinos, Higgsinos, or heavy neutrinos, then WIMPs are annihilating all about us ($(\sigma v)_{ann} \simeq 10^{-26}$ cm^3 sec^{-1}) and some of the annihilation products may be detectable. Silk and Srednicki (1984) discussed the possibility that photino annihilations in the halo might produce enough low-energy antiprotons to explain the anomalously high flux of low energy antiprotons detected by Buffington and Schindler (1981). Stecker etal. (1985) have gone a step further and calculated the expected spectrum of antiprotons for this scenario.

Press and Spergel (1985) have pointed that the sun will capture significant numbers of WIMPs if the halo is WIMP-dominated, and Olive and Silk (1985) have discussed the possibility that neutrinos and antineutrinos produced by WIMPs annihilating in the sun might be detectable in large, underground detectors (such as those used to search for proton decay). Freese (1986) and Krauss, Srednicki and Wilczek (1985) have pointed out that the annihilations of those WIMPs captured by the earth might also be detectable in large, underground detectors.

In a very recent paper Srednicki, Theisen, and Silk (1986) have proposed an even more intriguing way of detecting WIMP annihilations in the halo, through the gamma ray lines they produce when they annihilate into a bound quark-antiquark state (such as ψ/J, T, etc.) and a monoenergetic photon. Photinos, sneutrinos, heavy neutrinos, and Higgsinos in the halo could be directly detected this way. Furthermore, if such gamma ray lines are detected, not only could the mass of the halo WIMPs be directly determined, but also the mass distribution of the halo could be probed by the directional dependence of the strength of the line (Turner 1986).

Monopole searches — A variety of induction and energy loss searches are presently ongoing (for a recent review, see Groom 1986). Unfortunately, there has been no additional confirming evidence for the famous Valentine's Day event of 1982. The current level of sensitivity is about 10^{-13} cm^{-2} sr^{-1} sec^{-1}, and experiments are being designed and constructed at the sensitivity level of 10^{-16} cm^{-2} sr^{-1} sec^{-1}, which is a factor of 10 below the Parker bound (for monopoles lighter than 10^{17} GeV). Needless to say the discovery of monopoles as the dark matter would not only be a boon to astrophysicists, but also to particle physicists as it would be a confirmation of the idea of grand unification and to early Universe cosmologists as they would represent relics from the earliest moments of the Universe ($t \lesssim 10^{-34}$ sec).

Laboratory experiments -- Many of the dark matter candidates could have their existence confirmed in the laboratory. For example, spartners may be produced at the CERN SppS collider (although they have not yet), the Tevatron $p\bar{p}$ collider at Fermilab (which comes online in fall 1986), SLC at SLAC, and if they exist (and if it exists) at the SSC for certain. Discovering the LSP and determining its properties (mass, interaction cross section, etc.) would allow one to reliably calculate its relic abundance and settle the issue of whether or not it could be the dark matter. Needless to say, knowing the dark matter constituent's properties would make direct searches much more straightforward.

Experiments to directly measure the mass of the electron neutrino continue. A confirmation of the result of the ITEP group (Lubimov etal. 1981) would be strong evidence that the Universe is neutrino-dominated (and cosmologists and structure simulators would have to adjust accordingly!). The continuing neutrino oscillation experiments also bear on this issue.

5. A HOST OF DARK MATTER PROBLEMS

As a number of authors have emphasized there are several dark matter problems (see, e.g., Freese and Schramm 1984, Schramm 1986, Carr 1986, and Bahcall 1984). Bahcall (1984) has made a convincing case that there is an unaccounted for, dark disk component in our galaxy, with mass density comparable to the seen component (stars, gas, dust, etc.). If typical of spiral galaxies in general, this dark disk component corresponds to $\Omega \approx 0.005$. Since the formation of the disk involved dissipation, it is unlikely that this component is comprised of WIMPs. In all likelihood it is baryonic (which does not conflict with the nucleosynthesis bounds and in fact receives weak confirmation as primordial nucleosynthesis suggests that $\Omega_{baryon} \gtrsim 0.014$, which is about twice that seen in luminous matter).

As we have heard at this symposium there is very good evidence for the existence of a dark halo component in spiral galaxies. Since there is no convincing evidence as of yet for a rotation curve which 'turns over', at present we only have a lower bound to the amount of dark matter in the halos of spiral galaxies, something like $\Omega \gtrsim 0.05 - 0.10$.

There is also good evidence for dark matter in clusters of galaxies and (somewhat weaker evidence) for dark matter in small groups of galaxies. In the case of clusters, some of this matter is only dark as far as the optical astronomer is concerned, since it is X-ray bright. The uncertainties here are much greater, but the amount of dark matter in clusters probably corresponds to $\Omega \approx 0.1 - 0.3$.

Finally, there is the ultimate dark matter problem. The 'light' of theory casts a strong beam on $\Omega = 1.0$, the flat, Einstein-deSitter model. While the 'shadow' of observation is cast on the value $\Omega_{obs} \approx 0.2 \pm 0.1$ (where ± 0.1 is not meant to be a formal error bar, but rather a theorist's estimate of the spread of current determinations). To be sure, the observational determinations only apply to the matter which clusters with the visible matter on scales less than 10-30 Mpc. A component which is smoothly distributed on these scales would thus far have gone undetected. In order to reconcile theory with observation, this smooth, dark component would need to contribute $\Omega_{SM} = 1 - \Omega_{obs} \approx 0.8 \pm 0.1$.

5.1 The Ω Problem

This discrepancy between theory and observation has come to be known as the Ω problem. A variety of ideas have been put forth to save the flat, $\Omega = 1$ (more precisely k=0) Einstein-deSitter model. They are all based upon the same principle, a smooth component contributing about

$$\Omega_{SM} \simeq 1 - \Omega_{obs} \simeq 0.8 \pm 0.1$$

Suggestions for the smooth component include: failed galaxies (Kaiser 1984); a relic cosmological term (Peebles 1984, Turner, Steigman, and Krauss 1984); fast-moving, light strings or a network of light strings (Vilenkin 1985; 'light' here means a symmetry breaking scale $<<$ 10^{16} GeV, the canonical scale for cosmic strings which would lead to interesting isothermal perturbations); and relativistic particles produced by decaying WIMPs (Turner, Steigman, and Krauss 1984, Dicus, Kolb, and Teplitz 1978, Gelmini, Schramm, and Valle 1985, Olive, Seckel, and Vishniac 1985), I will briefly discuss the exotic (even by present standards) scenario of relic WIMPs decaying into WIRPs (Weakly-Interacting Relativistic Particles).

[A brief comment with regard to a relic cosmological term. A cosmological term corresponds to a uniform energy density and has exactly the same form as the vacuum energy density associated with a quantum field theory (Zel'dovich 1968). For this reason one might look to particle physics for a prediction for any relic cosmological term. The quantum contribution to the vacuum energy (or zero point energy) for a renormalizable theory is formally infinite, one of the infinities which is swept under the rug by renormalization. Forgetting about the infinity for a moment, there is no symmetry in the theory which excludes a vacuum energy as large as m_{pl}^4, and each stage of SSB should change the vacuum energy by order M^4; recall the present expansion rate of the Universe sets an upper bound to the vacuum energy density of about 10^{-46} GeV^4. The present theoretical situation then is somewhat discouraging; first one has to throw away an infinite contribution, then one has to fine-tune away terms in the theory which are permitted by all the symmetries of the theory (and if not put in ab initio would arise due to quantum corrections anyway) and which are 122 orders-of-magnitude larger than the present upper limit to Λ -- not much theoretical guidance here! There is one tiny ray of hope though. In supersymmetric theories, the fermionic and bosonic contributions to the vacuum energy cancel and the cosmological constant is zero. However, once SUSY is broken this cancellation no longer occurs and the vacuum energy becomes of order $\langle\phi\rangle^4$ or about 10^{10} GeV^4. Unless the Universe is really supersymmetric and we don't realize it, which doesn't seem likely, the apparent smallness of the cosmological constant, relative to what it has every right to be, is a very fundamental problem. That doesn't, however, preclude a solution (e.g., an axion-like mechanism) which leaves a tiny relic cosmological constant, say 0.8 of critical density or so.]

5.2 Structure Formation with 'Rotting Particles'

The basic idea of this scenario is that today the Universe consists of two components: NR particles which account for the mass which clusters, i.e., $\Omega \approx 0.2 \pm 0.1$; and R particles, which by the virtue of their high speed cannot cluster and account for the remaining $\Omega \approx 0.8 \pm 0.1$. (For simplicity's sake, it would be preferrable for the NR component to be baryons only.)

It has been long realized that density perturbations cannot grow in a Universe which is radiation-dominated, and so it is necessary that the R particles have a recent origin. In the rotting scenario, they are born from the non-radiative decays of unstable WIMPs. In order to account for the observed structure in the Universe the decays occur rather recently, redshift z_d of order 3-10 (corresponding to a WIMP lifetime of around $10^8 - 10^9$ yrs).

[In some particle physics models which address the so-called family problem, why are there 3 families, etc., there are new interactions which would allow a heavy neutrino to decay into a light neutrino and a massless, very weakly-interacting (essentially invisible) Nambu-Goldstone boson, with a lifetime which is about right for this scenario.]

For definiteness, let's suppose that the unstable WIMP is a neutrino and that $h = 0.5$. If the neutrino were stable then its mass would have to be about 24 eV to achieve $\Omega = 1$. Because the mass density in WIRPs scales like R^{-4} since the decay epoch (say redshift z_d), the neutrino mass required in the decaying scenario is about $24z_d$ eV. This in turn means that in the decaying scenario the Universe becomes matter-dominated earlier than in the non-decaying case, by a factor of z_d. Hence, density perturbations start to grow earlier; of course once the Universe becomes radiation-dominated they cease growing. They start growing earlier but stop growing before the present epoch -- the net effect is that they undergo the same amount of growth (actually, slightly more growth; after the decays, perturbations continue to grow logarithmically). In the case of a decaying neutrino the damping scale is smaller because the mass of the neutrino is larger. As White has discussed (1986a,b), this helps the viability of the neutrino scenario. The so-called 'rotting' particle scenario is summarized in Fig. 8.

What can one say about z_d? Since perturbations undergo about the same amount of growth, independent of z_d, perturbations at the epoch of decoupling are larger for larger z_d. This in turn implies larger anisotropies in the microwave background. Present observations probably constrain z_d to be less than about 5-10 (Vittorio and Silk 1985, Turner 1985).

Before I forget, the rotting particle scenario does have one drawback. It predicts a very youthful Universe, typically $H_0 t_0 \approx 0.53 -$ 0.58. Unless the Hubble constant is in the range 40-55 km sec^{-1} Mpc^{-1}, rotting particles are in deep trouble.

5.3 Testing Oddball Cosmological Models

Is it possible to use observational data to discriminte between the

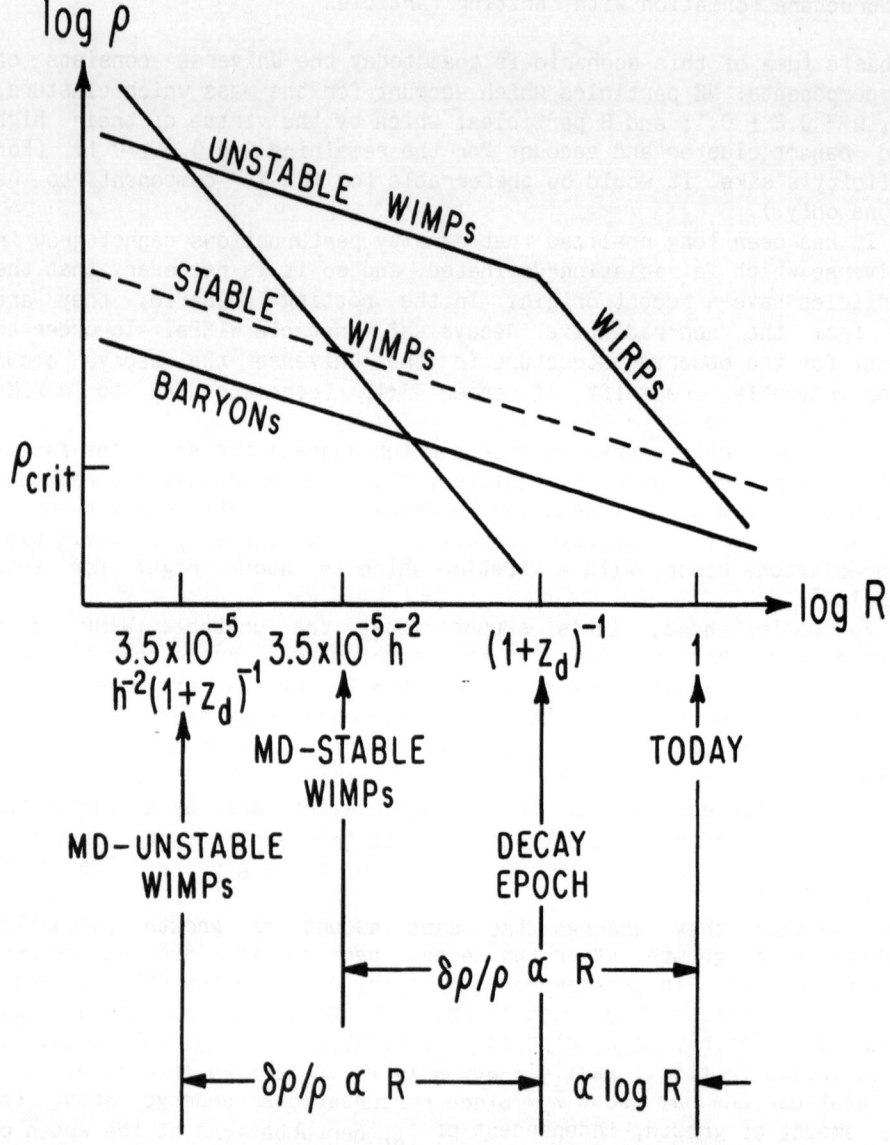

Figure 8 – Schematic summary of structure formation with 'rotting particles' (and the stable WIMP scenario for comparison).

theoretical ideas which have been put forth to save the flat Universe? [I have nightmares of a modern day Christopher Columbus proving all of us theorists wrong.] A graduate student working with me thinks the answer might be yes. Since the evolution of the cosmic scale factor in these models (rotting particles, relic cosmological term, fast moving strings) is very different from the usual $R(t) \propto t^{2/3}$, in a flat, matter-dominated Universe, one might expect that some of the usual cosmological tests might be good discriminators. She has recently calculated the magnitude vs. redshift, angular size vs. redshift, lookback time vs. redshift, and differential comoving volume element vs. redshift diagrams for these models. The two diagrams which look to be particularly useful are shown as Figs. 9,10 (from Charlton and Turner 1986).

6. PROGNOSTICATIONS AND CONCLUDING REMARKS

6.1 Great Dark Hopes

What are the most likely candidates from the new particle zoo for the dark matter? Unfair question -- but I'll answer it anyway. I would say the axion, the light neutrino, and the photino (or another LSP candidate).

Axion -- The Peccei-Quinn mechanism continues to be the most attractive solution to the only woe of QCD, the strong CP problem. Many SUSY models automatically have a PQ symmetry and indications are that the field theory limit of the superstring also has a PQ symmetry. The PQ symmetry breaking scale required for axions to dominate the Universe, an energy greater than about 10^{12} GeV, is an interesting scale. Finally, it seems possible that halo axions could be detected (especially if theorists could predict their mass more precisely).

Light Neutrino -- The neutrino is actually known to exist! Almost all theories beyond the standard model predict that neutrinos should have masses (albeit very small). With the neutrino one gets at least 3 (and possibly 4) shots at being right (I'll bet on the τ-neutrino). What about the experimental prospects? A recent paper by Bergkvist (1985) raises some serious questions about the validity of the ITEP experiment. [Bergkvist has shown that the line used to calibrate the ITEP detector has a non-Lorentzian tail, which he claims would give rise to the non-zero result they obtain for the neutrino mass.] In any case enough different types of experiments to determine the electron neutrino mass are now in progress that we should have a definitive answer soon. With regard to neutrino oscillation experiments, Boehm and Vogel (1984) have recently reviewed the experimental situation and find no conclusive evidence for the existence of neutrino oscillations. The experimental effort in this direction, however, continues.

John Simpson (1985) has recently caused some excitement with the results of his tritium endpoint experiment (which employs a Si(Li) detector). His data indicate a kink in the Kurie plot, which could be explained by the existence of a 17.1 keV neutrino mass eigenstate with about 3% mixing to the electron neutrino weak eigenstate.

The theoretical implications are very exciting. In order for his

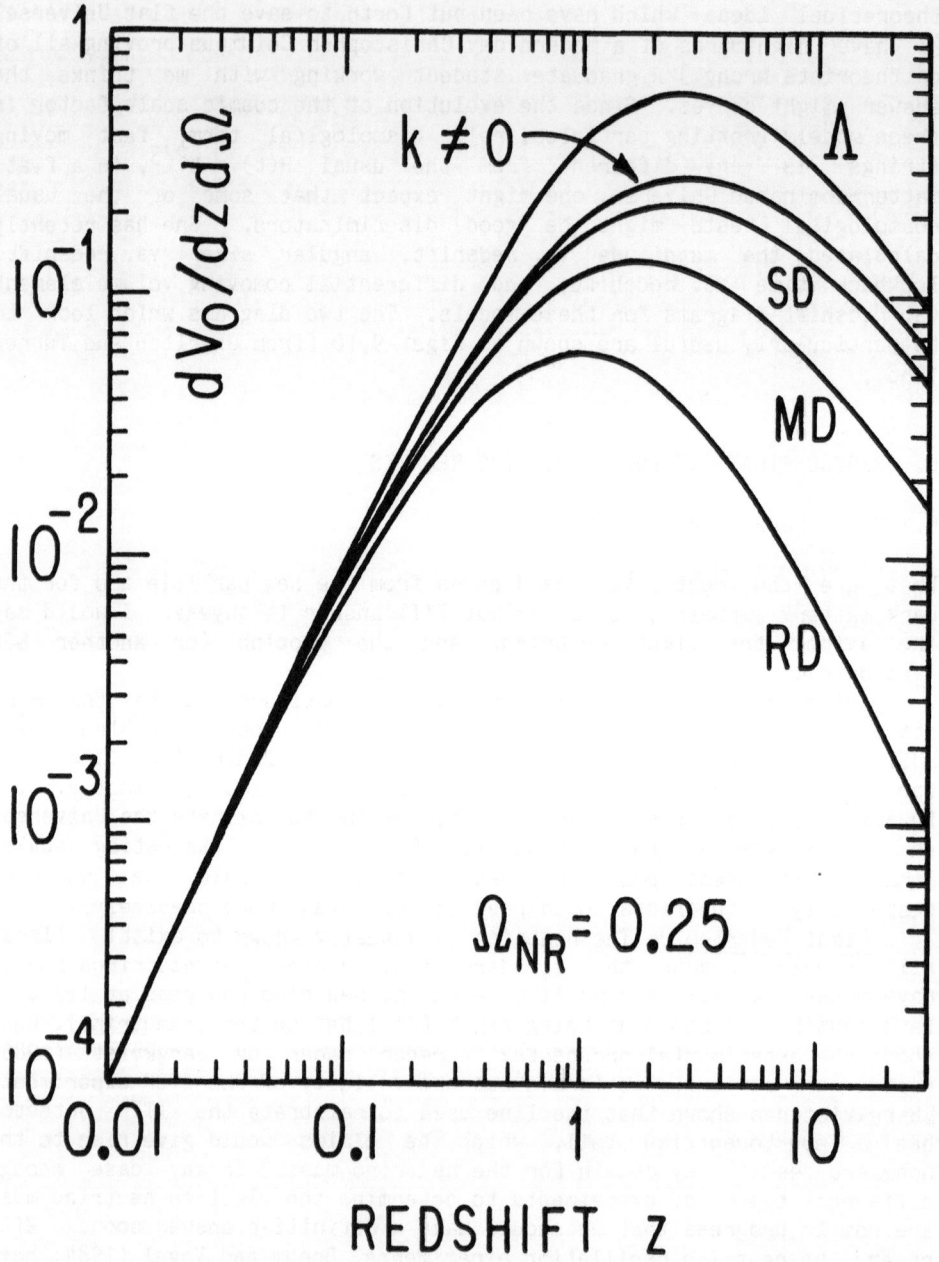

Figure 9 - The differential comoving volume, $dV_0/dZd\Omega$, vs. redshift z, for model universes with $\Omega_{NR} = 0.25$ and $k = 0$ ($\Lambda \neq 0$, smooth component of matter, fast strings, and relativistic particles); also shown for comparison is the $k \neq 0$ model (from Charlton and Turner 1986).

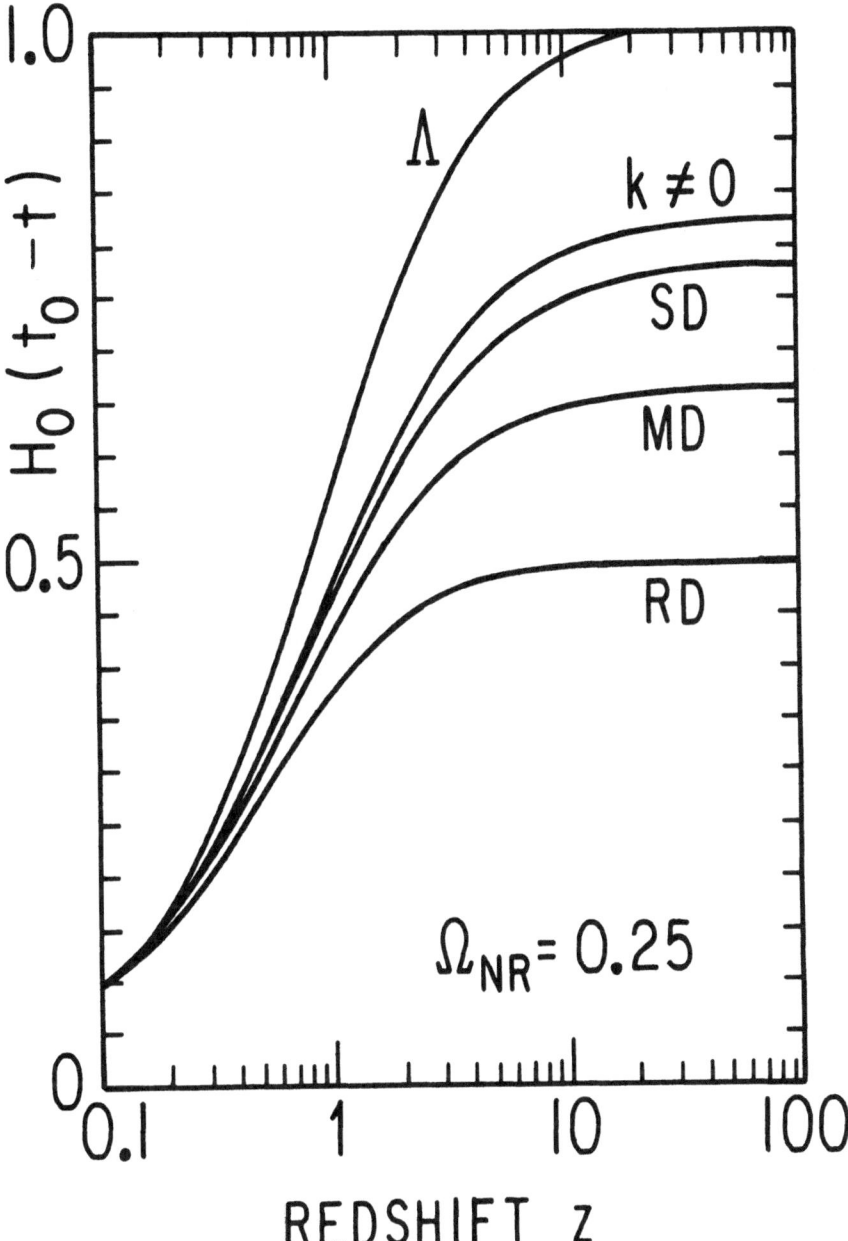

Figure 10 - Lookback time, $H(t_0 - t)$, vs. redshift z, for the same models as in Figure 9 (from Charlton and Turner 1986).

result to be consistent with other experimental data, there must be two 17.1 keV mass eigenstates (Dugan etal. 1985). Cosmology tells us that such a neutrino(s) cannot be stable and have the canonical abundance. Either it must have a smaller relic abundance or have a lifetime of less than about a year in order to avoid interfering with structure formation in the Universe (Steigman and Turner 1985). Either alternative implies that neutrinos must be endowed with interactions other than the usual electroweak interactions (e.g., as in the majoron model of Gelmini and Roncadelli 1981). If this is the case, neutrino annihilations will be more effective, keeping them in equilibrium until much lower temperatures, which results in their having a much smaller relic abundance (cf, Eqn. (3) or Kolb and Turner 1985). In fact, if they have these additional 'stronger than weak' interactions and are stable, their relic abundance could be such that they are the dark matter. It is interesting to note that in this case they would behave like cold dark matter.

Hold everything! Two other groups have now looked for the same effect in the decays of ^{35}S (Q value of 166.8 keV) and at the 90% confidence level set a limit on the mixing of a mass eigenstate greater than about a few keV of less than 1% (Altzitzoglou etal. 1985, Markey and Boehm 1986).

Simpson has worried that the kink might be due to a solid state effect (or Coulomb effects) since the kink occurs so near threshold and he plans to look at the tritium β-decay spectrum with a Ge(Li) in the near future.

Photino (or another LSP candidate) -- SUSY is a very attractive theoretical idea and just as importantly it makes predictions which can be tested in the forseeable future. Thus far, none of the experimental data provide any unambiguous evidence for SUSY. However, all of the SUSY candidates suggested for the dark matter should be able to be produced at CERN, Fermilab or the SSC. We will have an answer, maybe not tomorrow, but before the turn of the century. In addition, there is the very real possibility that if relic LSPs are the halo dark matter, they or their effects can be detected. Of all the spartners, a 1-10 GeV photino seems to be the most likely LSP candidate.

My Favorite Dark Horse Candidates -- A planck mass monopole which does not catalyze nucleon decay could provide the halo mass density, close the Universe and safely elude all the astrophysical bounds (see Fig.6). Not only that, but its flux ($\simeq 10^{-13}$ cm^{-2} sr^{-1} sec^{-1}) is such that detection is just around the corner! For personal reasons as well as for the novelty of it I also include shadow matter in my dark horse list. Quark nuggets are so attractive that they too have to be included as a dark horse possibility.

6.2 Concluding Remarks

So much for prognostications (the majority of which must necessarily be wrong!). The organizers of this symposium have flattered me by asking that I review all the exciting dark matter candidates from the new particle zoo. Now for the harsh realities. God forbid, but it is very possible that we live in a low Ω, baryon-dominated Universe. After all, we are only compelled to appeal to non-baryonic dark matter if Ω is greater than about 0.15 (higher if we are foolish enough to ignore the primordial nucleosynthesis constraint of $\Omega_{baryon} \leq 0.035h^{-2}$). Theoretical prejudice aside, there is no convincing evidence (or even unconvincing evidence for that matter) that Ω is any larger than about 0.2 ± 0.1. History repeats itself; once again we have convinced each other that there are only two possible stories of structure formation — cold and hot dark matter with $\Omega = 1$. We may be in for some real surprises. Fortunately, it's not surprises that puts theorists out of work, rather more often it's the lack of surprises. [There is an old saying which dates back to the early days of experimental physics; theorists untethered by experimental data are doomed to rise in their own hot air never to be seen again.] Perhaps it will be a scenario based upon cosmic strings or the role of astrophysical fireworks (see Ostriker's contribution to these proceedings) that will eventually prevail; then again, it could be an $\Omega = 1$ WIMP-dominated Universe. Now I'm covered either way!

Acknowledgements. I am grateful to a multitude of colleagues for enlightening discussions. This work was supported in part by the Department of Energy (at Chicago and at Fermilab), by the National Aeronautics and Space Administration (at Fermilab), and by an Alfred P. Sloan Fellowship.

REFERENCES

Abbott, L. and Sikivie, P. 1983, Phys. Lett., **120B**, 133.
Albrecht, A. and Steinhardt, P. 1982, Phys. Rev. Lett., **48**, 1220.
Altzitzoglou, T., Calaprice, F., Dewey, M., Lowry, M., Piilonen, L., Brorson, J., Hagen, S., and Loeser, F. 1985, Phys. Rev. Lett., **55**, 799.
Alcock, C. and Farhi, E. 1985, Phys. Rev., **D32**, 1273.
Applegate, J. and Hogan, C. 1985, Phys. Rev., **D31**, 3037.
Audouze, J. 1986, in these proceedings.
Bahcall, J. 1984, Astrophys. J., **287**, 926.
Bardeen, J., Steinhardt, P., and Turner, M. S. 1983, Phys. Rev., **D28**, 679.
Barrow, J. and Turner, M. S. 1981, Nature, **291**, 469.
Bergkvist, K. 1985, Phys. Lett., **154B**, 224.
Blumenthal, G. and Primack, J. 1984, unpublished.
Boehm, F. and Vogel, P. 1984, Ann. Rev. Nucl. Part. Sci., **34**, 125.
Boesgaard, A. and Steigman, G. 1985, Ann. Rev. Astron. Astrophys., in press.
Bond, J., Kolb, E., and Silk, J. 1982, Astrophys. J., **255**, 341.
Bond, J. and Szalay, A. 1983, Astrophys. J., **276**, 443.
Bond, J., Szalay, A., and Turner, M. 1982, Phys. Rev. Lett., **48**, 1036.
Buffington, A. and Schindler, S. 1981, Astrophys. J., **247**, L105.
Cabrera, B., Krauss, L. L., and Wilczek, F. 1985, Phys. Rev. Lett., **55**, 25.
Carr, B. J. 1986, in Inner Space/Outer Space, eds. E. Kolb etal. (Univ. of Chicago Press, Chicago).

Charlton, J. and Turner, M. 1986, Univ. of Chicago preprint (submitted to Astrophys. J.).

Davis, M., Lecar, M., Pryor, C., and Witten, E. 1981, Astrophys. J., 250, 423.

Degrand, T. and Kajantie, K. 1984, Phys. Lett., 147B, 273.

Dicus, D., Kolb, E., and Teplitz, V. 1978, Astrophys. J., 223, 327.

Dine, M. and Fischler, W. 1983, Phys. Lett. 120B, 139.

Drukier, A., Freese, K., and Spergel, D. 1986, Phys. Rev. D, in press.

Dugan, M., Gelmini, G., Georgi, H., and Hall, L. J. 1985, Phys. Rev. Lett., 54, 2302.

Efstathiou, G. and Silk, J. 1983, Fund. Cosmic Phys., 9, 1.

Ellis, J., Hagelin, J. S., Nanopoulos, D. V., Olive, K., and Srednicki, M. 1984, Nucl. Phys., B238, 453.

Freese, K. 1986, Phys. Lett. B, in press.

Freese, K. and Schramm, D. N. 1984, Nucl. Phys., B233, 167.

Gelmini, G. and Roncadelli, M. 1981, Phys. Lett., 99B, 411.

Gelmini, G., Schramm, D. N., and Valle, J. 1984, Phys. Lett., 146B, 386.

Georgi, H. 1984, Weak Interactions in Modern Particle Theory (Benjamin/Cummings, Menlo Park).

Goodman, M. and Witten, E. 1985, Phys. Rev., D31, 3059.

Green, M. B. 1985, Nature, 314, 409.

Greenberg, O. 1985, Physics Today, 38 (9), 22 (Sept.).

Groom, D. 1986, Phys. Rep., in press.

Guth, A. 1981, Phys. Rev., D23, 347.

Guth, A. and Pi, S.-Y. 1982, Phys. Rev. Lett., 49, 1110.

Hawking, S. 1982, Phys. Lett., 115B, 295.

Kaiser, N. 1984, Astrophys. J., 284, L9.

Kibble, T. 1976, J. Phys., A9, 1387.

Kolb, E. 1984, in Proc. XIth Int. Conf. ν Phys. Astrophys., eds. K. Kleinknecht and E. Paschos (World Scientific, Singapore).

Kolb, E. and Turner, M. 1983, Ann. Rev. Nucl. Part. Sci., 33, 645.

Kolb, E. and Turner, M. 1985, Phys. Lett., 159B, 102.

Kolb, E., Seckel, D., and Turner, M. 1985, Nature, 314, 415.

Krauss, L. L., Srednicki, M., and Wilczek 1985, preprint.

Langacker, P. 1981, Phys. Rep., 72, 185.

Lee, B. W. and Weinberg, S. 1977, Phys. Rev. Lett., 39, 165.

Linde, A. 1982, Phys. Lett., 108B, 389.

Linde, A. 1985, Phys. Lett., 158B, 375.

Lubimov, V., Novikov, E., Nozik, V., Tretyakov, E., and Kosik, V. 1981, Phys. Lett., 94B, 266; JETP, 54, 616.

Madsen, J. and Epstein, R. 1985, Phys. Rev. Lett., 54, 2720.

Markey, J. and Boehm, F. 1986, Phys. Rev. Lett., in press.

Olive, K. and Silk, J. 1985, Phys. Rev. Lett., 55, 257.

Olive, K., Seckel, D., and Vishniac, E. 1985, Astrophys. J., 292, 1.

Peccei, R. and Quinn, H. 1977, Phys. Rev. Lett., 38, 1440.

Peebles, P. J. E. 1982, Astrophys. J., 263, L1.

Peebles, P. J. E. 1984, Astrophys. J., 284, 439.

Perkins, D. 1984, Ann. Rev. Nucl. Part. Sci., 34, 1.

Preskill, J. 1984, Ann. Rev. Nucl. Part. Sci., 34, 461.

Preskill, J., Wise, M., and Wilczek, F. 1983, Phys. Lett., 120B, 127.

Press, W. and Spergel, D. 1985, Astrophys. J., 296, 679.

Quigg, C. 1983, Gauge Theories of the Strong, Weak, and EM Interactions (Benjamin/Cummings, Menlo Park).

Ramond, P. 1983, Ann. Rev. Nucl. Part. Sci., **33**, 31.

Ross, G. G. 1984, Grand Unified Theories (Benjamin/Cummings, Menlo Park).

Scherrer, R. and Turner, M. 1986, Phys. Rev. D, in press.

Schramm, D. N. 1985, Nucl. Phys., **B252**, 53.

Schramm, D. N. and Steigman 1984, Phys. Lett., **141**B 337.

Seckel, D. and Turner, M. 1985, Phys. Rev., **D32**, 3178.

Sikivie, P. 1983, Phys. Rev. Lett., **51**, 1415.

Silk, J. and Srednicki, M. 1984, Phys. Rev. Lett., **53**, 624.

Simpson, J. 1985, Phys. Rev. Lett., **54**, 1891.

Srednicki, M., Theisen, S., and Silk, J. 1986, Phys. Rev. Lett., in press.

Starobinsky, A. 1982, Phys. Lett., **117B**, 175.

Stecker, F., Rudaz, S., and Walsh, T. 1985, Phys. Rev. Lett., **55**, 2622.

Steigman, G. 1986, in Inner Space/Outer Space, eds. E. Kolb etal. (Univ. of Chicago Press, Chicago).

Steigman, G. and Turner, M. 1985, Nucl. Phys., **B253**, 375.

Steinhardt, P. and Turner, M. 1983, Phys. Lett., **129B**, 51.

Tremaine, S. and Gunn, J. 1979, Phys. Rev. Lett., **42**, 407.

Turner, M. 1985, Phys. Rev. Lett., **55**, 549.

Turner, M. 1986a, Phys. Rev. D, in press.

Turner, M. 1986b, Ann. NY Acad. Sci., in press.

Turner, M. 1986c, Univ. of Chicago/Fermilab preprint (submitted to Phys. Rev. D).

Turner, M. and Schramm, D. 1979, Nature, **279**, 1979.

Vilenkin, A. 1984, Phys. Rev. Lett., **53**, 1016.

Vilenkin, A. 1985, Phys. Rep., **121**, 263.

Weinberg, S. 1978, Phys. Rev. Lett., **40**, 223.

White, S. D. M. 1986a, in Inner Space/Outer Space, eds. E. Kolb etal, (Univ. of Chicago Press, Chicago).

White, S. D. M. 1986b, in these proceedings.

Wilczek, F. 1978, Phys. Rev. Lett., **40**, 279.

Witten, E. 1984, Phys. Rev., **D30**, 272.

Zel'dovich, Ya B. 1986, Sov. Phys. Uspekhi, **11**, 381.

DISCUSSION

DAVIS: You concluded your talk by discussing cosmic strings. A number
of us who do simulations are aware that they would make beautiful
structures and have non-random phases that might explain enhanced galaxy
clustering on large scales. But we've been inhibited from doing
simulations by the fact that they are laid down non-randomly. The
predictions are not yet very specific as to how to lay them down, how to
form the loops, how to set up the velocities, and so on. We need more
work in this field before it is reasonable to do simulations. Is
progress being made on these problems?

M. TURNER: Since we are treating this like a presidential press
conference, I guess I don't have to answer that question but can answer
some other question that I know the answer to (laughter and applause).
Seriously, I think progress is being made. I'd particularly like to
emphasize the work of Alex Vilenkin, who calculated the spectrum one
would expect. But that's not the only important thing; there is the
distribution of loop sizes and the question of non-random phases. I
believe that Neil Turok and collaborators are doing simulations. And
Bob Scherrer at Chicago is also trying to do a simulation, actually
laying down the loops and trying to get initial conditions for
simulations that one could run.

DATTA: Do superstring theories either specify or constrain the
cosmological constant?

M. TURNER: There is the hope that superstring theories will explain
everything, including the cosmological constant, but they haven't done
so yet. I would make one comment along these lines. The superstring
theories are supersymmetric theories. It has been known for a while
that in a supersymmetric theory, if you set the cosmological constant
equal to zero, you won't get ugly radiative corrections. Unfortunately,
that doesn't solve the whole problem. We know the world isn't
supersymmetric today, and the lowest supersymmetry-breaking scale is
such that you would still have an enormous cosmological constant. But
it is certainly one of our hopes that we will be able to explain the
cosmological constant.

PRIMACK: I don't know how seriously one should take this - I haven't
had a chance yet to absorb their paper - but Antoniadis, Kounnas and
Nanopoulos claim to prove that in their standard supersymmetry theory
there is an automatic cancellation that results in a cosmological
constant of zero. The theory allows inflation, and then resets the
cosmological constant to zero automatically. It looks like it is right.

SILK: Do you have any "warm" particle candidates? Cold dark matter
results in initial structure on scales much smaller than galaxies, and
hot dark matter on scales much larger. For some aspects of galaxy
formation theory it would be attractive to have a dark matter candidate
that yielded a scale for massive halos that was just right.

M. TURNER: Well, people have suggested that right-handed neutrinos might have masses in the keV range.

STEIGMAN: I can answer that. I was at CERN a few months ago; you can find many things in the drawers of offices at CERN. There is the axino, which is the superpartner of the axion. Also, in the no-scale supersymmetry theories of the CERN group the gravitino could have a mass of order a few hundred eV and so could be an ideal warm particle candidate.

GOTT: From the point of view of galaxy formation, it is worth emphasizing that cold dark matter particles are more useful to you than neutrinos. For instance, White's simulations with $\Omega = 0.2$ in dark matter explained a lot of things with very few parameters. In the paper on inflation that you wrote with Bardeen and Steinhardt you found that fluctuations coming into the horizon at a radiation-dominated epoch were a factor of ten larger than ones coming in when matter dominated. Then the extra logarithmic growth factors found by Jim Peebles allowed you to live with very low microwave background fluctuations. So you have an extra factor of almost a hundred in the growth of fluctuations in the cold dark matter scenario that you don't have in the neutrino models.

ALCOCK: Just a comment on the quark nugget hypothesis. The primary difficulty with quark nuggets is not the uncertainty of whether they form or not, but the fact that they evaporate when the Universe cools below 20 MeV unless they have planetary mass. That mass is out of line with any natural mass scale in that phase transition. So I wonder if you have any idea of how you might make very large nuggets.

M. TURNER: No, I don't. The natural scale at that phase transition is much smaller than the size of a planet. Unless you change the dynamics of the early Universe, slowing down the expansion to make the natural scale larger, it seems implausible to me that you would get boulder-sized quark nuggets.

SPERGEL: How much freedom is there in the axion mass if you want axions to close the Universe?

M. TURNER: If you express the axion mass in terms of $h = H_0/100$ km s^{-1} Mpc^{-1}, the microwave temperature, and the initial misalignment angle, you have at least a factor of two uncertainty in addition due to the finite temperature behavior of the axion mass. If you remember that the mass is related to the symmetry-breaking scale, you can see that there is quite a lot of uncertainty floating around before you even worry about the angle. After all, we don't know H_0 to better than a factor of two, while the microwave background temperature is known to a factor of ~ 1.1. Now throw in that factor of two uncertainty that you have from how the oscillation got started, and you're probably getting close to a factor of ten uncertainty before you even worry about what the angle was initially. And in an inflationary Universe it has different values in different bubbles. Just to give credit where credit is due, So-Young Pi was the first person to emphasize this very important point. You can, of course, push all the parameters in one direction and set an upper limit to how big the axion mass could be. If my numbers are correct, this is 5×10^{-5} eV to have $\Omega = 1$.

YAHIL: You mentioned the possibility that the decay of WIMPs may solve the problem of having $\Omega = 1$ despite the small infall velocity toward Virgo and the cosmic virial theorem. There's a very nice paper by George Efstathiou (1985, M.N.R.A.S., 213, 29P) that shows that this doesn't work, because the growth of velocity perturbations during the epoch before the decay is sufficiently large that WIMPs don't decay early enough to resolve the discrepancy.

M. TURNER: That's right. If you want our infall velocity toward Virgo to be small enough, then you derive a lower bound on the redshift of decay of $z_d > 10$. Now, Nicola Vittorio tells me that the microwave fluctuations in the decaying neutrino scenario give an upper bound of $z_d < 4$. So I asked George whether he could bring his limit down to four. He said that if pressed, maybe he could. So there is a conflict, but it is not clear that the conflict is big enough to rule out this scenario. Others, like Olive, Seckel and Vishniac, have also emphasized the possibility of decaying cold dark matter.

CARR: This morning, Martin Rees noted the slightly unsatisfactory feature of WIMP models that it requires a coincidence that the WIMP density is so close to the baryon density. The two candidates you mentioned that might obviate this difficulty are quark nuggets (although Alcock and others suggest that this scenario is unlikely), and shadow matter. For these it might be fairly natural that the baryon and WIMP densities are comparable. Can you comment?

M. TURNER: I certainly agree with Martin's concern about why the ratio of exotic stuff to ordinary matter is close to 1 and not $10^{\pm 122}$. But we have this problem: All we are left with now is debris, and we are trying to figure out how it all happened. We have to decide which facts are important and which are not. So it is not clear how seriously we should take this coincidence.

I'm not sure that shadow matter is an easy way out. The most intriguing possibility was that there exists a shadow world that's identical to ours. But when we wrote our paper on shadow matter, we ruled out this possibility on the basis of primordial nucleosynthesis.

REES: There is another cosmological number which is just as important as the photon-to-baryon ratio and Ω. That is the fluctuation amplitude, which is $\sim 10^{-5}$ in natural units. Now of course strings with the appropriate μ are a way of accounting for this value. Otherwise, it is presumably discussed by most people in the context of fluctuations during the inflationary phase. It is my impression that no-one yet has a model which naturally gives you the right value, although the models do give a scale-independent spectrum. Could you comment?

M. TURNER: I would echo what you've just said, that inflation predicts the Zel'dovich spectrum – that's a very general property – with slight deviations that don't appear to be very important. While you can concoct models to give you an amplitude of 10^{-5}, none of them is particularly compelling. In other words, there are proof-of-existence models, but none which jump out as being very, very pretty.

In this regard, I might just emphasize that I think inflation should be elevated from a scenario to a paradigm. In the past few years inflation has been found to be rather more general than symmetry-breaking phase transitions. Linde has emphasized that inflation need not be connected with a phase transition at all; you could just have some potential that starts away from its minimum. And people who look at compactification transitions, the transition from more than four dimensions to four dimensions, find that the transition can be inflationary. A student and I looked at an induced gravity model and found that that model inflates. In fact, inflation seems to be a rather general phenomenon. It may be so general that it had to happen. Maybe it has nothing to do with a symmetry-breaking phase transition. In that regard, if there are extra dimensions, and some are very small and others very large, then I find very compelling the idea that inflation might be responsible. That is, the Universe could have started out with all those extra dimensions, and then inflation could have made some of them much bigger than the others.

E. TURNER: There is an aphorism that says it is as important to know what you don't know as it is to know what you know. You drew a useful line, with which I roughly agree, on the astrophysical side, when you said that we had no compelling evidence for $\Omega > 0.15$ and no compelling astronomical need for anything other than baryons. I wonder if you could also draw a similar line for us from a physicist's point of view. If we came back in 30 years, what would it surprise you to have lost? Is Weinberg-Salam secure? Are GUTs secure? Is Kaluza-Klein secure?

M. TURNER: (Puts up his viewgraph of conclusions and indicates that nothing is secure. Loud laughter.) I would say that the standard SU(3) × SU(2) × U(1) model is as well tested as the standard cosmology back to $\sim 10^{-2}$ seconds. We have the W and Z bosons – everything's checking out just right. QCD is rather well checked out; we would like to be able to predict the masses of the proton, neutron and so on, but that has eluded us because of terrible non-linearities in the theory. But even within QCD and the Weinberg-Salam model, there are nagging questions. Why are there these two sets of particles, quarks and leptons, and why are they patched together with multiplication signs, rather than in some nicer way? I think something more is called for, and grand unification is a compelling idea. One of the great successes of grand unification is its ability to predict the so-called Weinberg angle – this happens in the SU(5) model. So I think most people believe that something like this must happen. Then the picture gets a bit hazier, when we try to unify the gauge forces that we already know. And the logical progression from these is to try to patch in gravity.

I'd like to make a comment about superstring theory, because it represents a real philosophical jump in thinking such as hasn't been made since Einstein. That jump is to try to guess the theory of the world. Instead of working your way up from the bottom, we guess from the top. We guess the theory of everything at energies of 10^{19} GeV and above, in spite of the fact that we don't live there, and then work everything down to where we live. People think this is very promising. But maybe it doesn't have anything to do with reality. There was one possibility that I didn't mention, and that is that quarks and leptons are themselves not fundamental. It may be that the path to unification proceeds that way.

BURKE: What does no proton decay threaten?

M. TURNER: It doesn't really threaten a whole lot.

BURKE: They said it would! (laughter)

M. TURNER: Well, proton decay is a generic prediction of grand unified theories. The lifetime of the proton depends, among other things, on the fourth power of a symmetry-breaking scale. Now it turns out that in the simplest grand unified theory, SU(5), you can calculate that scale rather accurately, so that the lifetime is predicted to within an order of magnitude or two. Everyone was very excited, because the answer fell in the window between experiments that had already been done and experiments that could never be done. Today this simplest grand unified theory is probably at a point where 50% of the community would say it has been falsified. But the long proton lifetime doesn't have much effect on more general grand unification, because pushing the symmetry-breaking scale up by a factor of three raises the proton lifetime by a factor of 100 and puts it into the inaccessible region. And when you construct a supersymmetric grand unified theory, the unification scale usually goes up automatically. So if we didn't observe proton decay, it would just be rotten luck at this point; it wouldn't say that there is no grand unification.

THE EVAPORATION OF STRANGE MATTER IN THE EARLY UNIVERSE

Charles Alcock and Edward Farhi
Massachusetts Institute of Technology

A new candidate for the dark matter of the universe is strange matter.[1] This substance consists of roughly equal numbers of up, down and strange quarks confined in a quark phase which is <u>conjectured</u> to have a lower energy per baryon number than ordinary nuclei. Strange matter is absolutely stable, has a density comparable to that of nuclei and can exist in lumps ranging in size from a few fermis to ~ 10 km. If it is distributed in space in lumps larger than ~ 1 cm, it could close the universe without ever encountering the earth and would be astronomically unobservable.[2]

A lump of strange matter contains ordinary quarks (up, down and strange) and gluons plus a small component of electrons to guarantee charge neutrality. The hadronic material is in a "quark phase" in which nucleons and mesons do not exist and the quarks are free to roam within the lump.[3] Witten[1] suggested that this form of matter could be absolutely stable. A detailed study[3] has shown that within the uncertainties inherent in a strong interaction calculation, the existence of stable strange matter is reasonable.

Witten[1] also outlined a scenario for the production of strange matter in the early universe. The production occurred when the universe cooled through the QCD phase transition at a temperature T_C (roughly, 100-200 MeV). Witten's scenario has been criticized[4], but the physics involved is sufficiently complex that a clear determination of the outcome of the QCD phase transition is unlikely to appear soon. We avoided these uncertainties by examining the fate of lumps of strange matter at temperatures below 100 MeV, assuming that strange matter is formed during the transition.

The strange matter efficiently evaporates neutrons and some protons. The evaporation rate is very efficient, being limited primarily by the rate of heating by thermal neutrinos. We are able to strongly conclude that a lump of strange matter can only survive if its baryon number $A \geq 10^{52}$.

This number is large in an important sense cosmologically. If we assume the universe is closed by baryons in either the strange or normal phase, then the mean baryon number in the particle horizon at these early epochs is ~ $10^{55} (1 \text{ MeV}/T_u)^3$. At T_u = 50 MeV this number is ~ 8×10^{49}, much smaller than the minimum baryon number of a lump which could survive. This means that the process that leads to the formation of strange matter lumps must involve large perturbations in the baryon number on the horizon scale. A mechanism for producing this perturbation is not known to us.

1. E. Witten, *Phys. Rev.* <u>D30</u>, 272 (1984).
2. A. DeRujula and S. Glashow, *Nature* <u>312</u>, 734 (1984).
3. E. Farhi and R. L. Jaffe, *Phys. Rev.* <u>D30</u>, 2379 (1984).
4. J. Applegate and C. Hogan, Cal Tech preprint GRP032 (1984).

J. Kormendy and G. R. Knapp (eds.), Dark Matter in the Universe, 489.

DETECTING COLD DARK MATTER CANDIDATES

A.K. Drukier, K. Freese and D.N. Spergel
Department of Astronomy, Harvard University
60 Garden Street
Cambridge, Massachusetts 02138
U.S.A.

ABSTRACT. We consider the use of superheated superconducting colloids as detectors of weakly interacting galactic halo candidate particles (e.g. photinos, massive neutrinos, and scalar neutrinos). These low temperature detectors are sensitive to the deposition of a few hundreds of eV's. The recoil of a dark matter particle off of a superheated superconducting grain in the detector causes the grain to make a transition to the normal state.[1] Their low energy threshold makes this class of detectors ideal for detecting massive weakly interacting halo particles.[2]
    We discuss realistic models for the detector and for the galactic halo. We show that the expected count rate ($\approx 10^3$ count/day for scalar and massive neutrinos) exceeds the expected background by several orders of magnitude. For photinos, we expect ≈ 1 count/day, more than 100 times the predicted background rate. We find that if the detector temperature is maintained at 50 mK and the system noise is reduced below 5×10^{-4} flux quanta, particles with mass as low as 2 GeV can be detected. We show that the earth's motion around the Sun can produce a significant annual modulation in the signal.[3]

References

[1]Drukier, A.K. and Stodolsky, L. Phys.Rev.D, 30, 2295 (1984).
[2]Goodman, M.W. and Witter, E. Phys.Rev.D, 30, 3059 (1958).
[3]Drukier, A.K., Freese, K., and Spergel, D., submitted to Phys.Rev.D.

J. Kormendy and G. R. Knapp (eds.), Dark Matter in the Universe, 490.

CONSTRAINTS ON DARK MATTER DENSITY AND AXION MASS FROM THE LARGE-SCALE STRUCTURE OF SPACETIME

P.S. Joshi* and B. Datta**
* Tata Institute of Fundamental Research, Bombay, India
** Indian Institute of Astrophysics, Bangalore, India

On the basis of general properties of the large-scale structure of spacetime, we present new and general theoretical upper limits on the density of dark matter in the Universe, assuming a 90% content for the dark matter, and lower limits on the mass of the axion, assuming the dark matter to be made up of axions. These limits are derived in terms of the possible lower limits to the age of the Universe and the Hubble parameter. We find that for the age in the range $(8 - 24) \times 10^9$yr, the maximum density of dark matter is in the range $(1.25 \times 10^{-28} - 1.38 \times 10^{-29})$ g cm^{-3} and the minimum value of axion mass in the ranges $(0.36 - 2.39) \times 10^{-5}$ eV and $(1.44 - 9.51) \times 10^{-5}$ eV.

J. Kormendy and G. R. Knapp (eds.), Dark Matter in the Universe, 491.

BARYOGENESIS IN THE INFLATIONARY UNIVERSE

Hideo Kodama, Katsuhiko Sato and Nobuaki Sato[*]
Department of Physics, University of Tokyo, Japan
*)Department of Astronomy, University of Tokyo, Japan

As is known well, the inflationary universe model resolves most of the fundamental problems concerning the large scale structure of the universe and is now becoming a standard model for the early universe. However, there is one important problem yet to be made clear. In this model the number density of particles effectively goes to zero during the inflation and everything is created after the universe is heated up again at the end of inflation. Since the reheating temperature is much lower than the GUT temperature in general, however, it is not clear whether the observed baryon asymmetry is generated in this process.

In order to make clear this problem we investigated how baryogenesis depends on the reheating temperature T_i in the SU(5) GUT model. We numerically traced the development of the abundance of heavy bosons and the asymmetry in quarks and leptons after reheating treating T_i and the superheavy Higgs boson mass M_H as free parameters. The calculation is carried out until the asymmetry in quarks and lepton are frozen. We found that T_i should be greater than 5×10^{10} GeV for the observed asymmetry to be generated.

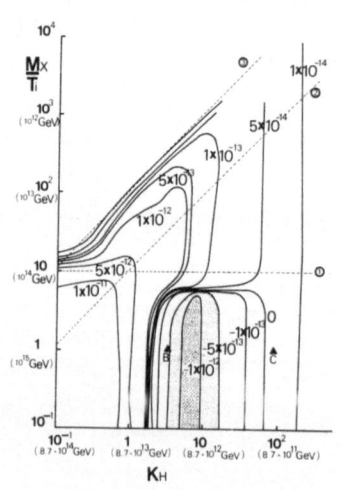

Fig. 1 The contour map of the final baryon/entropy ratio on M_H-M_X/T_i plane. The sign of n_B/s is negative in the dotted region.

Besides this general conclusion we found an interesting phenomenon. For some range of CP-breaking parameters there appear regions in M_X/T_i-M_H plane(M_X is the superheavy gauge boson mass) in which the baryon asymmetry has different signs, as shown in Fig.1. Since the boundary layer of these regions is very narrow, small fluctuations of the reheating temperature are expected to produce large fluctuations of baryon/entropy ratio. This yields a possible mechanics of generating isothermal density perturbations, and dividing the Universe into domains of baryons and antibaryons.

Reference

N.Sato, H.Kodama and K.Sato, Prog. Theor. Phys. **74**(1985)no.2.

J. Kormendy and G. R. Knapp (eds.), Dark Matter in the Universe, 492.

COSMOLOGICAL CONSEQUENCES OF AN UNSTABLE HEAVY NEUTRINO

T. Padmanabhan and M.M. Vasanthi
Tata Institute of Fundamental Research, Bombay

In a recent paper Simpson has reported evidence for a heavy neutrino of about 17.1 kev mass. Cosmological bounds on stable neutrino species imply that this neutrino [ν_H] must be unstable. The most likely decay mode $\nu_H \rightarrow \nu_L + f$ where ν_L is a light neutrino and f is a scalar boson leads to a cosmological scenario which is quite different from the conventional picture. In this scenario universe becomes <u>matter dominated</u> at a redshift of z ~ 10^7 and becomes <u>radiation dominated</u> [by the decay product ν_L of ν_H] at z ~ 310. The kinematic constraints on the lifetime of ν_H do not lead to any contradictions. On the other hand, growth of baryonic perturbations is severely limited in this model due to two reasons: (i) virtually no growth can take place in the radiation dominated region z < 310.   (ii) Decay of ν_H is likely to disrupt and smoothen out past growth by a large factor. It is doubtful whether a simple way out of this difficulty exists.

Reference.

J.J. Simpson, <u>Phys. Rev. Letts.</u>, <u>54</u>, 1891 (1985).

J. Kormendy and G. R. Knapp (eds.), Dark Matter in the Universe, 493.
© *1987 by the IAU.*

PROSPECTS FOR AXION DETECTION

John Moody

Institute for Theoretical Physics, UCSB

Axions[1] are a natural consequence of the only known solution of the "Strong CP Problem."[2] The axion represents a minimal $U(1)$ extension to the standard $SU(3) \times SU(2) \times U(1)$ model of the strong and electroweak interactions and furthermore appears very frequently in the contexts of SUSY GUTs, Kaluza-Klein, Superstring, Technicolor, and other unified models. As such, the axion is perhaps the best motivated conjectured particle.

Constraints from nuclear and particle physics experiments[3] and stellar evolution[4] have suggested that the axion must have a very low mass and must interact extremely weakly with ordinary matter. Such weakly-interacting axions are known as "invisible axions";[5] they are predicted to have masses of less than 10^{-2} eV. The fact that these axions are so weakly interacting means it is not possible to detect them using the traditional techniques of particle physics.

Invisible axions may play a central role in cosmology. If the mass of the axion is of order $\sim 10^{-5}$ eV, then standard axion cosmology scenarios predict that a primordial background of axions provides the closure density.[6] This axion background is formed while the universe is at a temperature of $\sim 10^{12}$ GeV; since the axion background interacts very weakly with ordinary matter, it rapidly departs from thermal equilibrium as the universe expands. The axion background behaves as a high occupation number (classical), coherent, oscillating pseudoscalar field, and has been rightfully dubbed the "cosmic harmonic oscillator." Because axions departed from thermal equilibrium at a very early epoch, they provide a cold dark matter background for galaxy formation. If our galactic halo is composed of axions, the local number density is expected to be $\sim 10^{13}$ per cm^3.

The hope that axions **can be detected** is based not upon single particle interactions but rather upon newly proposed macroscopic techniques in which large numbers of axions and conventional particles interact coherently. There are two classes of experiments: those which search for galactic halo axions and those which attempt to measure macroscopic forces mediated by virtual axions.

Galactic halo axions can be detected by converting them to microwave photons. Sikivie first suggested that this be accomplished in a microwave cavity via the axion-photon-photon coupling ($a\mathbf{E} \cdot \mathbf{B}$) by using a background magnetic field as a catalyst.[7]

In such an experiment, non-relativistic axions, say of mass 10^{-5} eV, are converted coherently into a microwave signal of frequency 2.4 GHz which is observed classically. The difficulty in such experiments stems from our ignorance of the precise value of the axion mass, which requires that several orders of magnitude in frequency be searched. Several authors have analyzed in detail the sensitivity required for such search experiments.[8,9] Given present technology, a halo density of axions may be detectable, but these experiments will not be sensitive enough to categorically rule-out axions as being important for cosmology. At this writing,

494

J. Kormendy and G. R. Knapp (eds.), Dark Matter in the Universe, 494–495.

three experimental groups are building magnetic field coupled detectors,[10] and certain improvements in the original designs of Sikivie have been suggested.[11]

The inherent difficulty of these experiments suggests the need to examine other possibilities. It was recently proposed that larger conversion rates could be achieved in principle by using aligned electron spins as the catalyst.[12,8] Experiments based upon electron spins are difficult to analyze, however, because the conversion rates depend sensitively upon material properties.[13] Various possibilities have been suggested including the use of exotic ferrites, exotic paramagnetic salts, spin-polarized conductors, ferromagnetic resonance in YIG, and so on, but no workable schemes have emerged as yet.[14]

Experiments to detect axions via the macroscopic forces which they mediate[15] provide an independent test for the existence of those axions which happen to be of the greatest interest cosmologically, those with masses in the range around 10^{-5} eV. Such axions are expected to mediate forces over a range given by their Compton wavelength $\lambda \sim 2cm$. Three types of forces are possible: a Yukawa force between nucleons, a tensor force between spins, and a very novel P- and T-violating force between nucleons and spins. The presence of these forces can be probed by using the extremely sensitive techniques of experimental gravity.

REFERENCES

1. S. Weinberg, Phys. Rev. Lett. **40**, 223 (1977); F. Wilczek, Phys. Rev. Lett. **40**, 279 (1977).

2. R.D. Peccei and H. Quinn, Phys. Rev. Lett. **38**, 1440 (1977).

3. T. Donnelly, et al. , Phys. Rev. **D18**, 1607 (1978).

4. M. Fukugita, S. Watamura, M. Yoshimura, Phys. Rev. Lett. **48**, 1522 (1982), and Phys. Rev. **D26**, 1840 (1982); L. Krauss, J. Moody, F. Wilczek, Phys. Lett. **144B**, 391 (1984); N. Iwamoto, Phys. Rev. Lett. **53**, 1198 (1984); D. Morris (LBL report 18690, 1984).

5. J.E. Kim, Phys. Rev. Lett. **43**, 103 (1979); M. Dine, W. Fischler, and M. Srednicki, Phys. Lett. **104B**, 199 (1981); M. Shifman, A. Vainshtein, and V. Zakharov, Nucl. Phys. **B166**, 493 (1980).

6. J. Preskill, M.B. Wise, and F. Wilczek, Phys. Lett. **120B**, 127 (1983); L.F. Abbott and P. Sikivie, Phys. Lett. **120B**, 133 (1983); M. Dine and W. Fischler, Phys. Lett. **120B**, 137 (1983).

7. P. Sikivie, Phys. Rev. Lett. **51**, 1415 (1983), and **52**, 695, (1984).

8. L. Krauss, J. Moody, F. Wilczek, and D. Morris, ITP preprint NSF-ITP-85-76.

9. P. Sikivie, University of Florida preprint (1985).

10. A. Melissinos, et al. , private communication (1985); P. Sikivie, N. Sullivan, D. Tanner, Experimental proposal (1984); P. Lubin, D. Morris and C. Pennypacker, Experimental Proposal (1985).

11. D. Morris, LBL Report LBL-17915 (1984).

12. L. Krauss, J. Moody, F. Wilczek, and D. Morris, ITP preprint NSF-ITP-85-08.

13. J. Slonczewski, IBM preprint (1985); private communications of C. Bennett, S. Coleman and N. Fortson.

14. Private communications of C. Bennett, G. Chapline, N. Fortson, J. Moody, D. Morris, J. Slonczewski, N. Snyderman, and F. Wilczek.

15. J. Moody and F. Wilczek, Phys. Rev. **D30**, 130 (1984).

THE MISSING MASS AND THE SOLAR NEUTRINO PROBLEM

David Spergel
Department of Astronomy, Harvard University
60 Garden Street
Cambridge, Massachusetts 02138
U.S.A.

ABSTRACT. If the halo of our galaxy is composed of weakly interacting particles, they will be captured by the sun.[1] If the mass of these particles exceed 5 proton masses, they will remain in the Sun where they will serve as an effective means of transporting energy in the solar core. They will make the Sun's core more nearly isothermal, thus decreasing the rate of the PPIII reaction.[2,3] If the halo is composed of particles with masses between 5 and 10 GeV and cross section between 10^{-34} and $10^{-37} cm^2$, this mechanism could resolve the solar neutrino problem.[4] If these particles exist, they could be detected by a low temperature detector.[5] However, if the particles annihilate in the Sun, (e.g. Photinos or Scalar Neutrinos), their number density will be too low.[6]

References

[1]Press, W.H. and Spergel, D.N. Ap.J., 296, 000 (1985).
[2]Spergel, D.N. and Press, W.H. Ap.J., 294, 000 (1985).
[3]Gilliland, R. and Faulkner, J. Ap.J., 299, 000 (1985).
[4]Gilliland, R., Faulkner, J., Press, W.H., and Spergel, D.N., in preparation.
[5]Drukier, A.K., Freese, K. and Spergel, D.N., submitted to Phys.Rev.D.
[6]Krauss, L.M., Freese, K., Spergel, D.N. and Press, W.H. Ap.J., 299, 000 (1985).

J. Kormendy and G. R. Knapp (eds.), Dark Matter in the Universe, 496.

UNSTABLE DARK MATTER AND GALAXY FORMATION

Yasushi Suto, Hideo Kodama and Katsuhiko Sato
Department of Physics, The University of Tokyo, Japan

Recently cosmology with unstable particles has attracted much attention as a possible solution to several important problems in the present universe. In this scenario, however, nonlinear structures in the universe would not easily form since their binding energy would decrease gradually in the course of the decay of unstable particles. Thus this scenario is stringently constrained by the galaxy formation problem. In order to obtain the constraints quantitatively, we carried out numerical calculations of the evolution of density perturbations after the time of recombination. Universe is assumed to consist of three components; unstable X-particles, its non-radiative decay product (massless particles) and baryons. Initial conditions are specified by the density contrast of X-particles $\delta_{X,i}$ and the baryon density parameter $\Omega_{b,i}$. Resultant baryon density contrast at $Z = 4$ (epoch of galaxy formation) and $Z = 0$ (present) is shown below for $\Omega_{b,i} = 0.005$.

The amplitude of $\delta_{X,i}$ at recombination is constrained from an isotropy of cosmic microwave background radiation. The resultant constraints for adiabatic perturbations are as follows; $\delta_{X,i} < (10^{-3} \sim 10^{-1})$ for galaxies and clusters of galaxies and $\delta_{X,i} < (10^{-4} \sim 10^{-3})$ for superclusters. Thus from the results shown in the figures, we conclude that the adiabatic perturbations can account for the formation of galaxies and clusters of galaxies if τ_X(decay life of X-particles) $> 10^{15}$ sec. This is based on the rapid growth of density perturbations in a nonlinear stage. The formation of supercluster size objects, however, is very difficult in this scenario for any values of τ_X.

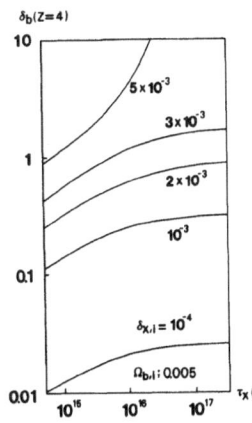 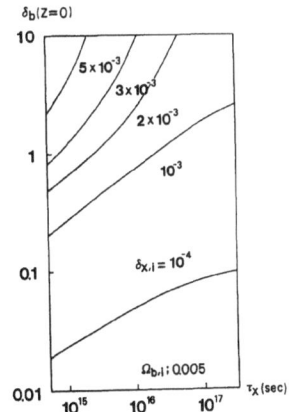

References

Suto,Y., Kodama,H. and Sato,K.: 1985, Phys. Lett. 157B,259.
Suto,Y., Kodama,H. and Sato,K.: 1985, submitted to MNRAS.

J. Kormendy and G. R. Knapp (eds.), Dark Matter in the Universe, 497.

DO BARYON STRUCTURES SURVIVE AFTER THE DECAY OF UNSTABLE PARTICLES ?

Yasushi Suto[1] and Masafumi Noguchi[2]
1 Department of Physics, The University of Tokyo, Japan
2 Tokyo Astronomical Obs., The University of Tokyo, Japan

A possibility of a universe dominated by unstable particles has been suggested recently as an interesting solution to various cosmological problems. In this scenario, however, nonlinear structures may be disrupted or even smoothed out during the decay of unstable particles. In order to obtain the survival condition of baryon structures, we studied evolution of systems composed of baryon and unstable particles by the numerical integration of the Vlasov equation.

The initial behavior of the system depends on the ratio of the lifetime of the unstable particles τ_X to the initial free-fall time scale $t_{ff,i}$ of the system; if $\alpha \equiv \tau_X/t_{ff,i} \lesssim 1$, then the system has been smoothed out without reaching the equilibrium state (see Fig.1; the numbers in the figure denote the time in units of $t_{ff,i}$). On the contrary, the system with $\alpha \gg 1$ once reaches the nearly equilibrium configuration and afterwards evolves adiabatically (Fig.2). In the latter case, it is easy to show that the characteristic radius $R_c(t)$ and velocity dispersion $\sigma_c(t)$ of the system change as $M_{tot}^{-1}(t)$ and $M_{tot}(t)$, respectively. The survival condition of structures is found as $\alpha > 1/(3\beta^2)$, where β is the ratio of the energy densities of baryons and unstable particles before decay.

For cosmologically plausible ranges of τ_X and β, galaxies satisfy this condition and their evolution is understood very well by an adiabatic behaviour. On the other hand, clusters of galaxies have a possibility of disruption in future due to the decay of unstable particles.

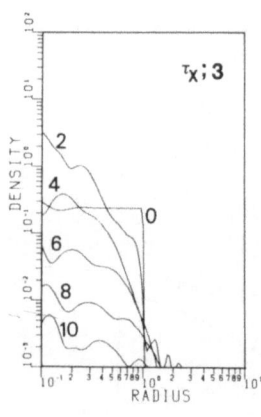

Fig.1

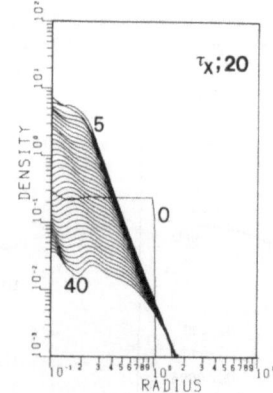

Fig.2

References
Suto,Y., Kodama,H. and Sato,K.: 1985, Phys. Lett. 157B,259.
Suto,Y. and Noguchi,M.: 1985, submitted to MNRAS.

J. Kormendy and G. R. Knapp (eds.), Dark Matter in the Universe, 498.

CONSTRAINTS ON DARK MATTER FROM PRIMORDIAL NUCLEOSYNTHESIS

Jean AUDOUZE

Institut d'Astrophysique du CNRS, Paris, France
and Laboratoire René Bernas, Orsay, France.

ABSTRACT. Primordial nucleosynthesis which is responsible for the formation of the lightest elements (D, ^{3}He, ^{4}He and ^{7}Li) provides a unique way to determine the present baryon density ρ_B in the Universe and therefore the corresponding cosmological parameter Ω_B. After a brief summary of the relevant abundance determinations and of the consequences of the Standard Big Bang nucleosynthesis, it is argued that one needs to call for specific models of chemical evolution of the Galaxy in order to reconcile the observations with the predictions of this model. In this context the predicted values for Ω_B should range from $4\ 10^{-3}$ to $6\ 10^{-2}$. These values are significantly lower than those deduced from current M/L determinations.
   In order to reconcile the early nucleosynthesis with larger values of Ω_B (i.e. with the presence of dark matter) two scenarios departing from the Standard Big Bang models are presented : they are (i) the possible partial photofission of ^{4}He (and ^{7}Li) into D and ^{3}He induced by energetic photons coming from the decay of massive (500 MeV) neutrinos and/or gravitinos (ii) the existence of some semi baryonic form of matter referred to as "quark nuggets". In these two cases the formation of the very light elements could be consistent with values of Ω as large as 1, i.e. similar to those suggested by some of the determinations collected in this book and which are favoured by the current models of inflationary Universes.

1. INTRODUCTION

The paramount importance of the formation processes of the very light elements (D, ^{3}He, ^{4}He and ^{7}Li) in cosmology and in particular in the determination of the baryonic density of the Universe (and also of the maximum number of neutrino families) has been pointed out by a large number of papers, including e.g. Yang et al. 1984, Boesgaard and Steigman 1985, Audouze 1984. Primordial nucleosynthesis occurs at a time of about 100 sec after the critical event from which Big Bang

J. Kormendy and G. R. Knapp (eds.), Dark Matter in the Universe, 499–523.

originates and might constitute one of the most severe constraints regarding the presence of dark matter in the Universe.·

This presentation starts with a brief review concerning the current status of the abundance determinations of the relevant elements. Then it is argued that if the primordial nucleosynthesis occurs according to the standard Big Bang model and is followed by the most classical galactic evolution models such as those used by Audouze and Tinsley (1974), the baryonic density deduced from the primordial ^{4}He abundance is significantly lower than that which come from the inferred primordial D. This why Delbourgo-Salvador et al. (1985) (see also Gry et al. 1983) have considered further galactic evolution models which lead to primordial abundances of D and ^{3}He in agreement with that of ^{4}He and provide a consistent range of values for the cosmological parameter Ω. However one should draw attention on the fact that the deduced value of Ω is significantly lower than that inferred in other papers. For instance Boesgaard and Steigman (1985) quote $0.014 < \Omega < 0.19$. After a discussion concerning this discrepancy, this presentation ends with two proposals for reconciling the early nucleosynthesis with larger values of Ω. The first one concerns some possible partial photofission of ^{4}He (and ^{7}Li) into D and ^{3}He which could be triggered by the decay of massive (500 MeV) neutrinos and/or gravitinos (as suggested by Audouze et al., 1985). The second calls attention on the possible existence of heavy "quark nuggets" which may come naturally from quantum chromodynamic theories (Witten 1984) and which could constitute the most abundant form of existing matter (Schaeffer et al., 1985).

2. ABUNDANCE DETERMINATIONS OF THE LIGHT ELEMENTS

The abundances of D, ^{3}He, ^{4}He and ^{7}Li have been thoroughly measured and analyzed in a large number of articles. References can be found in Audouze 1982 and 1984, Boesgaard and Steigman 1985 and in the conference proceedings edited by Shaver et al. in 1984.

2.1. Deuterium

The deuterium abundance can be determined either in the interstellar medium from its UV (950 Å) Lyman line (see e.g. Vidal Madjar et al., 1984) or in the solar system mainly through the determination of the ^{3}He/^{4}He ratio in the solar wind (Geiss and Reeves 1972). The interstellar D/H ratio is about $(1 \pm 0.5) \ 10^{-5}$. These abundances are therefore still quite uncertain mainly because of the complexity of the interstellar lines of sight. Moreover because D is a very fragile nuclear species (it is transformed into ^{3}He in H rich regions at T > a few 10^5K) its abundance is very dependent on the stellar and galactic evolution.

2.2. ^{3}He

The ^{3}He abundance is determined in the solar system also in the solar wind and in some gas rich meteorites ^{3}He/H~1.4±0.4 10^{-5} (D+^{3}He/H~ 3.6±0.6 10^{-5}). Rood et al. 1984 have attempted to determine the interstellar ^{3}He$^+$/^{4}He abundance by observing the 8.7 GHz line of ^{3}He$^+$ in a few galactic H II regions. The corresponding ^{3}He/H ratios range from less than 2 10^{-5} (for W49 and M17A) to 5 10^{-4} for W3. This very large variation of the ^{3}He/H ratio from one H II region to another clearly shows that this interstellar abundance is still badly determined. Because ^{3}He is the out-product of D, its abundance depends also much on the stellar and galactic evolution.

2.3. ^{4}He

The ^{4}He abundance has been observed in astrophysical sites as different as the Sun and Jupiter, the solar wind, the stellar atmospheres, planetary nebulae, the globular clusters and the HII regions of blue compact galaxies (see e.g. the book of Shaver et al., 1984). The most often adopted primordial ^{4}He abundance comes from the analysis of Kunth and Sargent (1984), i.e. Y~0.245±0.005. Given the cosmological importance of an accurate determination of this abundance (see following sections) two remarks should be made at this point : (i) Given the large spread of the abundances observed in blue compact objects, the uncertainty on the primordial He abundance deduced by Kunth and Sargent 1984 might be larger than that quoted by these authors (ii) Davidson and Kinman (1985) have made a careful analysis of the He abundance in IZW18 and deduce from their analysis a value Y~0.23±0.02 (iii) Vigroux et al. (1985) have shown recently that one could deduce a primordial $\overline{Y_p}$ value of 0.24 from the He/H versus O/H correlation but a much lower value of 0.20 from the He/H, N/O correlation. I do not claim here that Y_p has such a low value but I do want to call attention on the fact that this primordial value is still uncertain (at least more than what is quoted in the current literature).

2.4. Lithium 7

Spite and Spite (1983) who analyzed the Li abundance in population II stars deduce from it a primordial Li abundance of (^{7}Li/H)~10^{-10}, i.e. ten times lower than its value in some young F stars and in the solar system. Given the fact that convection and diffusion processes which could affect the atmosphere composition of such stars are not yet properly understood it might be still possible that the primordial Li/H ratio is as high as 10^{-9} as previously thought e.g. by Reeves 1974.

Table 1 provides our estimates of the interstellar, solar system and primordial abundances of the relevant light elements. As said above, the primordial abundances of D and ^{3}He are very dependent on the galactic evolution models and will be discussed in section 4.

Element	Primordial Abundance	Solar system Abundance	Interstellar abundance
D	$3\ 10^{-5}$–$3\ 10^{-4}$	$(3\pm1)10^{-5}$	$3\ 10^{-6}$–$2\ 10^{-5}$
^{3}He	$2\ 10^{-5}$–$6\ 10^{-5}$	$(4\pm2)10^{-5}$	$4\ 10^{-5}$–$2\ 10^{-4}$
^{4}He	$0.22 - 0.25$	0.15–0.24	$0.22 - 0.30$
^{7}Li	$(6\pm0.3)10^{-10}$	$\sim10^{-8}$	$7\ 10^{-10}$–$2\ 10^{-9}$

TABLE 1.
Abundances (by mass) of the light element produced
by the primordial nucleosynthesis

3. THE CONSEQUENCES OF THE STANDARD MODEL

The primordial nucleosynthesis occuring under the assumptions made by the Standard (also called canonical) model has been thoroughly discussed in the literature (see e.g. Audouze 1984, Yang et al. 1984, Boesgaard and Steigman 1985). Figure 1 taken from Yang et al. (1984) displays the well known dependence of the primordial abundances of D, ^{3}He, ^{4}He and ^{7}Li with the baryonic density $\eta=n_{B}/n_{\gamma}$ (where n_{B} and n_{γ} are respectively the baryon and photon densities). I will only remind that two important parameters can be deduced from this model :
(i) the maximum number of neutrino (or lepton) families especially sensitive to the primordial abundance of ^{4}He (Y_{p}) : an increase (or a decrease) of Y_{p} by 1% corresponds to an increase (or a decrease) of this number by one unit : If $Y_{p}<0.23$ and $N_{\nu}<2$, one can see on Figure 2 that such a low Y_{p} (which cannot be ruled out by the present observations) would imply very large D_{p} values (and therefore very low baryonic densities as we will see below) and that neutrinos can exist only at most in two separate components (the tau neutrino ν_{τ} should therefore be considered as a mixture of the ν_{e} and ν_{μ} states). If $Y\sim0.24\pm0.010$ there is of course a very good agreement between the "canonical" primordial nucleosynthesis and the three families of neutrinos. This agreement is invoked as an argument in favour of Grand Unification Theory schemes which relate the three lepton families to the three quark families (see e.g. Fayet 1984).
(ii) We are especially concerned here by the constraints put by the canonical primordial nucleosynthesis on the present baryonic

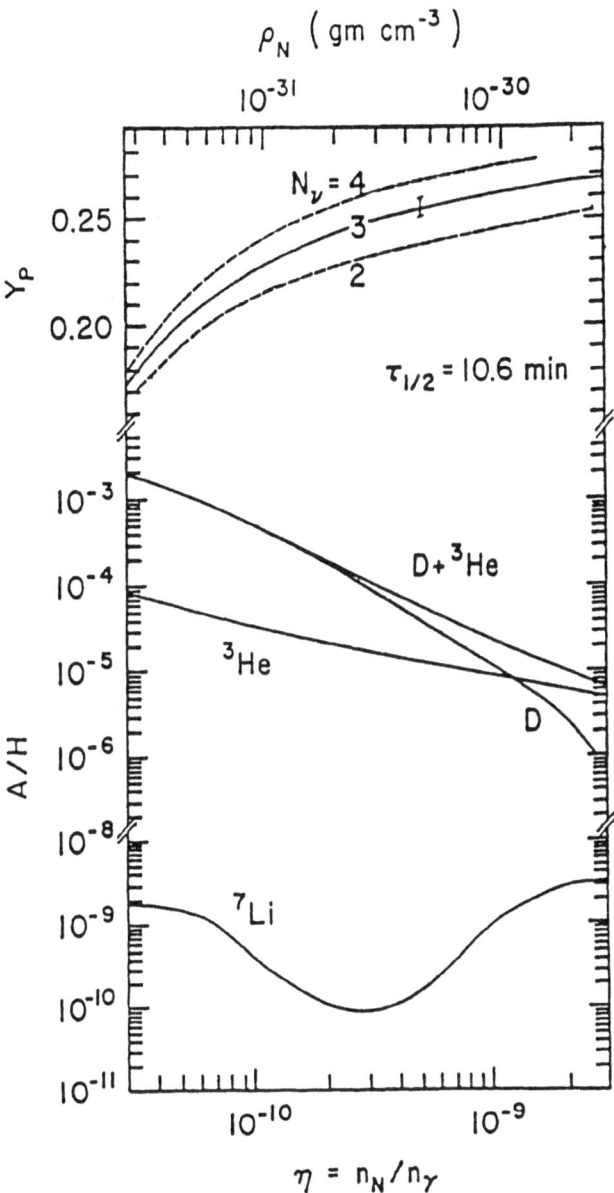

Figure 1
      Abundances of D, ^{3}He, ^{4}He and ^{7}Li predicted by the Standard Model
(see text) against the baryon to photon ratio η. The three curves for
the ^{4}He abundance (Y) correspond respectively to 2, 3 and 4 different
neutrino families (from Yang et al., 1984).

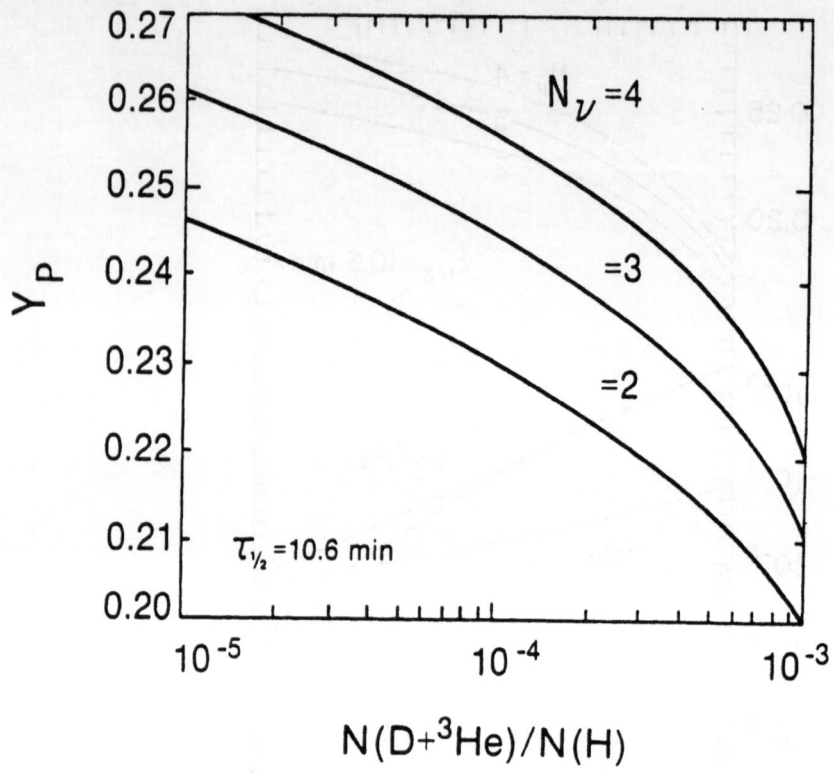

<u>Figure 2</u>
    Predicted [4]He abundance (Y_p) against the predicted $N(D + {}^3He)$
abundance for 2, 3 and 4 different neutrino families (from Yang <u>et
al</u>., 1984). This diagram shows that if $Y_p < 0.22$ the Standard Model
runs into difficulty.

density of the Universe and therefore on its overall dynamics : given the critical density $\rho_c = 3H_0^2/8\pi G = 1.88 \ 10^{-29} h_0^2 gcm^{-3}$ (h_0 is the Hubble constant H_0 expressed in units of 100 $kms^{-1} Mpc^{-1}$) corresponding to a flat Universe and to a cosmological parameter $\Omega=1$, Ω_B (the baryonic cosmological parameter)$=3.53 \ 10^{-3} \ h_0^2 \ \theta^3 \ \eta_{10}$ ($\theta=T/2.7K$, $\eta_{10} \ 10^{10} \ \eta$). Boesgaard and Steigman 1985 following Yang et al. (1984) consider that there is a good agreement between the primordial abundances deduced from observations and those calculated for $3 < \eta_{10} < 10$ which leads to $0.01 < \Omega_B < 0.19$. This estimate is clearly less than that inferred from the dynamics of large structures $\Omega \geqslant 0.3$ but is consistent with the values of Ω deduced from the dynamics of small groups of galaxies. Should the cosmological parameter Ω be larger than 0.2 either to account for large M/L ratio determinations or to satisfy the inflationary scenarios according which $\Omega=1$, there should be a significant fraction of the matter in non-baryonic form.

In contrast with the optimistic view presented by Yang et al. (1984) who argue in favour of such a good agreement, one may give credit to the point made by Vidal-Madjar and Gry 1984 who claim that the comparison between the observational deductions and the calculations regarding D and ^{4}He respectively may lead to discrepant values of η. This is the case if Y_p ranges between 0.22 and 0.25 and if only 30% to 70% of D is destroyed during the galactic history as it has been evaluated by Audouze and Tinsley 1974 in the frame of simple models of chemical evolution of galaxies. This is the reason why we have examined somewhat different models of that type in order to examine what could be the necessary conditions to restore such an agreement.

4. CHEMICAL EVOLUTION OF D AND ^{3}He AND PRIMORDIAL NUCLEOSYNTHESIS

This section is a brief account of the analysis of some models of galactic evolution concerning D and ^{3}He recently proposed by Delbourgo-Salvador et al. 1985. While ^{7}Li and especially ^{4}He are not as significantly affected by such processes, it is possible to design some models leading to the destruction of most of the primordial D during the galactic history.

Two types of models have been considered :
(i) the possibility of admixture (by infall or inflow) of stellar processed (i.e. D free) material in the considered galactic zone.
(ii) the mass loss of D free material released during the pre main sequence phase,
(i) Figures 3a and 3b shows the galactic evolution of D, ^{3}He and ^{4}He computed with type (i) models in which the astration rate $\nu=0.45$ (expressed in units of 10^9 $year^{-1}$), the admixture rate of D free material $\delta=0.12$ (Mo per unit of 10^9 Mo) and the rate of ^{3}He stellar production being respectively $5 \ 10^{-5}$ $(M/Mo)^{-4}$ (fig. 3a) and $5 \ 10^{-4}$ $(M/Mo)^{-4}$ (fig. 3b) : this parametrization takes into account the fact that more massive stars are likely to have hotter inner regions inside

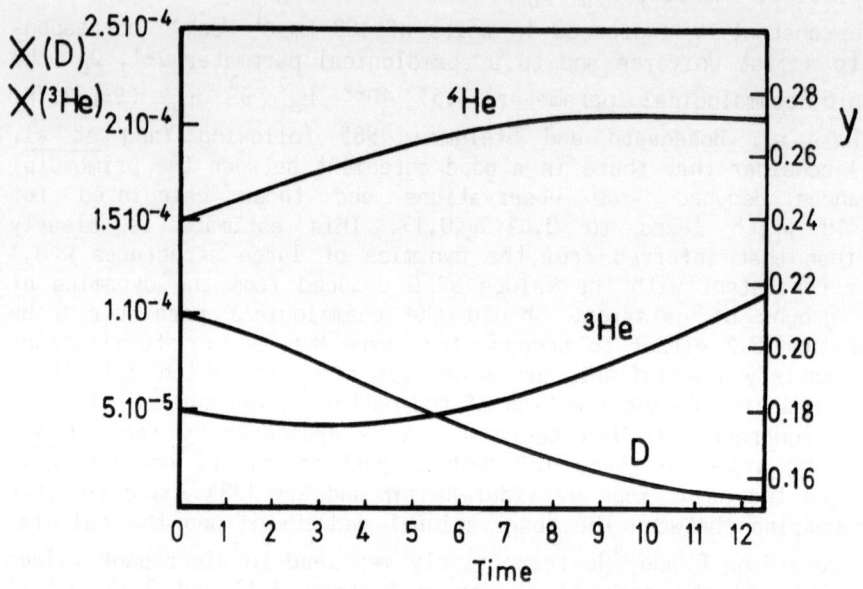

<u>Figure 3a</u>

Abundances (by mass) of D, ^{3}He and ^{4}He as a function of time (in 10^9 years units) calculated with type (i) (see text) models where there is some infall of processed material such that the ^{3}He production rate is 5 10^{-5} (M/Mo)$^{-4}$ (from Delbourgo-Salvador <u>et al.</u>, 1985).

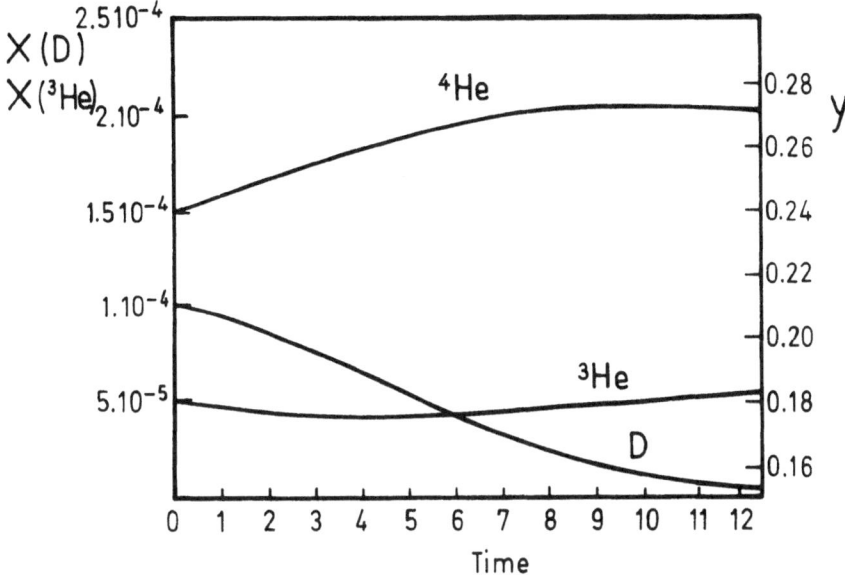

Figure 3b

Abundances (by mass) of D, ^{3}He and ^{4}He as a function of time (in 10^9 years units) calculated with type (i) (see text) models where there is some infall of processed material such that the ^{3}He production rate is 5 10^{-4} (M/Mo)$^{-4}$ (from Delbourgo-Salvador et al., 1985).

which ^{3}He can itself be transformed into ^{4}He. With such models $D^{Now}/D_{prim} \sim 15$, a present interstellar value of D/H$\sim$5 10^{-6} as found by Vidal-Madjar et al., (1983) would be consistent with a solar system value of $\sim 10^{-5}$ and a primordial value of $\sim 10^{-4}$ while ^{3}He/H which could be 5 10^{-5} at the beginning of the galactic life would be about 5 10^{-5} in the interstellar medium in case (a) and 1.2 10^{-4} in case (b). Large values of the rate of stellar production of ^{3}He might explain the large ^{3}He interstellar abundances observed by Rood et al., 1984.

(ii) If stars suffer large mass losses during their pre main sequence phase they could release significant amounts of D free-^{3}He rich material which could then affect the galactic evolution of these two isotopes. The temperature reached by such material is indeed sufficient to destroy D into ^{3}He. Figure 4 shows the evolution of D, ^{3}He and ^{4}He computed with a model where we assume that stars in ave- rage lose about 20% of their mass during the pre main sequence phase and where the rate of stellar ^{3}He production is 5 10^{-5} (M/M$\odot$)$^{-4}$. In such models D evolves in a quite similar fashion as the previous ones while the present (interstellar) ^{3}He/H abundance is about the same as that computed in model (i) with a rate of 5 10^{-4} (M/M$\odot$)$^{-4}$.

These two types of models alleviate the difficulty encountered in simple models which imply primordial D abundances which could be inconsistent with the primordial ^{4}He abundance.

If we adopt such models of galactic evolution together with the standard primordial nucleosynthesis scheme, we can find a range of baryonic density η consistent with the primordial abundances deduced from these models. The primordial abundance ranges of D, ^{3}He, ^{4}He and ^{7}Li are respectively 3 $10^{-5} < X(D) < 3$ 10^{-4}, 3 $10^{-5} < X(^3He) < 6.10^{-5}$, $0.235 < T < 0.255$ and 5 $10^{-10} < X(^7Li) < 2$ 10^{-9} (where the abundances are expressed by mass). Figure 5 shows that η_{10} can range from 1.2 to 4.5 which means that $0.004 < \Omega_B < 0.06$. The baryonic cosmological parameter deduced from our analysis is significantly smaller than the one of Yang et al., 1984 who derive $0.014 < \Omega_B < 0.19$. Our Ω_B value falls about also below the Ω values deduced from large scale dynamics. The present derivation implies that a large fraction of the matter present in the Universe should be in a non baryonic form.

5. PARTIAL PHOTODISINTEGRATION OF ^{4}He (AND ^{7}Li) BY ENERGETIC PHOTONS

As said before the standard primordial nucleosynthesis with or without galactic evolution effects puts severe constraints on the baryonic density and possibly on the cosmological parameter Ω and therefore on the overall evolution of the Universe. This constraint comes mainly from the fact that D and ^{3}He are very underabundant in Universes such that $\rho_{present} \approx \rho_{critical}$. In this section, we review the first scenario that we (i.e. David Lindley, Jo Silk and myself)[1] have

[1] More detail can be found in Audouze, Lindley and Silk, 1985.

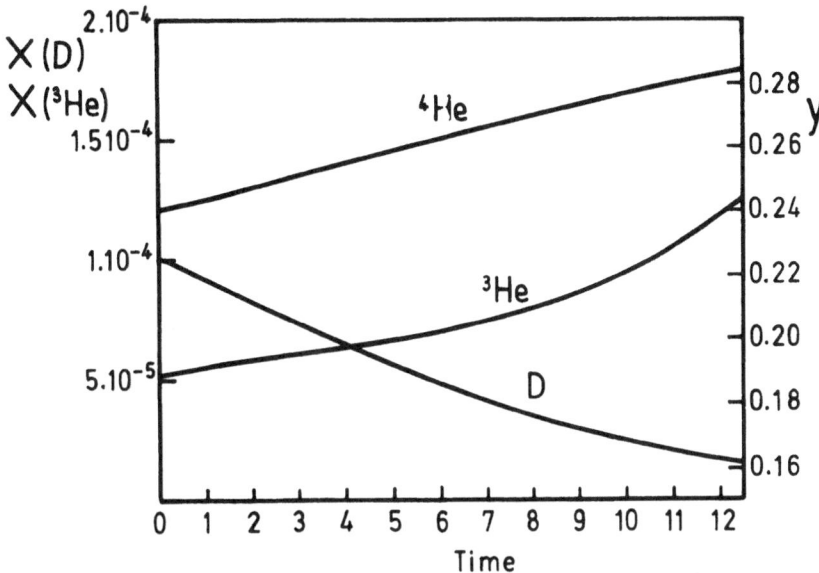

Figure 4

Abundances (by mass) of D, ^{3}He and ^{4}He as a function of time (in 10^9 years units) calculated with type (ii) (see text) models where mass loss processes suffered by pre main sequence stars which release D free, ^{3}He rich material are taken into account. Here 20% of the stellar mass is involved in this process (from Delbourgo-Salvador et al., 1985).

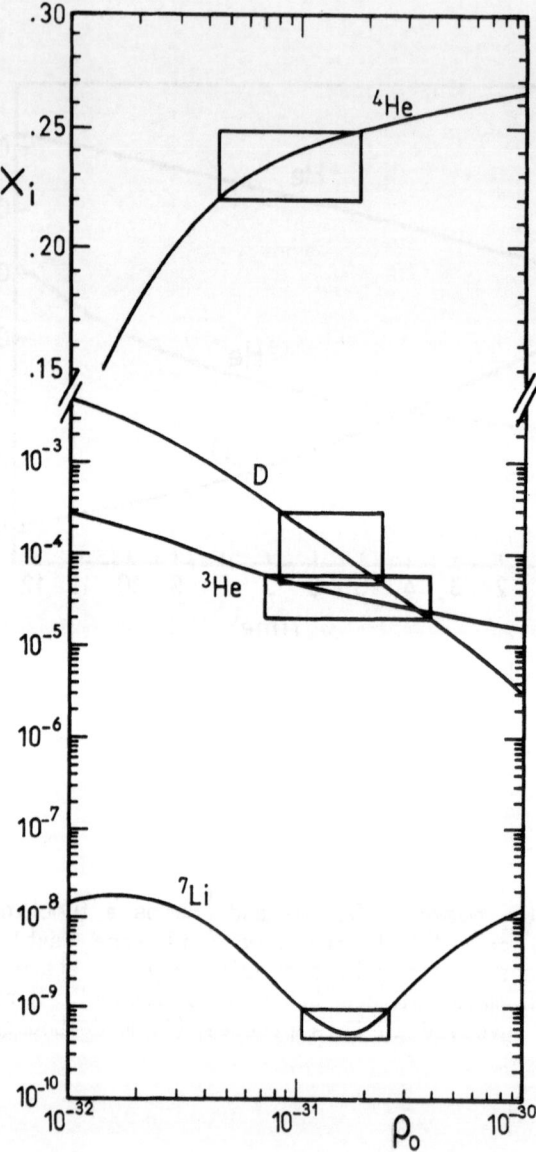

<u>Figure 5</u> Comparison between the light element abundances computed by
Delbourgo-Salvador <u>et al.</u>, 1985 and the primordial abundances of D,
^{3}He and ^{4}He taking into account the effects of galactic evolution
estimated by these authors. One deduces from this comparison that
$4 \ 10^{-3} < \Omega_B < 6 \ 10^{-2}$ (from Delbourgo-Salvador <u>et al.</u>, 1985).

sketched in order to reconcile the production of these elements with a large baryonic density. For that purpose we assume that ^{4}He (and possibly also ^{7}Li) are partially photodisintegrated by energetic photons coming from massive unstable particles.

Two different hypothetical candidates have been considered :

(i)   massive neutrinos with a mass $\geqslant 500$ MeV as already envisaged by Audouze and Silk (1984) (see also Lindley, 1979 and Hut and White, 1984),

(ii) gravitinos which are the leptons of spin 3/2 associated to the gravitons (of spin 2) in the supersymetric theories (see e.g. Fayet 1984) and the life time of which $\tau_{3/2} \sim m_{pl}^2 / m_{3/2}^3$ i.e. $\sim 10^8$ (100 GeV^{-3}/m$_{3/2}$)$^{-3}$ sec where m_{pl} is the Planck mass and $m_{3/2}$ the mass of the gravitino.

The high energy photons which are produced by the decay of such particles can either (a) scatter on thermal photons (i.e. the photons which constitute the background cosmological radiation) and they can produce e^-e^+ pairs if the product of thermal energy by the energy of the photons is higher than a threshold value computed by Lindley (1985) such that

$$E_\gamma \cdot kT = \frac{1}{50} \text{ MeV}^2 \qquad (1)$$

(b) if the product is lower than this threshold value the energetic photons coming from the decay can suffer some Compton scattering on electrons, induce pair production by interaction with nuclei or induce some photofission. In situation (a) thermalization is too rapid for photofission to take place. This implies that kT should be at least lower than 10^{-3} MeV for the decay photons to have an energy >20 MeV (which is the threshold energy of partial photofission of ^{4}He). The consequence is that the life time of these gravitinos or massive neutrinos should be $\geqslant 10^5 - 10^6$ sec.

The evolution of the abundances of the very light elements can be written as :

$$dN_4 = -\Sigma_4 N_4 \, dn_E / n_e$$
$$dN_3 = (-\Sigma_3 N_3 + f_{43} \Sigma_4 N_4) dn_E / n_e \qquad (2)$$
$$dN_2 = (-\Sigma_2 N_2 + f_{32} \Sigma_3 N_3 + f_{42} \Sigma_4 N_4) dn_E / n_e$$

N_2, N_3 and N_4 are respectively the abundances of D, ^{3}He and ^{4}He, the Σ are the corresponding rates, the f$_{ij}$ are the branching ratios, n_E is the density of energetic electrons. These electrons are the secondary particles coming from the energetic photons and Lindley (1985) has shown that it is equivalent to write these equations with electrons since photons behave like two electrons of half the energy. Finally n_e is the thermal electron density.

Figures 6 and 7 show respectively the effect of gravitinos and of massive unstable neutrinos on the light element abundances. One can see that there is a range of mass and lifetime for such particles

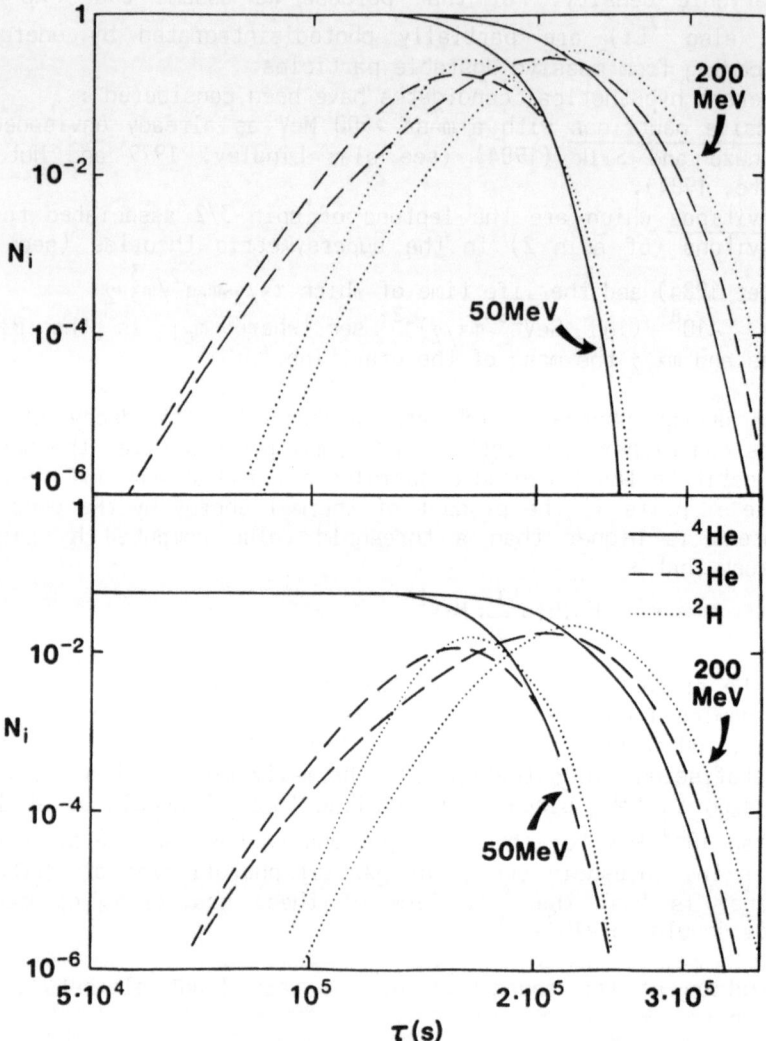

<u>Figure 6</u>
        Effect of decaying gravitinos inducing partial photofission of
^{4}He (and also ^{3}He and D). The upper panel corresponds to $Y_{initial}=1$
while the lower panel corresponds to $Y_{initial}=0.24$ with no initial D
and ^{3}He. Calculations have been performed per 50 MeV and 200 MeV
electrons. The resulting abundances are plotted against the gravitino
lifetime (from Audouze <u>et al.</u>, 1985).

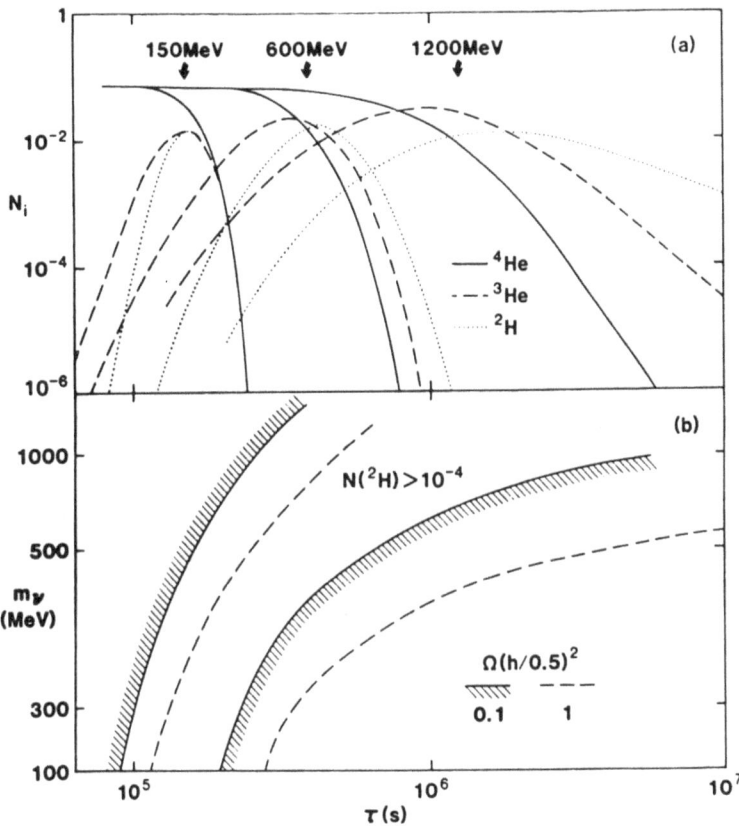

Figure 7

Same calculations for decaying massive neutrinos. The upper panel shows the resulting D, ^{3}He and ^{4}He abundances from 3 different values of the neutrino mass (150 MeV, 600 MeV and 1200 MeV) and $Y_{initial}$=0.28. The lower panel shows the domain neutrino mass (m_ν) neutrino lifetime (τ_s) inside which significant D abundances can be reached (from Audouze et al. 1985).

which induce photofission processes in order that the predicted D, ^{3}He and ^{4}He abundances are consistent with their primordial abundances (figure 6 and 7). The existence of these particles at the beginning of the evolution of the Universe may reconcile a dense (probably closed) Universe with the observed abundances of the light elements.

6. EFFECT OF QUARK NUGGETS ON THE PRIMORDIAL NUCLEOSYNTHESIS

Among the many hypotheses which have followed the irruption of particle physics in cosmology and especially the occurence of the unification theories, it is interesting to consider the scenario proposed by Witten (1984) who argued in favour of the existence of "quark nuggets" aside from that of nuclei. These particles would be made from an equal number of u, d and s quarks. They could have appeared at a time when the temperature of the Universe was T~100 MeV by a first order quark-hadron phase transition : if this transition occurs smoothly, these nuggets could coexist with nucleons after the nucleation (see Witten, 1984, for details). For a quark nugget made of 3A quarks (A being the atomic mass of such particles), its radius is $r=r_0$ $A^{1/3}$, its volumic mass is comparable to that of the nuclear matter and its electric charge is $Z\sim5$ $A^{1/3}$. These particles, if they exist, have interesting properties : (i) they could populate the "nuclear desert" which ranges from $10^3 < A < 10^{57}$ (10^3 corresponding to the largest atomic mass for nucleons and 10^{57} to that of a neutron star considered as a single particle, (see e.g. de Rujula 1984) ; (ii) when they are accelerated they might trigger "Centauro" events : these events which have been detected in nuclear emulsion chambers at high altitude are very energetic $10^{15\pm1}$ ev, they produce many (70-100) secondary particles but no (or very few) π_0 (Kazanas et al. 1985). They might be present in the very high energy component of the cosmic rays (Audouze et al. 1985b); (iii) the properties of the high energy cosmic rays coming from the direction of Cygnus X3 might be due to the presence of such still hypothetical particles see e.g. Barnhill et al., 1985.

Here we are concerned by their possible effect on the primordial nucleosynthesis. In other words, if such particles exist it possible to be in at flat ($\Omega=1$) or a close ($\Omega>1$) Universe and still make predictions on the production of D, ^{3}He, ^{4}He and ^{7}Li consistent with their presently adopted primordial abundances. As analysed by Schaeffer et al., 1985, these nuggets s can interact with nucleons n and p through reactions n+s⇄s and p+s⇄s within specific conditions which depend on the stability of these nuggets and their absorption and emission properties. The nucleon absorption rates are respectively $\lambda(n+s\rightarrow s)=\tau_0\langle v\rangle$ $\rho_B 1/A$ and $\overline{\lambda(p+S\rightarrow 1)=\tau_0\langle v\rangle\rho_B}$ $f_c 1/A$ for the neutrons and the protons. In these expressions τ_0 is the geometrical cross section $\tau_0=\pi R^2$; $\langle v\rangle$ is the relative velocity of the nuggets and the nucleons (averaged over the Maxwell-Boltzmann distribution of these particles), ρ_B is the baryonic density, A the atomic mass of the nugget and f_c the Coulomb barrier that a proton has to overcome before its absorption (f_c=exp $-\epsilon c/kT$ where ϵ_c is the Coulomb energy of the proton inside the nugget electrical field). These expressions hold for stable nuggets. If these particles are unstable these rates should be multiplied by a factor f_B which is a barrier, the height of which is fixed by the binding energy ϵ of the nucleon inside the nugget such that f_B= exp(-ϵ/kT) (ϵ~10-100 MeV).

To summarize : (3)

λ_n stable=$\tau_0\langle v\rangle\rho_B 1/A$ $\qquad$ λ_n unstable=$\tau_0\langle v\rangle\rho_B f_B 1/A$

λ_p stable=$\tau_0\langle v\rangle\rho_B f_c 1/A$ $\qquad$ λ_p unstable=$\tau_0\langle v\rangle\rho_B f_c f_b 1/A$

For the <u>emission of nucleon</u> s→n+s and s+p+s the corresponding rates are :

$$\lambda_n \text{ stable} = \frac{\eta c}{R} f_B \frac{1}{A} \qquad\qquad \lambda_n \text{ unstable} = \frac{\eta c}{R} \frac{1}{A}$$

$$\lambda_p \text{ stable} = \frac{\eta c}{R} f_B f_c \frac{1}{A} \qquad\qquad \lambda_p \text{ unstable} = \frac{\eta c}{R} \frac{1}{A} f_c$$

(4)

In these expressions, the term $\eta c/R$ is similar to a fission barrier : it represents the frequency of attempts for the nucleon (proton or neutron) to go out of the nugget multiplied by an energy barrier factor.

After the phase transition which took place at T=100 MeV between nucleons and nuggets, the relative density of these two classes of particles is governed by

$$\frac{d}{dt} \frac{\Omega_B}{\Omega} = \frac{\Omega_S}{\Omega} \lambda_{em} - \frac{\Omega_B}{\Omega} \lambda_{abs}$$

(5)

($\Omega=\Omega_B+\Omega_S$) : Ω_B is the mass fraction made of baryons while Ω_S is the one made of nuggets from the above expressions, λ_{em}= n $\frac{c}{R} \frac{1}{A}$ ~ n_Q $A^{-4/3}$ where n_Q is the baryon number density within a nugget (n_Q~10^{39} cm^{-3}). λ_{abs}~Π R^2 n_S $\langle v\rangle$~Ω_{nug} $n_c A^{-1/3}$ where n_c is the critical density of the Universe (~ 10^{23} cm^{-3}) at the time of nucleosynthesis. The comparison of these two terms in (5) clearly shows that the process of baryon emission is much more important than that of absorption for not too large values of A :

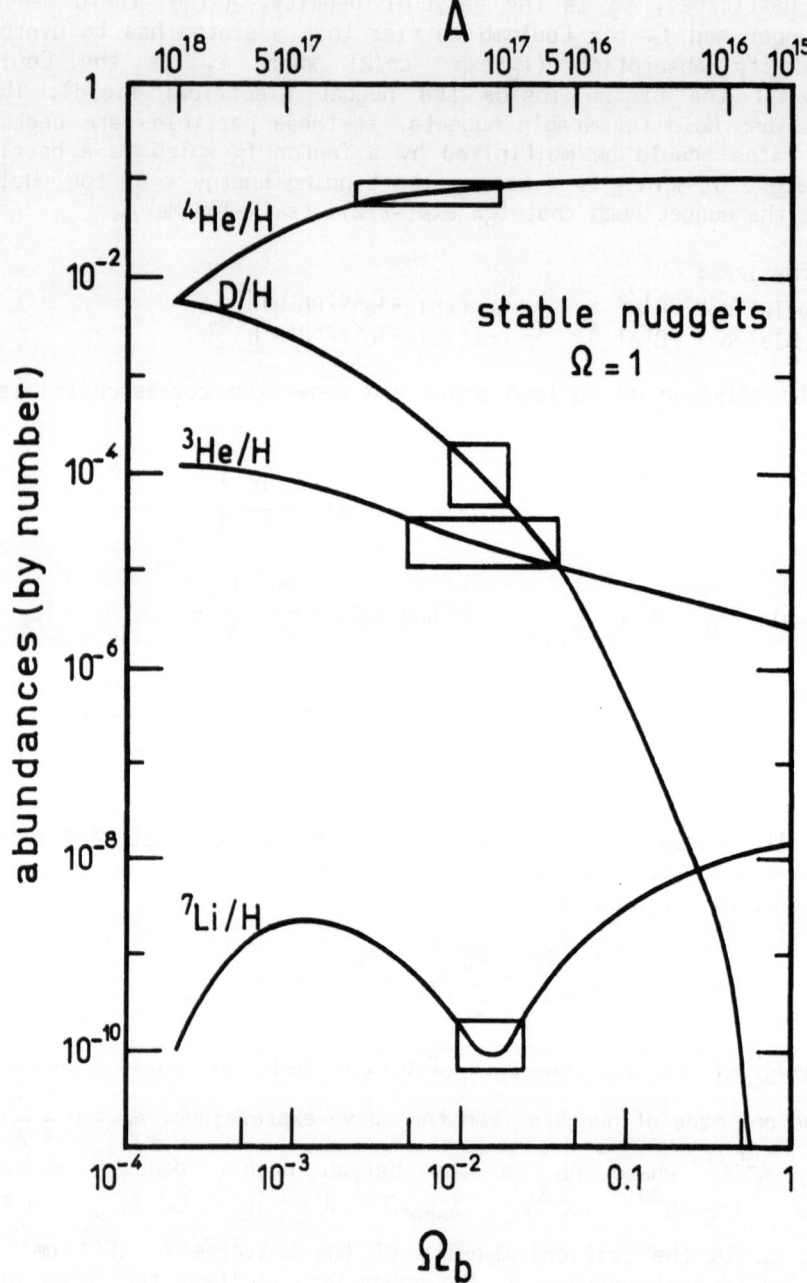

Figure 8
     Primordial abundances calculated from a model for which Ω=1 and
including stable quark nuggets. If these nuggets have atomic masses
~ 10^{17}, the results of the Big Bang nucleosynthesis can be
consistent with a closed universe (from Schaeffer et al., 1985).

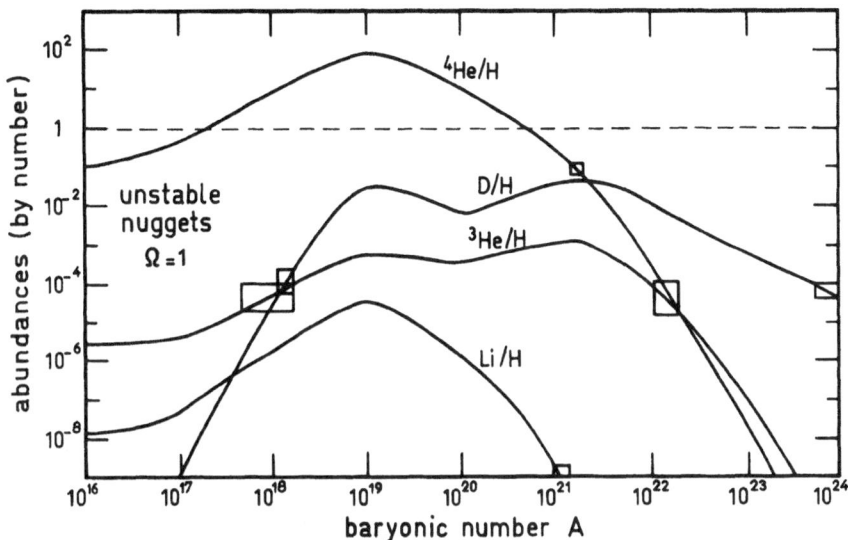

Figure 9
    Primordial abundances calculated from a model for which $\Omega=1$ including instable quark nuggets. The agreement between observations and calculations which is achieved for stable nuggets does not exist in this case (from Schaeffer et al., 1985).

(1) In the case of <u>stable or metastable</u> quark nuggets the solution of
(5) can be written as :

$$\frac{\Omega_b}{\Omega} = 1-e^{-x}$$

with $x = \int_{0}^{\infty} \lambda_{em}\, dt = (\frac{A}{A_{tran}})^{-4/3}$

with A_{tran} (the transition atomic mass) such that

$A_{tran} = 1.6\ 10^{16}\ (\frac{\varepsilon}{10\ MeV})^{-3/2}\ \eta^{3/4}$

for example $A_{tran} = 2.10^{15}$ if $\varepsilon = 10$ MeV and $\eta = 0.1$

For $A < A_{tran}$ the nuggets are rapidly transformed into nucleons by the
emission process while the number of nucleons become insignificant at
$A > 10^{18}$. Figure 8 shows the outcome of the primordial nucleosynthesis
calculated in a model of Universe for which one assumes that $\Omega = 1$ (flat
Universe) and where the stable quark nuggets exist with an atomic mass
A which is the free parameter. One notices that if $A \sim 10^{17}$ there is a
good agreement between the prediction of this model and the selected
primordial abundances of D, ^{3}He, ^{4}He and ^{7}Li.

(2) If the quark nuggets are <u>unstable</u>, the rates of emission and
absorption are of course modified along the lines described above. The
equilibrium between the nuggets and the nucleons remain qualitatively
the same. Namely the nucleons overcome the nuggets at low A and the
situation is reversed at high A : for $A < 10^{16}$, the nucleons dominate
and the nucleosynthesis occurs like in the standard model; for
$10^{16} < A < 10^{22}$ there is a significant emission of nucleons coming from
the decay of nuggets during the nucleosynthesis phase (at times
corresponding to $10^8 < T < 10^9$ K) which increase significantly the light
element production (fig. 9); for $A > 10^{22}$ the Universe is made only from
nuggets the neutron release occur ar lower temperatures than in the
case considered by the standard model. There is a large difference
between this case and the one concerning the stable nuggets. Because
the emission rate is significantly larger and leads to a huge neutron
release which is responsible for an overproduction of D, ^{3}He and ^{4}He
(fig 9) there is no more agreement between the observations and the
predictions of this model. Therefore if quark nuggets exist (see
Alcock and Farhi, these proceedings for a rather pessimistic view
regarding their evolution) they should be stable and have an atomic
mass $A \sim 10^{17}$ in order to reconcile very satisfactorily the standard
Big Bang nucleosynthesis with a significant presence of dark matter in
the Universe.

7. SUMMARY AND CONCLUSIONS

Although the determinations of the primordial abundances of the very light elements D, ^{3}He, ^{4}He and ^{7}Li are quite uncertain for various reasons, current models of primordial nucleosynthesis are able to fix important cosmological constraints especially on the present <u>baryonic</u> density of the Universe and also on the number of different existing neutrino (and lepton) families. By using models of chemical evolution of our Galaxy like those with inflow of processed matter or with significant stellar mass losses occuring during the pre-main sequence phase (as designed by Delbourgo-Salvador <u>et al.</u>, 1985) we were able to set conditions by which the $D_{primordial}/D_{present}$ ratio can be as large as ~10 and therefore find a range of baryonic densities consistent with all the present determinations of primordial abundances. We have deduced a range of baryonic cosmological parameter $4 \ 10^{-3} \leqslant \Omega_B \leqslant 0.06$ significantly smaller than that one deduced e.g. by Yang <u>et al.</u>, 1984 for whom $0.01 \leqslant \Omega_B \leqslant 0.19$. If our conclusions are confirmed it would mean that non baryonic matter should be invoked to explain large scale dynamic features such as those governing large clusters. Although one can still play with the uncertainties on the primordial abundances, on the actual value of the Hubble constant, on the neutron life time or on the values of the cosmological constant, the standard Big Bang model of primordial nucleosynthesis does imply a quite severe limitation on the baryonic dark matter.

We have also examined a few hypotheses such that Universes with large Ω parameter could be consistent with the primordial abundances of D, ^{3}He, ^{4}He and ^{7}Li. The two scenarios reviewed here which fulfill that purpose invoke respectively the photofission of ^{4}He (and ^{7}Li) induced by energetic photons coming from the decay of gravitinos and/or massive neutrinos (Audouze <u>et al.</u>, 1985a) or the presence of stable quark nuggets of atomic stars A~10^{17}. In the first case (gravitinos and/or massive neutrinos) the dark matter which could lead to large Ω values may be consistent with current inflationary schemes) could be baryonic while in the second case most of it would be made of these still hypothetical quasi baryonic particles. In any case these two scenarios do not exhaust other possibilities to escape the strong constraint on the present baryonic density of the Universe coming from the Standard Big Bang nucleosynthesis. Cosmologists have enough imagination to design other elaborate and ingenious ways to solve the problems which are presently put to us by the observable Universe.

This review is based on investigations performed with several collaborators whom I am enjoying to work with. They are Pascale Delbourgo-Salvador, Cécile Gry, David Lindley, Guy Malinie, Richard Schaeffer and Jo Silk. I would like to thank also Elisabeth Vangioni-Flam and Alfred Vidal-Madjar for many interesting discussions on these topics. Most of the writing of this review presentation has been made at the Astronomy Department of U.C. Berkeley. I thank Jo Silk for his hospitality. I want also to express my gratitude to M.C. Pelletan for her careful typing of this paper.

REFERENCES

Alcock, C., Farhi, E., 1985, Phys. Rev. D., submitted and these
    proceedings
Audouze, J., 1982, in Astrophysical Cosmology, Eds H.A. Brück, G.V.
    Coyne and M.S. Longair, Pontificiae Academical Scientiarum
    Scripta Varia, p.395
Audouze, J., 1984, in Large Scale Structure of the Universe, Cosmology
    and Fundamental Physics, Eds G. Setti and L. Van Hove (Geneva
    CERN), p.293
Audouze, J., Lindley, D., and Silk, J., 1985, Ap.J. Letters, 293, L53
Audouze, J., and Silk, J., 1983, in Primordial Helium, Eds P. Shaver,
    D. Kunth and K. Kjär (ESO Garching) p.71
Audouze, J., Schaeffer, R. and Silk, J., 1985, 19th ICRC La Jolla USA,
    8, 290
Audouze, J. and Tinsley, B.M., 1974, Ap.J., 192, 487
Barnhill, M.V., Gaisser, T.K., Stanev, T. and Halzen, F., 19th ICRC
    La Jolla, USA, 1, 99
Boesgaard-Merchant, A. and Steigman, G., 1985, Ann. Rev. Astron.
    Astrophys. in press
Davidson, K. and Kinman, T.D., 1985, Ap.J. Suppl., 58, 321
Delbourgo-Salvador, P., Gry, C., Malinie, G. and Audouze, J., 1985,
    Astron. Astrophys., 150, 53
De Rujula, A., 1984, CERN-Th 3980-84 preprint
Fayet, P., 1984, in Large Structure of the Universe, Cosmology and
    Fundamental Physics, Eds G. Setti and L. Van Hove (Geneva CERN)
    p.33
Geiss, J. and Reeves, H., 1972, Astron. Astrophys., 18, 126
Gry, C., Malinie, G., Audouze, J. and Vidal-Madjar, A., 1983, in
    Formation and Evolution of Galaxies and Large Structures in the
    Universe, Eds J. Audouze and J. Tran Than Van, Reidel Dordrecht,
    p.279
Hut, P. and White, S.D.M., 1984, Nature, 310, 637
Kazanas, D., Balashbhramanian, V.K., Streitmatter, R.E., 1985, Phys.
    Letters B, in press
Kunth, D., and Sargent, W.L.W., 1984, Ap.J., 273, 81
Lindley, D., 1979, M.N.R.A.S., 188, 159
Lindley, D., 1985, Ap.J., 294, 1
Reeves, H., 1974, Ann. Rev. Astron. Astrophys., 12, 437
Rood, R.T., Bania, T.M. and Wilson T.L., 1984, Ap.J., 280, 629
Schaeffer, R., Delbourgo-Salvador, P. and Audouze, J., 1985, Nature,
    in press
Shaver, P. Kunth, D. and Kjär, K. (Eds), 1984, ESO workshop on
    Primordial Helium (ESO, Garching)
Spite, F. and Spite, M., 1982, Astron. Astrophys., 115, 357
Vidal-Madjar, A. and Gry, C., 1984, Astron. Astrophys., 138, 285
Vidal-Madjar, A., Laurent, C., Gry, C., Bruston, P., Ferlet, R. and
    York, D.G., 1983, Astron. Astrophys., 120, 58

Vigroux, L., Stasinska, G., and Comte, G., 1985, Astron. Astrophys.
    submitted
Witten, E., 1984, Phys. Rev. D30, 272
Yang, J., Turner, M.S., Steigman, G., Schramm, D.N. and Olive, K.A.,
    1984, Astrophys.J., 281, 493

DISCUSSION

REES: I'd like to ask for clarification of the limits on Ω. As I understand it, a spread of at least a factor of four or five is due to the uncertainty in the Hubble constant squared. So how can you have only a factor of five spread altogether?

AUDOUZE: That's why I was questioning my value.

STEIGMAN: The uncertainty in the microwave background temperature increases the spread to a factor of five or six.

AUDOUZE: Our chemical evolution models give $\eta < 3$, which gives $\Omega < 0.06$ if we want to have good agreement with the He and D abundances.

ALCOCK: I just want to point out that if quark nuggets existed then they could have had a powerful effect on nucleosynthesis. If you form these lumps and they then evaporate, most of your baryons are condensed into very high-density regions. So at the epoch of nucleosynthesis the photon-to-baryon ratio where the baryons are is much lower than the mean photon-to-baryon ratio in the Universe. That is probably how the signature of strange matter on nucleosynthesis came about.

DAVIS: Do you include that?

AUDOUZE: No, because we did our computation ignoring the evaporation of the nuggets. If we include the evaporation of the nuggets, it does not work, because too many neutrons are released in the evaporation and we get too much D and ^{4}He. In fact, we get more ^{4}He than H.

MADSEN: The calculation of He synthesis with strange nuggets by Riisager and myself assumes that nuggets survive to the era of nucleosynthesis. In that case we find a lower limit to the baryon number of nuggets of order $A > 10^{20} \, \Omega_s{}^3$. This limit of course disappears if nuggets evaporate before that, as suggested by Alcock and Farhi.

AUDOUZE: The calculations concerning the four light elements (D, ^{3}He, ^{4}He and ^{7}Li) performed by R. Schaeffer, P. Delbourgo-Salvador and myself are in fair agreement with your conclusions given the uncertainties in the interaction between nuggets and nucleons.

FABER: I'm a little confused as to why your lower limit on Ω_b is smaller than that derived by other people.

AUDOUZE: There are two reasons. First, we took a low ^{4}He abundance – 23% – and second, we assume that we have a lot of deuterium at the very beginning.

FABER: Could you not have used ^{3}He + D to estimate the primordial D abundance?

AUDOUZE: If you want to make ^{3}He + D consistent with ^{4}He, you need to have a fairly low value of η if you want to have a value of the ^{4}He/H ratio of 23 - 24%.

STEIGMAN: I believe that that answer to Sandy Faber's question is correct. Audouze's constraint on ^{3}He + D is different from ours because he uses these models of infall and chemical evolution to permit him to have more ^{3}He + D than we've estimated from the destruction of ^{3}He. And that's the contradiction: we believe that there is an upper limit on ^{3}He + D which is violated in these models where Audouze has a factor of 15 destruction of deuterium.

FABER: Is it possible in your deuterium destruction models to significantly affect ^{3}He + D as well?

AUDOUZE: Producing too much ^{3}He kills the model, of course.

STEIGMAN: It is unfortunate that our knowledge of the abundances important for cosmology comes largely from our own local swimming hole, from the Solar System and the local interstellar medium. In writing the review article with Ann Boesgaard, I was impressed by the fact that if you make a plot (which I've never seen made before) of the column density of H versus that of D from local interstellar medium data, the H column density has a range of three orders of magnitude but correlates linearly with D with deviations of less than a factor of two. In the past I've seen D/H plotted as a ratio versus distance, and then the scatter looks large. Now I think that the deviation in the deuterium abundance is remarkably small, with D/H $\simeq$ 2$\times$10^{-5}.

AUDOUZE: Yes, but as I said, maybe the deuterium abundance is not much of a problem. Maybe the problem comes from infall.

STEIGMAN: We are in perfect agreement that if the ^{4}He abundance is determined to some level of confidence to be 23% or less, the standard model is in serious trouble.

J. BAHCALL: If you believe despite the neutrino problem that we understand the Sun, then the lower limit on ^{4}He/H from standard models is 0.235. That is, if you take the standard model of the Sun and vary every uncertainty, as we've tried to do in looking at the problem with the solar neutrinos, you find ^{4}He/H = 0.245 ± 0.01.

AUDOUZE: But you know that the Sun is not made of primordial material. So something could have happened between the beginning and the formation of the Sun.

STEIGMAN: The helium abundance in the Sun worries me a bit because of this upper limit of 25%. Why doesn't the Sun have a solar helium abundance? The interstellar medium has a helium abundance of 28 -32 % measured in HII regions. Is the Sun discrepant, or are the uncertainties larger than people have claimed?

C. NORMAN, the session chairman, invited J. Bardeen, P. Hut, R. Larson,
D. Lynden-Bell, B. Paczynski, M. Rees and E. Salpeter to introduce subjects
covered in the posters. Each of their comments was followed by a brief
general discussion.

HUT: One of the most important problems which we face in this meeting is
the question: What is the composition of the most abundant form of matter
in the Universe? We have heard from many speakers that the observed types
of matter contribute only a small fraction of the gravitationally inferred
average matter density. Whether the main type of matter consists of exotic
elementary particles or plain old baryons, we do not know.

The dark matter problem is most dramatic on the scale of clusters
of galaxies, where the ratio of unseen to seen matter is largest. Some
inferences can be made from the distribution of galaxies on these and
larger scales. For example, we heard that the observed distribution can be
reproduced much more easily with "cold" elementary particles than with "hot"
particles such as neutrinos.

However, there are advantages to starting closer to home. Here the
observations and theoretical inferences are more direct, even though the
dark matter problem is less dramatic. Near the Sun, we have dark matter in
the galactic disk with a density comparable to that of the observed matter.
I would like to discuss three poster papers which contain hints about the
composition of this local dark matter.

Recently an upper limit has been put on the average mass M of the
constituents of the local dark matter, $M < 2 M_\odot$ (Bahcall et al. 1985,
Ap. J., **290**, 15). Because of the large population of wide and therefore
fragile binaries, heavy black holes are excluded. Neutron stars are barely
allowable on dynamical grounds (the upper limit quoted is rather generous),
but pose severe problems for the metallicity evolution of the Galaxy.
Elementary particles which are not dissipative would not have gathered in
the galactic disk and are therefore also excluded.

Of the remaining candidates the only promising ones are white dwarfs and
brown dwarfs. However, straightforward extrapolations of local observations
of either type of star give densities far below what is needed. We have to
investigate where we can reasonably deviate from these extrapolations.

One poster paper is clearly in favor of the first solution. In his
paper on bimodal star formation, Larson suggests a deviation from a simple
power law for the initial mass function (IMF) of star formation. He shows
how a bimodal IMF can explain several problems in the chemical evolution
of galaxies, and at the same time naturally produce enough white dwarfs to
provide the local dark matter. His approach is interesting in that it is
not entirely ad hoc; it is plausible in the wider context of observations of
galactic evolution. The problem, of course, is the question of why we have
not seen these many white dwarfs. Perhaps they cool fast enough to have
escaped detection? If so, Larson's model has an extra advantage, since it
produces most of the white dwarfs early in the history of the Galaxy.

Two poster papers contain arguments restricting the viability of the
second solution. Boeshaar, Tyson and Seitzer provide upper limits on the

J. Kormendy and G. R. Knapp (eds.), Dark Matter in the Universe, 525–536.

density of the brighter type of brown dwarfs in the solar neighborhood.
Mathieu shows that observations of open star clusters are compatible with
the absence of a significant population of brown dwarfs, which might suggest
that they do not play a dynamically important role anywhere in the disk.
But neither paper contains strong constraints, as the authors point out.

If we combine the arguments of these papers, it seems that white dwarfs
are favored as candidates for the local dark matter, but brown dwarfs are
not ruled out. So we have to resign ourselves to the fact that we still
do not know the nature of half of the matter even in our own galactic back
yard.

ALCOCK: A comment on the white dwarf situation. It is beginning to look
as though the absence of very low-luminosity white dwarfs means that the
observations are inconsistent with a constant rate of white dwarf formation.
So in order to have a lot of unseen matter in degenerate stars, we could add
an initial burst of white dwarf formation around seven billion years ago.

J. BAHCALL (to Richard Larson): Do you think that there is still a mystery
in the observed numbers of white dwarfs?

LARSON: I suggested several ways around it. (1) The scale height of the
oldest white dwarfs may be larger than we think. (2) White dwarf cooling
theory may not be quite right. (3) As Alcock just pointed out, if all of
the star formation occurs early enough, the white dwarfs are now all older
than a cooling time. (4) If you allow me more parameters in my initial mass
function, I can arrange to make the typical white dwarf mass be $\geq 1 \, M_\odot$, in
which case the cooling time is almost certainly shorter than 10^{10} years.

PEEBLES: Could the formation of the disk have compressed the halo material
enough to account for the missing mass in the disk?

HUT: I don't think so, because of the high velocities in the halo. How
would you trap it?

PEEBLES: Since you are deep in the potential well of the disk, you trap the
low-velocity tail of the distribution of high-velocity objects.

HUT: I haven't looked at it, but you would have to increase the halo density
by a factor of ten to account for the local disk density.

BARNES: I have done N-body models in which I start with a spheroid or cloud
which represents the halo and then impose the disk field. You don't flatten
the halo very much; you can't flatten it enough to get a local dark-matter
density equal to that of the luminous matter.

BINNEY: I don't agree. How much you flatten the halo depends on how
radial a velocity distribution you are willing to have in the initial
configuration. The halo becomes squishable as you take away its tangential
velocity dispersion. I think you could find a model; whether or not you
would believe it is another matter.

SCHECHTER: You hang a lot on the fact that the wide binaries seem to

survive. There was a question the other day (and I wasn't quite satisfied
with the answer) about whether the wide binaries could be stabilized by a
third unseen companion.

HUT: It doesn't work. If the companion is far away from both visible stars,
the binary is broken up anyway. If it is close to one of them, you will
notice the very different radial velocities of the visible stars.

GUNN: There is a range of separations that would allow you to hide a lot of
mass in a wide binary. If you were seeing two planets that orbit at $\gtrsim 0.01$
pc around something dark that weighs 100 $M_\odot$, they would be very hard to
break up and you would be none the wiser from the dynamics.

J. BAHCALL: That's right, but the frequency of wide binaries is such that
you can't afford to have a very massive black hole for every wide binary.

LARSON: Several posters address the question of whether the ratio of dark
to luminous matter is the same in all galaxies. The situation is not simple
or clear. On the one hand there are at least two posters, by Carignan and
by Casertano and Bahcall, which show that when you fit rotation curves with
models, the similarity in the rotation curves demands a ratio of disk mass
to halo mass which is remarkably close to one for all cases. This suggests
that the disk and halo matter are closely related, perhaps even made of the
same kind of material (as Casertano and Bahcall suggest). On the other
hand, a poster by Kent finds more variability in results of this kind.
    Several posters address the question of how M/L ratios correlate with
galaxy properties. One by Vader studies the dependence of M/L on color for
both visual and IR colors. She confirms a conclusion obtained by Tinsley
several years ago that the variation of M/L with color disagrees with the
predictions of "standard" models in the sense that the bluer galaxies have
much more mass than those models would predict. This suggests that the
luminous and dark matter don't vary together, at least as a function of
color. Then there is a poster by Athanassoula, Bosma and Papaioannou, in
which they determine M/L ratios by fitting multiple-component disk and
halo models. They show a plot of M/L *versus* color which to my eye looks
almost like a scatter diagram. There does appear to be a trend, and they
plot the Larson-Tinsley models, but there is more than an order of magnitude
of scatter. There is also a poster by Bosma, Athanassoula and van der Hulst
about disk galaxies with very low surface brightnesses. Dynamically these
are giant spirals, but in terms of light they are very dim. Again something
funny is going on with the mass-to-light ratios.
    This gives me a chance to make a comment on mass-to-light ratios and the
Larson-Tinsley models. The predictions of models are critically dependent
on the initial mass function. These and nearly all other models assume for
the initial mass function a power law or something similar that resembles
the original Salpeter function. I would contend that the Salpeter power
law is remarkably poorly constrained by existing observations except at the
high-mass end. The relationship between the high- and low-mass ends, which
is crucial here, is hardly constrained at all. The conventional continuity
constraint is nearly always assumed. If you drop it, all hell breaks loose.

You can make models with mass-to-light ratios ranging from a few tenths to infinity for a population of stars and their remnants. So looking at the light tells you almost nothing about the mass of the associated stellar population except that it is not zero. It could be essentially infinity for possible assumptions about the initial mass function and the time dependence of the star formation rate. Because of this, I would suggest that the concept of luminous mass is a myth; the term should either be dropped or defined very carefully.

BURSTEIN: I would like to emphasize a point that has been made before. I am impressed with the lack of constraint placed on dark matter in spirals by comparing luminosity profiles and rotation curves. For each galaxy observed, there is a wide range of ratios of luminous to dark matter that produces acceptable fits. The arbitrariness in the luminosity-dependent approach is consistent with what Vera Rubin and I find for the systematics of rotation curves: they *do not* reflect the Hubble type of the galaxy.

SANDERS: I want to make a similar point. Everyone who plays the game of using rotation curves to make models basically assumes that the mass-to-light ratio of every component in a galaxy is independent of radius. Can you think of any reason why that should be true?

LARSON: No.

BARDEEN: I want to raise some questions about the dynamics of the formation of halos and elliptical galaxies. There are quite a few poster papers on this subject. We have seen suggestions that luminous ellipticals form mainly from mergers while spirals and small ellipticals form from more isolated density perturbations. This raises the question: To what extent can we make more-or-less spherical accretion models *versus* something more lumpy and complicated? Also, some recent studies, e. g., by Primack, Blumenthal and Faber, of the violent relaxation of collapsing collisionless systems suggest that it is difficult to get an extended flat rotation curve. On the other hand, the simulations by Frenk suggest that an isolated lump which is not formed by merging gets a rotation curve that is flat out to quite a large distance. And the final question is: What is the time of galaxy formation? In the case of accretion, this is not a particularly well-defined concept; in the cold dark matter scenarios, galaxy formation is fairly recent, and in the numerical simulations by Frenk and collaborators, the galaxies are still forming now. To what extent might these predictions be in conflict with attempts to find young galaxies? These are the questions I'd like to open up for discussion.

PRIMACK: You suggest that the simulations by Blumenthal *et al.* of halo formation are in conflict with those of Frenk. But actually, we are only trying to simulate the inner parts; our outer boundary conditions are not realistic. And Frenk *et al.* are simulating the outer parts; their resolution is not adequate to study the inner regions. My impression is that the models fit together rather nicely. The characteristic features of rotation curves are that they rise rapidly and within a couple of exponential scale

lengths are essentially at their asymptotic value. This is very hard to
understand without having the baryons modify the dark matter distribution.
This happens automatically, it is a dynamical fact of life. It is shown
by several simulations, e. g. by Barnes and by Ryden and Gunn. And it
is rather nice that we automatically get the kinds of spectra that are
predicted by the cold dark matter model. So we get rotation curves which
are fairly flat even beyond where you can measure them in real galaxies.

KAISER: Isn't it true, though, that if we could measure rotation curves for
rich clusters, we would find that they also are flat? But the slope of the
spectrum on cluster scales is rather different from the slope on galaxy
scales. So it is not clear that flat rotation curves are the result of a
"proper" choice of spectrum.

BINNEY: A few years ago a paper by Simon White demonstrated that merger
remnants always have r^{-3} density profiles, in agreement with the Hubble-
Reynolds formula. At this meeting, on the other hand, White told us that
in simulations of the clustering of cold dark matter, galaxies predominantly
form through mergers and have flat rotation curves. What has changed?

DEKEL: That's a very important question. Those simulations showed that,
with almost any initial condition, a finite system which collapses and
violently relaxes ends up with a de Vaucouleurs-law profile. Now the
question is: What happens when you allow secondary infall after the center
has relaxed? The results turn out to be very similar to the predictions of
the self-similarity solutions of Goldreich and Fillmore, of Bertschinger,
and of Gunn and Ryden. If the fluctuation spectrum is steeper than $n = -1$,
you don't have enough secondary infall to change the original de Vaucouleurs
profile. But if you have a lot of power on small scales, e. g., if on
galactic scales the cold dark matter spectrum has $n = -2$, then you have
enough secondary infall to produce the r^{-2} profile. So the theoretical
solution says that if $n \leq -1$ you end up with an r^{-2} density profile.

WHITE: Let me make a comment in response to James Binney. People who read
other people's papers should be careful. Like salesmen of any kind, the
people who do N-body work always present their results in the way that best
makes the point. If you want to show that your model has an r^{-3} luminosity
profile like in an elliptical galaxy, you plot the logarithm of density
versus the logarithm of radius and find a slope of -3. If you want to
demonstrate that you have a flat rotation curve, you take the mass within a
radius, divide it by r and plot the result as a function of radius. Now, if
the models are well defined over only a factor of four or five in radius,
you can very well have something whose density looks like r^{-3} but whose
rotation curve looks flat. (laughter, catcalls, uproar)

FRENK: There are two generic types of rotation curves that arise in our
high-resolution N-body simulations of a universe dominated by cold dark
matter. Those clumps which remain relatively isolated for long periods of
time (and 8 of our 10 largest clumps do so) develop rotation curves which
are essentially flat from at least 10 kpc outward. In contrast, clumps
which form by merging of similar-size sub-clumps have rotation curves which

rise slowly out to $\sim$ 50 - 80 kpc and only then become flat. These are the
objects that may be identified with the halos of ellipticals. It is not
clear to what extent flat rotation curves are specific to the cold dark
matter model. I suspect that the substantial large-scale power in this
model is responsible and may explain the difference between our results and
those obtained earlier from studies of the collapse of an isolated system or
the merger of two isolated ones.

BARDEEN: I think it is fair to say that in these models you expect a rapid
burst of star formation at some fairly early time, like at a redshift of
five or six, and then a more gradual star formation after that. This might
not violate the limits set by searches for primeval galaxies.

LYNDEN-BELL: None of us knows what dark matter is. It is now so fashionable
to think that it is some exotic and unknown type of elementary particle
that many people refer to the observed matter as the "baryonic density".
However, there are perfectly good baryonic candidates for dark matter, such
as giant planets.

    We now have rather good evidence that around a number of giant
elliptical galaxies baryonic matter is disappearing from hot, X-ray-emitting
gas. The place where it disappears is right for the making of dark halos;
the rate of its disappearance would build a halo in 10^{10} years. If we want
to believe the observations rather than our prejudices, we should take as
our best bet that dark halos are baryonic and made from cooling flows. In
a subject where observations are few, theorists have great freedom to build
and are loath to abandon their castles in the sky. But if the disappearing
hot gas does not make dark matter, how else can we get rid of it? When
exotic neutral particles have been found in laboratories, I shall be happy
to postulate them in the cosmos, but until then, let us use the observations
and not the prejudices.

DRESSLER: The talks on cooling flows suggest a scenario for making both
relatively high-mass ($\sim$ 2 $M_\odot$) stars in the centers of galaxies and very
low-mass stars which cannot be seen forming in their halos. Has anyone
thought about what might control the mass function in these two regimes?

FABIAN: The initial mass function is probably pressure-dependent. If high
pressures give low-mass stars, then most of the matter in cooling flows
turns into such stars. But it is possible to produce transient, low-
pressure regions from large blobs cooling in the flows, giving rise to Hα
filaments. If there are stars forming in those regions, they will form
at pressures similar to those in the disk of our own Galaxy, and so will
have higher masses. But I agree entirely that we need more work on star
formation.

J. JONES: You don't need to form as many Jupiters as you might believe.
In the models of elliptical galaxy evolution I have been making, I find
that, provided you are considering mostly low- to intermediate-mass stars
with lifetimes of $\sim$ 10^9 to a few x 10^{10} years, these stars eject enough gas
when they die that, if it falls into the center, it will provide a means of

condensing the galaxy in a way that agrees with the observations. You don't
actually need to put the gas into extremely low-mass stars, provided you are
not making ultra high-mass stars, i. e., stars with masses $> 10 \ M_{\odot}$.

STEIGMAN: There is a worse problem than understanding star formation that
makes only low-mass stars. You have to turn *all* of the gas into stars.
This is not observed in the Galaxy: molecular clouds certainly don't turn
completely into stars.

LYNDEN-BELL: We don't know that no gas is left over. We see gas in lots of
elliptical galaxies, for example in the form of filaments.

SALPETER: A poster by Trinchieri and Fabbiano describes the extensive X-ray
emission from elliptical galaxies which are not in cluster cores. I want
to discuss a controversy raised by such data--cooling flows *versus* galactic
winds -- and a "diplomatic compromise" -- galactic fountains. The data
in the poster refer directly only to isolated galaxies, but they also have
an indirect bearing on rampressure stripping for elliptical galaxies in a
cluster environment.

   (a) X-Ray Halos and Galactic Fountains: The X-ray emission from large,
isolated elliptical galaxies has two properties which suggest thermal
emission from hot gas that was ejected in star deaths and is now in a
cooling flow: (i) The total X-ray luminosity L_x of a galaxy with optical
luminosity L_{opt} scales approximately as $L_x \propto L_{opt}^{1.6}$. The rate of gas mass
loss by stars scales as L_{opt} and the kinetic energy per particle scales as
the square of the velocity dispersion, $\sigma^2 \propto L_{opt}^{2/3}$. The observed L_x thus
scales correctly as the rate of (thermal plus gravitational) energy release.
(ii) For a given galaxy, the observed X-ray emissivity $f_x(r)$, as a function
of distance r from the galaxy center, is approximately proportional to
the star density $\rho_s(r)$. The emissivity, proportional to the square of gas
density $\rho_g(r)$, scales correctly with gas release rate but the density *ratio*
$\rho_g(r)/\rho_s(r)$ *increases* with increasing r, roughly as $\rho_s^{-1/2}$. Thus, most of the
total gas mass resides in the outer parts of the halo where the X-ray data
are poorest. Furthermore, the empirical relation (ii) breaks down in the
outer parts of the better-studied elliptical galaxies: (iii) The average
$\rho_g(r)$ decreases even more slowly with increasing r at large r, giving the
appearance of a "broken ring" or an outer halo of gas.

   The "orthodox" theoretical view (e. g., Forman, Jones and Tucker 1985,
Ap. J., **293**, 102) invokes supernovae only for mild stirring of the gas;
the X-ray emission leads to inward cooling flows and is powered by the
gravitational energy released in the flow. According to the orthodox view,
a complete galactic wind, i. e., immediate ejection of gas from the galaxy
by supernova explosion, would give X-ray emission many orders of magnitude
less than the slow cooling flows because of the very disparate flow times.
Ostriker, on the other hand, was arguing that galactic winds are necessary
to inject the heavy elements into the intergalactic gas. This sounds like a
controversy. But I want to argue that both sides are right, as follows.

   The radiative cooling that feeds the inward flow is not steady. Small
density fluctuations are magnified into strong cooling instabilities.
This must lead to an irregularly shaped gas surface even in the absence of

supernovae. Sometimes supernovae will erupt close to a "dipping part" of the gas surface; with little overburden of gas, these supernova will give a *temporary* and *local* galactic wind carrying heavy elements outward. While this local wind produces negligible X-ray emission itself, it does *not* decrease the emission from the other inward-flowing regions appreciably: the inward mass flow rate is decreased only slightly, and even this is offset partly by compression and heating from inward-directed parts of supernova explosions. Furthermore, in this more complex picture (coupled with an assumed dark halo which increases the escape velocity V_{esc} slightly) one should also get a third phenomenon, namely galactic fountains: The cooling instability and the increased V_{esc} lead to blobs cooling and condensing far out, falling inward, clashing with upwelling gas, etc. This process will puff up the gas surface on the average, but will be highly irregular since it is highly dynamic. Although these fountains have been invoked mainly for spiral galaxies (e. g., Bregman 1980, *Ap. J.*, **236**, 577), I consider the broken rings of gas surrounding ellipticals (point iii, above) as the most direct visualization of the fountain predictions.

   (b) Rampressure Stripping of a Puffed Up Galaxy: The above internal effects for an isolated elliptical galaxy must also interact with rampressure effects in a cluster environment. There are two kinds of interplay. The simpler of the two has just been calculated in a Cornell Ph. D. thesis by Terry Gaetz. He calculates rampressure heating and stripping by external gas flowing with velocity V_f relative to a spherical elliptical galaxy, with inward radiative cooling flows and central star formation included. His results differ from previous, simpler calculations in several ways. The increased cross section of the "puffed up" galactic gas increases the efficiency of rampressure stripping. Also, the stripping efficiency depends on a larger power of the ratio (V_f/V_{esc}) than the previously assumed $(V_f/V_{esc})^2$, because heating effects are as important as momentum effects.

   A second kind of interplay has not been calculated yet, but the external flow must also affect the internal gas dynamics. Consider a very massive elliptical galaxy which, when isolated, has galactic fountains but very little galactic wind. The same galaxy in a dense cluster has its outermost internal gas stripped by the external rampressure. The decreased density far out could then convert a galactic fountain into a partial galactic wind, resulting in an even larger mass loss. I therefore conjecture that a galaxy's environment can have a strong influence on the mass inflow rate in a cooling flow and on the resulting central star formation rate. However, the effect on the X-ray luminosity is weaker and more complicated, because the decreased mass flow is partially compensated by the increased heating from the external flow.

TUCKER: Do you predict that X-ray halos should become dimmer as you move closer to the core of a cluster?

SALPETER: For a galaxy with low mass and large velocity (large V_f/V_{esc}), the X-ray halo should shrink in size with little change in central surface brightness. However, for massive galaxies moving slowly in the cluster core $(V_f/V_{esc} \lesssim 1)$, the X-ray halos should brighten instead of dimming (accretion heating instead of stripping).

FABIAN: We find that, in order to fit the data for (say) NGC 4472, the gas must cool out of the flow as fast as it appears. If you strip the outer layers of gas, you stop star formation in the outskirts of the galaxy. But you don't strip the center, so you don't stop star formation there. The central star formation shouldn't depend very much on stripping.

SALPETER: My opinions are half way between yours and Jerry Ostriker's. I think the supernova rate is always enough to lift the gas produced by planetary nebulae to about twice its initial radius, while Jerry says the factor is infinity.

SHAPIRO: The success of a galactic fountain usually depends on the dimensionless ratio of the cooling time for the hot gas to the dynamical flow time, and is thus a reasonably sensitive function of metallicity. I therefore suspect that there is an additional correlation between the star formation rate, which affects the metallicity, and the ability to strip the galaxy *versus* the production of a galactic fountain.

EKERS: I'd like to show a slide and describe an observation of a galaxy we've heard quite a lot about, NGC 4472. The slide shows NGC 4472 and the X-ray contours. Also shown are HI contours from observations by Sancisi, Carignan and myself. There is a small HI cloud between NGC 4472 and its dwarf companion, containing about as much HI as you'd expect for a dwarf of that brightness. But the HI is no longer in the dwarf galaxy. So the suggestion is that we are actually seeing the ram-pressure stripping of HI from the dwarf by NGC 4472's X-ray halo. This observation is also of interest because we seem to have HI right inside a very hot corona, and we have to worry about how it survives.

PACZYNSKI: The beautiful maps of the six known gravitational lenses were shown by Ed Turner. These maps convinced me that in two cases (Huchra's lens and 0957+561AB) there is an obvious galaxy that does most or all of the lensing. In the other four cases there is no galaxy which looks important. In fact, in some cases the faint candidates are not within, but outside of the group of observed images. The image splittings are too large for galaxies and too small for the cores of clusters of galaxies. I think we should very seriously contemplate the possibility that most of these lenses are due to unknown dark objects.

SCHECHTER: Can you put limits on the surface density of the lensing object?

PACZYNSKI: It must be higher than some critical value which depends on the distance. In typical cosmological models, the central density of a rich cluster of galaxies is not quite sufficient, although not by a large factor. At the optimum distance, the center of a large Abell cluster has $\sim$ 60 - 70 % of the critical density. So some clusters might be able to do the overfocussing. In such cases you would expect to see an additional image at some large distance of the order of a few arcminutes or greater.

EKERS (to Ed Turner): What is the evidence that the galaxies found in the vicinity of gravitational lenses are anything but chance coincidences?

E. TURNER: The sizes of the lens systems are typically only a few arcsec, around galaxies whose typical magnitudes are $\gtrsim 20$. The density of galaxies is not high enough for such a coincidence to happen easily. But that's not the way to look at it. The whole phenomenon is due to a chance coincidence. We see that the galaxy is there, it is at a smaller distance than the quasar, and it has a gravitational field. So it must be involved in the lensing.

PACZYNSKI: It might be involved, but it is not necessarily making the largest contribution.

E. TURNER: I agree.

TYSON: In cases of "missing" lens galaxies: As an alternative to over-focussing, the multiplicity observed (even instead of odd) would also be consistent with a dark, compact lensing object (an idea originally due to Press and Gunn).

PACZYNSKI: I would expect a galaxy to produce an even number of images, because nuclei are very compact. Whenever the central density in a lensing galaxy is very high, the image formed near the center is very faint. One can show that the intensity for a typical magnification goes like (surface density)$^{-2}$. You would expect that the central image should be demagnified by a factor of 100 for a typical galaxy having a core radius of a few kpc.

BURKE: I think it is important to use all of the available information. Unfortunately, only 0957+561 has a relative wealth of information available. In this case, one should note that the VLBI jet shows the parity reversal expected for the pair of images. Any of these other suggestions should be able to demonstrate the proper parity pair.

REES: I would like to focus on a topic that has attracted surprisingly few poster papers, i. e., the implications of background radiation measurements and upper limits in various wavebands. Most of the baryons in the Universe could be uniformly spread in a diffuse intercluster gas. It is well known that such gas, if ionized, yields a UV and X-ray background. About ten years ago Field and Perrenod, Boldt and others considered whether the hard X-ray background could be due to such gas at temperatures $\sim 40(1 + z)$ keV. Recent work by Guilbert and Fabian renders this idea implausible on energetic and other grounds. Nevertheless, the constraints on gas at more modest temperatures ($10^4 - 10^7$ K) aren't very strong. The background from such a gas with a density sufficient to contribute $\Omega = 1$ would not be detectable. The best constraints on such gas come from considering the pressure confinement of the clouds that cause QSO absorption lines, but these are rather model-dependent. It may be possible to reconcile the data even with $\Omega = 1$ with a carefully contrived thermal history and a low Hubble constant.

Turning now to longer wavelengths, the detection of distortions in the microwave background spectrum or in an infrared background would tell us about early galactic history and about the energy production associated with

Population III star formation. Maybe someone who has thought about this
would care to comment?

CARR: I would like to comment about the importance of background light
constraints. Many processes in the period after decoupling would be
expected to produce radiation, including primeval galaxies, population
III stars, pregalactic explosions and black hole accretion. Therefore,
observational upper limits on the background radiation density place
interesting constraints on these processes, especially if the radiation
presently resides in the optical and UV, where the observational limits
are strong. Probably, however, the radiation will have been reprocessed
by dust. Dust absorption may occur near the sources or in the background
Universe; dusty galaxies could provide the background absorption if they
cover the sky. In these situations one would expect to see a far-infrared
background peaking at 200 - 500 μ, a wavelength which depends only weakly on
the grain characteristics and cosmological parameters (Bond, Carr and Hogan,
preprint). Such a background and its anisotropies could be detectable
with future space experiments and may already have been seen at 100 μ by
IRAS (Rowan-Robinson, preprint). Jonathan McDowell has calculated the
constraints which background light limits already impose on astrophysical
processes in the early universe.

MCDOWELL: I think the interesting data have not yet arrived. For a
significant contribution to Ω from pregalactic very massive objects, the
models I have made generally predict backgrounds that are close to current
observational limits. We should have interesting constraints in both the
near and far infrared in the coming decade.

DAVIS: I am surprised that Joe Silk hasn't mentioned the particle
background. If they formed the halo of the Galaxy, some candidates for cold
dark matter would annihilate there. In a paper with Mark Srednicki, Silk
finds that such annihilation can provide a natural explanation for the local
low-energy antiproton flux, which is otherwise very difficult to understand.
Perhaps Silk would like to talk about this?

SILK: Suppose the dark halo consists of any Majorana-type massive fermions.
A particularly attractive candidate is the supersymmetric partner to the
photon, the photino. Srednicki and I have shown that annihilations of these
particles in the halo produce an observable flux of low-energy cosmic-ray
antiprotons. Once the photino mass is specified, the prediction is quite
specific, and the photino mass is restricted to a narrow range by the
value of Ω. The annihilation products include p-$\bar{p}$ pairs of energy several
hundred MeV, which accumulate in the halo over a typical leakage time of $\sim$
10^8 yr. There are essentially no secondary cosmic-ray antiprotons produced
in this energy range by the interactions of high-energy primary cosmic rays
with the standard grammage of interstellar matter assumed in cosmic-ray
confinement models. We concluded that this could provide a unique signature
of a not-improbable form of halo dark matter. One experiment has reported
detection of low-energy cosmic-ray antiprotons, but this result has to be
confirmed before any conclusion can be reached about the antiproton source.

STEIGMAN: If the density in ordinary baryons is, say, 10 % of the critical density, and most of it doesn't turn into galaxies, can you hide it?

REES: Gas can be collisionally ionized if it is hotter than 10^6 K and it can be photoionized by an ultraviolet background if it is at $\sim 10^4$ K. I don't think there is any objection, if $H_0 = 50$ km s^{-1} Mpc^{-1}, to having an Ω of at least 0.1 or 0.2 in uniformly distributed baryons in that temperature range.

SHAPIRO: I think you can probably have a hotter gas at a higher density and still hide it, with the possible exception of the constraint imposed by the pressure inferred for the metal-free Lyα clouds. If the gas were between 5×10^6 and 10^7 K with an Ω of 0.3, I doubt we could see it. It would have too low a temperature to be seen as the X-ray background and too low a density to compete with the local soft X-ray contribution of our Galaxy. But it would overconfine the Lyα clouds.

REES: We don't know the pressure of the Lyα clouds unless we think they are spherical and unless we also know the UV background that is ionizing them.

SHAPIRO: Then it is even easier to hide this gas.

CONFERENCE SUMMARY

James E. Gunn
Department of Astrophysical Sciences
Princeton University
Princeton, NJ 08544

I will attempt in this talk to address the questions that Drs.
Schwarzschild and Faber posed at the beginning, in the light of what
has been said in the last few days and a few things that were not but
perhaps should have been.

1. DO WE REALLY NEED IT?

Do we really need dark matter to explain the world as we see it and
presently understand it? The certainty of the "yes" answer to this
question varies rather a lot depending on where it is that one thinks
one needs it. We have heard conflicting views about the necessity for
it in the solar neighborhood, but I think it is safe to say that there
is very little remaining doubt about scales from the outer parts of
galaxies on up. At the very largest scales the question is "how much",
but even if Ω is as small as the dynamics of relatively well-
understood structures will allow, there must still be a large amount
compared to "luminous" matter, a loose term I shall use in this summary
to mean the stars, gas, dust, remnants, and other objects we
"understand". We might use "ordinary" or "baryonic", but those terms
are predjudicial and probably not quite correct either.
It should be remarked at the outset that all that I shall say is
predicated upon the physics we know and love being reasonably correct.
Many of the phenomena which dark matter is invoked to explain can be,
at least qualitatively, explained by some variation on Newtonian or
Einsteinian gravitation, perhaps the best developed of which (but by no
means the only one) is the theory of Milgrom which was described here.
Let us look briefly at the question of dark matter in the solar
neighborhood, where the results discussed here by Bahcall seem to be in
conflict with those of Kuzmin. It should be realized that the two
determinations of the local mass density refer to two rather different
populations, the Bahcall results on F dwarfs and K giants for a rather
old, large scale-height system and the Kuzmin ones for a very young
population confined very closely to the plane. It is thus not obvious
that the discrepancy is real. Taken at face value, the results might
suggest that the old disk population has something like half its total

537

J. Kormendy and G. R. Knapp (eds.), Dark Matter in the Universe, 537–546.

mass in unknown form, while we understand the young population reasonably well.

If we are forced to admit the existence of dark matter in the solar neighborhood and are led by economy of hypotheses to identify it with dark matter elsewhere, the consequences are striking. Bahcall has shown quite convincingly that its scale height cannot be more than double that of the old disk, which means that it currently occupies a very thin disk; there is little doubt that the bulk of it farther out is more-or-less spherically distributed, but if there is a concentration in a disk, it presumably must be dissipative and therefore presumably baryonic. Furthermore, since the bulk of the evidence suggests that the disk formed comparatively recently, the dark matter presumably formed out of gas in much the same manner as stars form.

We are not, of course, forced to assume that this stuff, if it exists, is the same stuff out of which halos are made and that accounts for the virial mass in great clusters, but if not we have two mysteries instead of one. It therefore seems prudent to ask how firmly one is required to believe that there is missing mass in the disk. I would here only like to issue a caution or two. Both of Bahcall's tracer populations have a potential problem brought about by the realities of stellar evolution. In the F dwarfs, slightly evolved subgiants which are photometrically and spectroscopically very similar to main-sequence stars are very common, and have luminosities which are a fraction of a magnitude to a couple of magnitudes above the main sequence at a given color. Since the derived density goes as the inverse luminosity of the population, an admixture of these class IV stars will lower the derived density. The K giants are in some senses even worse. One knows that stars from many populations funnel into the K giant region, and that the luminosity at a given color is very metallicity dependent. If there were, for instance, a strong inverse metallicity-height correlation, which there may well be, the effect would be that more distant stars would be more luminous and the density would be overestimated. One needs quantitative narrow-band photometry to investigate both of these effects in the samples used; that is hardly a difficult job today and needs badly to be done.

2. HOW MUCH IS THERE?

The total amount of dark matter associated with luminous matter, in galaxies, binary galaxies, groups, clusters, and stuctures like the Local Supercluster is in principle amenable to determination by dynamical investigations. Such investigations have, of course, been in progress by many workers for some years now, and the answers seem now to be converging, a phenomenon which is comforting but should not necessarily inspire confidence in the result. Marc Davis reviewed this subject ably here; it would appear from many independent lines of argument that Ω is about 0.2 under the assumption that the matter on scales larger than galaxies is distributed like galaxies are. For most of the arguments that go into determining that value the uncertain Hubble constant drops out. As interesting or perhaps more so is the question of the ratio of luminous to dark matter. The determination of that quantity and its possible variation from system to system and from one kind of system to another depends upon knowing mass-to-light ratios well as a function of color and population type. There seems no longer to be any strong evidence for variation with scale, or, indeed, any real variation at all once one is speaking of scales larger than the main bodies of galaxies. Values of $f=\rho_{\ell}/\rho_d$ of .05-.07 are obtained from arguments ranging from the dynamics of the Local Group to binary galaxies to the dynamics of other small groups and clusters to "global" measurements which in fact are only superpositions of the aforementioned results. The poster by Quinn here suggests that the ratio is about 0.1 for elliptical galaxies based on the behavior of the Malin-Carter shells. It would appear that a ratio f of 0.07 is unlikely to be wrong by a factor of two. In most formation scenarios this ratio should be universal; separation of dark and visible matter should occur only on small scales where dissipation can occur on the timescales of interest. (It should be noted that this situation need not prevail in the neutrino pancake picture and would not be expected to prevail in explosive scenarios like that of Ostriker and collaborators.) In biassing schemes, which are almost certainly necessary if one is to believe that Ω is unity, galaxies are not formed efficiently in low-density regions, but if the dark matter is primordial the ratio of matter which would have become the luminous matter in galaxies to the dark stuff should presumably be the same number. The constancy of f over observed systems may eventually, with somewhat better data than we possess currently, put strong constraints on biassed galaxy formation. Peebles has here and elsewhere strongly stressed the point that even if the biassing is a strict threshold phenomenon, heirarchical clustering will cause some mixing of the stuff in which no galaxy formation has occurred with that in which it has, giving apparent abnormally low values of f there. This phenomenon has not really been addressed in any of the n-body studies to date in which crude biassing schemes have been used.

It is worth at this point screaming about something about which it will do no good to scream, viz. the poor state of our knowledge of the Hubble constant. While it does fortuituously drop out or nearly drop out of the Ω determinations via dynamics, for almost all other questions it is of crucial importance. All inferences about the present universe from nucleosynthesis, from the processing of the perturbation spectrum through the early universe, from relic blackbody background fluctuations, depend sensitively on its value, and our ignorance of its value is the limiting factor in the application of many of these arguments. The obvious exhortation is clear: go do a better job.

3. WHERE IS IT?

That is, what is its distribution relative to that of luminous matter? As we have argued earlier, the existence of local dark matter distributed more-or-less like the stars in the disk would argue for dissipation in that component. An amusing possibility is the existence

of a shadow galaxy made of shadow matter coincident with our own; if
the physics of the shadow world were exactly like that in our own, this
would be likely if not inevitable, and would neatly explain the
required factor of two. The constraints on primordial helium
production, however, rule out this possibility, since there would
always be as much shadow stuff as "real" stuff, and the universe would
expand too rapidly through the nucleosynthesis era, in precisely the
same way it would if there were too many families of leptons with their
associated neutrinos.

In galaxies, we know now that the rotation curves at large radii
are dominated by dark material, a point made clear here by van der
Kruit, Rubin, and Sancisi. The striking demonstration by Kalnajs four
years ago at Besançon that rotation curves can be explained by constant
M/L disks seems not to be tenable when applied to galaxies with modern
rotation curves which extend to very large radii. The demonstrations
we have heard here from Freeman and from Kormendy that dwarf disk
systems also have dark halos is very important, and may, through the
phase-space constraints developed by Tremaine and myself, eventually
rule out neutrinos of a few tens of ev as candidates for the dark
matter. A very interesting question discussed here is the dynamical
importance of dark matter in the inner parts of galaxies. Since one
does not know a priori the distribution of dark matter, there are no
"maximal halo" models without a lower bound on the M/L for the disk
material. The Galaxy is the only system for which we have direct
evidence on this question, but the uncertainty in the disk mass is at
least a factor of two and perhaps more. Here the disk is certainly a
major contributor to the rotation curve interior to the sun, but it may
not dominate. An exciting recent development is the Athanassoula-Bosma
work on the multiplicity of the spiral pattern as influenced by the
halo-to-disk mass ratio; minimum disk masses can in principle be
determined by their technique.

We may eventually know the answer to this by the existence of
rotation curves which can clearly not be explained in their inner parts
by any reasonable M/L value for the visible mass, but in all the cases
discussed so far there are reasonable doubts about the distribution of
the gas (in edge-on galaxies with 21-cm rotation curves) and the
existience of noncircular motions (in optical measurements in systems
like NGC5194). If it turns out to be the way that current measurments
suggest, i.e., that some galaxies require dark matter in their centers
and others do not, a ready explanation may be found in the varying
clumpiness of the initial perturbation. All workers who have
investigated the effects of initial clumpiness have found that clumpy
initial conditions lead to deVaucouleurs-law like systems with strong
central concentrations, while smoother initial distributions result in
less centrally condensed final configurations. Thus most galaxies have
deVaucouleurs bulges, and a few, those with especially messy inital
conditions, might have deVaucouleurs halos.

It would appear, then, that all big galaxies have halos, the
evidence for spirals coming directly from the rotation curves, that for
ellipticals in rather more indirect fashiion from group M/L's, shell
geometry, and, for some, rather directly from X-ray data. A strong

possibility exists that all little galaxies also have halos; certainly some do, as we have seen here.

The evidence from binary galaxies concerning the existence of dark matter is quite compelling, but the evidence on its distribution is very confusing. It would appear from current data and their interpretation that the total masses of binary systems are weakly or not at all correlated with their luminosities. It is yet unclear how uncertainties and systematic selection (physical as well as observational) for orbital eccentricities and phases influence these results, but if taken at face value, they imply that f is highly variable on the scale of individual galaxies. These results may or may not be related to the apparent large dispersion in M/L for small groups, which seemed some years ago to be adequately explained by the work of Gott and Turner in terms of a combination of contamination and statistical fluctuations in the virial ratio for small systems. It is an extremely important result if true, simply because it says that there can be variations, whatever the mechanism; biassing demands such variations systematically with density. Thus the study of binary galaxies would seem to bear importantly on this crucial issue.

On yet larger scales, from cluster dynamics, local group infall, and the correlation structure in velocity space, there is unequivocal evidence for large amounts of dark matter, and again there seems to be slow convergence in the consensus from these data to a value of Ω in the vicinity of 0.2.

What then, are we to do if our prejudices demand that Ω be unity? In this connection it should be noted that not all inflationary scenarios demand simultaneous flatness and homogeneity, and that at least one primordial inflationary model, that of Gott, produces negatively curved homogeneous models naturally. It may be that the currently fashionable GUT phase transition models which demand that k=0 if the universe has inflated enough to be reasonably homogeneous are not correct, or that quantum effects earlier had already established homogeneity prior to the GUT inflation.

If Ω must be unity, it would appear that most of the mass cannot be where most of the galaxies are. Where, then, is it? The presumption of the biassing picture is that the efficiency of galaxy formation is high in high-density regions and low in low-density regions. In great clusters the gas mass is of the same order as the stellar mass (though there is some dispute about that) so the efficiency is high, of order unity, in dense regions. The constancy of f in smaller systems argues that the biassing is not a strong function of density when the density is high, and a simple picture in which there is a threshold below which the efficiency is low or zero, and above which it is high, suggests itself. It is important to note that essentially all our information about dynamics comes from places where the galaxy density is at least an order of magnitude above the mean, except for the local supercluster infall, in which case we deal with a density only about three times the mean. The fact that f is of the same order in the local supercluster as it is in very much denser structures is a little disconcerting. If the supercluster kinematics could be traced with high accuracy to greater distances and lower

densities yet, severe limits might be placed on biassing; the data,
however, will be long in coming.

One possibility which may or may not be relevant, but may again
bear on the biassing quesion, is whether the dark matter exists in
large black lumps of galaxy or even higher mass. This does not by
itself settle the question of where it is, because one must still
arrange for the lumps not to associate with galaxies, but their
existence would clearly enhance the biassing cause: collapsed
structures of dark matter without associated luminous mass are earmarks
of all the biassing schemes discussed so far. Clues about this matter
may be crying for attention in most of the gravitational lens systems
so far discovered, in which no plausible lens galaxies are found even
in quite sensitive searches. Even the beautiful data Tyson showed us
for Q2345 may turn out to be the best case; the galaxy between the
images is so faint/and/or distant that it is very implausible that it
can be the lens.

The distribution of mass on very large scales thus holds the key
to the crucial connection between dark matter and cosmology, however it
may be distributed on the scales of galaxies and clusters. Almost the
only handle we have on this is the dipole anisotropy in the microwave
background, which we believe must arise from our peculiar velocity,
which must in turn have arisen from peculiar gravitational acceleration
due to the inhomogeneous distribution of mass in our neighborhood. We
have heard here both from Yahil and Davis of the application of Gott's
luminosity/force technique to the IRAS catalog, and the derivation
thereby of fairly large Ω, implying that the mass is distributed
much more smoothly than the light. These results are very interesting,
but clearly some caution must be exercised in their interpretation.
The derived M/L's of the IRAS sources vary enormously, over at least a
couple of orders of magnitude, whereas optical M/L's vary probably over
no more than a factor of 5 or so (VISIBLE mass/light) Thus slight
environmental biasses in the M/L's can produce any effect one wants,
and it would be surprising to me if there were not in fact fairly
strong such effects.

Another matter related to the large-scale distribution is the
question of the cluster-cluster correlation function, the scale of
which has profound consequences for currently fashionable ideas about
the nature of the dark matter, but we will defer the discussion until
we take up that question, which we do next.

4. WHAT IS IT?

First of all, is there more than one kind? There is, I think, no
convincing evidence on this point as yet. If the dark material in the
disk of the Galaxy cannot be explained by astrophysical processes, or
if global cosmological tests indicate that the universe is currently
radiation-dominated (both of which I find unlikely), then the question
will have to be faced squarely, but now I think we cannot intelligently
address the question.

The current fashion in certainly to ascribe the dark matter to
some new stable or nearly stable neutral particle, among which the

favorites are GUT axions and the lightest supersymmetric partner to ordinary particles, probably the photino or gravitino, but perhaps something more exotic. The current state of this rapidly evolving scene was reviewed beautifully here by Mike Turner. It would be very surprising if it stabilizes anytime soon.

There are arguments for baryonic dark matter beyond the wish not to invent arbitrary solutions to our problems. There is evidence that the Population II mass function is very steep in the halo field as well as in massive globular clusters, and an extension at the low-mass end to quite plausible masses leads to very large mass-to-light ratios (and, incidentally, to very low heavy-element yields). A picture in which the low-mass cutoff progresses smoothly from masses like 0.1 solar mass to perhaps 10^{-3} solar masses as one goes from the central regions of a galaxy out makes a qualitatively plausible model which explains rotation curves quite handily. It entails no mystery as to why the amount of dark matter is within an order of magnitude or so of the amount of visible matter, and at least makes plausible the observed fact that rotation curves are flatish from regions where galaxies are almost certainly dominated by visible mass out to regions in which they are clearly dominated by dark matter. There is evidence from cooling flows, as we have heard, that such baryonic dark matter is being formed by some process before our eyes in a few particularly spectacular flows, and may be a general feature of such flows. It does seem a little strange that any such "extended bulge" halo population produces dark halos which seem to be completely uncorrelated with the amount of ordinary bulge population—it is true that large-bulge systems have, on average, higher rotation velocities, but large Sc's can have extensive halos and no bulge at all.

There are, of course, serious problems with baryonic dark matter, some of which were discussed here and others not. Understanding the absence of microwave background fluctuations on arcminute scales is very difficult whether the perturbations are adiabatic or isocurvature, and probably can be understood only if there is reionization, with its attendant energy problems. Primordial nucleosynthesis can be understood only if Ωh^{-2} is small, of order a few hundredths; the exact upper bound is a matter of some controversy, as we have heard discussed. It is unlikely that Ω is much smaller than its currently fashionable dynamical value, about 0.2, so a reliable value for the Hubble constant would have a large impact upon this question. With Ω of 0.2, the reionization must occur earlier than z=20, probably considerably earlier (for H=50, which probably is near the maximum value consistent with the nucleosynthesis data, the number is 32; these are all for unit scattering optical depth, and that is almost certainly insufficient—more realistic redshifts are about 1.6 times larger, corresponding to TAUs of about two. Thus 50 is a realistic minimum redshift). The fraction of matter now in known structures dense enough to have been formed then is very small, and furthermore the Compton cooling is very efficient, so it is not at all clear whether reionization can reasonably occur (and survive). We certainly know of no energy sources at all at those epochs, but that is probably irrelevant. It is worth noting that if the dark matter is like that claimed to be forming in the cooling flows at present, it is not a candidate for the ionizing energy, both because it emits none and because dynamical arguments rule out most of it having formed so early.

Rough estimates place the energy requirements at about 1 keV per nucleon to ionize and stay ionized long enough to create the desired optical depth, with perhaps one percent of the matter contributing energy, or 100 Kev per contributing nucleon in ionizing energy. These figures are not much in excess of supernova energies, but are far in excess of net supernova output per nucleon in stars with current mass functions. Thus something exotic will almost certainly have to be invoked to do the ionizing--but the energy requirements per se are not unreasonable.

Thus it might just be possible to have baryonic dark matter and not give up primordial nucleosynthesis. The notion is not outrageous, as Bernard Carr attempted to persuade us, but there are difficulties.

What about neutral particles? The first question is why f is about .07 and not a trillion or a trillionth. The coincidence is not actually so startling except perhaps for the axions, since one has no a priori notion of where the Peccei-Quinn symmetry should be broken. The expected masses for the neutrinos (if any) are of the correct order but too small by a few orders of magnitude, though perhaps the neutrinos belonging to the heaviest leptons should be heavier. The mass (and indeed the identity) of the lightest supersymmetric particle is likewise very uncertain, but favorite values are a few GeV, and if it interacts weakly, the mean density is about that required, as shown some time ago by Lee[1] and Weinberg.

The advantages of nonbaryonic dark matter for nucleosynthesis and for reducing the amplitude of the background fluctuations are well known; the former was discussed here in detail by Audouze, the latter unfortunately not discussed, though there is excellent recent published work by Vittorio and Silk[2,3] and by Bond and Efstathiou[4].

It is perhaps worth discussing briefly the matter of the flat rotation curves under the supposition that the dark matter is nonbaryonic. I think that the phenomenon is due to concidence, but one we are presented with entirely independently. It has been known for a long time, first from Peebles' analytical work and later from a long series of numerical experiments, that the mean value of the angular momentum parameter $\Lambda = J|E|^{1/2}M^{-5/2}G^{-1}$, which for a system in virial equilibrium is roughly the ratio of the mean rotation velocity to the total dynamical velocity, is about 0.07. I have argued that f, the ratio of visible to dark mass, is of that same order. For a disk, Λ is of order unity, and if the baryonic component cools and sinks in the roughly $1/r^{-2}$ halo of dark stuff until it is rotationally supported, it must thus contract by a factor of roughly the reciprocal of its original Λ, or about 15. It thus becomes 15^3 times denser. The dark halo at the new radius is 15^2 times denser, and since the original baryon density was f = 1/15 times the total dark density, it is now at about the same density as the dark material. Thus the baryons should not either swamp or be swamped by the dark material at their final equilibrium place, but should be comparable in dynamical importance. It is thus not surprising that the rotation curves are nearly flat;

detailed models by Ryden and myself, outlined in a poster here, show
that the idea works in detail.

If the dark matter is nonbaryonic, the inevitable question of
whether it is hot or cold, in the sense of whether its phase space
density is much lower than that in galactic halos (hot) or much higher
(cold). The original elementary particle candidate for the dark matter
was of course some heavy neutrino, of order 100 eV in mass, which is a
hot particle. From the very beginning considerable doubt was raised as
to whether such particles could partake of the known structure in the
universe on the scale of galaxies, both on account of their phase-space
densities, which in the coarse-grained sense can only decrease in the
absence of dissipation, and the even more serious trouble brought about
by the free-streaming erasure of small-scale perturbations early. The
recent simulations by Davis, Efstathiou, Frenk, and White and described
here by Davis illustrate the difficulty very strikingly. I think it
very unlikely that the dark matter is neutrinos, but the unequivocal
measurement of a neutrino mass would, of course, cause a quick retreat
from this view.

We have on the other hand the almost uncanny success of the cold
dark matter picture, documented most vividly in the work of Blumenthal
et al.[5]; its appeal is obvious from the sheer number of poster papers
here dealing with various aspects of the scenario. The spectrum, the
shape (but not yet the amplitude) of which can be understood on very
general grounds, accounts well for galaxy and cluster-sized structures
when the amplitude is normalized to fit the galaxy distribution on
scales of a few megaparsecs. Even the shapes of the rotation curves of
galaxies seem to be predicted. There is now considerable confusion
about the normalization, however, since it is clear that either Ω is
less than one, in which case it is not clear that the assumptions which
lead to the cold dark matter spectrum are valid (inflation), or that
there is biassing, in which case the amplitude of the matter
fluctuations now must be smaller than those in the galaxy distribution
by a factor of two or three at the normalizing scales. There is in
addition the difficulty that the cold dark matter correlation function
goes negative at about 20 $(\Omega h^2)^{-1}$ Mpc, and no positive correlations
would be expected on larger scales. There is, of course, the
suggestion that the cluster-cluster correlation function is
significantly nonzero and positive to larger distances, perhaps as
large as 100 Mpc. How seriously this should be taken is a matter of
some debate; certainly the catalogs from which these data come are not
satisfactory statistically, but it will be some years before properly
objectively prepared catalogs are available. There is, of course, the
related observation of holes larger than the correlation length. It is
in my opinion not completely clear yet what one expects with a given
fluctuation spectrum; the largest numerical experiments yet run are not
large enough to address that question with any certainty.

5. POSTLOGUE

There is the exciting possibility that the dark matter particle will be
discovered, either in accelerator experiments or in clever direct-

detection experiments; the discovery of a stable weakly-interacting fermion in the relevant mass range (a few GeV), or the confirmation of the existence of the axion, or (heaven forbid) the unequivocal measurement of a neutrino mass would make it difficult not to take nonbaryonic dark matter seriously. It seems unlikely that any of these things will happen soon.

It also seems unlikely that there will be an accurate, agreed-upon value for the Hubble constant very soon, which in my opinion is the single thing we need most at this point to constrain theoretical flight. Other important observational material will also be long in coming: to MEASURE finally a galaxy- and cluster-scale perturbation in the background; to have an objectively constructed galaxy and cluster catalog with well-planned redshift coverage to look at random velocities and large-scale correlations, just to name a couple. I think that we will be debating these same issues for a long while to come, and I do not look forward to decisive answers in the next few years. We have, I think, more-or-less agreed to the existence of dark matter and its importance in understanding the origin and dynamics of structure in the univese. An important first step, but there will be very many more to go.

REFERENCES (to work cited in the summary which was not discussed fully
                      at the conference)

1. Lee, B. and Weinberg, S., Phys. Rev. Letters 38, 1237 (1977).
2. Vittorio, N., and Silk, J., Ap. J. (Letters) 285, L39 (1984).
3. ibid, Ap. J. (Letters) 293, L1 (1985).
4. Bond, J. and Efstathiou, G., Ap. J. (Letters) 285, L45 (1984).
5. Blumenthal, G., Faber, S., Primack, J., and Rees, M., Nature
       311, 587 (1984).

A HISTORICAL PERSPECTIVE ON DARK MATTER

Scott Tremaine
Canadian Institute for Theoretical Astrophysics
McLennan Laboratories
University of Toronto
Toronto M5S 1A1
Canada

The dominant impression which I carry away from this meeting is that
extragalactic astronomy has reached a crisis ("a state of affairs in
which a decisive change for better or worse is imminent", according
to Fowler). The nature, origin and distribution of the dark matter
and its role in galaxy formation and dynamics are issues whose resolution is likely to determine the direction of studies in galactic
structure and cosmology for decades to come.

This is not the first crisis which science has encountereed,
and so I thought it might be useful to ask whether there are any
historical lessons from previous crises which might provide us with
some solace or guidance in our present perplexed state. Fortunately
this question has already been addressed, in a very influential book
called *The Structure of Scientific Revolutions* by Thomas Kuhn of MIT.

The central concept in Kuhn's work is the paradigm, which he
defines as a "body of intertwined theoretical and methodological
belief that permits selection, evaluation and criticism". Examples of paradigms would include Newton's laws, Maxwell's equations,
the caloric theory of heat, or Keynesian economics, although many
paradigms are more specialized than any of these. The paradigm provides a coherent framework which suggests what experiments are worth
doing, gives scientists confidence that they are on the right track,
and permits selection, evaluation and criticism of both theory and
experiment.

Kuhn argues that most scientists spend most of their careers doing what he calls "normal science". Normal science is work within
the framework of a paradigm. It consists, not of a search for new
theories—indeed, in normal science new theories are regularly
suppressed—but rather of the articulation and elaboration of the
paradigm. It can be thought of as puzzle-solving within a well-
defined set of rules, and the motivation for doing normal science
is not the desire to be useful or the drive for knowledge but rather
"the conviction that, if only he is skilful enough, he will suc-
ceed in solving a puzzle that no one before has solved or solved so

J. Kormendy and G. R. Knapp (eds.), Dark Matter in the Universe, 547–549.

well". There may be anomalies or apparent contradictions in normal science—in fact these provide the puzzles to be solved—but the expectation of the community is that these can be resolved within the paradigm.

I believe that for the last several decades extragalactic astronomy has been normal science. The corresponding paradigm is difficult to describe fully but would include the assumptions that Newton's laws are correct, that the mass in galaxies is mostly contained in visible stars, that the initial mass function is more or less a power law and more or less the same everywhere, that galaxy formation is unbiased, etc. Many of these assumptions are not usually stated explicitly, but this is merely a sign of how deeply they are woven into the paradigm.

Not all science is normal science. According to Kuhn the regular progress of normal science is occasionally interrupted by an event which he calls a "crisis". A crisis is simply a recognition by the community that their paradigm no longer works. Examples include the distressing state of Ptolemaic astronomy at the time of Copernicus and the late nineteenth century crisis in physics caused by the photoelectric effect and the black body radiation spectrum. I would like to argue that extragalactic astronomy is in the midst of a classic Kuhnian crisis. In support of this contention, let me briefly list some of the characteristics of a crisis, taken directly from Kuhn's book.

The first sign of the crisis is that some anomaly appears to be more than just a puzzle of normal science. (From my own perspective this occurred around 1974, with the Roberts-Whitehurst rotation curve of M31 and the Ostriker, Peebles and Yahil article proposing massive halos.) The awareness of the anomaly spreads gradually throughout the community, attracting more and more attention. Eventually the most eminent scientists in the field begin to concentrate on the anomaly. (I am sure we all agree that this has happened!) As work proceeds, the unspoken assumptions of the old paradigm become stated more explicitly. (Thus we now see papers raising issues like bimodal initial mass functions, biased galaxy formation and even non-Newtonian theories of gravity.) For many scientists, discouragement and pronounced professional insecurity set in, generated by the persistent failure of the puzzles of normal science to work out as they should. (Kuhn quotes Pauli: "In any case, it is too difficult for me, and I wish I had been a movie comedian or something of the sort and had never heard of physics.") Experimenters devote themselves to finding new ways of magnifying and making explicit the breakdown of the paradigm. (For example, we have the work of Rubin and her collaborators on spiral galaxy rotation curves, Aaronson's work on mass-to-light ratios of dwarf ellipticals, and various efforts to search for faint dwarf stars in halos.) Many new paradigms are proposed, but their proponents are generally not very interested in talking to one another, since they have abandoned the old paradigm which was what they once had in common.

If, then, we are in a Kuhnian crisis, we can use the same histor-

ical evidence to predict the resolution of the crisis. According to
Kuhn the time interval between the breakdown of the old paradigm and
the enunciation of the new is almost always one to two decades; if
we count the beginning of the present crisis from 1974 then we have
less than ten years to wait. The winning theory will be difficult
to recognize at first among the proliferation of new theories generated by the crisis. In particular, it usually will not explain the
experimental facts much better than its predecessors, just as Copernicus's theory was no more accurate than Ptolemy's. (Thus we should
not be disturbed at the shortcomings of, say, the cold dark matter scenario for galaxy formation, since it probably works about as
well as Kuhn says the winning theory ought to at this point.) The
new theory succeeds, not because it explains the facts so well, but
because it converts a few influential scientists who are attracted
for aesthetic reasons. These workers then develop and improve the
theory to the point where its successes multiply and more converts
are made. The final few holdouts will eventually die out, leaving
a community which is once again united. The crucial observations
which prove the correctness of the new theory will probably only be
made long after most of the community has already accepted it. The
final sign that the crisis has been resolved will be a series of
textbooks which incorporate the new paradigm.

    Finally, Kuhn stresses that the scientists who eventually devise
the successful new theory of dark matter will almost certainly be
either very young or very new to the field. However, this does
not excuse you from reading any future papers I may write on this
subject.

GENERAL DISCUSSION (CHAIRMEN J. PEEBLES AND S. TREMAINE)

PEEBLES: The first part of the general discussion focusses on four particularly active and important topics that can be associated with four of the speakers at this conference. To provoke discussion, I will ask:

(1) *Local K_z determination:* Does anyone not believe John Bahcall?

Dark matter in the disk of our Galaxy is particularly conveniently placed for the study of the nature and distribution of dark matter. It is difficult, although not impossible, to see how weakly interacting particles like axions or neutrinos could have become concentrated in a disk. If we could convince ourselves that the local dark matter is baryonic -- brown dwarfs or stellar remnants -- it would encourage studies of the possibility that dark matter found elsewhere is also baryonic. We would also have the challenge of understanding how an appreciable fraction of the local baryons were converted into a dark state after galaxies formed.

LYNDEN-BELL (to J. Bahcall): Do there exist suitable star-count data in the south, and do they agree with those in the north?

J. BAHCALL: The only appropriate samples at this stage are in the northern hemisphere. I think that it would be useful to hear about the new programs being carried out in the southern hemisphere by Paul Schechter on the K dwarfs and Ken Freeman on the K giants. They can't yet answer your question, but they can tell you how their samples are designed to answer all of the questions that have been raised.

FREEMAN: We are getting a complete sample of K giants near the south galactic pole; they are all bright and within about 1 kpc of the Sun. We are going to do DDO photometry on the whole lot, which will give us metallicities and luminosities. Then we really will have a pure sample of K giants. We will also get slit spectra for all the stars. I think the obvious thing to do will be to use the relatively metal-rich stars, which we can identify pretty reliably with the old thin disk.

PEEBLES: What is known about the proper motions of the M dwarfs? Can one properly infer a mass per unit area from a mass per unit volume at the low-mass end of the mass function?

J. BAHCALL: From what we know about dwarfs brighter than $M_V = 16$, I don't think that there is any hope that the M dwarfs contribute significantly to the density of observed matter. By any extrapolation, even a flat one, they contribute $< 0.01 \ M_\odot \ pc^{-3}$, and I think the observers here will call that an overestimate.

PEEBLES: You are quoting a mass per unit volume. How well can you get a mass per unit area?

J. BAHCALL: Even if you integrate over 1 kpc rather than 300 pc, you don't get a useful contribution. But Larson's remark that the luminosity function

551

J. Kormendy and G. R. Knapp (eds.), Dark Matter in the Universe, 551–565.
© *1987 by the IAU.*

may have two peaks could apply. He talks about a second peak in the white dwarf region, but it could just as well be at 0.03 $M_\odot$.

FABER: In light of Jim Gunn's remark about the possible spread in the absolute magnitudes of the F stars, what can you say to reassure us?

J. BAHCALL: I think he made a good point. All of the F dwarfs of Hill, Hilditch and Barnes were observed in Strömgren four-color photometry; they also had spectroscopy for a representative sample. I think that is the best that can be done, and it is sufficient for the F dwarfs. There is a bigger problem for the K giants. We have MK classifications only for the Upgren sample, not for the Oort sample. But the Oort and Upgren samples turn out to have the same densities, within the errors. I estimate that these errors contribute $\leq$ 20 % to the error in the total amount of matter. The real test of this work will come in 3 - 5 years, when we have new samples like the ones by Freeman and Schechter.

FREEMAN: I just want to remind you of the work on face-on galaxies that I and Piet van der Kruit discussed. This uses comparable dynamical techniques and gives M/L ratios similar to those found in the K_z analysis.

(2) *Dwarf Galaxies:* Does anyone not believe Marc Aaronson?

PEEBLES: Studies of extremely low-luminosity galaxies may reveal distinctive properties of dark matter. The Cowsik-McClelland-Tremaine-Gunn phase-space argument tells us that neutrinos with masses of a few tens of eV would have space distributions broader than the stellar distributions of some dwarfs. If these galaxies were dark because they lost most of their baryons, then they could be left with canonically deep potential wells of dark matter. On the other hand, if dwarfs formed by dissipation out of debris from large galaxies, they may have mass-to-light ratios characteristic of purely stellar systems.

J. BAHCALL: Aaronson removed some stars because of evidence that they are binaries. Suppose he observed a representative sample of stars. Are the velocity differences measured for the binaries sufficient to account for the entire observed velocity dispersions of the galaxies?

PEEBLES: Aaronson is no longer here.

MATHIEU: I have done Monte Carlo simulations to study this problem in open clusters, and Marc and I are doing them for his dwarf spheroidals. My gut feeling is that binaries will not account for observed dispersions as large as 1 or 2 km s^{-1}. But a problem that Aaronson has not taken into account is the possibility that the galaxies contain very massive objects that inflate the observed central velocity dispersions.

FABER: If you think about what types of binaries you need to give a dispersion of 9 km s^{-1} when the stars have masses of $\sim$ 1 $M_\odot$, you conclude that the separations are $\sim$ 25 AU and the periods are like that of Saturn around the Sun. So at the moment we might just be on the hairy edge of being able to rule out binaries. I would feel a lot better if we had followed these stars for ten years.

RICHSTONE: Two reasons to be skeptical: (1) The observed distribution of velocities looks flat, not Gaussian (although there are few points). (2) We know that giant ellipticals have anisotropic velocity dispersions and that velocity anisotropy can affect M/L determinations even if the velocity dispersion is known precisely.

OSTRIKER: Two comments on why the high values of M/L for dwarf spheroidals may be right. (1) White and Davis have suggested from binary-galaxy considerations that M/L varies as $L^{-1/4}$. Then it isn't surprising that when L is very small, M/L is very large. (2) If we know the rotation curve of the Galaxy at large radii, we can determine M/L for the dwarf spheroidals on the assumption that they are tidally limited. Faber and Lin have shown that this gives high values for M/L.

GUNN: There are several scenarios, some of them quite prosaic, in which one would expect very high M/L ratios in dwarf systems. For example, one can remove the metallicity, using the simple continuum model I talked about to reduce the yield by pushing the mass function to very low values. This can explain systems with a very low light density and high mass-to-light ratio. However, it is worth remembering that the dwarf spheroidals that have been measured show an enormous range in M/L, from values of about 10, perhaps a little higher than one would like to believe on the basis of population, to values like 100. Because of that it is difficult to believe that there is one simple picture for their formation. I have thought for a long time that the dwarf spheroidals are a key, a Rosetta stone, to galaxy formation. I can't see how any of the suggestions for how they form would introduce such an enormous dispersion in M/L. I'm a bit skeptical purely for this reason. Also, the two cases in which M/L is particularly high are the most difficult objects in the sample.

DEKEL: I believe that dwarf galaxies must have extended halos (but not necessarily very high M/L within the visible region). Self-gravitating gas-loss models fail to reproduce the observed relations between luminosity, radius and metallicity. On the other hand, the simplest model of substantial gas loss inside massive halos is very successful in reproducing the observed relations (e. g., Dekel and Silk, this volume). The observational constraints indicate further that the halos originate from cold dark matter perturbations. Our model provides a simple physical mechanism for biased formation of bright galaxies.

(3) *Primordial Nucleosynthesis:* Does anyone not believe Jean Audouze?

PEEBLES: Nucleosynthesis is advertised to provide a constraint on the mean mass density in baryons. For "reasonable" values of the Hubble constant, the density in baryons has to be less that $\sim$ 10 % of the critical Einstein - de Sitter density predicted by inflation with negligible present cosmological constant. As Audouze emphasizes, to justify this we need to

make a precise comparison of computed abundances with observed abundances
extrapolated back to primeval. Also, we need to consider alternatives
to the canonical paradigm, such as primeval non-linear isocurvature
fluctuations.

STEIGMAN: The difference between the results of Audouze and those summarized
by Boesgaard and Steigman (1985, *Ann. Rev. Astr. Ap.*, **23**, 319) is his use of
the *estimated* primordial abundance of ^{4}He to determine the nucleon density.
To use ^{4}He requires that the abundance be known to 3 decimal places. I
don't think that is presently possible. Boesgaard and Steigman suggest that
the primordial ^{4}He abundance was $Y_p = 0.24 \pm 0.02$.

AUDOUZE: With this value of Y_p the conclusions of our work do not change!
The reasons why I advocate that $\Omega_B \leq 0.06$ are the following. As I said
yesterday, the error bars concerning all the four elements D, ^{3}He, ^{4}He and
^{7}Li are still extremely high. But at this point I do not agree with the
statement of Gary Steigman that it is safer to start first to find a good
agreement from D, ^{3}He and ^{7}Li and *then* to check ^{4}He. The reason is that
^{4}He is much less affected by chemical evolution than D and ^{3}He. I want to
state again that in order to find an agreement between the predictions of
the standard Big Bang and our present knowledge regarding these elements,
one should invoke specific models of chemical evolution concerning D and
^{3}He. Moreover, we obtain a limit on Ω_B more stringent than the majority
of the participants may think. But I still believe that progress will take
place when better measurements of primordial ^{4}He are available.

PEEBLES: Jean, where do we go from here? What are the directions of
research that will lead us toward better answers?

AUDOUZE: Let me cite Kunth and others who try to measure the primordial He
abundance in blue compact galaxies and places where the metallicity is low.
Vigroux *et al.* find an abundance of $\sim$ 20 %. Kunth and Sargent find a very
large spread in Y, from 20 to 26 %, for just one value of [O]/[H]. My way of
getting consistency in all these numbers is to have $\Omega < 0.06$. I'm sorry if
that creates trouble.

PEEBLES: Any other forward-looking remarks?

PACZYNSKI: I would like to point out that there is a low-mass binary system,
CM Draconis, that offers a possibility to determine helium content in the
unevolved Population II stars. The binary is eclipsing, both masses,
radii and luminosities are known, and preliminary analysis (Paczynski and
Sienkiewicz 1984, *Ap. J.*, **286**, 332) implies $Y = 0.3 \pm 0.1$. This estimate
may be considerably improved in the near future.

OSTRIKER: One thing which would be very useful and which could be done
moderately soon is the measurement of galactic gradients in the light
elements. It would better tie down the galactic evolution model if we
knew, for example, whether deuterium increases or decreases with increasing
metallicity.

(4) *The Early Universe:* Does anyone not agree with Mike Turner?

PEEBLES: New ideas from particle physics have greatly stimulated some old speculations in cosmology. Inflation provides a beautiful explanation for the isotropy of the Universe and lends respectability to prejudices against cosmological models with non-zero space curvature. Phase transitions may produce computable density fluctuations that end up as black holes or superclusters. And we are presented with a rich catalogue of forms of matter, from gravitational waves to cosmic strings, whose abundances may be computable and whose physics may help to account for the properties of galaxies and superclusters. But apart from the general enthusiasm, are astronomers justified in accepting any of this as received knowledge?

J. BAHCALL: I have a question for Mike Turner about the shadow Universe. It was not clear to me in reading the particle physics scenarios that we get what we need from the shadow Universe, namely that the shadow matter is in the same place as the matter that we see. Could it be that there are shadow galaxies out there, but that they don't coincide with the ordinary matter?

M. TURNER: Well, if you are asking about primordial adiabatic perturbations, then the shadow and ordinary matter will both participate in gravity, and you can't separate them (assuming they were well mixed in the first place) until non-gravitational forces become important.

J. BAHCALL: Why would they be well mixed in the first place? Why do they have to be coincident even on a scale of 10 kpc? Why isn't it just as likely that shadow galaxies are out there where there is practically no normal material, while our Galaxy is made mostly out of non-shadow material?

M. TURNER: OK, it is possible that at the start they were not well mixed.

OSTRIKER (*sotto voce*): Are there statistical requirements necessitating a shadow - John Bahcall asking the same question at the same time? How close together do they have to be? (outbreak of mirth in Ostriker's vicinity)

GUNN: I think, John, that all you need is that both Universes were relatively homogeneous early on. Even the development of the perturbations doesn't require that they be adiabatic, as long as the shadow Universe had a decoupling phase. The perturbations are linked; you can show very easily that only the potential matters; this is described by them both; the matter follows the potential, and so the shadow matter will follow the ordinary matter. So one would expect a shadow galaxy here with more or less the same properties as ours, at least in the halo.

SPERGEL: If inflation is important and occurs after E8 x E8' breaks, the relative number density of E8' matter would be much lower. If E8' matter is the dark matter, the physics in its sector would have to be different from the physics in the E8 sector.

FELTEN: If "believing Mike Turner" includes believing that $\Omega = 1$, it would be wise to keep perspective by noting that there is as yet to my knowledge no confirmed prediction of any observable from the inflationary theory.

There are explanations for a few problems which had been noticed previously,
such as the horizon problem, but these are explanations after the fact, not
predictions. If I am wrong, maybe Mike, or Gary Steigman, would comment.
The observational evidence suggests, if anything, that $\Omega < 1$. The evidence
for the theory is not compelling.

M. TURNER: I think it is hard not to believe me, because I'm an honest guy
and I always tell the truth (laughter). But mainly, the message of my talk
was that the early Universe is just starting to come into focus. Most of us
believe in primordial nucleosynthesis. But at earlier times we are getting
only hints. Inflation is extraordinarily attractive: for the first time
there is a possible explanation of the origin of density inhomogeneities.
The idea of relics is very attractive. But there is nothing conclusive yet.
There are hints that might focus the effort in understanding how structure
and dark matter formed.

TREMAINE: In fairness, it may be too early to demand a confirmed prediction
from the inflationary model. The average rate of discoveries in cosmology
is about one per twenty years, and inflation has been around for only four
or five.

TREMAINE: OK, let's move on from character assassination. I want to ask a
number of questions. First, *what* is the dark matter? My understanding of
the consensus is that we certainly need some baryonic dark matter, given
that there is a problem of missing mass in the galactic disk. We don't *need*
non-baryonic dark matter, although it is very attractive. At the moment
we seem to need up to four kinds of dark matter, one for the disk, one for
the halo, one for rich clusters and one for Alan Guth. Does anyone believe
that the initial mass function of the stellar population is the same as a
function of position and time? Almost certainly not, but everybody assumes
it anyway. Does anyone believe estimates of the local initial mass function
for $M < 0.2\ M_\odot$? I have a question about that: How well is the main-
sequence mass-luminosity relation determined at very low masses, and how
much could it affect our understanding of the initial mass function? Does
anyone have any comments on that question, or on any of these others?

SILK: At the moment, we have at least two, if not three, dark matter
problems. I would like to propose a means of reducing our difficulties to
only one dark matter problem. The argument is as follows. The evidence for
more-or-less spherical halos is highly biased. It consists of polar ring
galaxies and of X-ray-emitting ellipticals. However, both are likely to
be ongoing mergers or merger products, and we would expect a merged halo to
be fairly round. However, isolated spirals may have very flattened halos.
In fact, in at least one scenario, this is highly probable. I suspect
that what I have to say would apply to any generic pancake scenario for
galaxy formation, but let me consider the particular example of warm dark
matter. All small-scale structure in the primordial fluctuation spectrum
is suppressed by free streaming, and the collapse on massive halo scales
happens very asymmetrically. The dark matter is most likely to form a
sandwich containing a layer of denser, dissipating baryons. No doubt the

dissipation will further help to drag in the dark matter. Direct formation of a filament is less probable, but the sheet should be unstable and should turn into a highly flattened triaxial halo. Within this, the galaxy forms in the usual way. I suspect that such a halo could simultaneously explain the disk dark matter and flat rotation curves. On larger scales, warm dark matter is indistinguishable from cold dark matter. One other implication is worth mentioning: A very prolate halo would have interesting consequences for the velocity ellipsoid of old stars, galaxy rotation curves and warps; it might even be desirable.

TREMAINE: There is yet another possible way to build a disk, which I think I first heard from Jim Peebles. In some scenarios you might form small clusters of cold dark matter. If they had some initial angular momentum, dynamical friction might drag some of them down into the disk, at which point they would be tidally shredded and would form a thin disk of non-baryonic matter. It may be worth investigating possibilities like this.

GUNN: There is another strong constraint on the shape of halos that hasn't been discussed. In the Galaxy at 3 or 4 kpc beyond the solar radius, the HI disk flares very strongly. It flares in precisely the way you would expect for a disk of velocity dispersion 7 - 10 km s^{-1} (as observed for the vertical velocity dispersion in other galaxies) if the disk became non-self-gravitating at that radius. If you assume that the disk falls off exponentially with radius, you can show that it *should* go non-self-gravitating at that radius. And from the behavior of the scale height in the gas disk *versus* radius you can put constraints on the ellipticity of the halo. It must have a flattening of < 2:1. Now at the time I made these suggestions, I don't think that everyone believed that the rotation curve of the Galaxy was flat. And if there is no halo, the disk still becomes non-self-gravitating and it still flares. But observations of other galaxies now suggest that our own would be very strange if it didn't have a flat rotation curve and therefore a halo. So now this argument based on the flaring of the disk can be taken seriously.

TREMAINE: But it doesn't rule out the possibility that the dark matter in the disk is non-baryonic.

GUNN: True. But it implies that most of the matter that supports the rotation curve cannot be in a flattened system.

SANCISI: It is true that the HI disk seems to flare in our Galaxy and in several others (e. g., NGC 891). But can you explain why, at a radius of three times the optical radius, 50 kpc or so, the gas is still close to the disk plane in NGC 891 and in NGC 5907? If disk formation is recent, how does the gas find the plane so quickly when there is no matter there?

GUNN: I have no answer. That is a very difficult problem.

SILK: It is an argument for dark matter in the plane of the galaxy.

OSTRIKER: A comment on Joe Silk's intriguing suggestion. A hot disk would stabilize things as well as a halo. And one would indeed expect it to be triaxial. But then I don't see why rotation curves are so symmetrical. There are a couple of cases of galaxies with unsymmetrical rotation curves at large radii, but if Silk's suggestion is right, they should be the rule rather than the exception.

SANCISI: I think that asymmetries *are* very likely the rule. When you go very far out in radius, at some point you don't believe most rotation curves any more because they are not symmetric. In some cases the rotation curve even turns down, so it becomes doubtful whether you are seeing circular motion.

SILK: Let it be said that the possibilities are rather broad. There is a certain probability of collapsing to a very thin sheet, but collapse could equally well be to other configurations. I can imagine a wide variety of complicated triaxial shapes.

VAN DER KRUIT: The kind of arguments just given by Gunn and Sancisi concerning flaring of HI layers, complemented by similar work on stellar populations, gives information on at least the positional dependence of the IMF. These data imply that M/L and therefore the general form of the IMF are not varying in disks. Also, I believe that these estimates of M/L (including Bahcall's K_z analysis) are the only ones that can tell us what a "reasonable" M/L is.

LAKE: Are halos spherical? This is an important question, but the work on polar rings has not answered it. In order to stabilize both rings, we must be looking down the intermediate axis of a triaxial mass distribution. The counteraligning of closed orbits relative to the potential will make the observed velocities equal in these flattened potentials.

One thing that has disappeared from the oral saga is that we know the core radii and asymptotic velocities of halos. Deconvolutions of rotation curves now yield a halo contribution to V that is linear in r. We don't know any velocity scales of halos from rotation curves. What information can we compare to binary studies?

PACZYNSKI: Next come the questions dealing with low-mass stars. There are lots of rumors that the velocity dispersion for low-mass, late-M dwarfs is smaller than that of the brighter stars. Where do we stand on that?

GILMORE: The status of current observational evidence that very low-mass luminous stars might provide a significant contribution to the total density of matter in the solar neighborhood is as follows. Recent automated red-sensitive photometric surveys over large areas to intermediate depth (Gilmore *et al.* 1985, *M. N. R. A. S.*, **213**, 257) and small areas to greater depth (Gilmore and Hewett 1983, *Nature*, **306**, 669; Boeshaar and Tyson 1985, *A. J.*, **90**, 817) have provided the first volume-limited samples of stars which are complete to the absolute magnitude at the theoretical minimum mass for hydrogen burning ($0.085 \, M_\odot$). These surveys are in excellent agreement, and show that the stellar luminosity function has a broad maximum near M_V = +12, and then a slow decline to M_V = +19. Conversion of this function to a mass function is hampered by the very small number of data

points available to calibrate the mass-luminosity relation below $\sim 0.25\ M_\odot$
(Fig. 3 of Gilmore and Reid 1983, *M. N. R. A. S*, **202**, 1025). The available
data show that the stellar mass function has a maximum near 0.25 $M_\odot$ and
then a decline at lower masses. The existence of a maximum ensures that
the integral of the mass function converges before the minimum mass for
hydrogen burning is reached. Stars with masses below 0.20 $M_\odot$ therefore
do not contribute more than about 0.005 $M_\odot$ pc^{-3} to the total mass density
near the Sun. There are two major caveats to this conclusion, each relating
to the possibility that the lowest-luminosity stars have short luminous
lifetimes. First, several stars are now known with reliable (trigonometric-
parallax) absolute magnitudes fainter than M_V = +16, which corresponds to
the theoretical minimum mass for hydrogen burning. If these stars really do
have masses below the theoretical limit, they will only briefly be luminous,
and their derived space density must be increased by the ratio of their
luminous lifetime due to the release of gravitational energy to the age
of the galactic disk. This could be a large factor, implying a very large
total mass in such "stars" and their remnants. Reid and Gilmore (1984,
M. N. R. A. S., **206**, 19) show that a large correction factor is unlikely
but not totally excluded. They show further that the observed faint stars
lie at or just below the hydrogen-burning main sequence in an $M_{bol} - \log T_e$
HR diagram, while none lies near a plausible gravitational cooling track.
The dispersion below the nominal minimum absolute magnitude is therefore
probably due to a combination of cosmic dispersion and observational and
theoretical uncertainties. The exception to this is the companion to
VB8, which is certainly well below the minimum mass for hydrogen burning
(McCarthy *et al.* 1985, *Ap. J. (Lett.)*, **290**, L9). However, it is not an isolated
star.

The second possible problem relates to the discovery by Poveda and Allen
(1985, *Ap. J.*, in press) that stars in the immediate solar neighborhood
with masses below $\sim 0.2\ M_\odot$ have a variety of properties consistent with
young age. The most compelling of these is their apparently small velocity
dispersion. If this result is valid for a larger sample, then the usual
correlation of velocity dispersion with age suggests that these stars are
young. The only sign that this result may not generally be valid comes
from the deep photometric surveys of Gilmore and Hewett (1983, *Nature*, **306**,
669) and Boeshaar and Tyson (1985, *A. J.*, **90**, 817). These show that the
space density of very low-mass stars found at substantial distances from
the galactic plane is consistent with that found in the solar neighborhood
convolved with an exponential decrease in density with a scale height of
a few hundred parsecs. This scale height implies a much larger velocity
dispersion for these stars than that found by Poveda and Allen. The
resolution of this paradox is not known.

While the data are not yet conclusive, it remains true that there is no
strong evidence that low-mass stars ($M < 0.2\ M_\odot$) provide more than about
0.005 $M_\odot$ pc^{-3} to the total mass density near the Sun.

LARSON: I agree that the biggest worry for the determination of the mass
function at the low end is the mass-luminosity relationship for very faint
stars. I'm fairly persuaded that the luminosity function drops off. Some
of the most impressive evidence is due to Frank Low, who showed that the
main sequence, while remaining well-defined, thins out very remarkably

toward the bottom end. Anything like a conventional mass-to-luminosity relationship translates this into a steep drop in the mass function. I have looked at the data available for very faint binaries, and so have Scalo and Gilmore. For what it's worth, these stars define a very nice linear relationship between absolute visual magnitude and the logarithm of the mass right down to $M < 0.1\ M_\odot$. So if you want to convert the falling luminosity function into a rising mass function, the mass-luminosity relationship has to do something strange, which is not suggested by the data.

One comment on the question of very young stars: I, also, am fairly persuaded by the evidence of Poveda and Allen that these very low-mass stars are young. There is not only the kinematic evidence, but also some spectroscopic evidence. If I recall correctly, there is a high abundance of flare stars among these very young stars. Their interpretation is that the hydrogen-burning main sequence ends at $0.2\ M_\odot$ and that these objects are on their way down to invisibility. That interpretation is ruled out immediately by the fact that the main sequence in the HR diagram remains well defined to much below $0.2\ M_\odot$. Anyway, I haven't heard any expert in stellar interiors suggest that the end of the hydrogen-burning main sequence occurs at $0.2\ M_\odot$. I would suggest an alternative interpretation, which is that the IMF and the characteristic mass have changed with time, and that you are indeed looking at recently formed objects.

TREMAINE: The next question is: *Where* is the dark matter? What is in the voids and what is not in the voids? What do we learn from gravitational lenses? And a point that is designed to be provocative: We have now observed a lot of rotation curves of galaxies, but have had little success in explaining them theoretically. So why should we observe any more rotation curves? Also, why are rotation curves so flat and so similar? Jim Gunn addressed this question, but it would be interesting to know if anyone else has any ideas. Are there any mass determinations that we should abandon? I was struck by the lack of argument about mass determinations based on tidal radii and the almost complete lack of discussion of binary galaxies. This suggests that people have given up on them. And: since gravitational lenses give you some evidence that you could have dark things that don't contain galaxies, by the same token, could you have a galaxy without a dark halo? If so, what would it look like? Are there any candidates? Any comments?

RUBIN: There are two situations in which rotation curves should be valuable: (1) Galaxies in binary samples: Linda Stryker and Kirk Borne have rotation curves for the binary galaxies studied by Linda Schweizer. These may help us to interpret the results on binary galaxies. (2) Galaxies in very dense environments: I hope that spectra of galaxies in the compact Hickson groups will teach us something about how halos are altered, or perhaps fail to form, in dense environments.

TULLY: I think it is worth while to continue to measure rotation curves. One very interesting point which is still unresolved is the presence of the dip that occurs in the inner parts of the rotation curves of some bulge-dominated spirals.

MILGROM (to Gunn): You gave an argument to explain why the contributions of the disk and halo to the rotation curve would be similar at the optical radius. The argument was based on the value of an angular momentum parameter. Can it also be applied to ellipticals? (We have some evidence that the two contributions are similar in ellipticals, too.)

GUNN: One can wave one's hands about elliptical galaxies, but much less convincingly than for spirals. We know almost nothing about the relative contributions of baryonic and halo matter in ellipticals, but such evidence as there is suggests that there is continuity with spirals. And there are theoretical arguments about the amount of halo matter, although they are qualitative. One such argument that I find fairly persuasive says that as long as the local density is strongly dominated by the non-dissipative halo, you can't form clumps in the baryons. So you can't begin to form the stars of which ellipticals are made until baryons begin to dominate. Qualitatively, this gives you the same kind of picture as for spirals. How strongly we should believe these arguments, I don't know.

GERHARD: A comment on the radius at which the dark matter begins to dominate. This is inferred from fitting maximum-disk rotation curves, and so depends heavily on a small inner part of the measured curve. I wonder if significant velocity dispersions and/or non-axisymmetries in the inner disk could result in rotation velocity measurements smaller than the true circular velocity. We may then underestimate the disk M/L and the radius at which the halo takes over.

OSTRIKER: Tremaine asked about the utility of binary-galaxy mass estimates. I find that, when scaled appropriately for starting assumptions, the studies agree well, and agree with extrapolated rotation curves. New larger samples could better determine such unknowns as orbital eccentricity distributions and whether or not mass and luminosity are correlated.

WHITE: A comment about Jim Gunn's Big Black Lumps. Although it is relatively easy to think of a way in which one of these might form with no associated galaxy, it is hard to see how it could manage to form with no ordinary matter at all. As Doroshkevich pointed out a couple of years ago, one might therefore hope to see such objects as weak, extended X-ray sources with no associated visible objects. Limits on such objects from Einstein could put useful constraints on the abundance of Big Black Lumps.

SCHECHTER: On the same point: We see a good number of galaxies which are interacting with other galaxies, and might expect to see examples of galaxies with tidal streams which are interacting with nothing or with something that doesn't look big enough to produce a significant effect. Maybe this is a case of suppressing something we don't like: we only pay attention to such galaxies when we see the companions.

WHITMORE: In gravitational lenses I don't see why we should expect the center of mass to be very near the center of light. For example, if the halo extends 120 kpc on one side, but only 80 kpc on the other, the center of mass could be displaced by about 20 kpc.

BURKE: A displacement of only 20 kpc won't work in 0957; the displacement of the unseen matter has to be greater than that to explain the data.

WHITMORE: Can you give us numbers? What sort of displacements are needed?

BURKE: The separation has to be $\geq$ 50 kpc. That's not an unique determination, of course. A recent paper by Gorenstein, Falco and Shapiro indicates the range of solutions that are acceptable.

TREMAINE: The question on gravitational lenses was partly designed to demonstrate that what we learn from them seems very unfocussed (laughter).

DEKEL: Some biased galaxy formation scenarios, (e. g., Faber, this volume; Dekel and Silk, this volume) predict that dwarf galaxies would not avoid the voids, but rather trace the mass. This is testable. There are already indications that dwarfs are clustered more weakly than bright galaxies (in the UGC, and in Perseus-Pisces). To make a more direct test, one should search for dwarfs in regions which are known to be void of bright galaxies. Oemler and I are currently making such a search.

REES: A comment on neutrinos and antibiasing. As I understand it, the neutrino model is consistent with observations if you can prevent galaxies from forming in regions where matter eventually accumulates. Can one rule out the possibility that there are, in those regions, the neutrinos and a lot of very hot gas that has not had time to cool down? The gas may not even have a tremendous overdensity because it can't cool down, and therefore it won't be too conspicuous in X-rays. This seems to be one way to salvage the neutrino model and also to get large dark objects for lenses.

FELTEN (to J. Gunn): I'd like to question your value of Ω = 0.2, for the following reason. It seems to me that within the context of your review, if you deny the scale dependence of M/L, the tendency is to push Ω down, maybe to 0.1 $\pm$ 0.05. To defend your Ω of 0.2, I wonder if you could quote for us two numbers. Pick your magnitude system and your Hubble constant, and then tell us the M/L required to close the Universe and the mean M/L for galaxies, weighted by luminosity. It seems to me that if Ω = 0.2, these two numbers should differ by a factor of 5. I would like to see what pair of numbers you assume, and whether that pair of numbers meets general assent.

GUNN: The value of Ω = 0.2 is a consensus arrived at by several people. I think that the scale dependence of M/L has disappeared because of the growing realization of several things which at this point can't entirely be quantified. There is the question of accounting for the gas in clusters, which is a large fraction of the total baryonic content. Also -- and because of uncertainties about the initial mass function, one should take this with a grain of salt -- we don't really know that the natural value of M/L for an elliptical is different from that for the disk of a spiral. Our beliefs involve an assumption about the location of most of the mass that we don't see. If we assume similar initial mass functions, then it is natural to assume that the value of M/L in an elliptical is three times the value

in a spiral, because the blue stars which are contributing most of the light
in spirals are absent in ellipticals. Thus the progression of M/L from 75
for groups to 200 - 300 for clusters can be explained by stellar population
differences and by the hot gas content of clusters. Now I'm not sure that
that is a quantitative justification for saying that the ratio of invisible
to visible matter is constant as a function of scale. I'm simply saying,
and I think Marc Davis would agree with me, that these points remove the
evidence for variation.

FELTEN: The point I'm trying to make is that galaxies in rich clusters
are not typical of galaxies in the field. To get Ω, you have to know
the luminosity-weighted average properties of galaxies in the field. And
it seems to me that the numbers are such that unless you can take the
absolutely largest M/L that you can get in rich clusters and apply that to
all galaxies, you come up with a number substantially smaller than Ω = 0.2.
I see that Jim Peebles is writing something relevant on the blackboard.
Perhaps he would like to comment.

PEEBLES: M/L for closure is 1500. Density parameter = 0.2. Ratio of total
to ordinary mass $\sim$ 15, which says that the mass-to-light ratio of ordinary
matter is 20. This is high, which I think is Felten's point.

Who wants to vote on what Ω will turn out to be? All in favor say "Aye".

THE MAJORITY: Aye!

RUBIN: No!

DEKEL: Write down their names! (much laughter)

PEEBLES: What will Ω turn out to be? We will count hands.

The Poll

Ω	votes
$1.001 < \Omega$	2
$0.999 < \Omega \leq 1.001$	28
$0.05 < \Omega \leq 0.999$	29
$\Omega \leq 0.05$	2
Don't know	71
Don't care	0

PEEBLES: It has been pointed out by Juan Uson that right here at Princeton
there is a remarkable experiment going on to measure the density parameter.
This is a new test by Ed Loh and Earl Spillar. In the usual test for Ω,
one plots a function such as magnitude *versus* redshift. The curvature is
then used to constrain the cosmology and evolution. If you can measure
redshifts wholesale, you can get a function of two parameters, redshift
and magnitude. A way to measure redshifts wholesale is to use the Baum

method, which many people have tried and found difficult. Loh and Spillar have improved this method to the point where they believe they have firm evidence that it works. With these data they can deduce both evolution and cosmology.

USON: This morning, Ed Loh told me that he gets $\Omega = 1.15 \pm 0.25$ (cheering and applause). Ed is getting the redshift using six wideband colors; the bandwidths are 1000 Å and they all have comparable sensitivity. This is a significant improvement over previous attempts. Ed's error bars are estimated on the assumption that 10 % of the redshifts are totally wrong. He is continuing to increase the size of the sample.

PEEBLES: Are there any final remarks?

TULLY: A comment about the large-scale structure of the Universe. Scott Tremaine instructed us that during crises we seek ways to magnify the breakdown in the paradigm. I would like to present a result which I think is important in this regard. A look at the distribution of Abell clusters with respect to supergalactic latitude shows that on a scale of a tenth of the event horizon, the Abell clusters lie in the very same (supergalactic) plane as nearby galaxies. The second interesting fact is that the nearby galaxies show secondary peaks in number density, suggesting that their distribution is stratified in layers parallel to the supergalactic plane. I don't think this was anticipated by any existing theory.

OSTRIKER: What did you plot to show this?

TULLY: The distance from the supergalactic plane in Mpc was plotted against number counts. And let me point out that the concentration to the supergalactic plane contains 100 Abell clusters of 10^{18} $M_\odot$.

LYNDEN-BELL: The objects are not in the same direction but are at the same displacement?

TULLY: Yes. They are distributed all over the sky, in both the northern and southern hemispheres.

KAISER: Could contamination by faint foreground galaxies increase the number counts near the supergalactic plane?

TULLY: I don't think the effect is statistically important, although it could creep in in a small way.

OSTRIKER: Isn't your effect due to the fact that there is more area near the equator than near the poles?

TULLY: That has been taken into account in the normalization of the data. There are also corrections for the way I sample the data and for the fact that there is some galactic obscuration.

TYSON: This result may tell us something when compared to an experiment that Seitzer and I have just completed. In 6 widely-spaced high-latitude fields we see a remarkably constant density of faint field galaxies to J = 27.

PEEBLES: These results needn't contradict each other. One could have a
situation where the ups and downs are more prominent in some regions than
others, while the mean is more nearly uniform. In fact that's what we find
when we compare the Lick sample, for example, with the Abell sample.

SANDERS: I want to address the remarks made by Scott Tremaine at the
beginning of this session. If we are really in a crisis in the sense
described by Kuhn, a crisis leading to a scientific revolution, then that
would seem to call for a revision or extension of physical laws. Now, is
a hypothesis like cold dark matter (which requires an undiscovered heavy
particle) actually revolutionary or is it just a patch-up of the existing
paradigm? It seems to me to be analogous to the 19th-century attempts to
explain the anomalous precession of the perihelion of Mercury by inserting
an unseen planet close to the Sun. The only truly revolutionary idea
discussed at this meeting is that of Milgrom and Bekenstein.

TREMAINE: I don't want to get into a long discussion of the sociology of
science, but Kuhn takes as a revolution any large change in the way in which
a community looks at a problem. A revolution doesn't *require* a fundamental
revision of physical laws.

J. BAHCALL: I would like to rephrase one of Tremaine's earlier questions
as a general impression of this conference and of the way things are
proceeding. I think you might want to ask: Is there any reason any of
us are doing anything *other* than measuring rotation curves? Because I'm
enormously impressed by the speed with which our ideas have clarified as
a result of rotation-curve measurements. Already everyone is taking NGC
3198 as a classic case, yet it appeared in preprint form only a month or two
ago. Vera Rubin and David Burstein's work in preprint showing that galaxies
of different Hubble types have the same rotation curve is believed to be
showing something very fundamental. The similarity of the rotation curves
of NGC 891 and NGC 7814 is very new. The work by Ken Freeman and Claude
Carignan on bulgeless systems, which also have flat rotation curves, is very
new. I think there is an enormously rich future for us in rotation-curve
measurements. Maybe we will find one which has all the symmetries that
Renzo Sancisi would like but which decreases at large radii. This would
solve a lot of problems. I am not convinced that any of us should be doing
anything other than measuring rotation curves.

TREMAINE: I want to close with a quotation, again from Kuhn's book. This
was produced by a frustrated monk called Alfonso in the 12th or 13th century.
He was trying to predict planetary positions using Ptolemaic theory, which
at that time was in a dreadful state. He was feeling very depressed, and
this was his comment: "Had I been present at the Creation, I would have
given some useful hints for the better ordering of the Universe".

(laughter and applause)

INDICES

In the Name Index, a page number that is underlined refers to the first page of a paper by the indicated author. Things named after people are listed in the Subject Index. "Dark matter" is frequently abbreviated as DM.

INDEX OF NAMES

INDEX OF OBJECTS

CLUSTERS OF GALAXIES

CLUSTERS OF STARS

GALAXIES

INDEX OF SUBJECTS